THE LEGO® MINDSTORMS® EV3 LABORATORY

THE LEGO® MINDSTORMS® EV3 LABORATORY

build, program, and experiment with five wicked cool robots!

daniele **benedettelli**

Printed in USA
Second printing

17 16 15 14 2 3 4 5 6 7 8 9

ISBN-10: 1-59327-533-1
ISBN-13: 978-1-59327-533-4

Publisher: William Pollock
Production Editor: Riley Hoffman
Cover Design: Tina Salameh
Interior Design: Octopod Studios
Cover Photograph: Francesco Rossi
Comic Illustrations: Arte Invisibile
Developmental Editor: William Pollock
Technical Reviewer: Claude Baumann
Copyeditor: Paula Fleming
Compositors: Riley Hoffman and Alison Law
Proofreaders: Emelie Burnette and Nancy Sixsmith

For information on distribution, translations, or bulk sales, please contact No Starch Press, Inc. directly:
No Starch Press, Inc.
245 8th Street, San Francisco, CA 94103
phone: 415.863.9900; fax: 415.863.9950; info@nostarch.com; www.nostarch.com

Library of Congress Cataloging-in-Publication Data
Benedettelli, Daniele, 1984-
The LEGO Mindstorms EV3 laboratory: build, program, and experiment with five wicked cool robots! / by Daniele Benedettelli.
pages cm
ISBN 978-1-59327-533-4 -- ISBN 1-59327-533-1
1. Robots--Design and construction--Amateurs' manuals. 2. Robots--Programming--Amateurs' manuals. 3. LEGO Mindstorms toys. I. Title.
TJ211.B46325 2013
629.8'9--dc23

2013030612

To the memory of Nari

about the author

Daniele Benedettelli is an Italian robotics engineer known worldwide for his LEGO MINDSTORMS creations, such as the LEGO Rubik Utopia (2007), Cyclops (2011), and LEGONARDO (2013). He prefers to be called Danny, mainly to avoid being mistaken for a girl named Danielle. (He once received an appreciation plaque in Bahrain "for sharing *her* expertise and knowledge in robotics.")

In 1992, his creation "Tom the cat" lost early in a LEGO competition held in a toy store in his hometown, but he didn't give up. He kept *playing well* until he hit the "dark age of LEGO"—a period in the life of an adult fan of LEGO (AFOL) when real-life interests (girls, in his case) replaced his passion for the plastic bricks. That is, until 2001—when he discovered the LEGO MINDSTORMS RCX.

Since 2006, Benedettelli has collaborated with The LEGO Group in testing and developing the LEGO MINDSTORMS products as a MINDSTORMS Community Partner (MCP). In 2012, he was hired as an external programmer for LEGO Education. He was selected as one of 12 experts to test the LEGO MINDSTORMS EV3, and he created the EL3CTRIC GUITAR, one of the bonus models for the Retail set 31313.

In 2012, with the help of the openPICUS team, Benedettelli crowdfunded and brought to market the NXT2WIFI, a Wi-Fi adaptor for the NXT, which makes it possible to control robots with any browser-enabled device (such as computers, Android smartphones, iPhones, and iPads) and to build huge swarms of networked robots.

He currently works as a high school electronics and systems teacher and as a freelance LEGO designer for the Bricks4Kidz franchise. He has participated in many LEGO events, and he has been invited to ICT Education conferences around the world as a keynote speaker and workshop facilitator. His YouTube channel has millions of views, and his creations have been featured in many TV shows worldwide. Benedettelli sometimes plays piano and composes music (mainly soundtracks for his videos). He likes origami and drawing comics, the latter a passion that proved useful when creating this book. He is the author of two previous books: *Creating Cool LEGO MINDSTORMS NXT Robots* (Apress, 2008) and *LEGO MINDSTORMS NXT Thinking Robots* (No Starch Press, 2009). You can learn more about him at these links:

http://robotics.benedettelli.com/
http://music.benedettelli.com/
http://www.facebook.com/robotics.benedettelli/
http://twitter.com/DBenedettelli

about the technical reviewer

Claude Baumann has taught advanced LEGO MINDSTORMS robotics in after-school classes for 15 years. He participated in beta testing for the ROBOLAB software developed at the Center for Engineering and Outreach (CEEO) at Tufts University (*http://ceeo.tufts.edu/*). He created ULTIMATE ROBOLAB, a cross-compiler environment that allowed graphical programming of LEGO RCX firmware, and with it conceived the world's only self-replicating program for the LEGO RCX (some call it a virus). Claude also served as a co-developer on the CEEO NXT Module team. More recently, he has participated as a MINDSTORMS Community Partner (MCP) during the development of the new EV3 Intelligent Brick. He has been the assessor of various high school robotics projects and is the author of *Eureka! Problem Solving with LEGO Robotics* (NTS Press, 2013), several articles, and conference presentations. His special interest is robotic sound localization. He is the head of a network of high-school boarding institutions in Luxembourg (EU), and he's married with three children and three grandchildren.

about the comic designers

Arte Invisibile or Invisible Art (*http://www.arteinvisibile.com/*; *http://www.facebook.com/AssociazioneArteInvisibile/*) is a nonprofit association formed in 2007 by young artists living in Tuscany, Italy. It has over 100 members.

Arte Invisibile coordinates courses for comic design, illustration, digital art, screenplay writing, and animation with the goal of bringing young people into the field of art and related professions. Over the years, the association has produced various publications and organized exhibits and workshops with world-famous artists. It also manages a well-supplied comic library with many rare comic books.

acknowledgments

I didn't remember how tough it was to write a LEGO MINDSTORMS book, especially with a comic story inside! There are many people I need to thank for making this possible. First, thanks to my family for their support and patience during this period: to my completely newbie **parents**, who tested the robots' building instructions and helped make them crystal clear; to my **brother**, who kept telling me to get a real job; to my grandparents, who were dismayed to see what toys have become (especially my ninja-like **grandma** who stealthily approached my LEGO work desk from behind and whispered "What's up? Are you working?" turning my hair a few shades grayer). And let's not forget the tiresomely affectionate family dog, who covered my floor with fur and drool and barked at my LEGO robots as if they were alive.

Thanks to the No Starch Press team, especially to **Bill** for believing in this project and for his reviews and suggestions, and to **Riley** for her tireless and kind support.

A big thanks to **Claude Baumann**, renowned school manager, teacher, and author, who punctually and meticulously reviewed the technical aspects of this book. Thanks to the **twelve monkeys** group, of which I am proud member, for their friendship and inspiration. Thanks to **John Hansen**, for his early EV3 screen capture tool. And thanks to the LEGO MINDSTORMS team, especially **Lee** (for hiring me as a programmer for LEGO Edu); **Steven** (for endorsing my projects); and **Camilla**, **Flemming B.**, **Henrik**, **Jesper**, **Lars Joe**, **Linda**, **Marie**, **Oliver**, **Pelle**, and **Peter**.

A huge thanks to all the LDraw community members who developed the bits and tools to create high-quality LEGO-like building instructions. Special thanks to master builder and book author **Philippe Hurbain** (Philo), a master in modeling 3D LEGO elements, and to **Kevin Clague**, developer of LPUB4.

Thanks to my dear photographer friend **Francesco Rossi** (*http://www.fr-ph.com/*) for the great photo on the book cover! He's helped me with the crazy and beautiful photos for my top models, like Cyclops and LEGONARDO. Girls don't believe it's me portrayed in his photos and want to meet him at once! True story.

Thanks to **Marco** and **Susanna** for helping me transform a first-draft script into a full-blown graphic narrative and to **Nicola** for his last-minute help.

And finally, thanks to the unaware **Eddie** for inspiring the character of Dexter. Regarding the other comic characters, *any resemblance to real persons is purely coincidental*. No apprentices were harmed in the making of this book.

brief contents

contents in detail

introduction

The idea for this book was born in 2012 during a sandstorm in Saudi Arabia. I was locked in my hotel room waiting for the weather to improve and listening to Paul Dukas's symphonic poem *The Sorcerer's Apprentice*. At that time, I was involved in helping The LEGO Group in developing and testing the new LEGO MINDSTORMS EV3. I began drafting the story of a kid given the chance to become the apprentice of a scientist; I wanted this story to be the background of my next book about LEGO MINDSTORMS.

playing without a computer

The LEGO Group's idea for the EV3 set is that it should be fully usable if you have a computer with a fast Internet connection. Unlike previous versions, the EV3 Software is available only as a download. You also won't find a printed user guide, only a booklet with partial building instructions for the simplest official robot, TRACK3R, and a few hints about on-brick programming.

Even if you don't have a computer available, however, you'll still be able to enjoy playing with the robots in this book, thanks to the new EV3 on-brick programming. This is an effective (if limited) way to program the robots using the EV3 Brick menu. Chapters 1 through 4 and 8 through 10 offer numerous ways to play with your robots without a computer.

whom is this book for?

This book is for anyone with a passion for robots! No matter your age, this book can teach you how to build and program robots using the LEGO MINDSTORMS EV3 Retail set 31313.

Besides specific building and programming instructions, you'll learn general LEGO building techniques, as well as basic and advanced concepts of computer programming. Experts and more experienced robot makers will find sections scattered throughout the book that delve into topics a bit more deeply.

what do I need to use this book?

To use this book, you will need the LEGO MINDSTORMS EV3 Retail set 31313. You'll also need an Internet-connected computer to download and install the EV3 programming software, including the tutorials to build and program the five official models in the set. If you are a teacher or a student building with LEGO MINDSTORMS Education Core set 45544, see Appendix B for a list of the additional LEGO elements you'll need to make your set equivalent to the Retail set 31313.

You should find a USB cable in your set that will allow you to connect the EV3 Brick to your computer. To connect with a Bluetooth connection, your computer should have a built-in or external Bluetooth dongle. To connect the EV3 Brick to the computer using Wi-Fi, you must purchase a USB Wi-Fi dongle separately. As of this writing, the only dongle known to work with the EV3 Brick is the NETGEAR WNA1100.

the EV3 software

The EV3 Software was developed by National Instruments, creators of the LabVIEW development environment. The EV3 language is based on its visual dataflow programming language called G. National Instruments also developed the NXT-G programming language for the previous LEGO MINDSTORMS NXT generation.

If you are an NXT user and you have programmed with NXT-G, you will find EV3 programming much clearer: Now all the blocks show all settings at once, and you don't have to select a block to see its settings in the Configuration Panel. The software allows you to zoom and pan through the programs to explore them more easily. You'll also find additional programming features.

the structure of this book

This book is both a manual and a workbook. It introduces new concepts as the story in the comics develops. (Look at the comics carefully, as they include hidden clues to downloading bonus material from the companion website, including AUDR3Y, the people-eating plant; the L3AVE-ME-ALONE box; and many more.) "Digging Deeper" sidebars explain some advanced topics in depth. If you're an expert, you may want to skip some of the more introductory chapters and go directly to Chapters 9 through 16, where you learn how to build and program the four main robots. The chapters also contain experiments to apply and deepen your knowledge. Here's what you'll find in each chapter:

* Chapter 1: Contents of the 31313 set; identifying LEGO Technic elements.
* Chapters 2, 3, and 4: Build ROV3R, a wheeled robot that can be built quickly and programmed without a computer.
* Chapters 5, 6, and 7: Introduction to EV3 programming using the computer.
* Chapter 8: LEGO building techniques.
* Chapters 9 and 10: Build and program WATCHGOOZ3, a walking robot that can be programmed with or without a computer.
* Chapters 11 and 12: Build and program the SUP3R CAR, a steering car.
* Chapters 13 and 14: Build and program the SENTIN3L, a walking defense robot.
* Chapters 15 and 16: Build and program the T-R3X, a fearsome walking dinosaur.

the companion website

The companion website, **http://EV3L.com/**, contains the EV3 projects for the robots, errata, additional tips and tricks, and the bonus models for this book.

let's start already!

Welcome to the journey! Follow Dexter and Danny through their adventure and become an EV3L scientist's apprentice!

The EV3L Scientist's Apprentice
BLUE N
IT SHOULD BE SOMEWHERE AROUND HERE...
58100
OOPH!
BUMP!
WATCH OUT!
EXCUSE ME, BUT I'M LOST. WHERE'S THE EVOLUTION 3 LAB?
DANNY - 2013
ARTE INVISIBILE
N. PAPPALEPORE
YOU MUST BE BLIND, BOY!
01X1

> Target acquired!
> Facial recognition init...
> Detecting features... 3%
209304354
490459434 0543
4059493823
984325482892
32119843902
2309485490
330498494
3004599
300459
2982
WHERE IS THE DOORBELL?
I challenge anyone to read what's written here
FREEZE!
?!
ACCESS GRANTED! COME IN!
VRRRRR
YOUR IDENTITY HAS BEEN CONFIRMED: DEXTER ZIFU, SECOND-BORN OF MR. LI ZIFU, PREEMINENT OWNER OF ZIFU INDUSTRIES, BACKER OF THE EVOLUTION 3 LAB...
I KNOW WHO I AM! HOW ABOUT YOURSELF?
UH...VERY FRIENDLY INDEED!
PROTOTYPE BDL104, SEMI-HUMANOID CLASS, FRIENDLY NAME: CYCLOPS MARK I.
VRRRR
NOW FOLLOW ME!
VRRRR
WHERE ARE WE GOING?
WHAT BRINGS *YOU* TO THE LAB?
MY FATHER ASSURED ME THAT MR. DANIEL WOULD TEACH ME...
0x000DA1AB
INDEED, I'M BRINGING YOU RIGHT TO HIM!
MKAY...
01X2

'MORNING, DOC...
HOP!
CLICK!
MY FATHER SAID...
AH, IT'S YOU! WELL, I CAN FORGET ABOUT GETTING ANY WORK DONE TODAY!
YOUR FATHER SAID... I KNOW, I KNOW. I'VE MADE A PROMISE!
01X3

SOME DAYS EARLIER...
SO YOU UNDERSTAND HOW IMPORTANT IT IS FOR DEXTER TO BECOME YOUR APPRENTICE.
SURE, I UNDERSTAND! BUT I HAVE RESEARCH OF MY OWN TO ATTEND TO!
I KNOW THAT ONLY TOO WELL! IT'S WORK THAT I'M FUNDING, REMEMBER?
YES, SIR. BUT NOT ONLY THAT, WE'VE HAD SOME, UH, SETBACKS...HERE AT EV3L...
YOUR PROBLEMS, NOT MINE! ALMOST ALL YOUR FUNDING COMES FROM MY GRANTS!
YES, SIR. I UNDERSTAND. I'LL TAKE THE BOY ON! I'LL WAIT FOR HIM SATURDAY MORNING, AT THE EV3L BUILDING.
LET'S MAKE THIS QUICK! DID YOU BRING WHAT I ASKED? DO YOU KNOW THE BASICS?
HOW ABOUT A LITTLE BRUSHUP ON BASICS?
01X4

your LEGO MINDSTORMS EV3 set

Your LEGO MINDSTORMS EV3 31313 set includes a collection of LEGO elements, a printed manual (with instructions for building the official robot, TRACK3R, and some hints about how to get started with the EV3 Intelligent Brick), a USB-to-miniUSB cable to connect the EV3 Brick to your computer, and a paper test pad (just unroll the sleeve surrounding the box)—but no software. Where is the software? You can download it from the Downloads section of the LEGO MINDSTORMS EV3 official website (*http://LEGO.com/mindstorms/*). The LEGO Technic elements in the box are beams, pins, gears, and wheels as well as electronic components like motors, sensors, cables, and the EV3 Intelligent Brick itself.

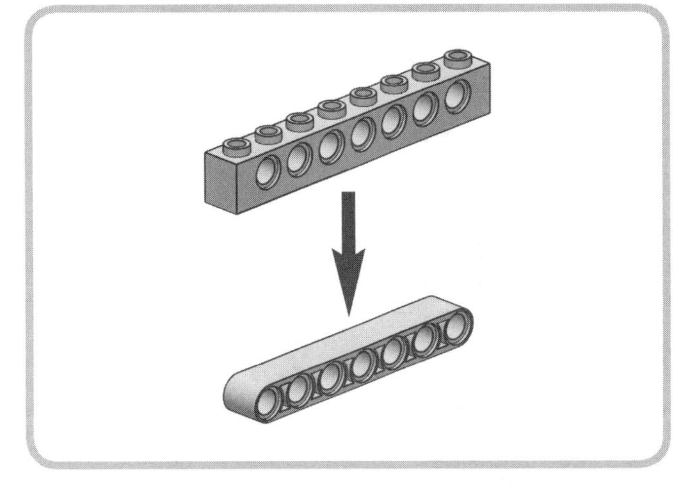

Figure 1-1: A classic 8M Technic brick compared to a 7M studless beam. Building with studless parts isn't always as intuitive as the classic way of building with LEGO by stacking bricks and plates from bottom to top. In fact, studless technique requires you to think three-dimensionally, from the inside out.

the studless way of building

As you may already know, there are no classic LEGO bricks in the EV3 box, and the beams don't have any studs. So how do you connect them?

Since 2000, LEGO Technic sets have been composed mainly of "studless" parts. The good old sharp-edged Technic bricks with studs (called *studded*) have slowly been replaced by smooth, studless Technic beams, which give the models a sleeker look (Figure 1-1).

I remember when I first switched from studded to studless building: Despite years of experience with "classic" LEGO Technic, I suddenly felt as though I could not build even the simplest thing. I was so frustrated! But as I took a close look at the official LEGO Technic models, I became more and more familiar with the parts. Sure, I had to learn a completely different way of building, but it was worth the effort. Studless building produces models that are lightweight, solid, and beautiful. Once you get started with studless building, you'll wonder how you could have lived without it!

studless vs. studded: the structural differences

Technic bricks have an even number of studs and an odd number of holes (a two-stud brick has one hole, a six-stud brick has five holes, and so on), and you measure and name them by counting their studs. Technic beams are like a minimalist, studless version of Technic bricks. Measure them by counting their holes, as shown in Figure 1-1. Like the studs on a LEGO brick, Technic pins act as the "glue" for your LEGO creations, as you can see in Figure 1-2.

The round ends of Technic beams allow you to build structures and mechanisms that are more compact and lighter than the ones you might build with standard LEGO bricks. For example, in order for two studded bricks to rotate next to each other on pegs, the pegs need to have two empty holes between them (see Figure 1-3). In contrast, the studless beams' pegs can be right next to each other.

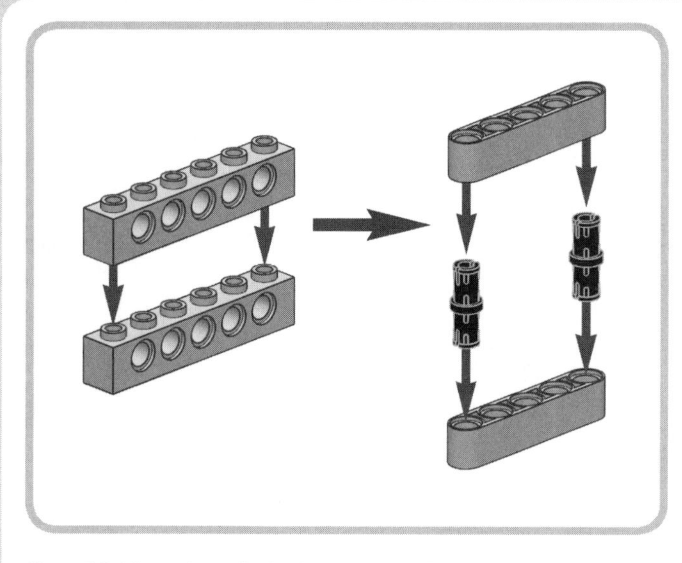

Figure 1-2: Like studs are for bricks, pins are the "glue" for studless beams.

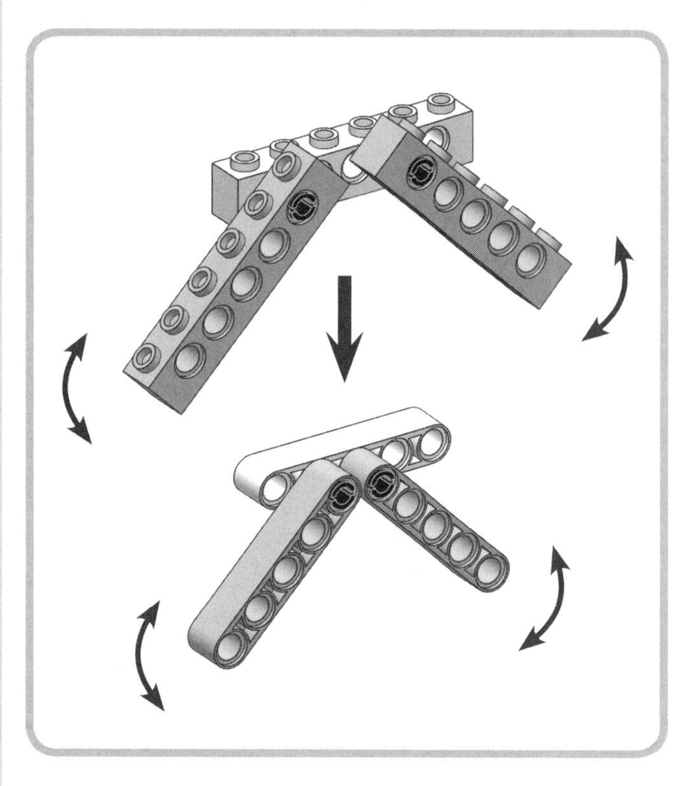

Figure 1-3: Technic beams occupy less space than bricks, allowing you to build more compact structures.

On the other hand, you can make sturdier, more rigid structures using standard bricks and plates. Depending on what you want to make, you might use studded, studless, or a combination of both techniques.

naming the pieces

Imagine that we're building a LEGO robot together and you find that you're missing a LEGO part. You ask me if I have one, but all you can muster is "Danny, would you pass me that . . . something, thingamajig, whatchamacallit, doodad, a habba whatsa?" and I don't understand what you need! Or worse, if you need to buy parts online (from a site like BrickLink; *http://www.bricklink.com/*) and you don't know how to refer to the parts, you'll be at a loss and unable to finish your robot.

Names are important. It's much easier to master LEGO building techniques if you know how to classify, name, and measure LEGO parts. You can't write a novel if you don't know grammar and vocabulary, and the same holds true for LEGO. You've got to know the parts.

The pieces in the EV3 31313 set can be divided into these categories:

beams straight beams, angular beams, frames, thin beams, and links
connectors pins, axles and bushes, axle and pin connectors, and cross blocks
gears spur gears, bevel gears, and worm gears
wheels and treads wheels, treads, and tires
decorative pieces panels, teeth, swords, and so on
miscellaneous pieces balls, ball magazine, ball shooter, rubber band
electronic pieces the EV3 Intelligent Brick, motors, sensors, and cables

NOTE For these categories, I've chosen to use the names that I think are the easiest to remember. For the official LEGO names, see Appendix A.

I'll describe the categories briefly, with a minimum of boring chatter.

beams

As mentioned earlier, *beams* are the studless equivalent of Technic bricks. This category includes straight beams, angular beams, and frames. We'll include thin beams and links in this category too. Beams can have round holes, which can fit pins, or cross holes, which can fit axles or axle pins. Links have ball sockets that fit pins with towballs.

straight beams

Figure 1-4 shows the straight beams; their names are listed in Table 1-1. The beams are measured by counting their holes. For example, a straight beam with three holes is a *3M beam* (and you can omit the adjective "straight"). The number of holes in a beam corresponds to the length of the beam as expressed in *Fundamental LEGO Units*, or *modules* (1M = about 8 mm). In all LEGO building instructions, you'll see a box for each building step that lists the parts needed in that step. The length of a beam is noted at its top-right corner.

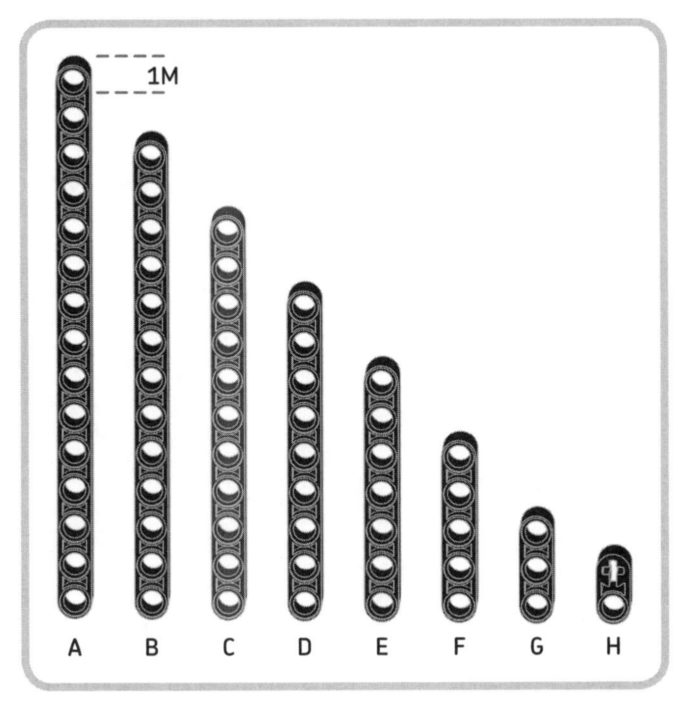

Figure 1-4: The straight beams

table 1-1: the straight beams

Label in Figure 1-4	Name	Color
A	15M beam	Black
B	13M beam	Black
C	11M beam	Red
D	9M beam	Black
E	7M beam	Black
F	5M beam	Black
G	3M beam	Black
H	2M beam with cross hole	Black

angular beams

Figure 1-5 and Table 1-2 show the angular beams and their names. An angular beam with three holes before and seven holes after the bend is a *3×7 angular beam*. The same naming pattern is used for the other angular beams. Notice that some angular beams have cross holes at their ends.

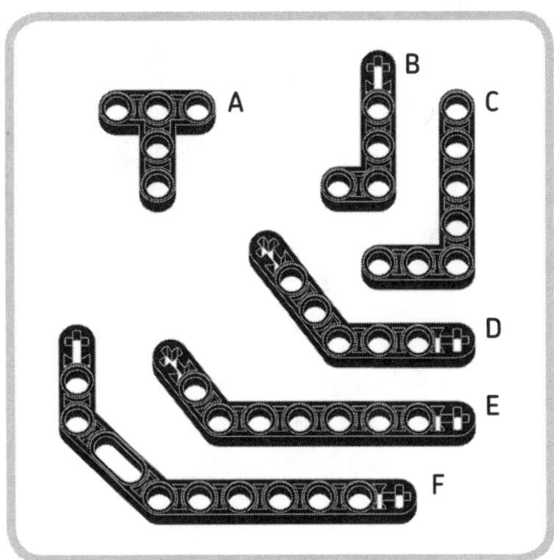

Figure 1-5: The angular beams

table 1-2: the angular beams

Label in Figure 1-5	Name	Color
A	T beam	Black
B	2×4 angular beam	Black
C	3×5 angular beam	Black
D	4×4 angular beam	Black
E	3×7 angular beam	Black
F	Double angular beam	Black

The angular beams labeled A, B, and C have right angles, while F has two 45-degree bends. But what about the others? What kind of strange angle is that, and how do you use it to build? You'll learn the secrets of working with the various angular beams in Chapter 8.

frames

We also have special beams called *frames*, as shown in Figure 1-6. We refer to these based on their shapes as *O-frames* (or simply *frames*) and *H-frames*. Once you know how to work with them, you'll find that they allow you to build rock-solid structures that will not come apart!

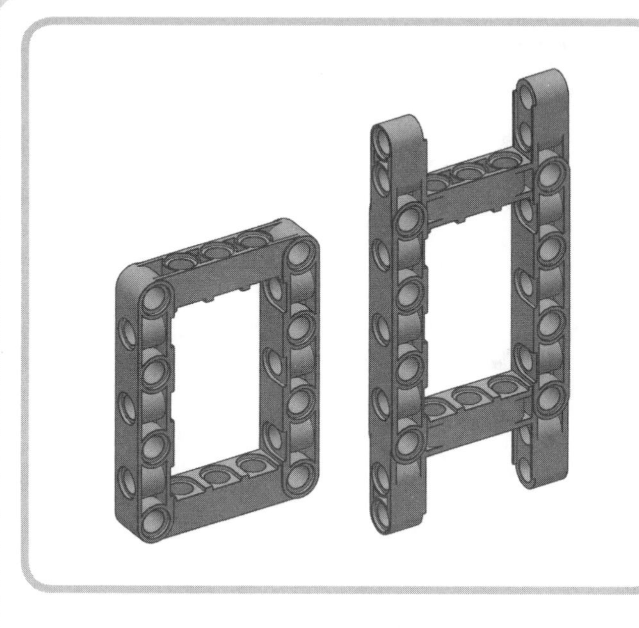

Figure 1-6: The O-frame and the H-frame

thin beams and links

The *thin beams* and *links* are shown in Figure 1-7 and Table 1-3. Thin beams have cross holes at each end, and they are one half-module thick. Think of the 6M and 9M links as beams with ball sockets at their ends. These fit pins with towballs (items D and H in Figure 1-8). Ball joints allow for a wide range of motion and rotation, similar to your shoulder or hip joints.

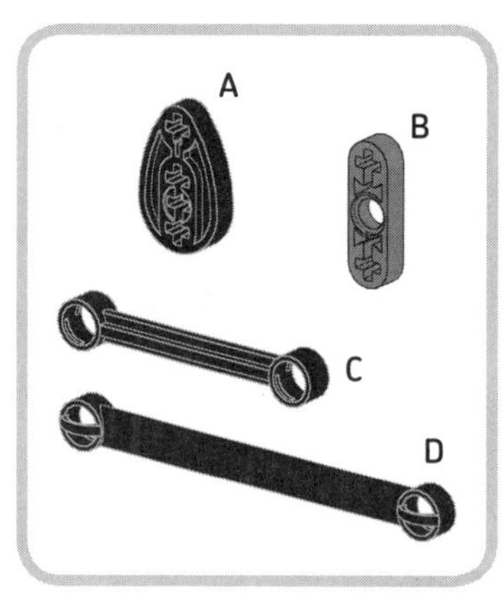

Figure 1-7: The thin beams and links

table 1-3: the thin beams and links

Label in Figure 1-7	Name	Color
A	Cam	Black
B	3M thin beam	Grey
C	6M link	Black
D	9M link	Black

connectors

Most parts in the EV3 set are connectors. When building with wood or metal, we use nails, glue, staples, screws, bolts, washers, and so on to connect the various pieces. In the wonderful world of LEGO Technic, we use pins, axles and bushes, axle connectors, and various cross blocks.

pins and axle pins

Pins hold beams together when fitted inside the beams' round holes. Pins are divided into two groups: pins with friction and pins without friction (also called *smooth pins*). Figure 1-8 and Table 1-4 show 2M and 3M pins, axle pins, pins with towballs, and a special 3M pin with a stop bush (also called a *bushing*).

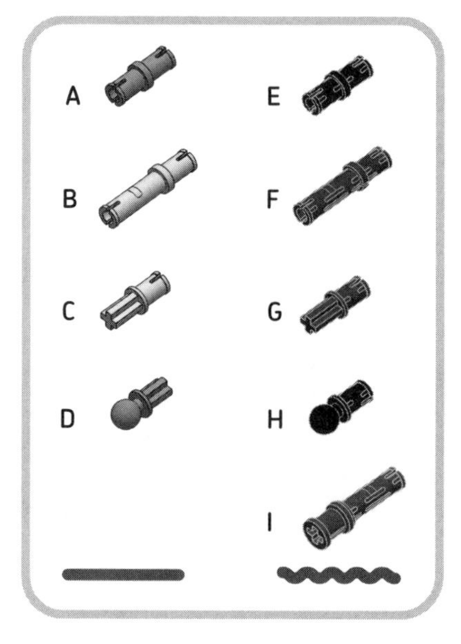

Figure 1-8: The renowned Technic pins. The straight line signifies the pins without friction (A–C); the wavy line indicates pins with friction (E–I). The axle pin with towball (D), while not technically a smooth pin, is listed here for comparison with the pin with towball (H).

table 1-4: the pins and axle pins

Label in Figure 1-8	Name	Color
A	Pin without friction	Grey
B	3M pin without friction	Tan
C	Axle pin without friction	Tan
D	Axle pin with towball	Grey
E	Pin with friction	Black
F	3M pin with friction	Blue
G	Axle pin with friction	Blue
H	Pin with towball	Black
I	3M pin with stop bush	Red

Pins without friction (labeled A, B, and C in Figure 1-8) turn smoothly and freely in the Technic holes. They are color coded: 2M pins are always grey, while 3M pins and axle pins are tan. Pins without friction are mainly used to connect moving beams.

NOTE The EV3 set has no axle pins without friction, labeled C in Figure 1-8, but I've included it here for the sake of completeness. Axle pins without friction can be used to hold a gear so that it can turn freely.

The *pins with friction* (labeled E, F, G, H, and I) have ridges that increase friction and make it harder for them to turn in the Technic holes. The ridges also prevent the pins from rattling. 2M pins with friction are always black, and 3M pins with friction and axle pins with friction are blue. 3M pins with stop bush come in many colors, but they're red in the EV3 set. The pins are color coded to help you identify their function at first sight.

Pins with friction are great for building stable structures because they hold beams together better than pins without friction. In the following chapters, you'll learn many ways to use pins and axle pins.

axles and bushes

Axles are designed to transfer rotational movement, for example from a motor shaft to a wheel. Axles can also be used to hold structures together. Their cross section looks like a cross (their complete name is actually *cross axle*), and they fit perfectly into parts that have cross holes, such as gears, angular beams, and cross blocks.

Like beams, axles come in many lengths. You can measure them by putting them next to a beam and counting the holes in the beam. Once you get used to working with them, you will be able to sort them by size at a glance, even without measuring them. This superpower really amazes people!

CROSSES AND HOLES

Build the following assemblies. Each one has a symbol to help you pick the right pieces. Wavy lines indicate a pin with friction (black or blue) and straight lines indicate smooth pins (grey or tan). A plus (+) indicates axle pins and a circle indicates round pins.

* Once you've built the assemblies, hold them and try to make the inclined beam swing. What happens in each case?
* In the rightmost assembly, which 2×4 angular beam is the easiest to turn?

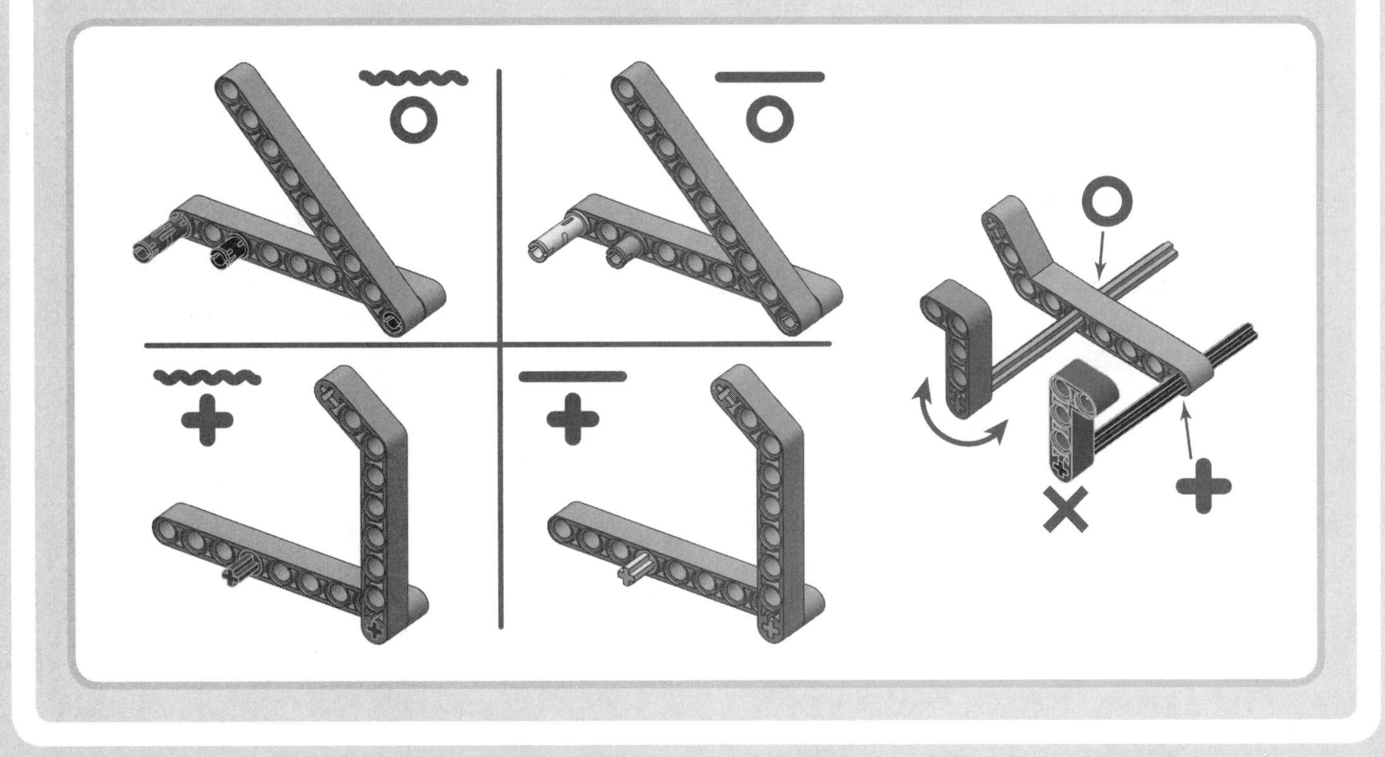

Like pins, axles are color coded, as shown in Figure 1-9: The 2M axle is red, odd-length axles are light grey (3M, 5M, 7M, 9M), and larger even-length axles are black (4M, 6M, 8M, 10M, 12M). The EV3 set doesn't have normal 4M and 8M axles; it includes a tan axle with a cylindrical stop in the middle (4c), as well as some axles with a stop at one end (3s, 4s, and 8s). Unlike the 4M and 8M axles with stop (4s and 8s), the 3M axle with stop (3s) has a protruding stud. In the 3s, 4s, and 8s axles, the stop looks like a built-in bush, and it *stops* the axle from passing through a hole or a cross hole. In the 4c axle, the stop in the middle stops the axle from passing completely through a cross hole.

Figure 1-9 also shows two bushes, labeled B1 (yellow, one half-module thick) and B2 (red, one module thick). You'll usually fit these bushes over axles to prevent the axles from coming out of holes, or you'll use them to keep space between two or more elements of a structure. Because the bushes are mainly used with axles, I've listed them together.

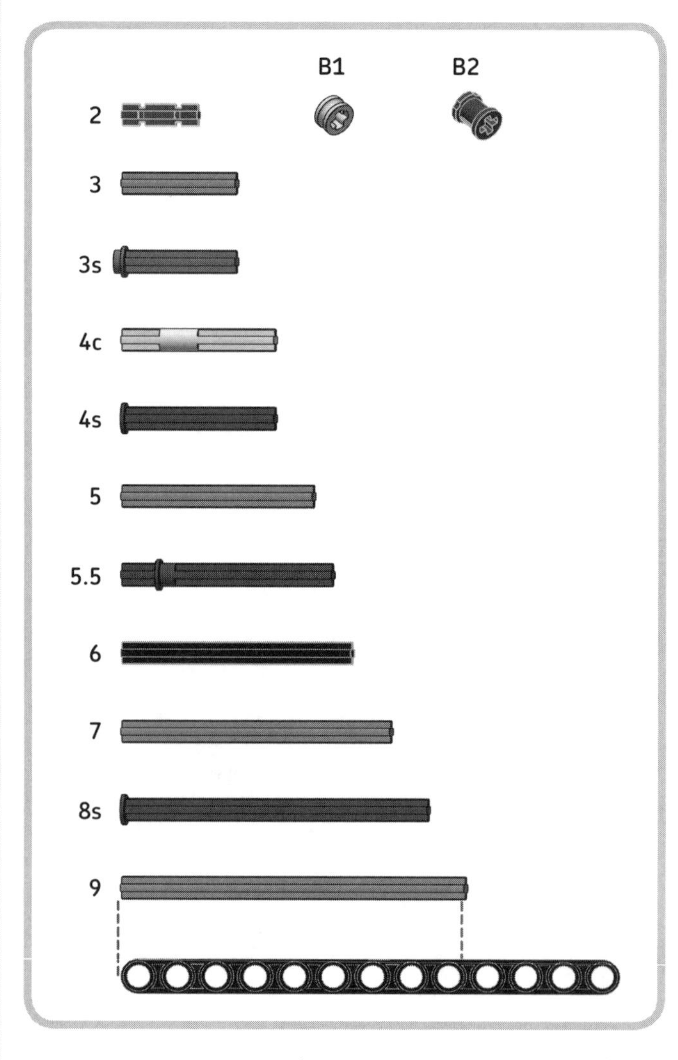

Figure 1-9: The axles and bushes (with 13M beam shown for comparison)

axle, pin, and angle connectors

Figure 1-10 shows the axle, pin, and angle connectors, and Table 1-5 lists their names. Each angle connector (those labeled E, F, G, and H) is identified by a number embossed on its body.

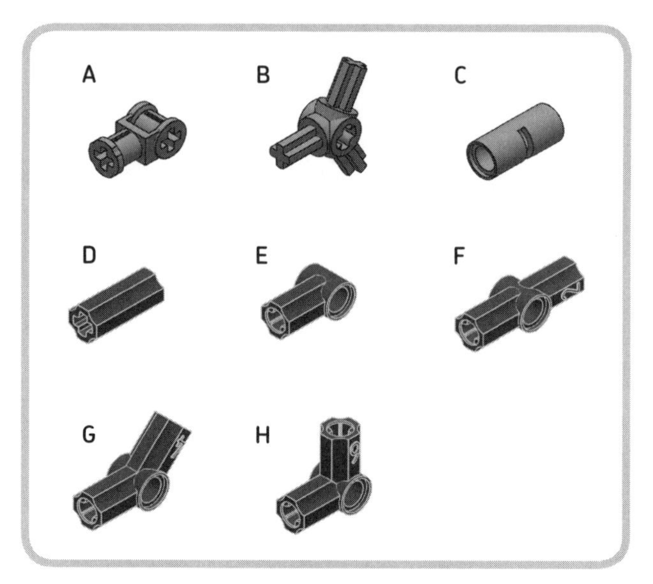

Figure 1-10: The axle, pin, and angle connectors

table 1-5: the axle, pin, and angle connectors

Label in Figure 1-10	Name	Color
A	Connector with axle holes	Grey
B	Connector hub with 3 axles	Grey
C	Pin connector	Grey
D	Axle connector	Red
E	Angle connector #1	Red
F	Angle connector #2	Red
G	Angle connector #4	Red
H	Angle connector #6	Red

cross blocks

Here comes the fun! Cross blocks are essential to studless building because they allow you to build—and think—in three dimensions. Remember, studless building isn't about simply stacking bricks; we're adding parts from all sides, as shown in Figure 1-11.

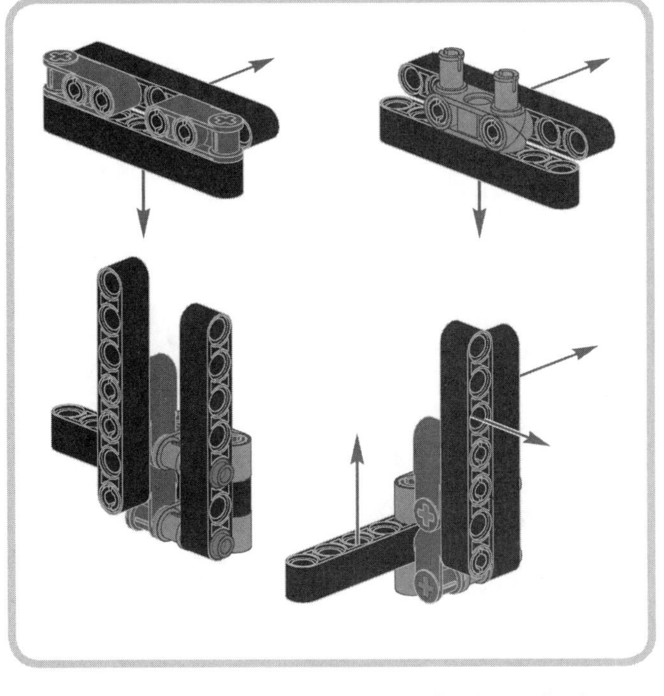

Figure 1-11: With cross blocks, you can build in any direction, not just from bottom to top.

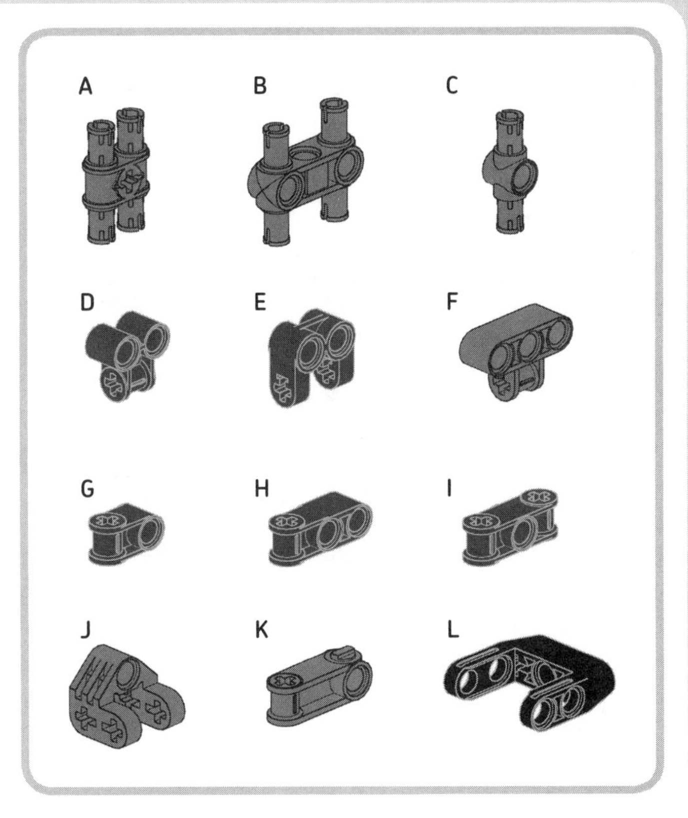

Figure 1-12: The cross blocks

Figure 1-12 shows the cross blocks in the EV3 set, and Table 1-6 lists their names. Some entries also list a nickname for the piece; for example, "Mickey" and "Minnie" are funny names for parts D and E, respectively. (Thanks to LEGO MINDSTORMS Education designer Lee Magpili for these nicknames.) The part labeled L can be used as a gearbox to hold 90-degree-coupled 12z and 20z bevel gears (for example, see "Medium Motor with Gearbox" on page 126).

It would be nearly impossible to show you all of the combinations you can build with cross blocks. The best way to learn how to use them is to draw your inspiration from the projects in this book and from the many Technic models in the wild.

table 1-6: the cross blocks

Label in Figure 1-12	Name	Color
A	2M beam with 4 pins	Grey
B	3M beam with 4 pins	Grey
C	3M pin with hole	Grey
D	2×1 cross block ("Mickey")	Red
E	2×2 fork cross block ("Minnie")	Red
F	3×2 cross block	Grey
G	2M cross block	Red
H	3M cross block	Red
I	Double cross block	Red
J	2×2×2 fork cross block	Grey
K	3M cross block, steering	Grey
L	Gearbox cross block	Black

gears

When people think of complicated machines, gears often pop into their minds, even if the machine is a computer with few moving parts! And when a machine stops working, people often blame its (sometimes imaginary) gears.

Gears are rotating wheels with teeth that mesh with other toothed parts (like gears, gear racks, and worm gears) to transmit movement. Figure 1-13 shows the gears included in the EV3 set, with their corresponding names in Table 1-7. LEGO gears are identified by number of teeth, as indicated in their name followed by *z*; for example, a 24-tooth gear is called a *24z gear*.

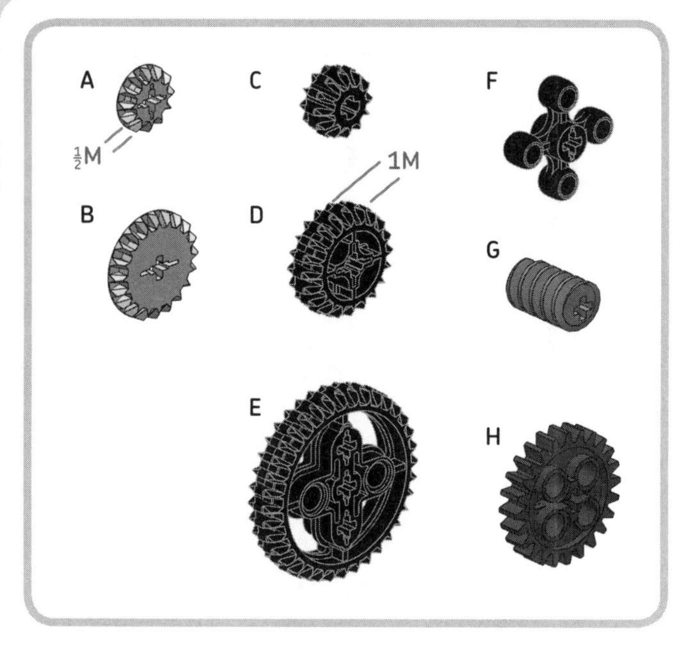

Figure 1-13: The gears

table 1-7: the gears

Label in Figure 1-13	Name	Color
A	12z bevel gear	Tan
B	20z bevel gear	Tan
C	12z double-bevel gear	Black
D	20z double-bevel gear	Black
E	36z double-bevel gear	Black
F	4z knob wheel	Black
G	Worm gear (1z)	Grey
H	24z gear	Dark Grey

Most gears are 1M thick, with the exception of the 12z and 20z bevel gears, which are both one half-module thick. The 24z gear (labeled H) is a spur gear, but "spur" can be omitted from the name (8z, 16z, and 40z spur gears also exist in the LEGO system). The worm gear is a particularly tough gear. You'll learn more about it and how to combine gears in Chapter 8 and while building the robots in this book.

wheels, tires, and treads

The simplest and most efficient way for your robots to move is on wheels. The EV3 set contains four large wheels with tires, three medium wheels with tires, four small wheels with two small tires, and two rubber treads. Figure 1-14 shows the various types of wheels, tires, and treads in the set, and Table 1-8 lists their names.

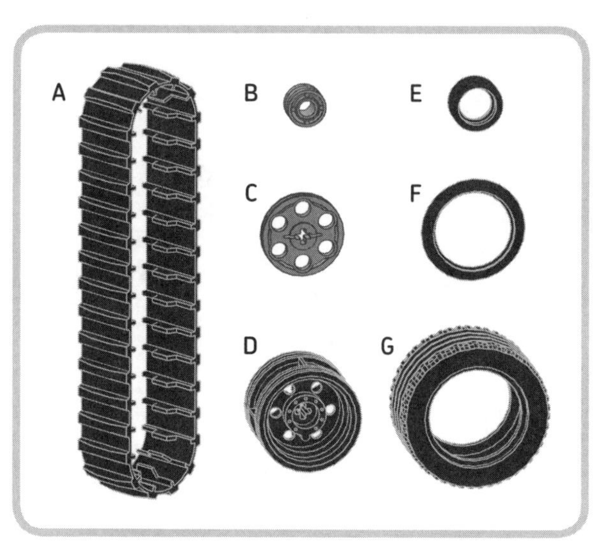

Figure 1-14: The wheels, tires, and treads

table 1-8: the wheels, tires, and treads

Label in Figure 1-14	Name	Color
A	Rubber tread	Black
B	Small wheel	Grey
C	Medium wheel	Grey
D	Large wheel	Black
E	Small tire	Black
F	Medium tire	Black
G	Large tire	Black

The large tires have their dimensions printed on their edge; for example, *43.2×22 ZR*. These measurements are in millimeters: In this example, 43.2 mm is the tire diameter and 22 mm is the width of the tire. The medium tire has a 30 mm diameter and is about 3 mm wide. The small tire has a 14 mm diameter and is 6 mm wide.

decorative pieces

The EV3 set contains several decorative pieces. In addition to the teeth with axle holes, there are many white panels, blades, and swords, as you can see in Figure 1-15. Their names are given in Table 1-9. The panels come in mirrored pairs and are identified by a number embossed on the concave side. Because these panels have many connection holes, you can even use them as large cross blocks when building.

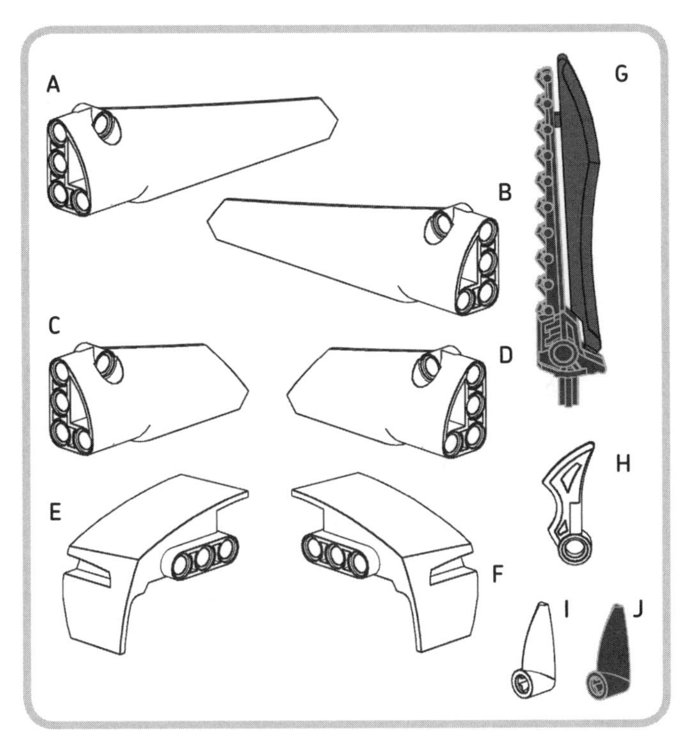

Figure 1-15: The decorative pieces

table 1-9: the decorative pieces

Label in Figure 1-15	Name	Color
A	Long panel #5	White
B	Long panel #6	White
C	Medium panel #3	White
D	Medium panel #4	White
E	Right mudguard	White
F	Left mudguard	White
G	Sword	Red/Grey
H	Curved blade	White
I	Tooth	White
J	Tooth	Red

miscellaneous pieces

The miscellaneous pieces are special elements: a ball magazine, a ball shooter, three balls, and a rubber band. LEGO rubber bands are color coded; the red one included in the EV3 set has a 24 mm diameter. These parts are shown in Figure 1-16 and listed in Table 1-10.

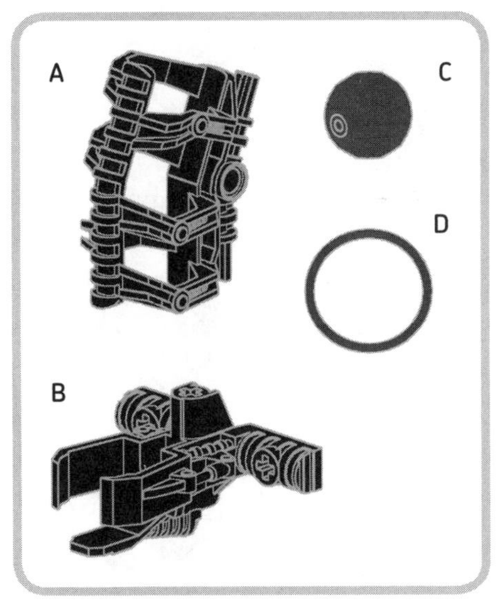

Figure 1-16: The miscellaneous pieces

table 1-10: the miscellaneous pieces

Label in Figure 1-16	Name	Color
A	Ball magazine	Black
B	Ball shooter	Black
C	Ball	Red
D	Rubber band	Red

electronic pieces

Finally, here's what makes a MINDSTORMS set a real robotics tool kit: the electronic pieces! These pieces are shown in Figure 1-17 and their names are listed in Table 1-11. The 31313 set contains two Large Servo Motors and seven cables: four 25 cm (10 in) cables, two 35 cm (14 in) cables, and one 50 cm (20 in) cable.

The EV3 Intelligent Brick is a microcomputer that acts as the brain for your robotic creations. It features the Linux operating system running on a 300 MHz ARM9 controller. It has 64MB of RAM and 16MB of flash memory, expandable with a microSD card up to 32GB! The screen resolution is 178×128 pixels (black and white).

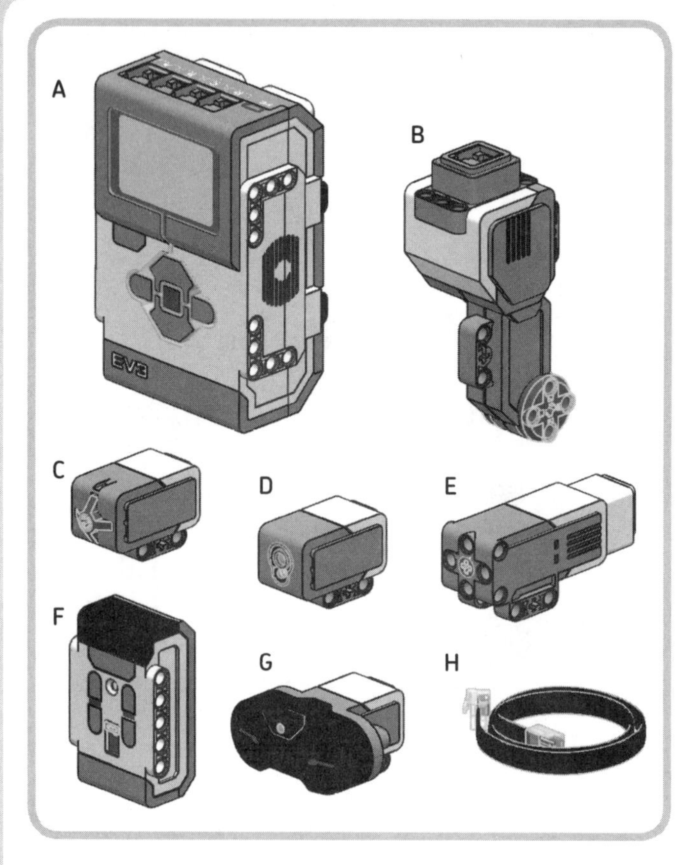

Figure 1-17: The electronic pieces

table 1-11: the electronic pieces

Label in Figure 1-17	Name
A	EV3 Intelligent Brick
B	Large Servo Motor
C	Touch Sensor
D	Color Sensor
E	Medium Servo Motor
F	Remote IR Beacon
G	IR Sensor
H	Connector cable

You can connect the EV3 Brick to a computer with a mini-USB 2.0 port, and you can connect other devices (daisy-chained EV3 Bricks or a Wi-Fi Dongle) to the USB 1.1 host port on its side. You can also connect up to four motors and four sensors to the Brick. The EV3 Brick can recognize which motor or sensor is attached to its ports, thanks to the Auto-ID feature. You can use 6 AA batteries to power it (LEGO Education sells a rechargeable battery; see Appendix B for details).

The EV3 Servo Motors are not plain LEGO motors: They have a built-in rotation sensor (1 degree resolution) to allow precise motion control. The Large Servo Motor runs at 160 to 170 rpm, with a running torque of 20 N·cm and a stall torque of 40 N·cm. The Large Servo Motor is slower but stronger than the Medium Servo Motor, which runs at 240 to 250 rpm with a running torque of 8 N·cm and a stall torque of 12 N·cm.

A motor is in stall (or is *stalled*) when it is commanded to turn but the shaft is blocked by some mechanical stop and is unable to move. This consumes a lot of battery power, and you should avoid this situation by, for example, turning the motor off before it gets stuck or removing the block that's preventing the shaft from turning freely.

The sensors give your robots the ability to touch and see. The Touch Sensor is basically a switch that your robot can use to detect contact with objects. The Color Sensor can measure ambient light, measure the amount of light reflected by objects, and recognize the color of objects. The IR Sensor can measure distance, detect the distance and the bearing to the Remote IR Beacon, and receive remote commands from the Remote IR Beacon. I'll describe the various sensors in detail later in the book.

the differences between the EV3 retail and education sets

The EV3 set comes in two versions: Retail set 31313 (the set used in this book) and Education Core set 45544. The sets have different assortments of parts, and they also differ in which EV3 sensors they include. While in the Retail version you have a Touch Sensor, an IR sensor with a Remote IR Beacon, and a Color Sensor, in the Education set you have two Touch Sensors, a Color Sensor, an Ultrasonic Sensor, and a Gyroscopic Sensor. The differences between the two sets are listed in detail in Appendix B.

conclusion

This chapter has provided an overview of the contents of the LEGO MINDSTORMS EV3 31313 set. You learned how to identify the various elements in the set. You've also been introduced to the unique aspects of studless Technic parts: round and cross holes, connection blocks, pins with and without friction, and so on. In Chapter 2, you'll build a simple mobile robot to start exploring the world of robotics with LEGO MINDSTORMS EV3.

GOOD! NOW, FOR YOUR FIRST JOB, HOW ABOUT YOU CLEAN UP AROUND HERE?
WHAT? BUT I CAME HERE TO LEARN, NOT TO CLEAN!
WELL, WHY DON'T WE BUILD AN EV3 ROBOT THAT CLEANS FOR YOU?
BUT I DON'T KNOW HOW TO *CREATE* ANYTHING...
SINCE YOU ARE A BEGINNER, I PREPARED A GUIDE FOR YOU!
LOOK AT THIS!
HERE YOU'LL FIND THE INSTRUCTIONS TO BUILD A MOBILE ROBOT, MADE WITH A MINIMUM OF PARTS!
0x0B1ESSED
WOW! THANKS!
HOW DOES IT WORK?
0x0B1ESSED
OZX1

0xB16B00B5
YOU'LL NEED TO INVENT THE CLEANING TOOL.
HOW DO I *COMMAND* IT?
FOR NOW, YOU WILL PROGRAM THE ROBOT WITHOUT A COMPUTER. WE WILL USE THE ***BRICK PROGRAM APP*** ON THE EV3 BRICK ITSELF!
NOW GET TO WORK!
WOW! LET'S START ALREADY!
THIS PLACE REALLY NEEDS A GOOD CLEANUP!
OZX2

building ROV3R

Now that you're familiar with the pieces that come in the EV3 set, it's time to build your first robot: ROV3R, a mobile robot that is built with just a few parts. Thanks to its modular design, reconfiguring it for various missions is a snap. In this chapter, I'll show you how to combine the wheeled version of ROV3R with different sensors and tools (see Figures 2-1 and 2-2), but you can easily swap out ROV3R's wheels for treads. In the chapters that follow, you'll learn how the added sensors and tools work and then program these ROV3Rs to accomplish various tasks.

Along with the building instructions that follow, you'll find many tips and tricks. As you read through the instructions, you'll learn various building techniques, tips for making good design choices, and some rules of thumb for building robots with studless LEGO Technic pieces.

Although this book is printed in grayscale, the contrast and readability of the images have been maximized. Almost all of the parts in the 31313 set are white, black, or red. When knowing the color of a certain element is important—for example, to distinguish pins with friction from pins without friction (or axle pins)—I've added labels to indicate the color (see Table 2-1).

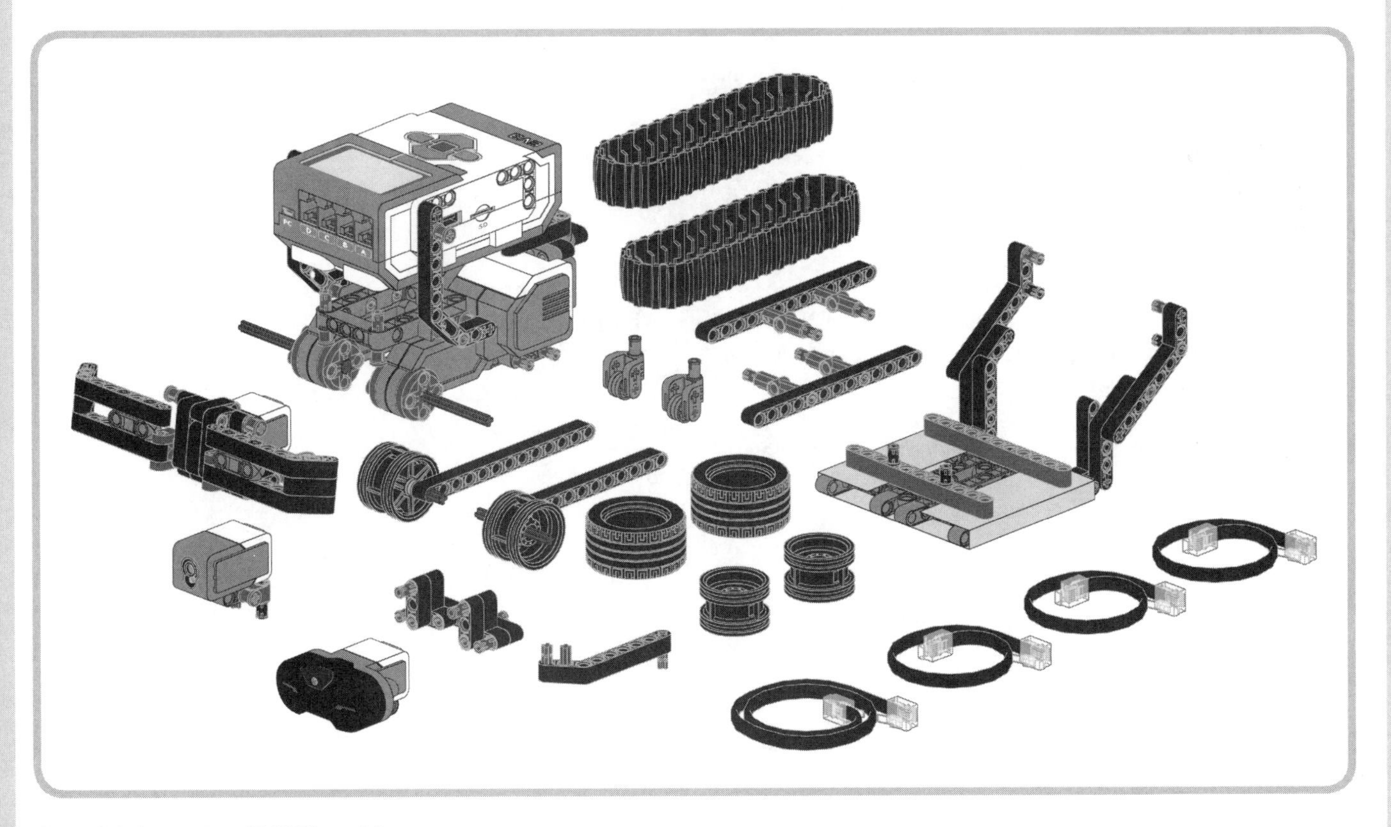

Figure 2-1: An overview of ROV3R's modules

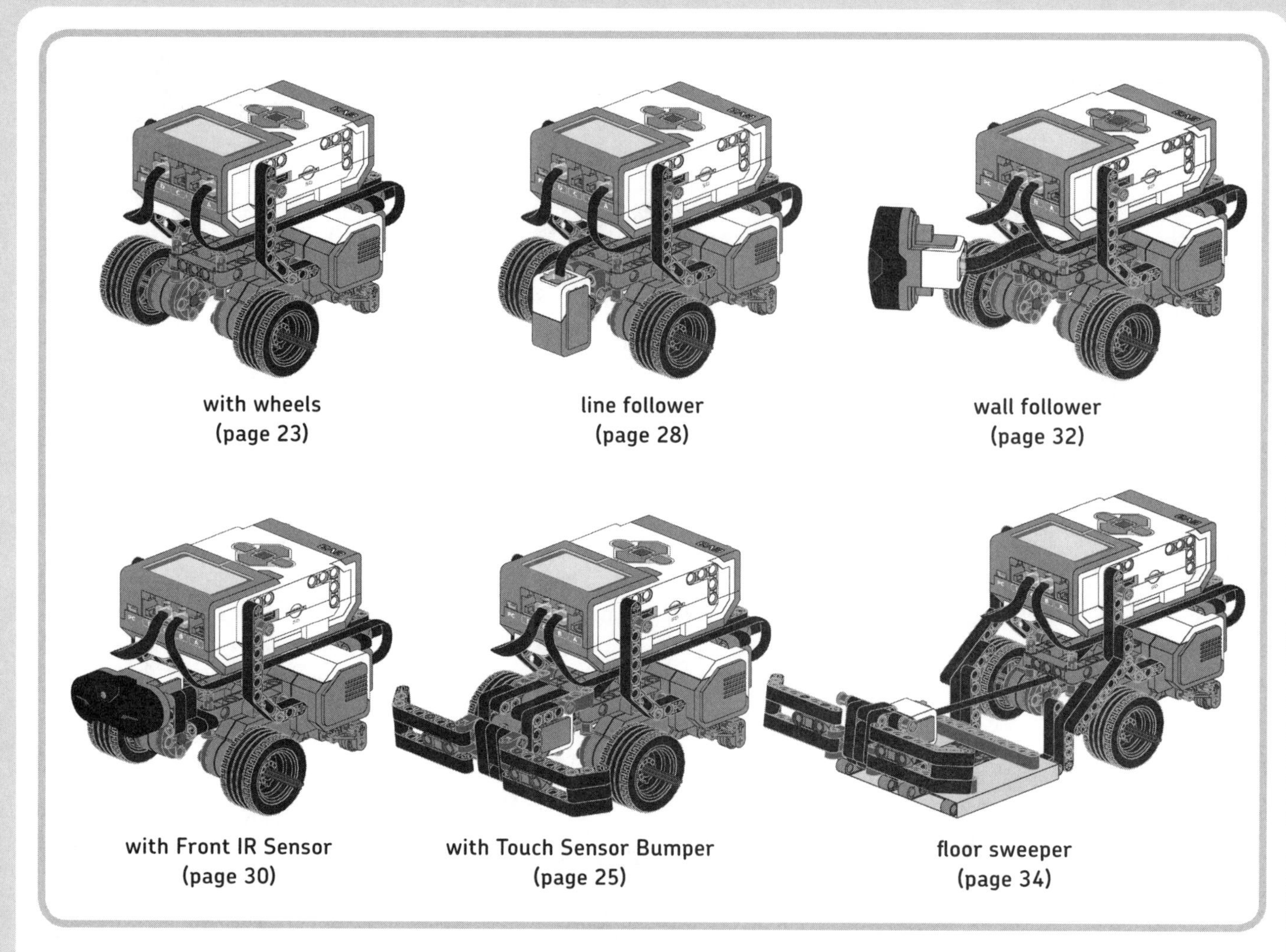

Figure 2-2: ROV3R can be reconfigured in many ways, thanks to its modular design. These are just a few of the possible combinations.

When not otherwise specified, pins are black pins with friction, axle pins are blue axle pins with friction, and 3M pins are long blue pins with friction. And remember that odd-length axles (3M, 5M, 7M, 9M) are light grey.

NOTE This color legend applies to all elements throughout this book.

table 2-1: labels used to designate colors in the building instructions

Color	Label
White	W
Grey	G
Dark grey	DG
Yellow	Y
Red	R
Blue	B
Tan	T

base module

First you'll need to build the Base Module, which can be used with wheels (see "ROV3R with Wheels" on page 23) or treads (see "ROV3R with Treads" on page 40).

At the top of an axle, you will see a number indicating its length. To determine the length of an axle, first place it next to a long beam and then count the holes in the beam that lie alongside the axle. You can also use these real-scale pictures to measure axles.

1

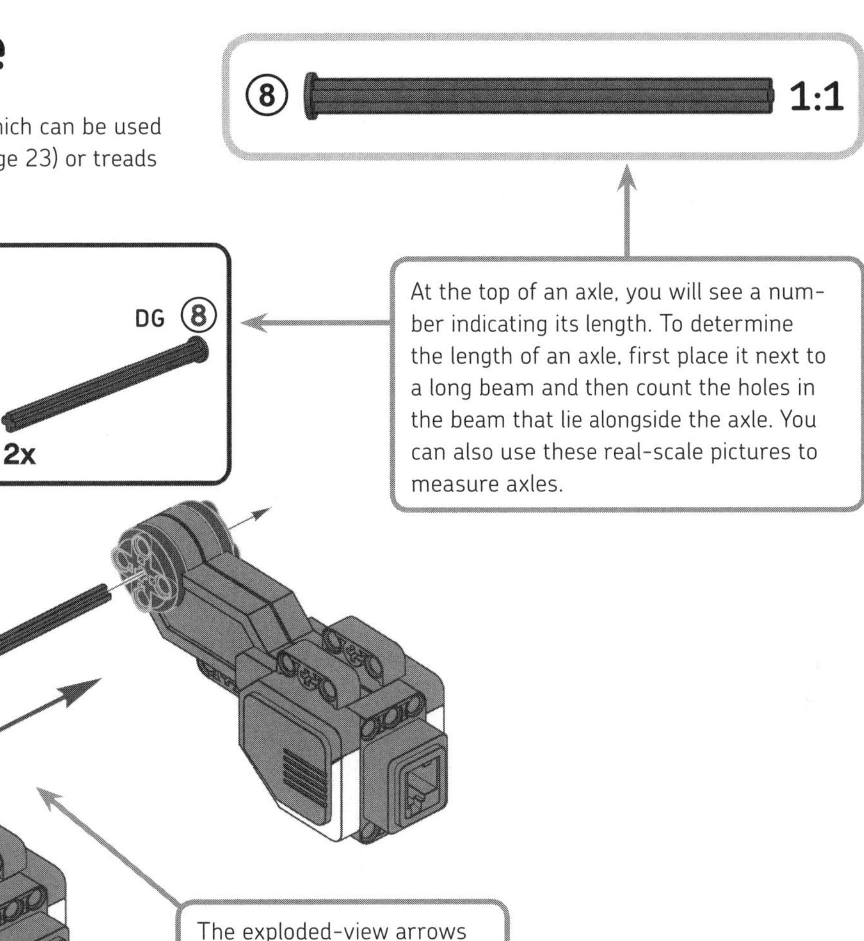

The exploded-view arrows show how the parts should be assembled.

The parts list box shows the elements you'll need for that particular step.

The EV3 Large Servo Motor is internally geared down, with a built-in one-degree-resolution rotation sensor. The Large Motor runs at 160 to 170 rpm, with a running torque of 20 N·cm and a stall torque of 40 N·cm.

2

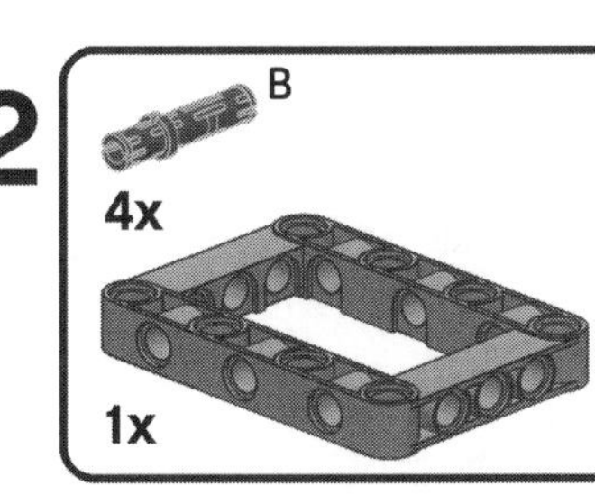

The O-frame holds the motors together. This technique is called *bracing*.

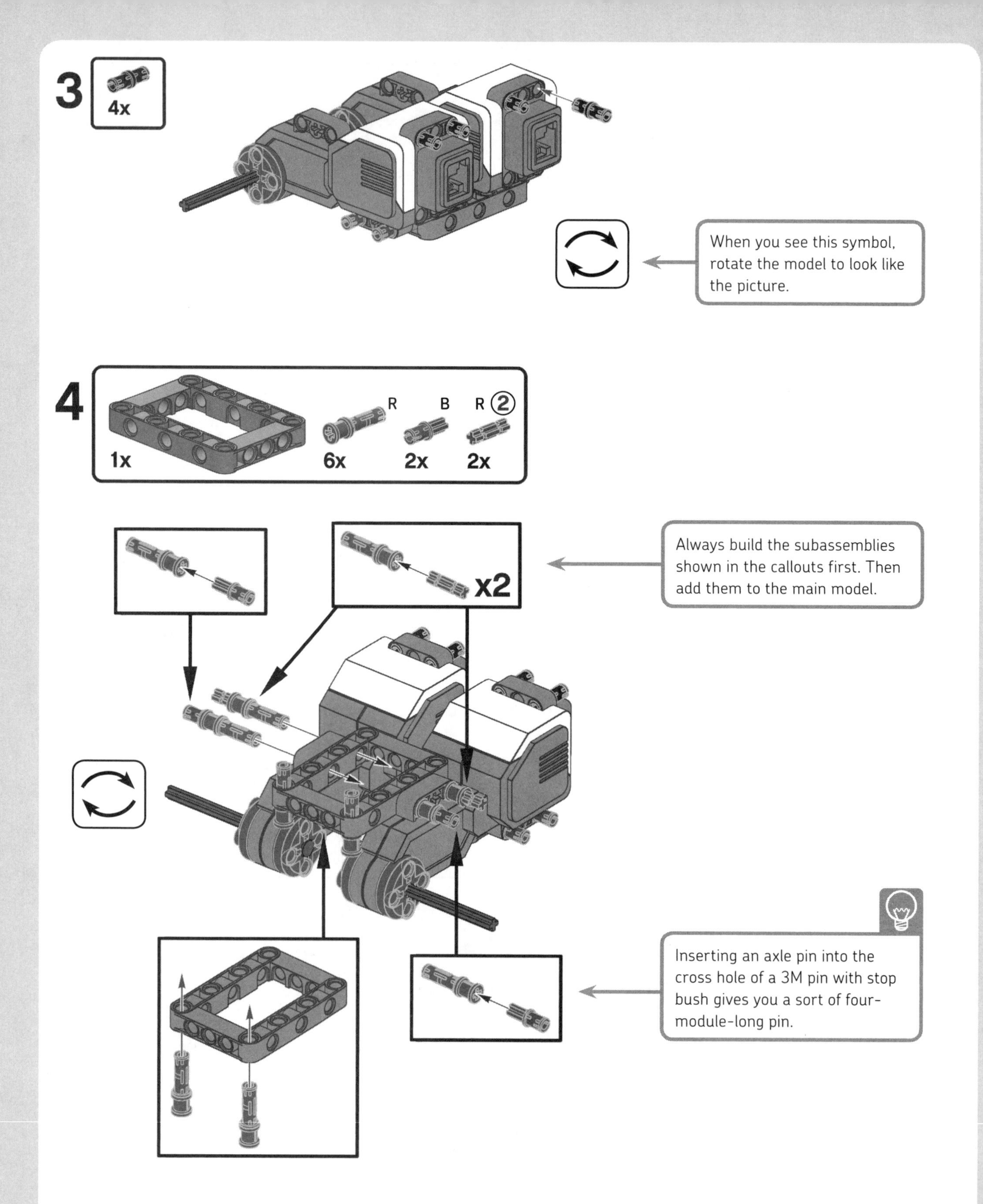
3
4x
When you see this symbol, rotate the model to look like the picture.
4
1x
R
6x
B
2x
R ②
2x
x2
Always build the subassemblies shown in the callouts first. Then add them to the main model.
Inserting an axle pin into the cross hole of a 3M pin with stop bush gives you a sort of four-module-long pin.

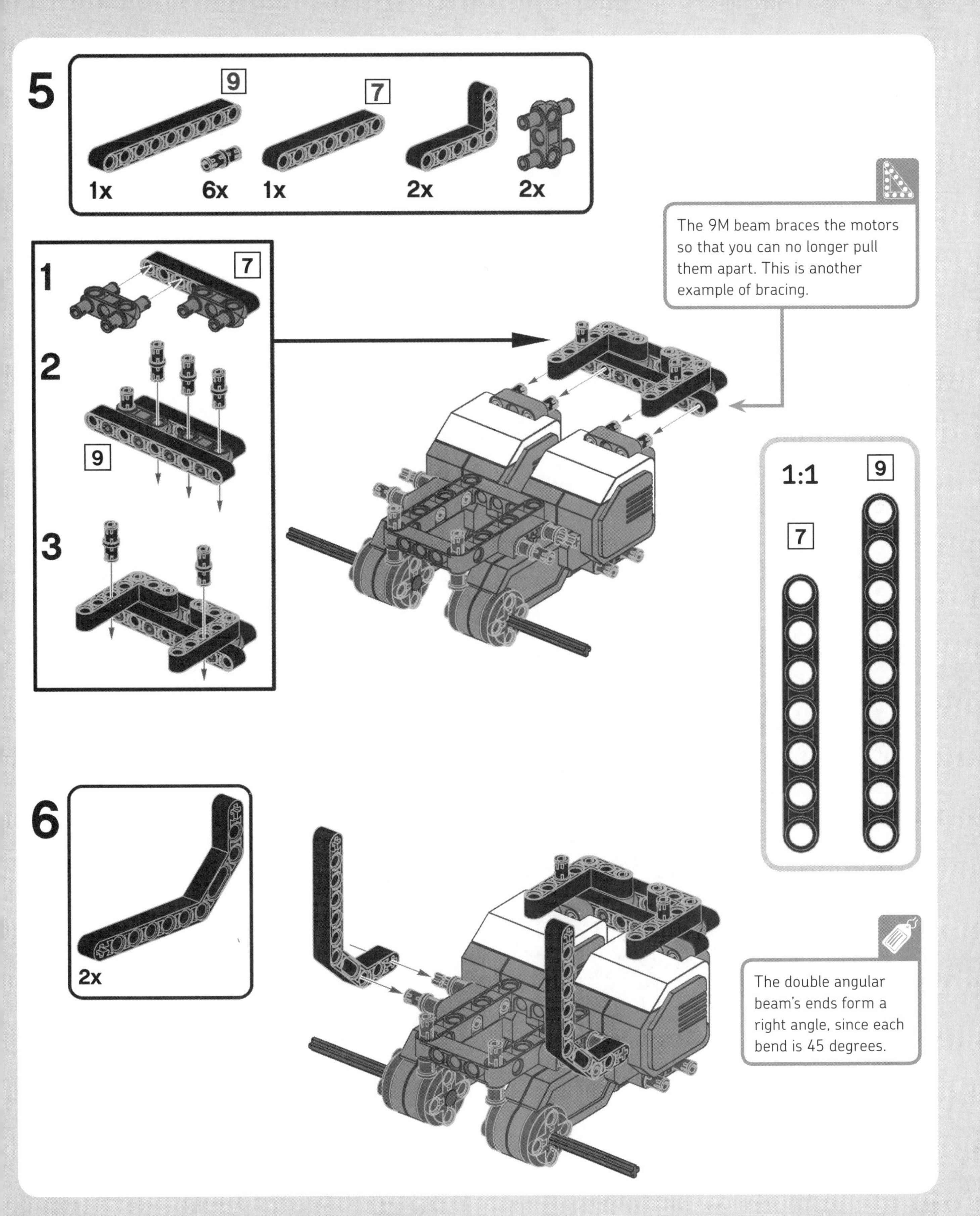

The 9M beam braces the motors so that you can no longer pull them apart. This is another example of bracing.

The double angular beam's ends form a right angle, since each bend is 45 degrees.

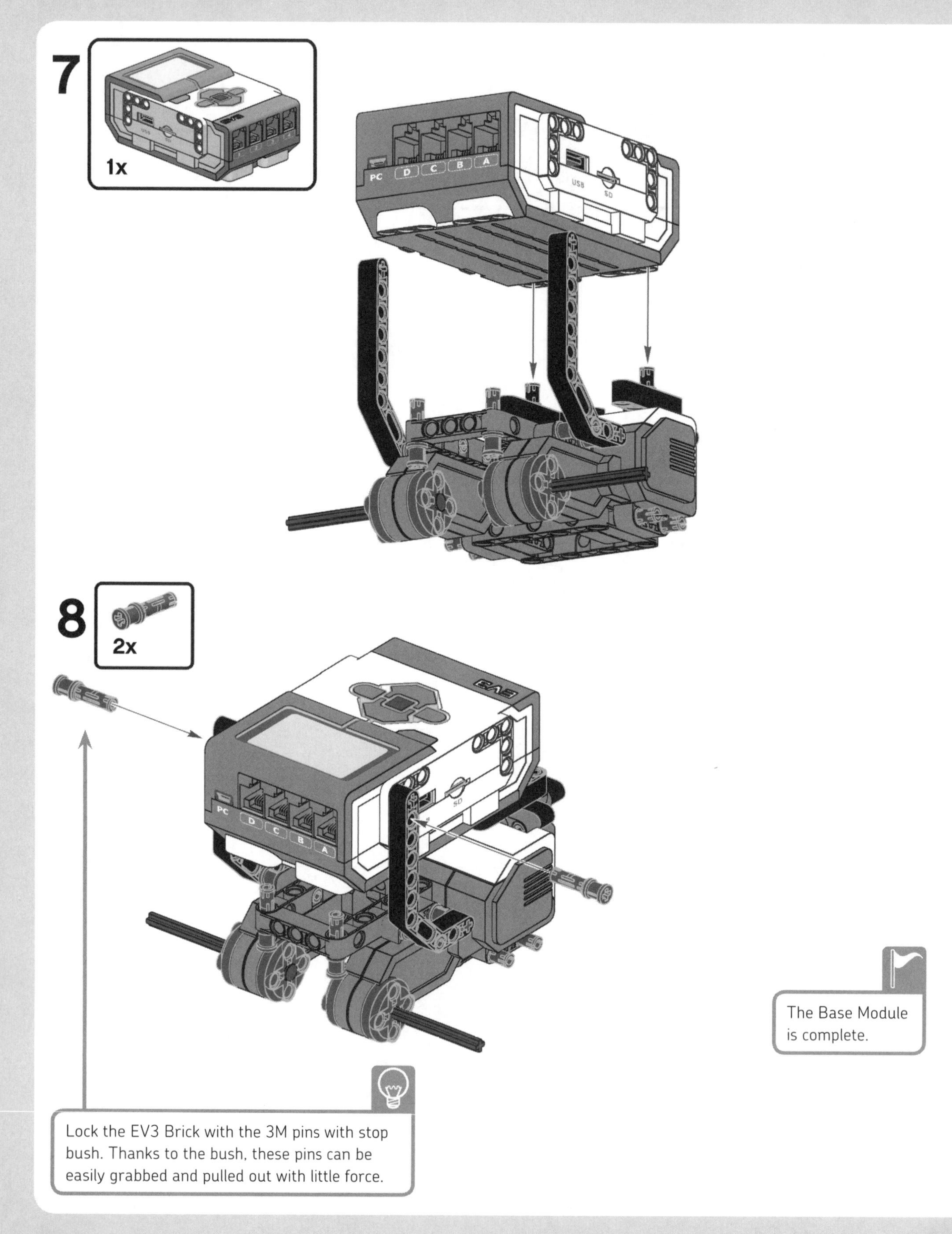

The Base Module is complete.

Lock the EV3 Brick with the 3M pins with stop bush. Thanks to the bush, these pins can be easily grabbed and pulled out with little force.

ROV3R with wheels

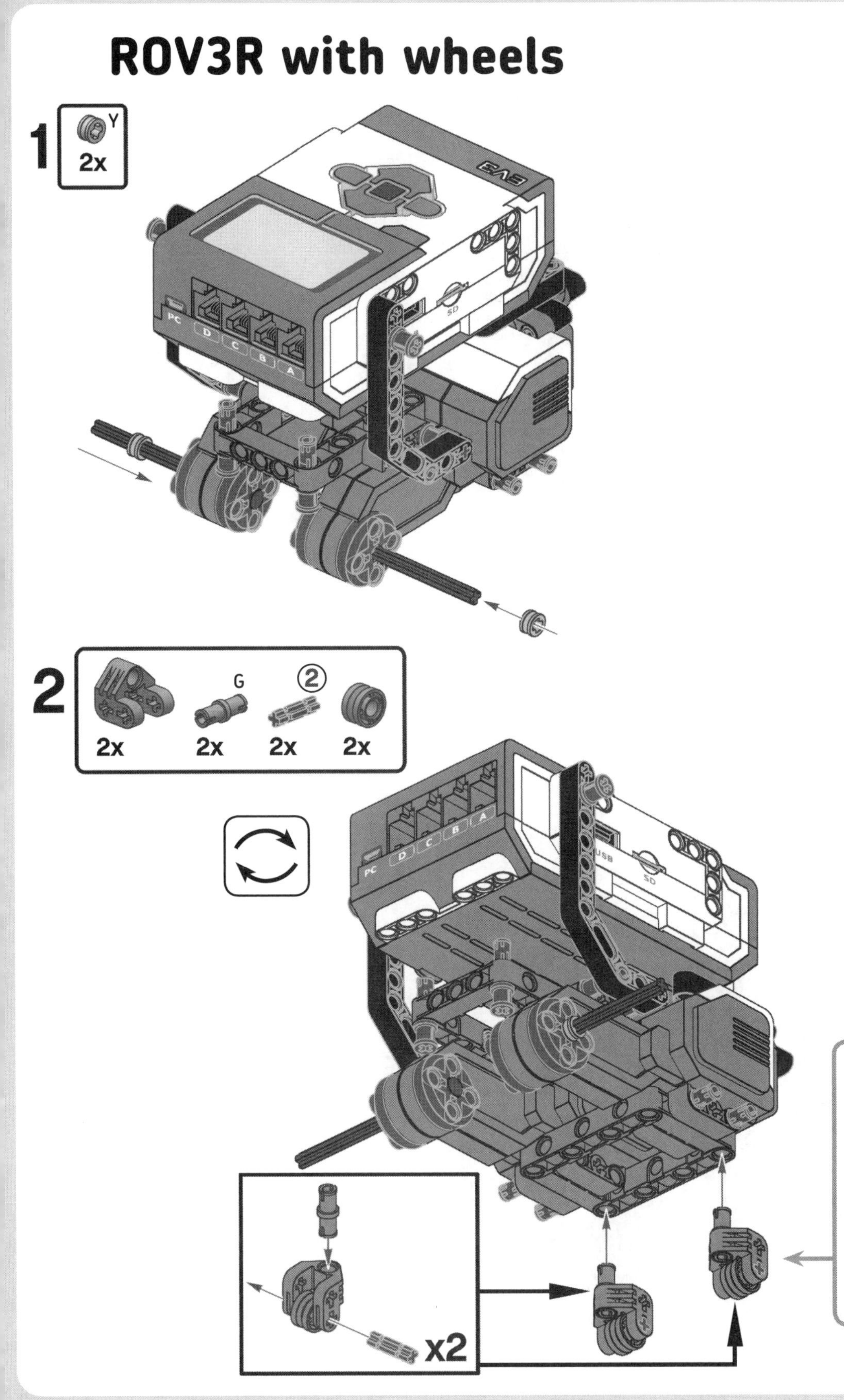

These small caster wheels support the robot. They are passive, meaning that they don't propel the robot but just follow the robot's motion. Like shopping cart wheels, these wheels sometimes swivel when the robot reverses direction, causing the robot to jiggle.

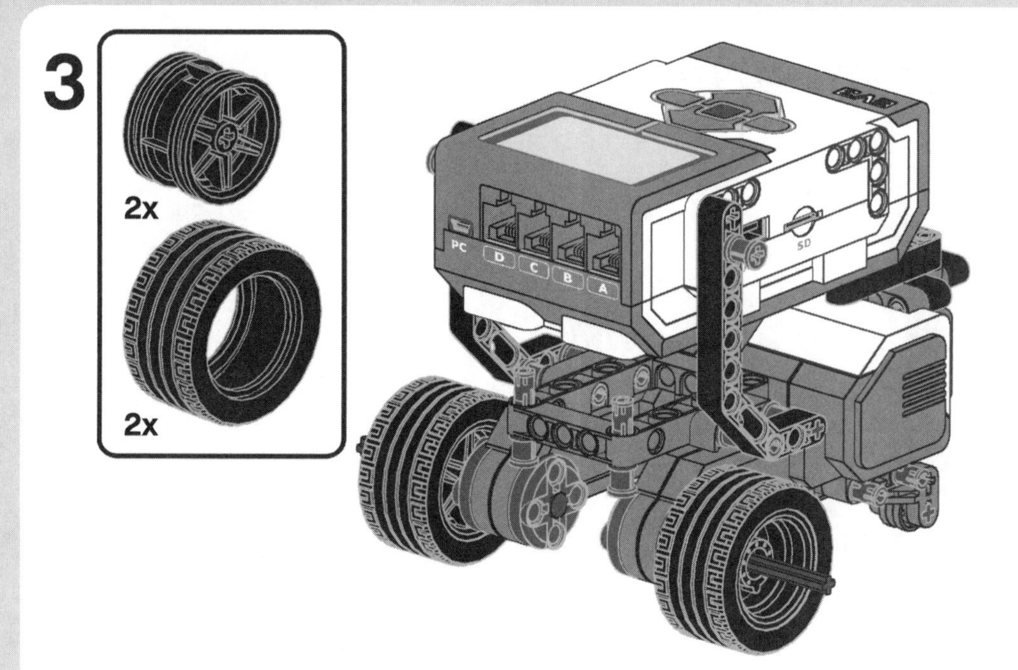

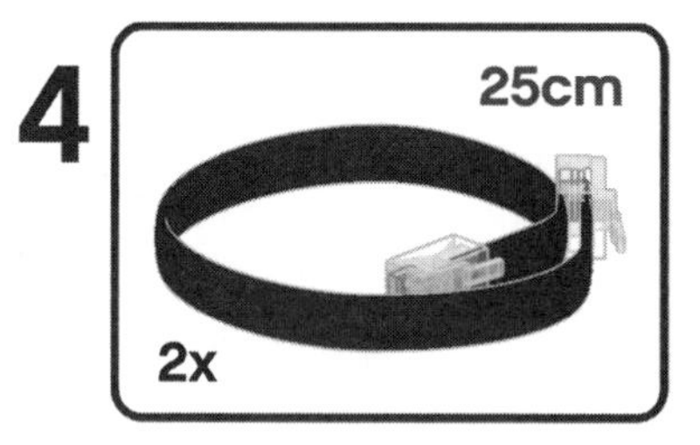

The EV3 Brick has four output ports, labeled A, B, C, and D. Use the short cables to connect the right driving motor to port C and the left driving motor to port B.

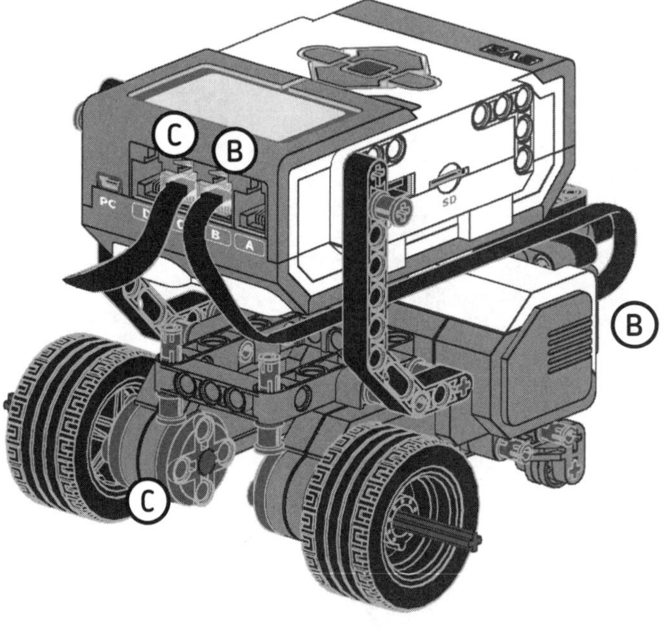

The ROV3R with Wheels is now complete. You can attach the modules in this chapter to this version of ROV3R *or* to the ROV3R with Treads.

touch sensor bumper

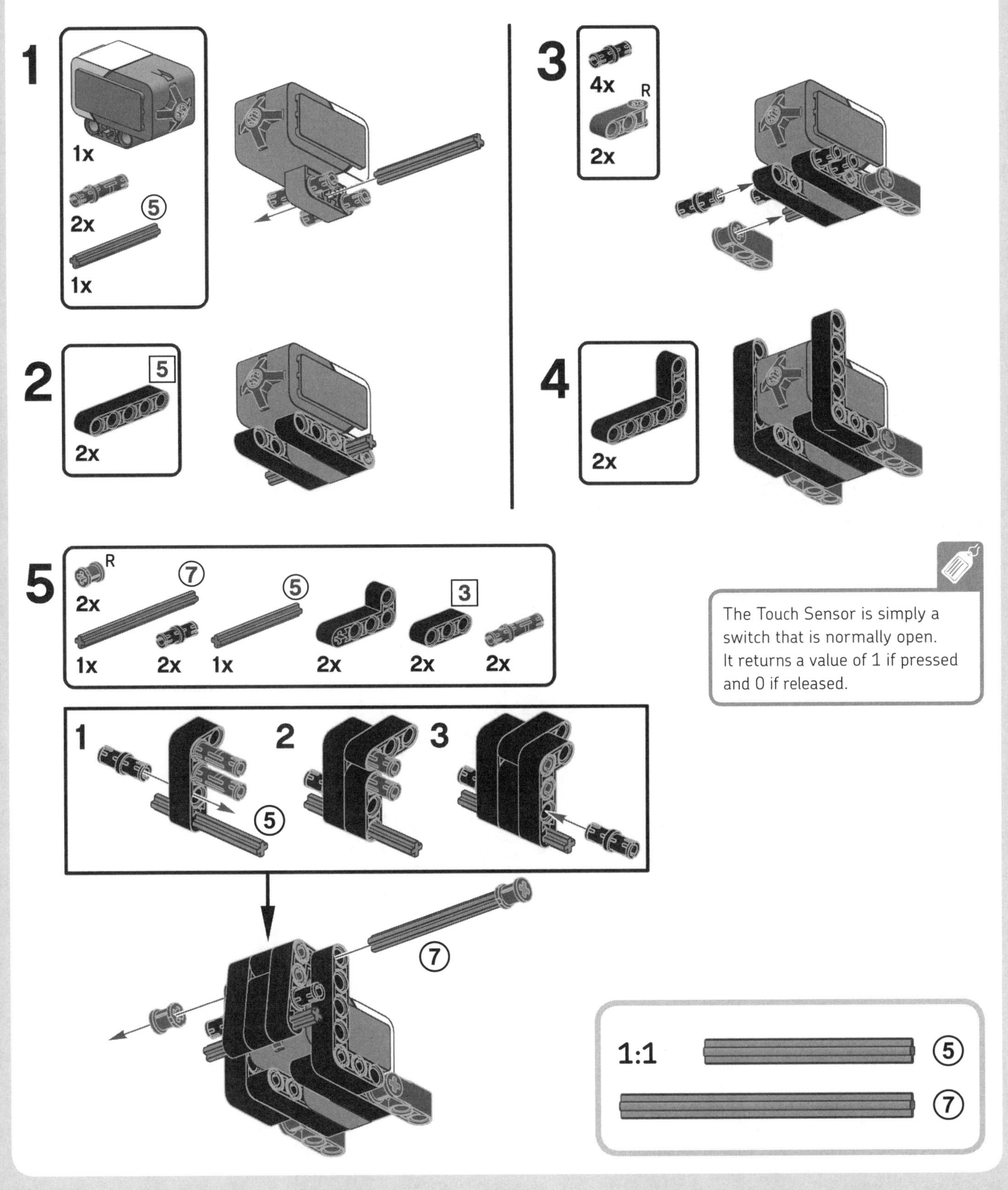

The Touch Sensor is simply a switch that is normally open. It returns a value of 1 if pressed and 0 if released.

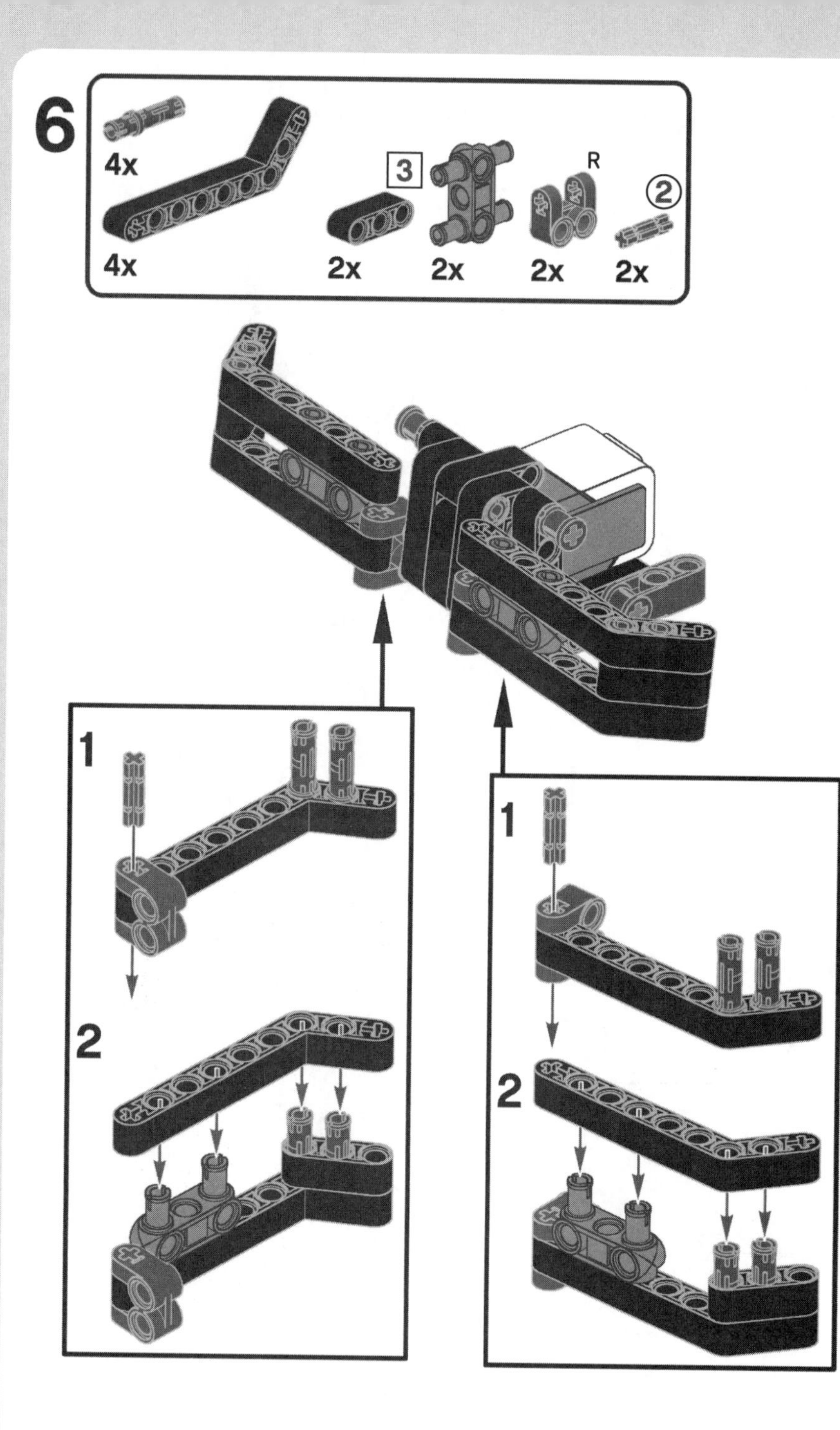

The Touch Sensor Bumper is now complete.

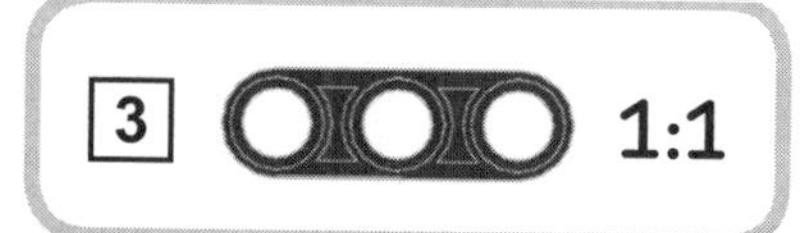

ROV3R with touch sensor bumper

1

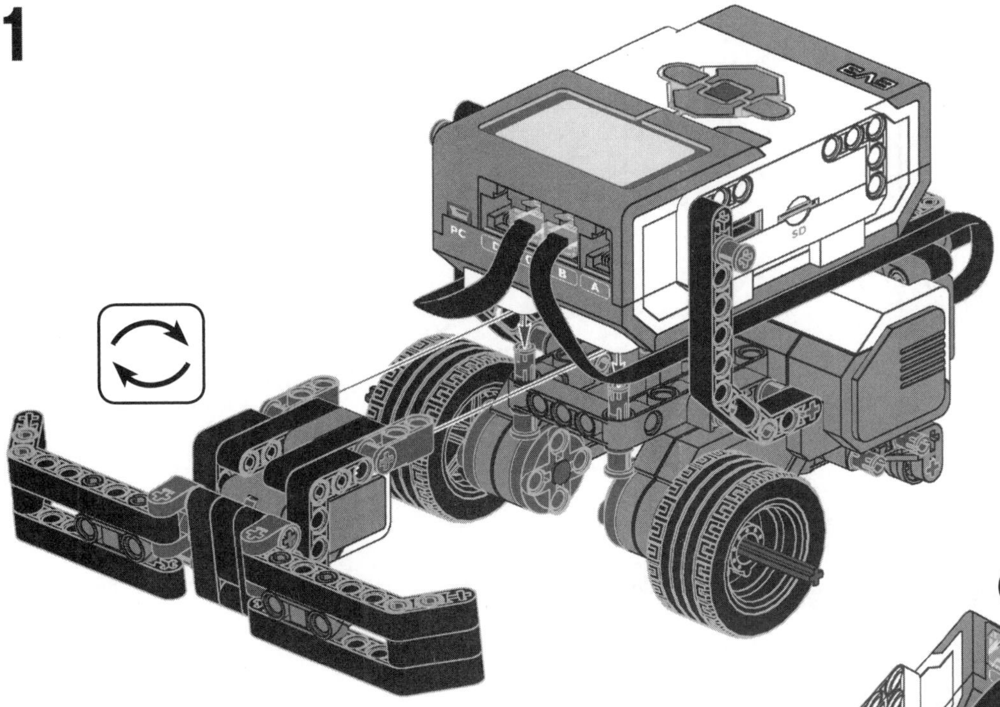

You can add the Touch Sensor Bumper to the ROV3R with Wheels or the ROV3R with Treads. The ROV3R with Wheels is shown.

2

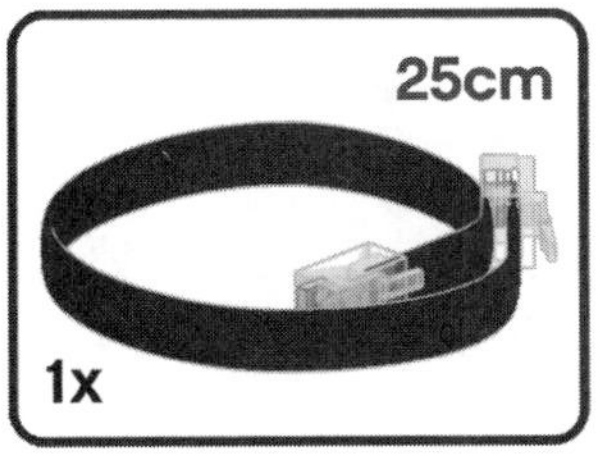

25cm

1x

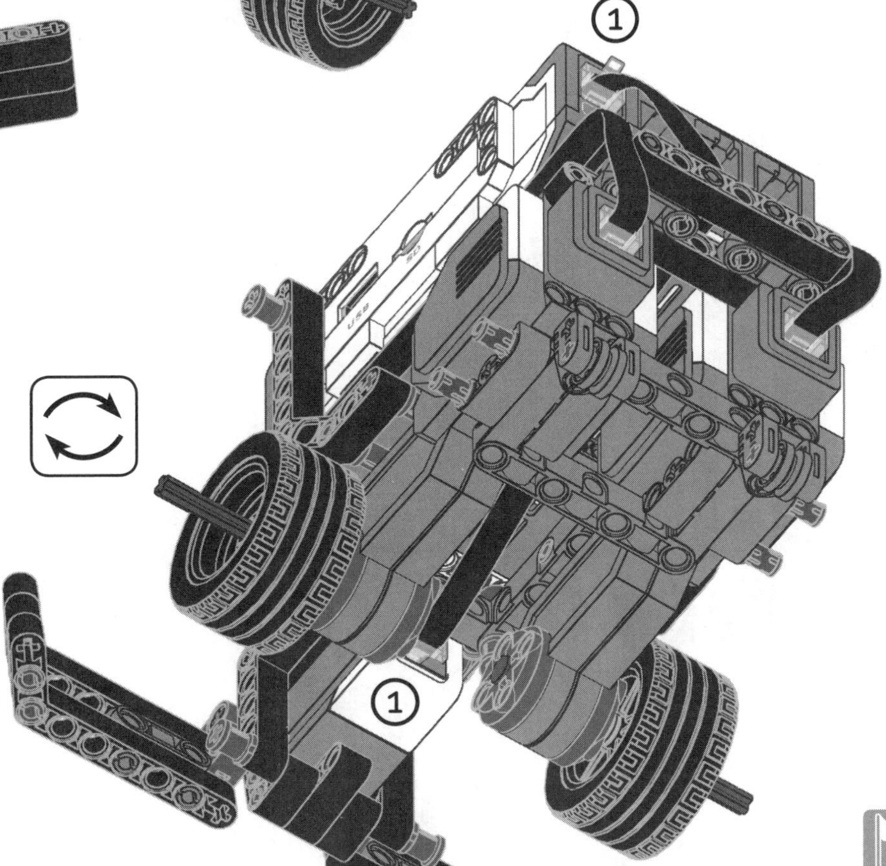

The Brick Program App allows you to program your robots without a computer by using the EV3 Brick menu. When using the Brick Program App, you must connect motors and sensors to the default ports. The Touch Sensor default port for the Brick Program App is 1. The EV3 system's Auto-ID feature allows the EV3 Brick to recognize the type of sensor attached to an input port. If you connect a motor to a sensor port, or vice versa, you'll see a warning.

The ROV3R with Touch Sensor Bumper is now complete.

line-following module

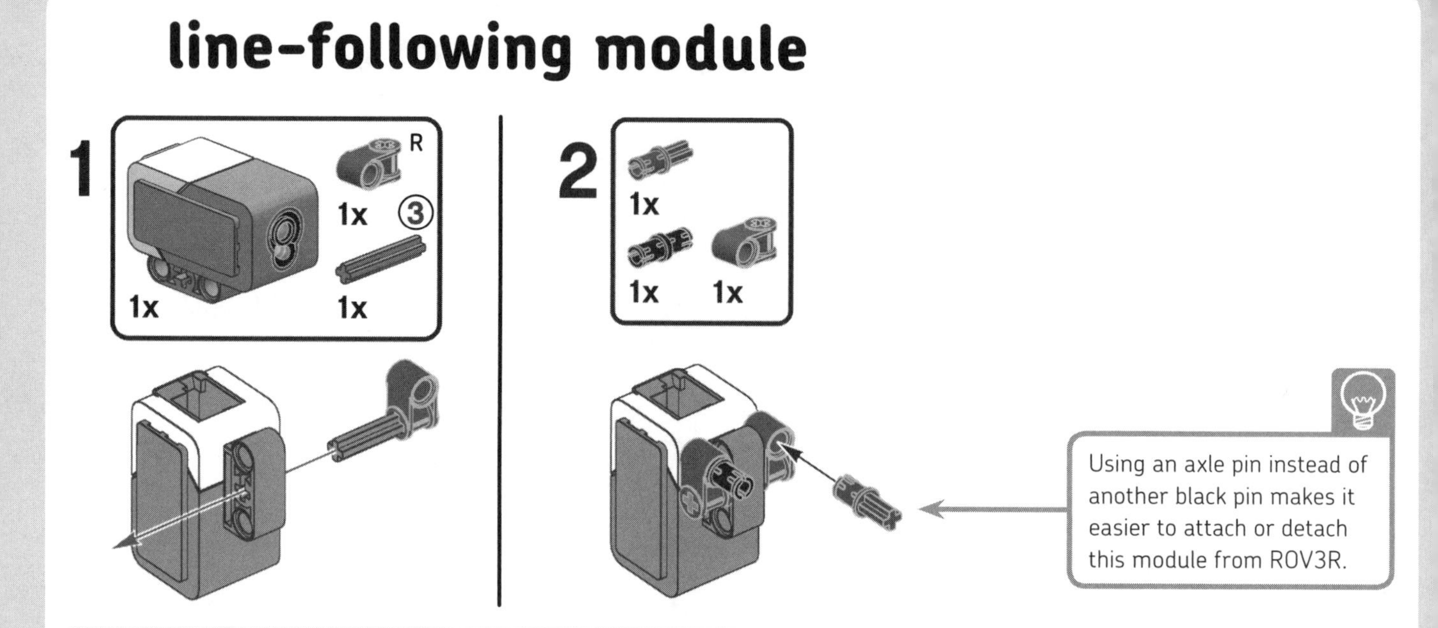

Using an axle pin instead of another black pin makes it easier to attach or detach this module from ROV3R.

line-following ROV3R

1

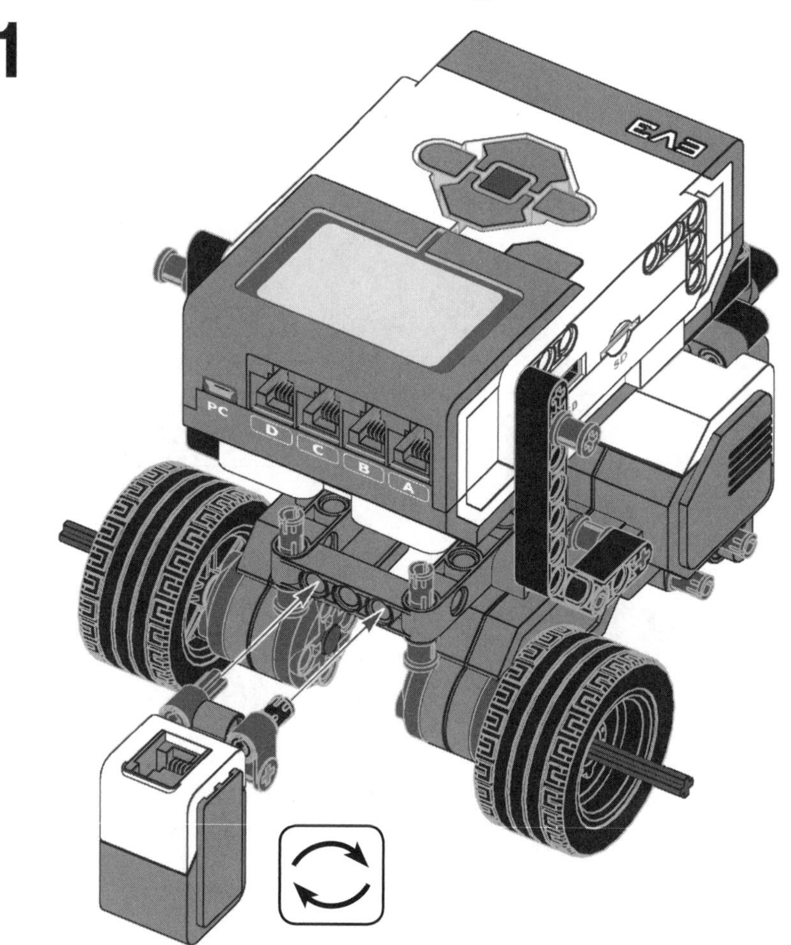

You can add the Line-Following Module to the ROV3R with Wheels or the ROV3R with Treads. The ROV3R with Wheels is shown.

The Color Sensor has a built-in RGB LED that can emit red, green, or blue light. The sensor detects different colors by flashing all three colors in a very fast loop and measuring the light returned by the surface being scanned. When the sensor is measuring the amount of reflected light, the LED glows red. You'll learn more about how ROV3R uses the Color Sensor to follow a line in Chapter 4.

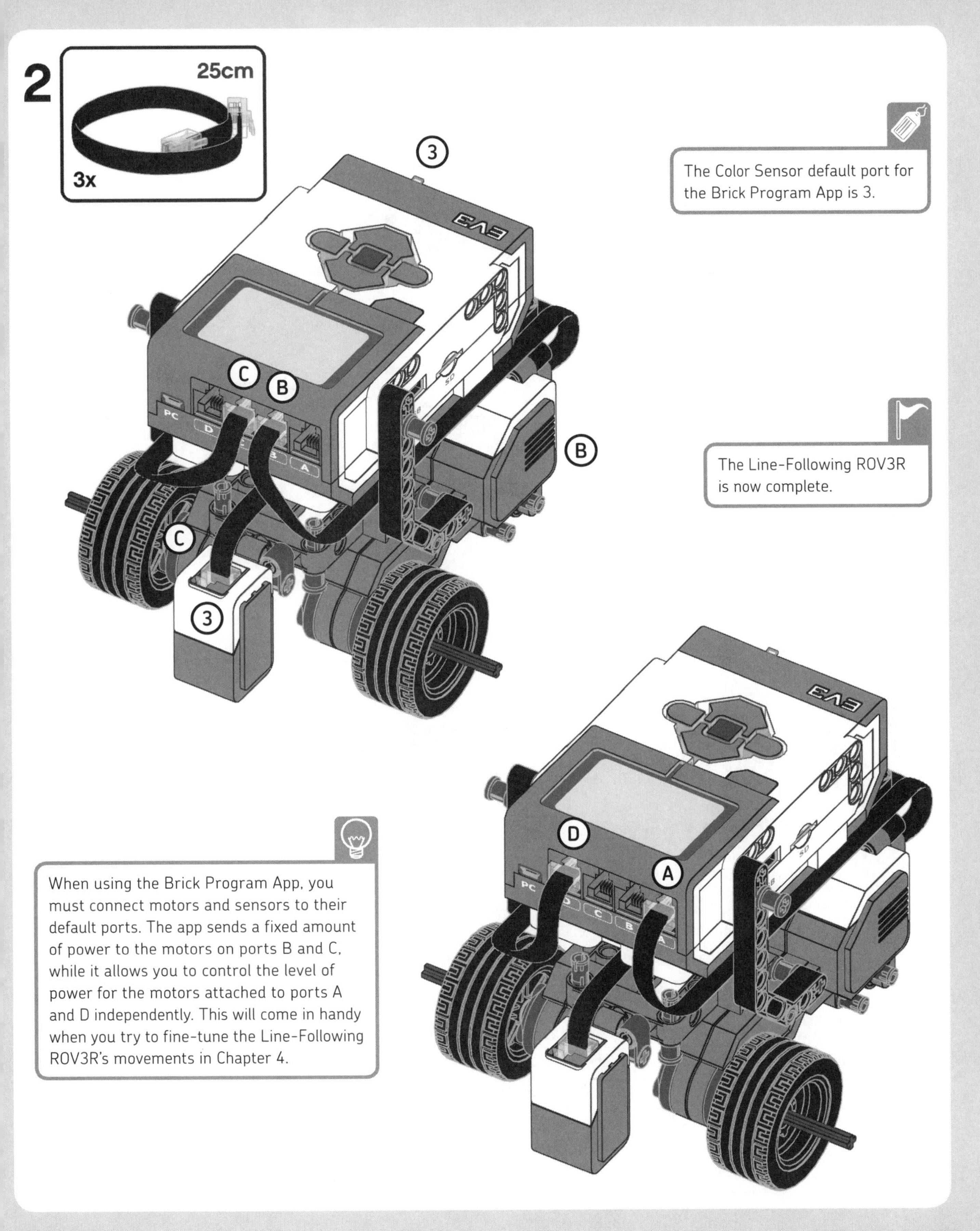
2
25cm
3x
3
C
B
B
C
3
D
A
The Color Sensor default port for the Brick Program App is 3.
The Line-Following ROV3R is now complete.
When using the Brick Program App, you must connect motors and sensors to their default ports. The app sends a fixed amount of power to the motors on ports B and C, while it allows you to control the level of power for the motors attached to ports A and D independently. This will come in handy when you try to fine-tune the Line-Following ROV3R's movements in Chapter 4.

front IR sensor

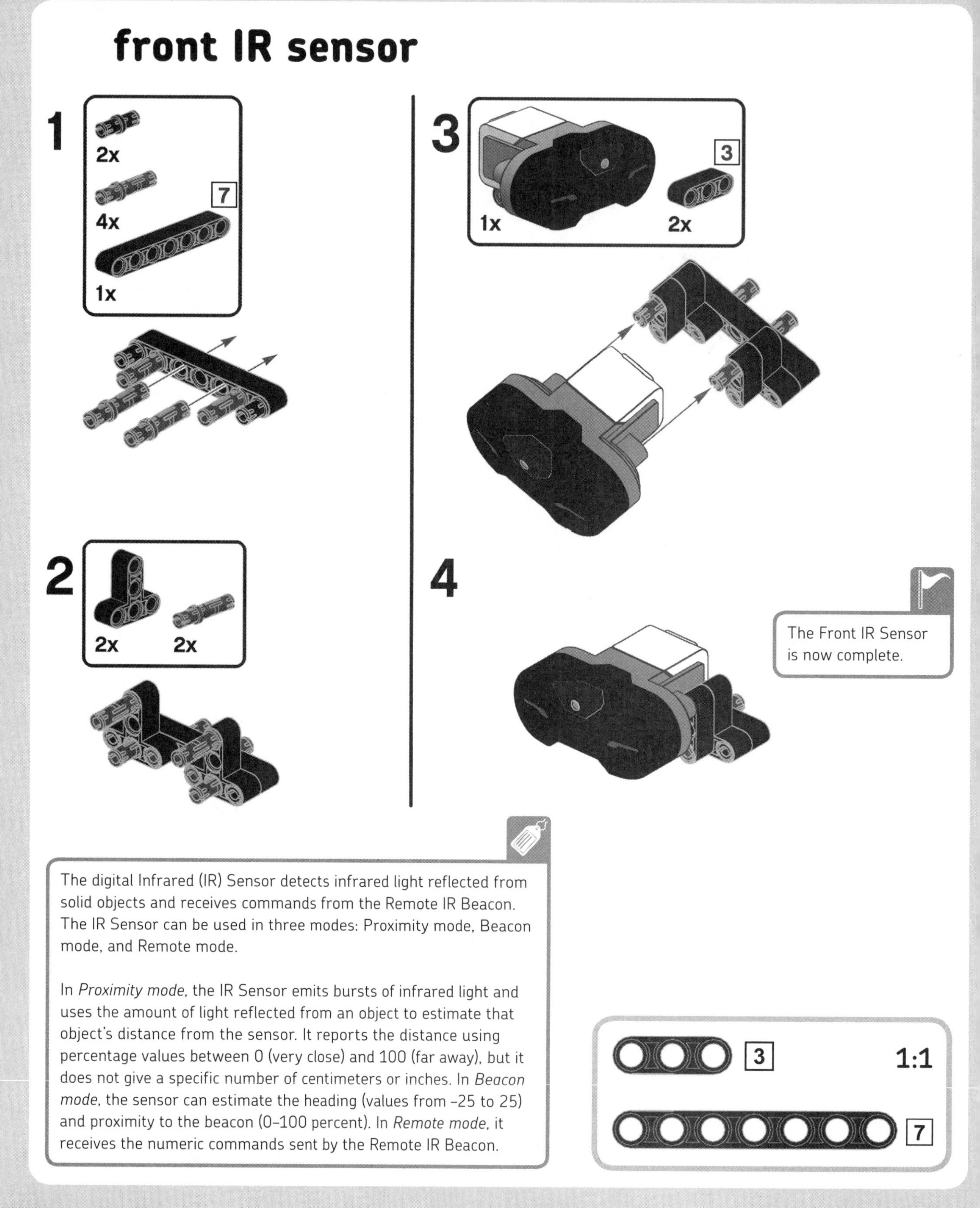

The digital Infrared (IR) Sensor detects infrared light reflected from solid objects and receives commands from the Remote IR Beacon. The IR Sensor can be used in three modes: Proximity mode, Beacon mode, and Remote mode.

In *Proximity mode*, the IR Sensor emits bursts of infrared light and uses the amount of light reflected from an object to estimate that object's distance from the sensor. It reports the distance using percentage values between 0 (very close) and 100 (far away), but it does not give a specific number of centimeters or inches. In *Beacon mode*, the sensor can estimate the heading (values from –25 to 25) and proximity to the beacon (0–100 percent). In *Remote mode*, it receives the numeric commands sent by the Remote IR Beacon.

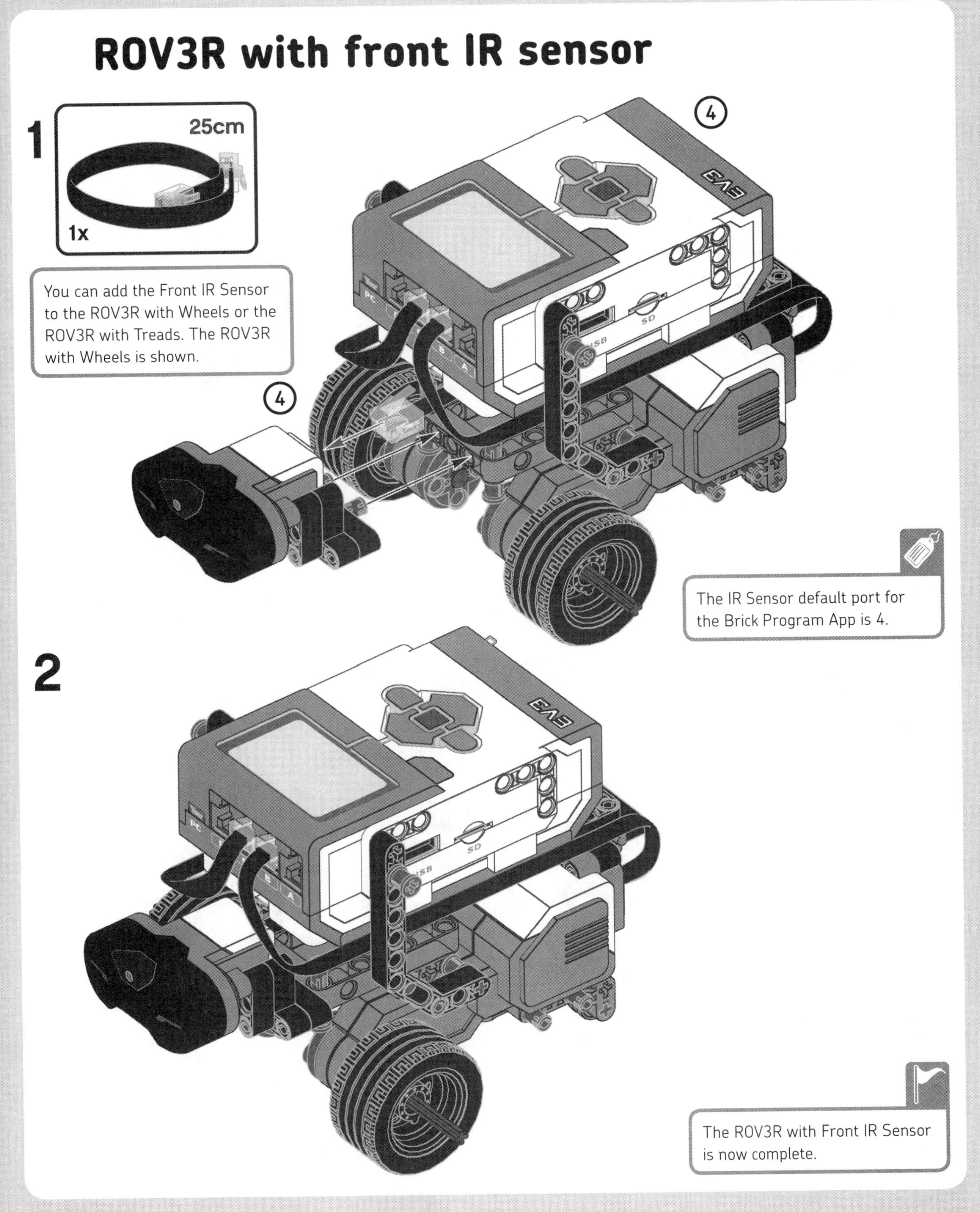
ROV3R with front IR sensor
1
25cm
1x
You can add the Front IR Sensor to the ROV3R with Wheels or the ROV3R with Treads. The ROV3R with Wheels is shown.
4
4
The IR Sensor default port for the Brick Program App is 4.
2
The ROV3R with Front IR Sensor is now complete.

wall-following module

1

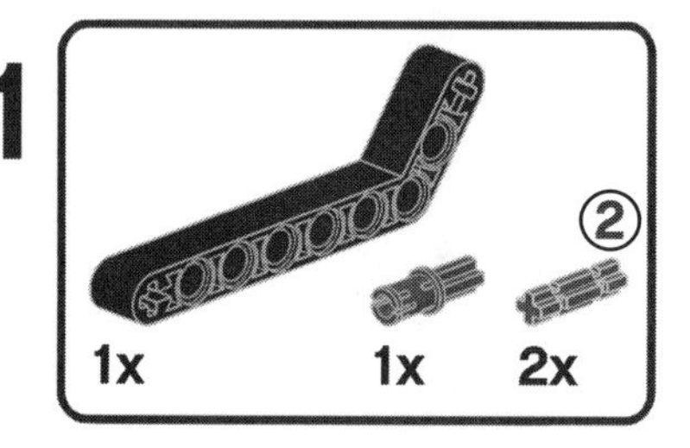

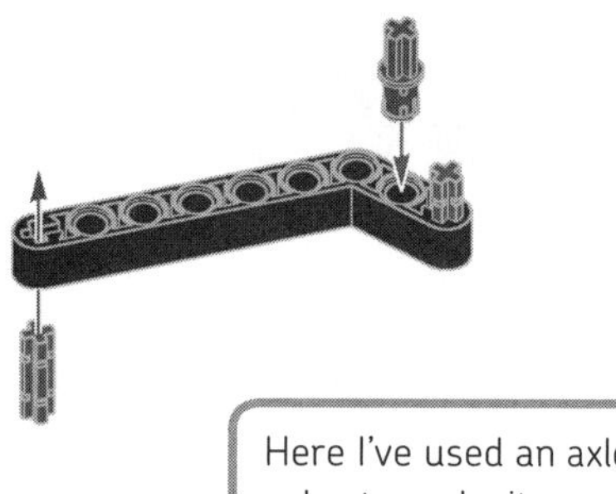

2

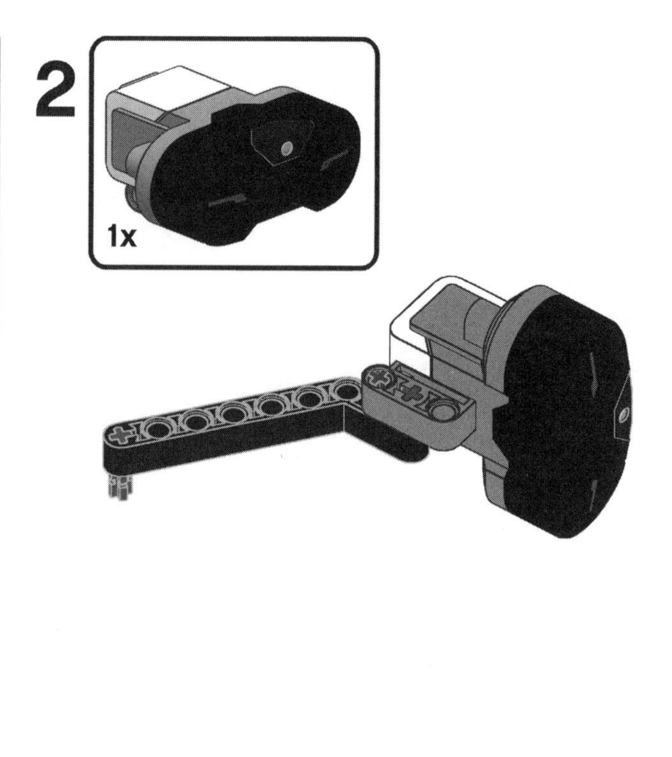

Here I've used an axle pin and 2M axles to make it easy to disassemble the IR Sensor and this module from ROV3R. For a sturdier assembly, you should use pins that snap into holes, like pins with friction or axle pins.

wall-following ROV3R

You can add the Wall-Following Module to the ROV3R with Wheels or the ROV3R with Treads. The ROV3R with Wheels is shown.

1

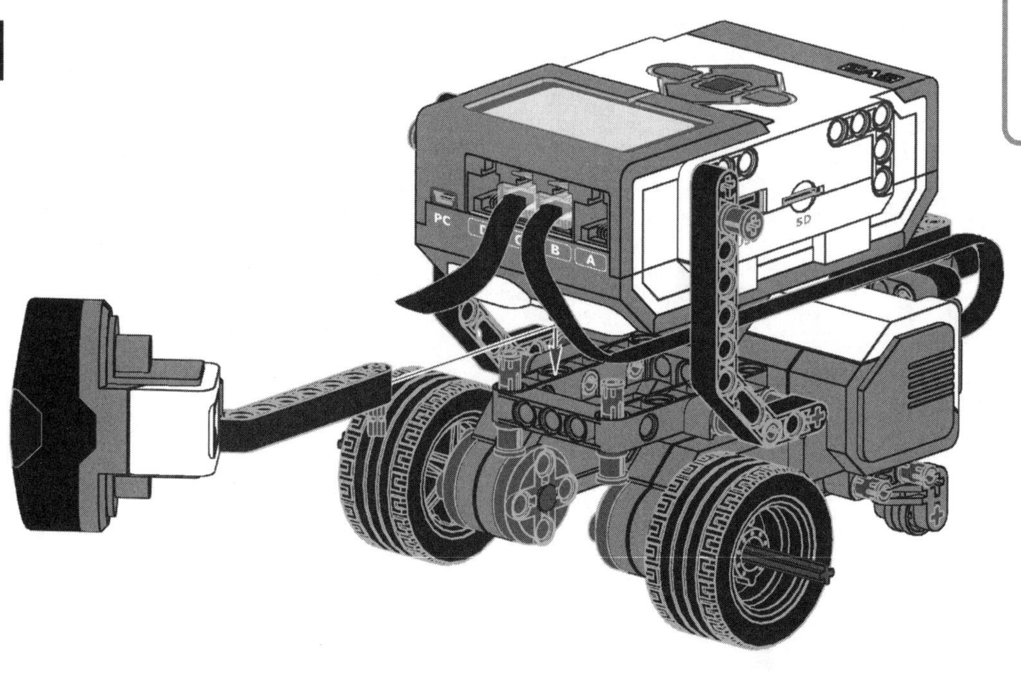

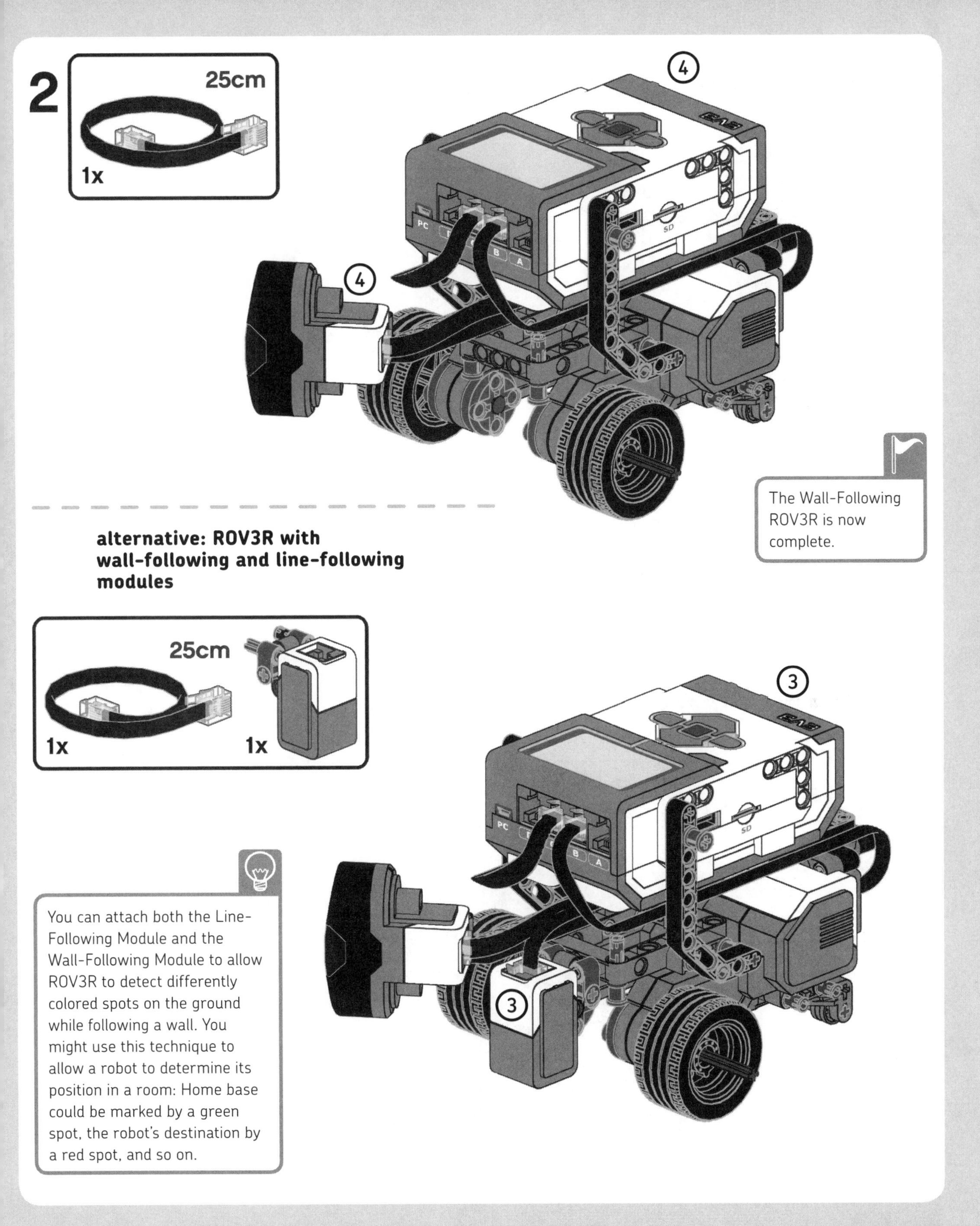

The Wall-Following ROV3R is now complete.

alternative: ROV3R with wall-following and line-following modules

You can attach both the Line-Following Module and the Wall-Following Module to allow ROV3R to detect differently colored spots on the ground while following a wall. You might use this technique to allow a robot to determine its position in a room: Home base could be marked by a green spot, the robot's destination by a red spot, and so on.

Dexter's cleaning tool

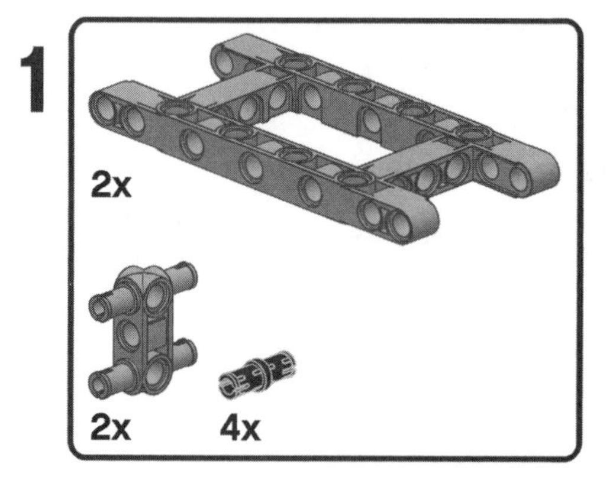

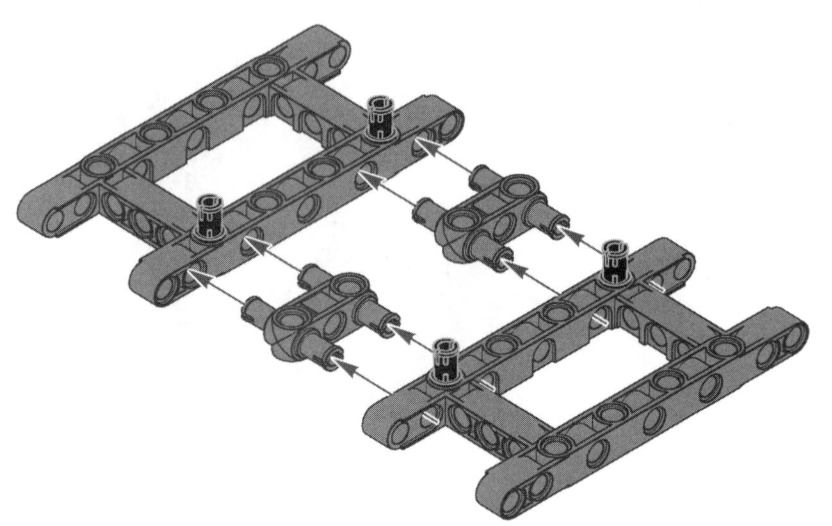

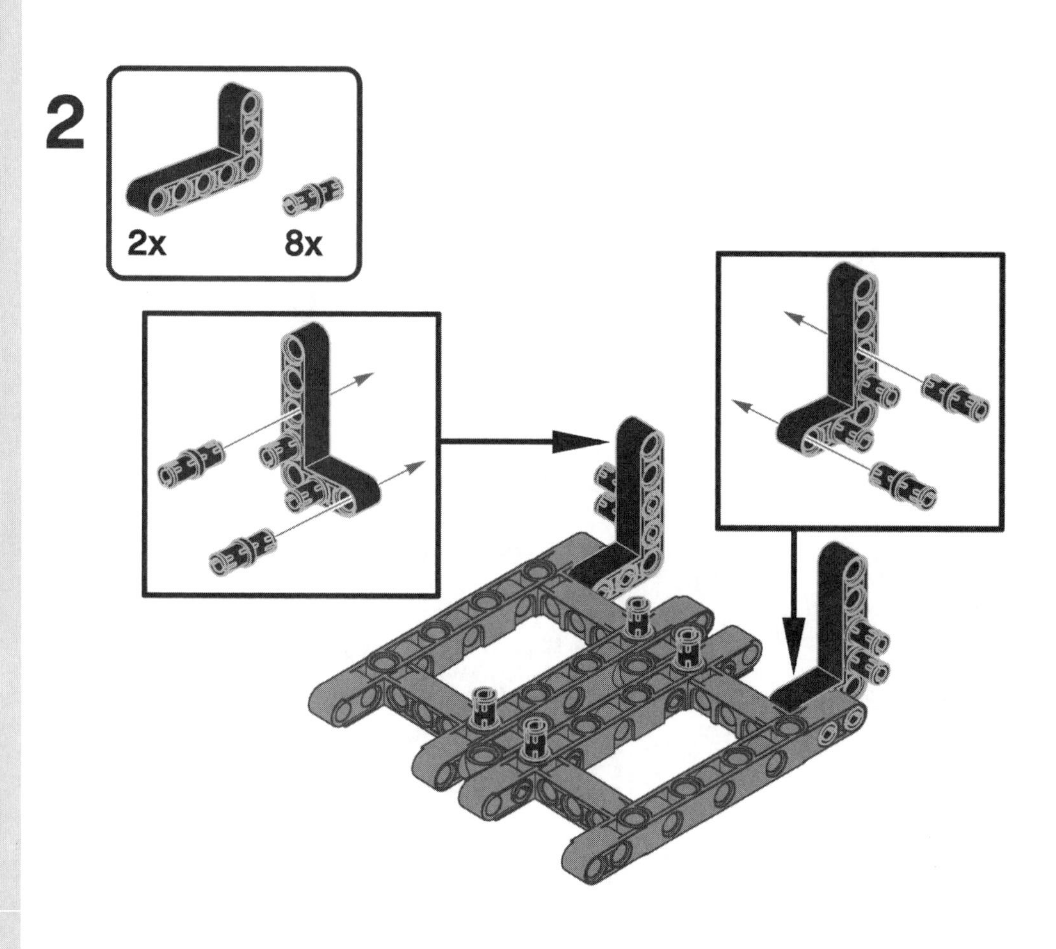

3

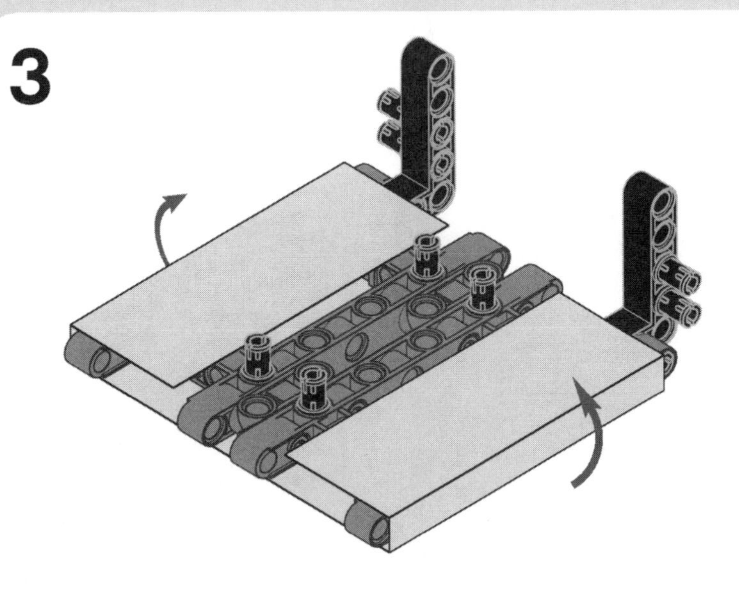

Wrap an electrostatic cleaning cloth around the frames. This allows you to pick up dirt and dust without using moving parts. Such cloths are typically treated with chemicals that give them a negative charge, which readily attracts dust particles as the robot moves around the room.

4

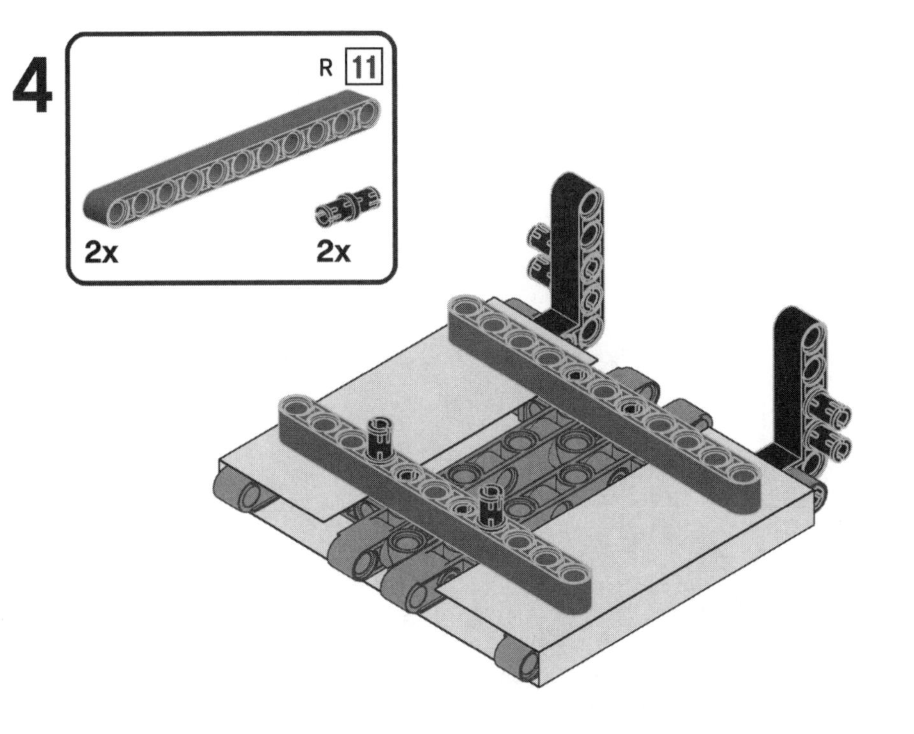

Lock the cloth into place using two 11M beams.

Dexter's Cleaning Tool is now complete.

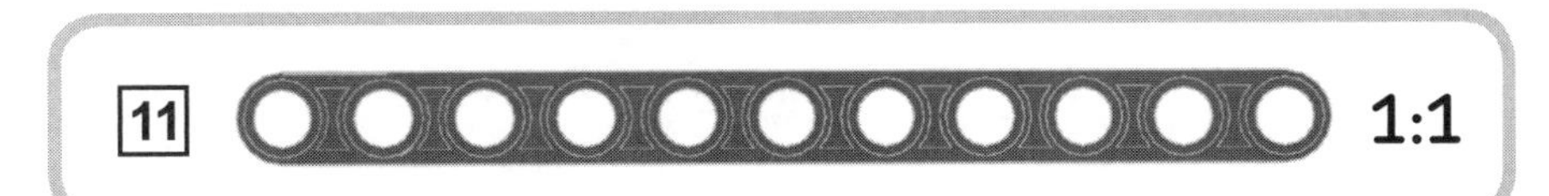

ROV3R with cleaning tool

You can add Dexter's Cleaning Tool to the ROV3R with Wheels or the ROV3R with Treads.

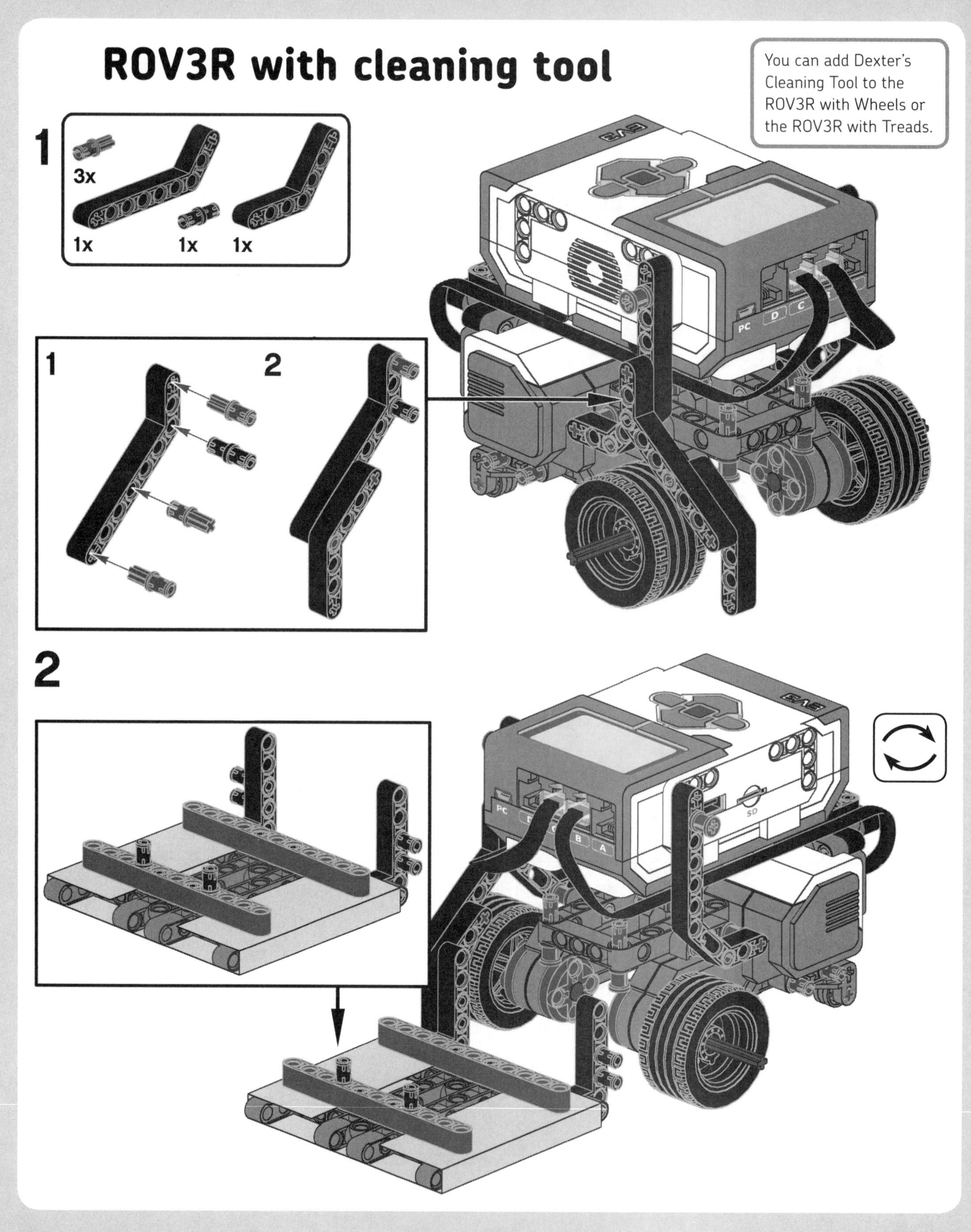

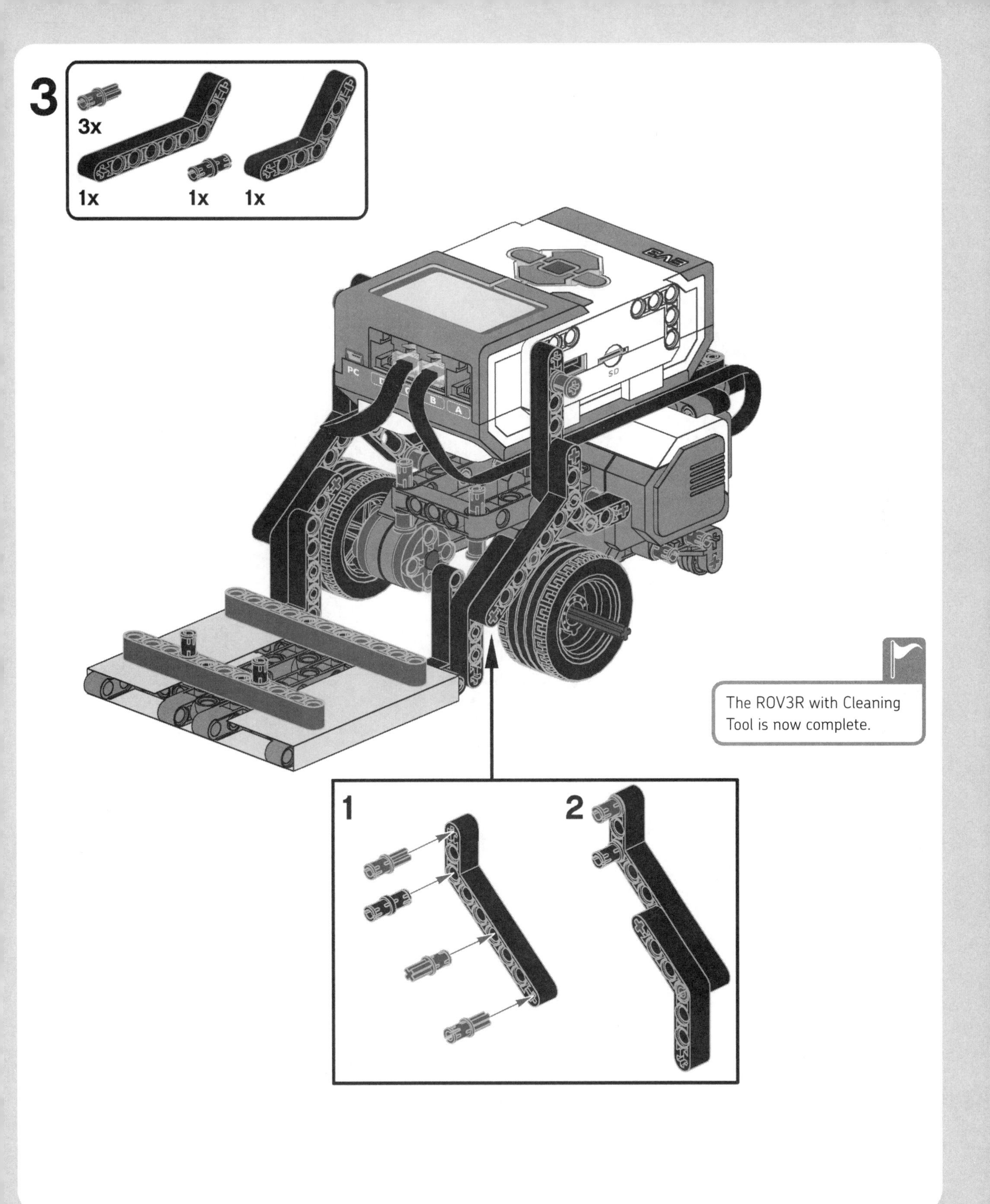

The ROV3R with Cleaning Tool is now complete.

alternative #1: ROV3R with cleaning tool and touch sensor bumper

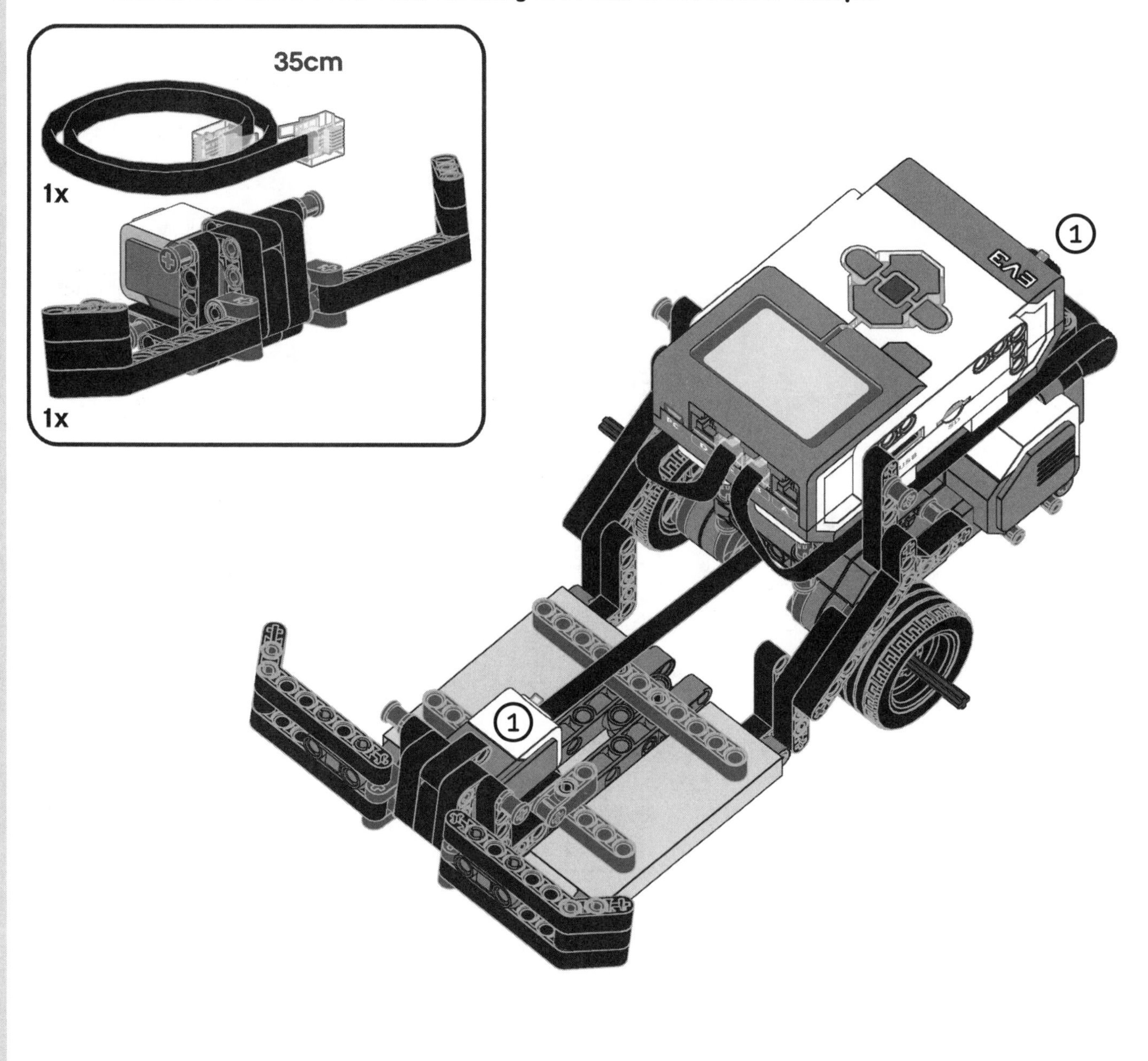

alternative #2: wall-following ROV3R with cleaning tool

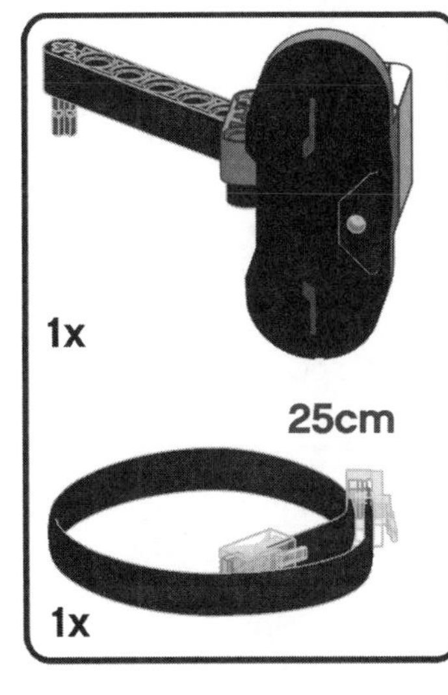

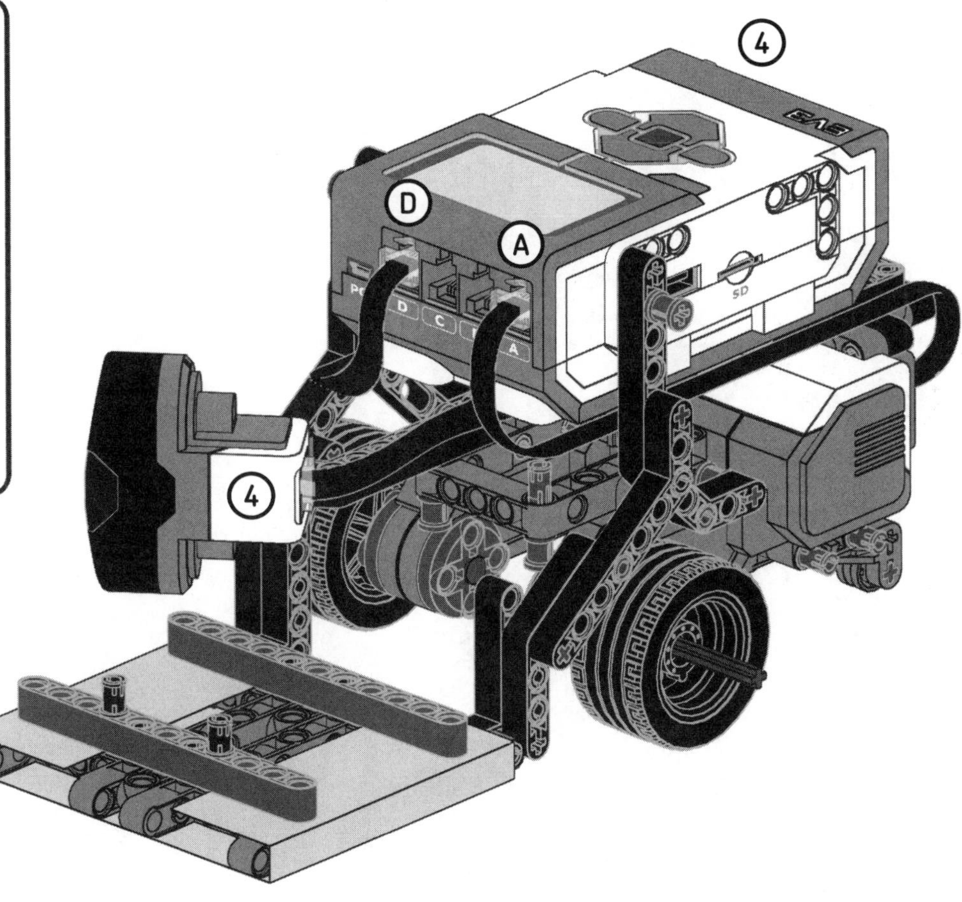

When using the Brick Program App, you must connect motors and sensors to their default ports. The app sends a fixed amount of power to the motors on ports B and C, while it allows you to control the level of power for the motors attached to ports A and D independently. This will come in handy when you try to fine-tune the Line-Following ROV3R's movements in Chapter 4.

ROV3R with treads

To build the ROV3R with Treads, start with the Base Module (step 8 on page 22).

You can attach any of the modules in this chapter either to the ROV3R with Wheels or the ROV3R with Treads.

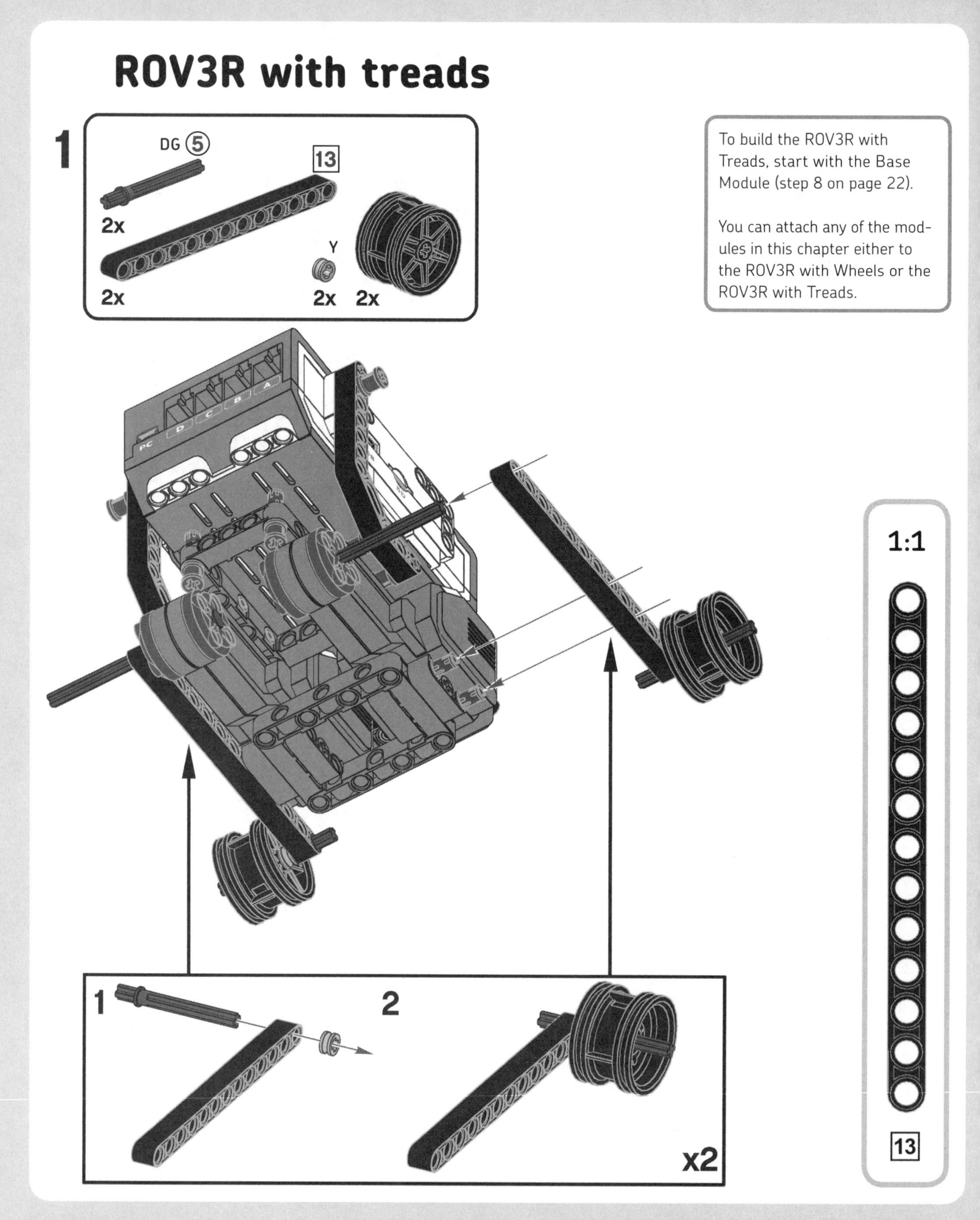

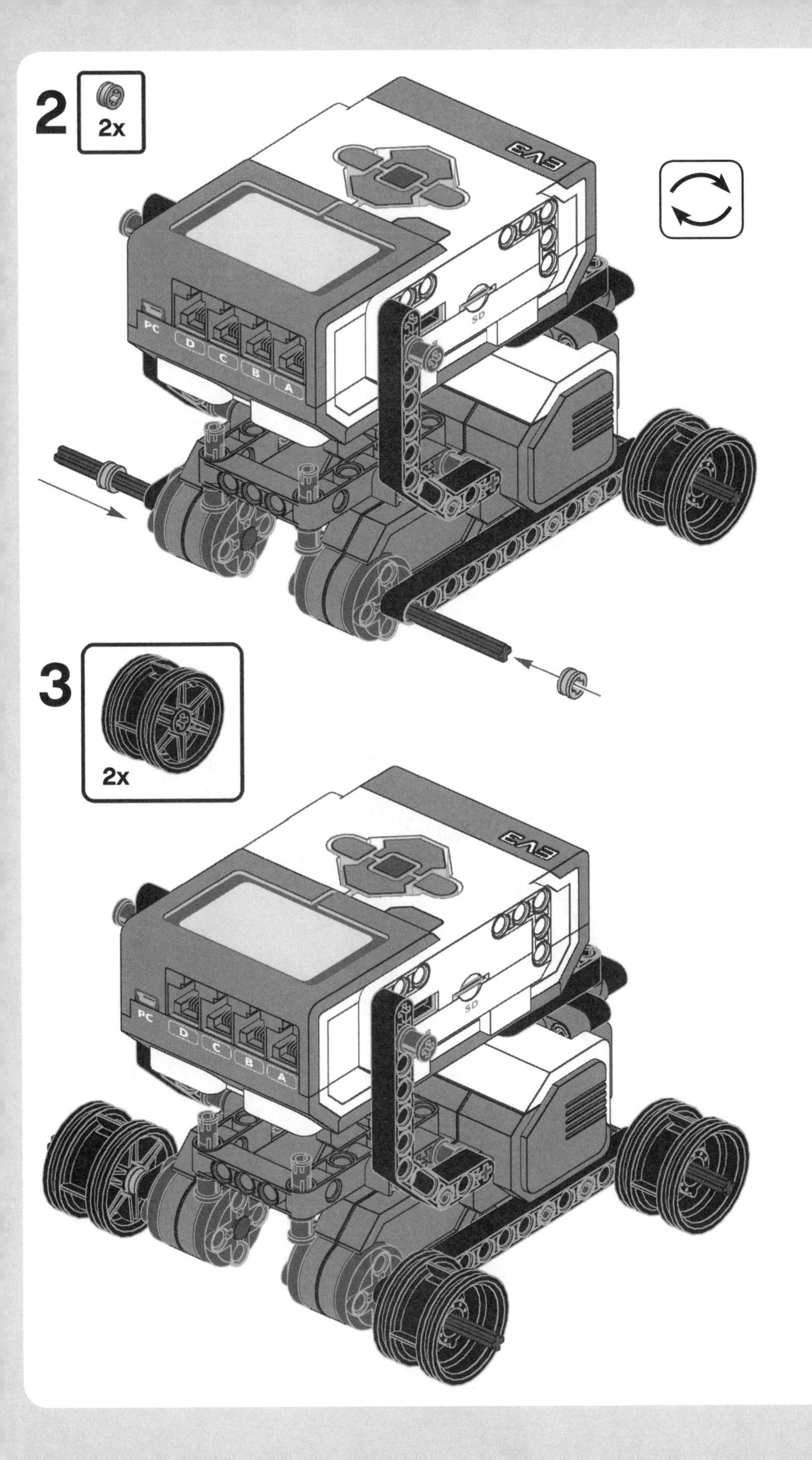
2
2x
3
2x
EV3
PC
D
C
B
A
SD

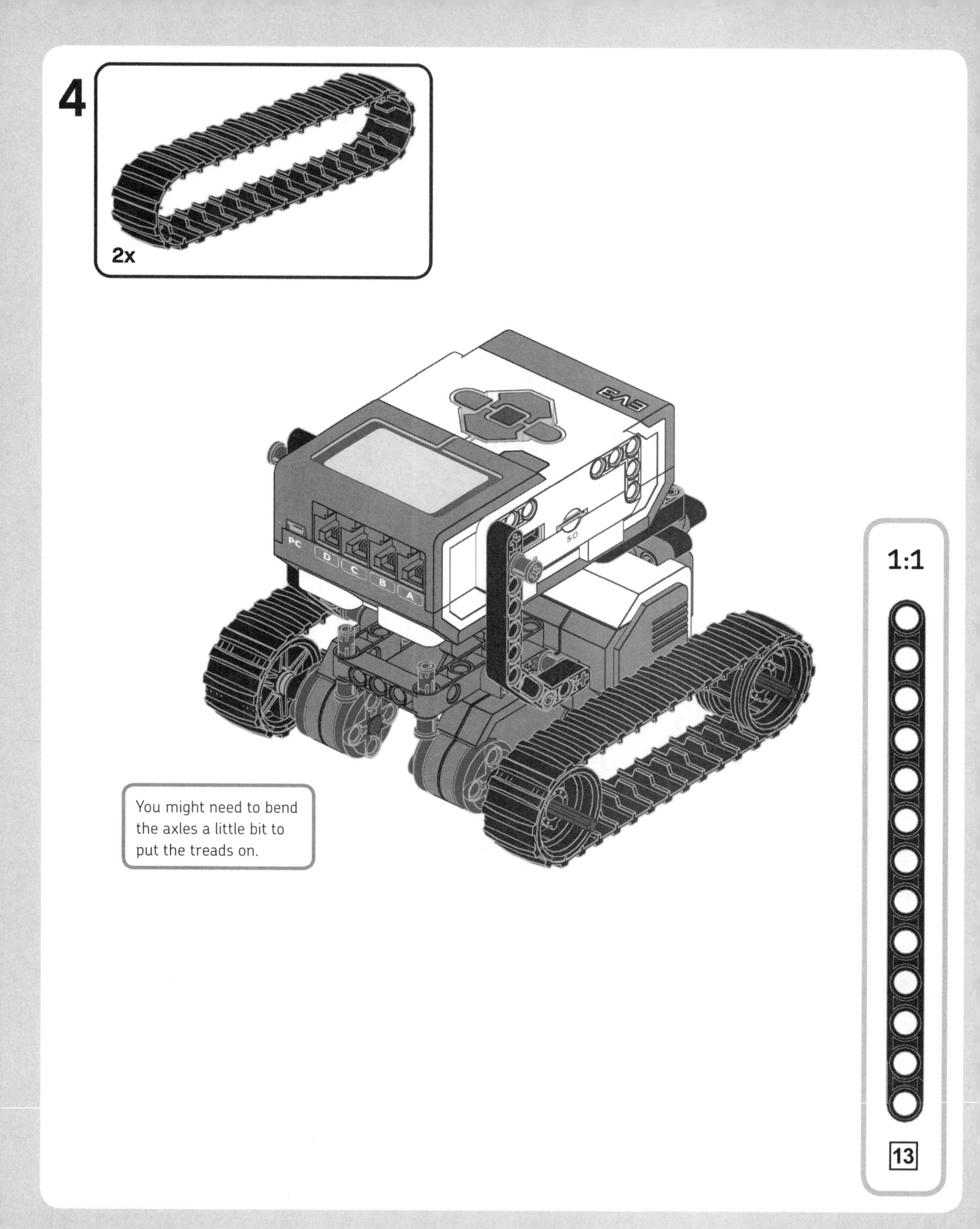
4
2x
PC
D
C
B
A
SD
1:1
13
You might need to bend
the axles a little bit to
put the treads on.

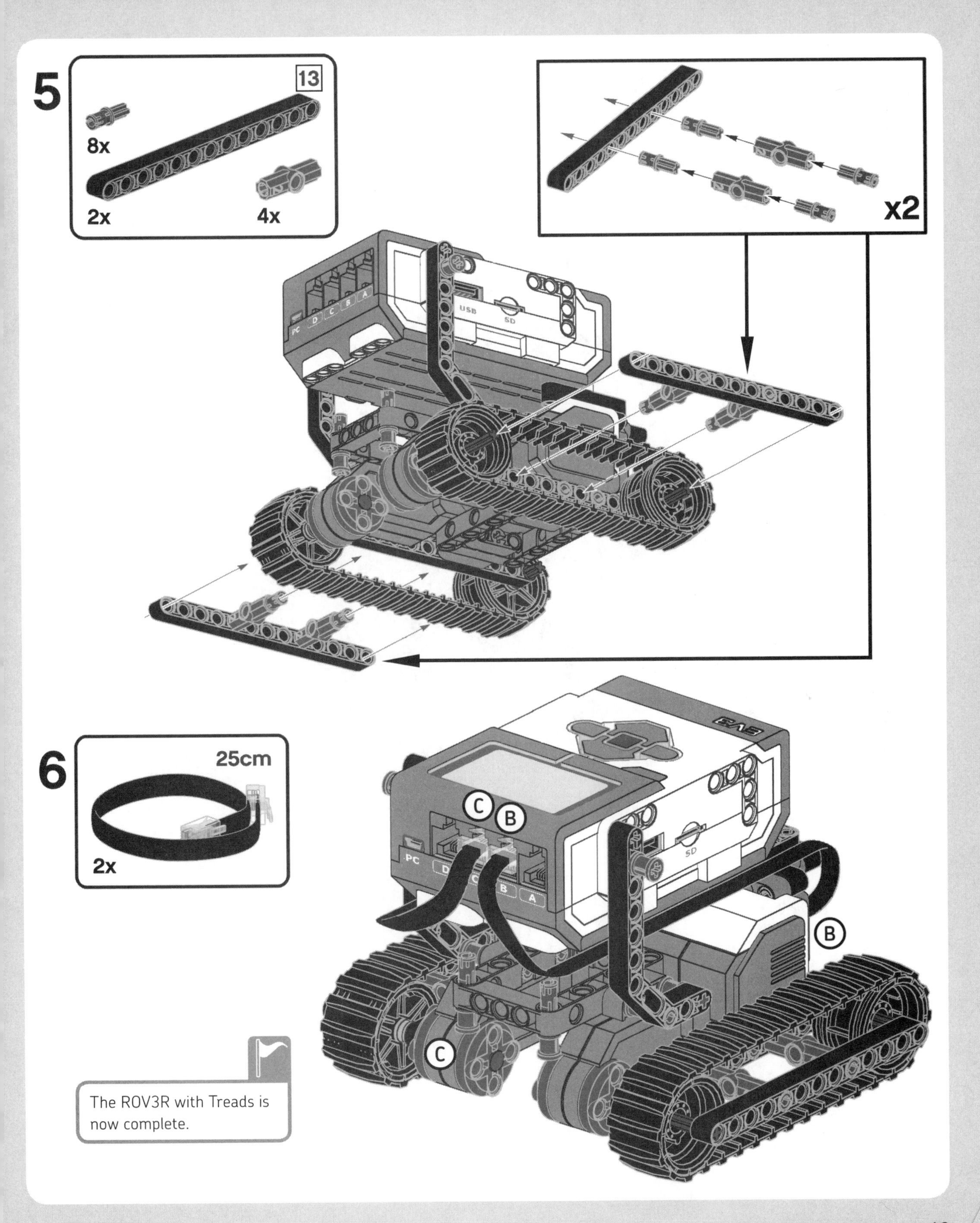
5
13
8x
2x
4x
x2
USB
SD
PC
D
C
B
A
6
25cm
2x
C
B
C
B
The ROV3R with Treads is now complete.

secret project: grabber module

This grabber module allows ROV3R to pick up and lift objects with a single Medium Motor.

Visit *http://EV3L.com/* for bonus material!

conclusion

In this chapter, you built ROV3R and equipped it with a variety of sensors and tools. As you followed the building instructions, you learned some fundamental building techniques, like bracing, and you got an overview of the LEGO MINDSTORMS EV3 system. Next you'll learn how to program ROV3R using the Brick Program App.

I'M NOT GREAT AT NAMING THINGS. WHAT WOULD YOU CALL IT?
SAME AS MY DOG: ROVER!
NO! ROV3R!
WHAT'S THE DIFFERENCE? IT SOUNDS THE SAME!
WELL, YEAH, BUT THE "E" IS REPLACED BY A "3"!
DON'T YOU H3AR THE DIFFER3NCE?
WHIIIRRRR
DOC, IT WORKS!
COUGH! COUGH!
OH, WHAT A *CHOKE* CLEVER SOLUTION, USING AN ELECTROSTATIC CLOTH TO PICK UP THE DIRT!
03X1

programming

To bring a robot to life, you need to tell it what to do by writing a computer program for it. A *program* is a step-by-step list of basic instructions designed to produce a result. These instructions are written in a *programming language*, an artificial language that a computer can understand.

For example, the instruction "prepare some tea" on its own would not be sufficient for a robotic butler. This would be considered a high-level instruction for that robot, meaning that it would need to be broken down into a sequence of lower-level instructions to actually control the robot.

The high-level instruction "prepare some tea" can be broken down into a list of well-defined, discrete, elemental steps like these:

* Pick up a pot.
* Fill it with water.
* Turn on the stove.
* Put the pot on the stove.
* Wait until the water temperature reaches 95°C (203°F).
* Turn off the stove.
* Pour 200 mL of water into a cup.
* Put the tea bag in the cup.
* Wait 4 minutes.
* Remove the tea bag.
* Add 4 teaspoons of sugar to the tea.
* Squeeze 10 drops of lemon juice into the tea.
* Wait until the tea temperature falls to 45°C (113°F).
* Place the cup in front of the person who will be drinking the tea.

All the above steps seem basic to us, but even they need to be broken down further for our robot. For example, "Pick up a pot" should be broken into lower-level commands like "Turn motor A 90 degrees clockwise."

Programs usually have inputs and outputs: In the case of a robot, basic outputs can be turning a motor on and off or rotating a shaft by a certain number of degrees. Inputs are readings from the sensors or commands coming from the master—that is, you.

the building blocks of any program

Even if you don't know a programming language, you can still design a program. How? You can use a *flowchart*, a type of diagram that represents a program by showing the steps as boxes connected with arrows.

You can describe the flow of any program using three basic structures: *sequences*, *choices*, and *loops*.

sequences

Sequences are lists of basic actions or commands—like "Turn that motor on" or "Go forward"—written inside rectangular boxes, like those in Figure 3-1.

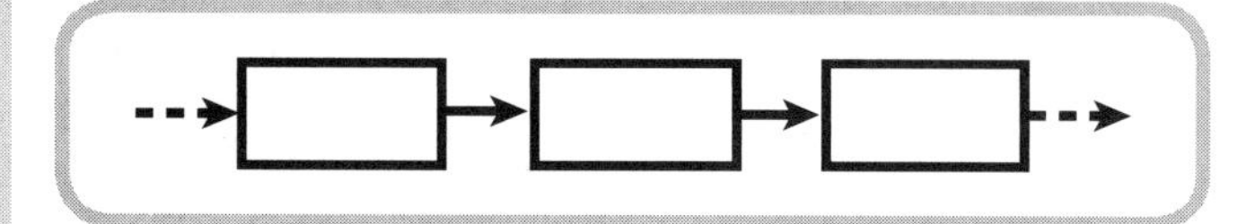

Figure 3-1: A flowchart sequence

choices

Choices are used to switch between different actions depending on a test (for example, a test might check an input value). In Figure 3-2, a choice is represented as a question mark inside a diamond. The program flow is switched to one of two branches, as indicated by arrows marked with an *X*, meaning the test failed, or a check mark, meaning the test succeeded. In the tea-making example, you could add a choice in which you tell the robot whether or not you want to add sugar.

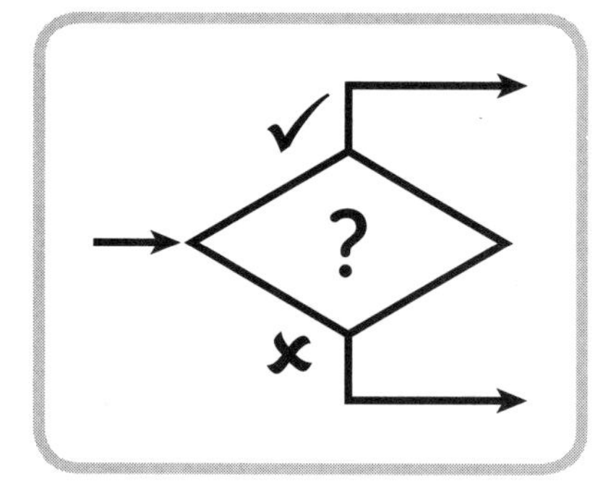

Figure 3-2: A flowchart choice

loops

Loops repeat a group of actions until a certain condition becomes true. That condition is represented in Figure 3-3(a) as a question mark inside a diamond. In Figure 3-3(b), you can see a loop that repeats a sequence of actions *N* times: *N* can be 1, 2, 3, or even infinity, meaning the sequence will be repeated forever. You can also use a loop to do nothing while waiting for a condition to become true, as in Figure 3-4. For example, in the program for the tea-making robot, you told it to wait for the tea temperature to reach 45°C before serving it.

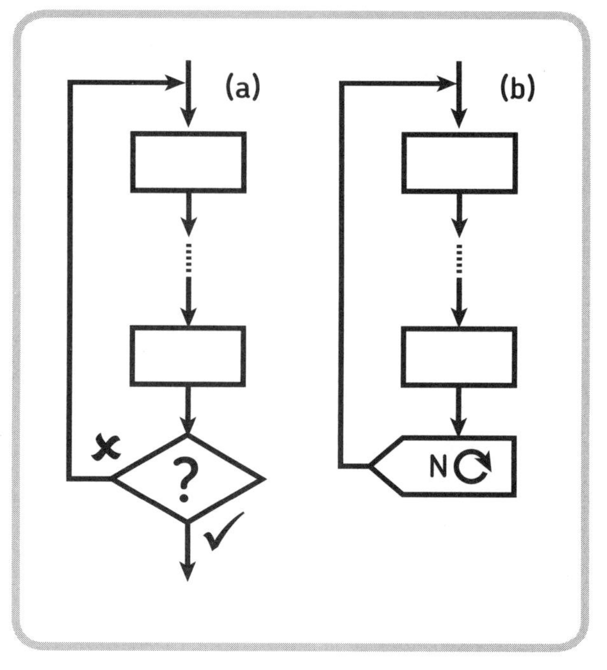

Figure 3-3: A conditional loop (a) and a loop that repeats a sequence N times (b)

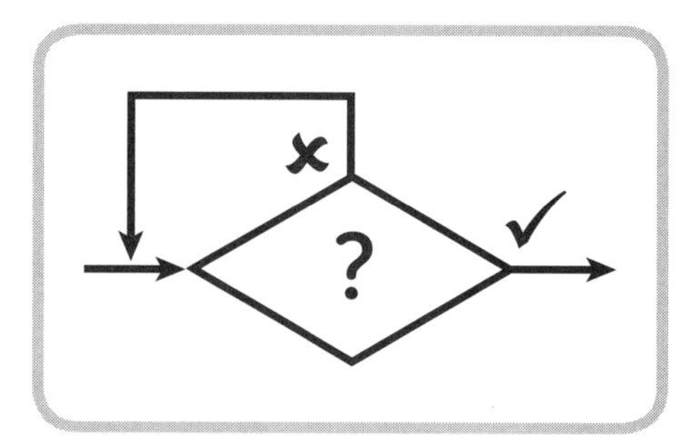

Figure 3-4: A waiting loop

programming with the brick program app

In the case of your EV3 robots, you won't actually write any code, but you will assemble the program just as you do LEGO elements! You can build a program for your robot using the EV3 Brick Program App (the BP App) directly on your EV3 Brick. The Brick Program App will make your programming experience easy and smooth. You don't always need a PC to have fun with robots!

NOTE **You can also tell your robot what to do using the EV3 graphical programming language (see Chapter 5).**

Programming with the Brick Program App is based on the idea that robots do nothing but perform certain basic actions and wait for certain things to happen. Even if somewhat limiting, this assumption works great! In the Brick Program App, you can do the following:

* Place a maximum of 16 blocks, each one commanding an action or waiting for a certain sensor reading.
* Have just one main loop: You can run the sequence once, twice, and so on. You can even run the sequence forever, batteries permitting.
* Build a single sequence of actions, such as starting and stopping motors, playing sounds, and displaying images on screen.
* Add Wait blocks like the loop in Figure 3-4.
* Customize one parameter for each programming block. A *parameter* is a configurable setting that changes the operation of a block.

But you can't make choices that split the program flow in two. That may seem limiting, but I assure you that for the robots we will program, you won't miss the choices. Even a robot that walks and avoids obstacles can be programmed to work without making any choices.

your first brick program

Let's see how to build a program using the Brick Program App. Figure 3-5 shows the EV3 Brick buttons:

* The Enter button allows you to accept changes or select options.
* The Escape button allows you to exit menus and discard options.
* The navigation buttons (Up, Down, Left, and Right) are used to navigate through the various menus.

If it's not already on, turn on your EV3 Brick by pressing the **Enter** button, and then wait for it to start up. Browse the EV3 Brick menu using the **Right** button, and go to the third tab (see Figure 3-6); this contains the Port View, Motor Control, IR Control, and Brick Program Apps. Select the **Brick Program App** using the **Down** button, and open it by pressing the **Enter** button. You should see the empty BP sequence, as shown on the right of Figure 3-6.

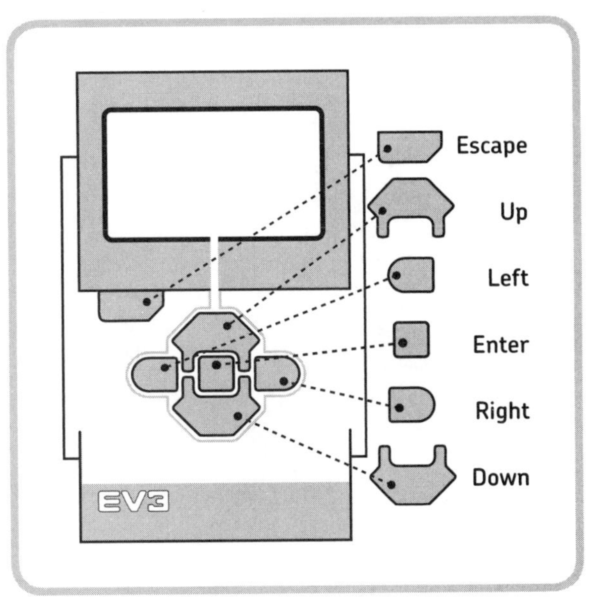

Figure 3-5: The EV3 Brick buttons

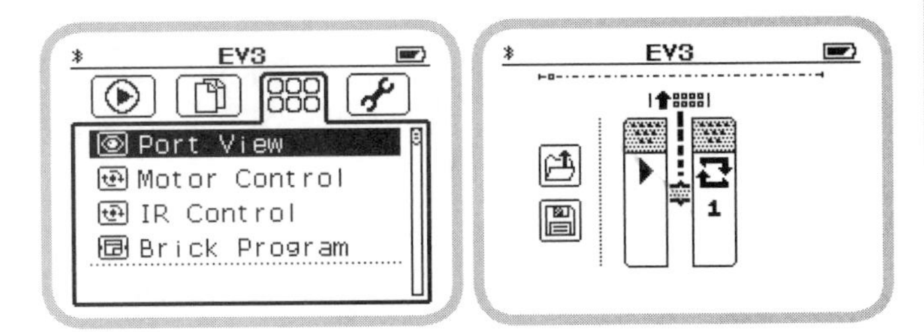

Figure 3-6: From the Apps tab in the EV3 menu (left), you can select the Brick Program App, which will show you the empty sequence on the right.

On the left of the programming sequence, you can see the icons to open or save a Brick Program. You can navigate through these choices using the four arrow buttons, the Enter button, and the Escape button. Depending on which element of the sequence you select, the buttons will have different functions.

To get started with the Brick Program App, let's give ROV3R the ability to use the IR Sensor to detect and then avoid the obstacles in a room. Build the version of ROV3R with the Front IR Sensor (as in Figure 3-7) according to the instructions in "ROV3R with Front IR Sensor" on page 31. Make sure the left motor is attached to port B, the right motor to port C, and the IR Sensor to port 4. Then try to sketch a flowchart for a program that makes ROV3R go straight until it sees an obstacle, drive backward on a curved line, and then go straight again. It should look similar to the one in Figure 3-8.

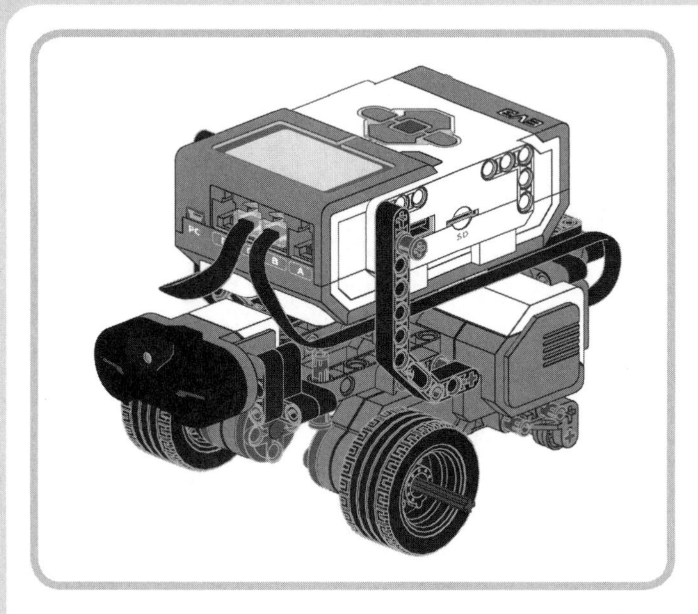

Figure 3-7: ROV3R with Front IR Sensor

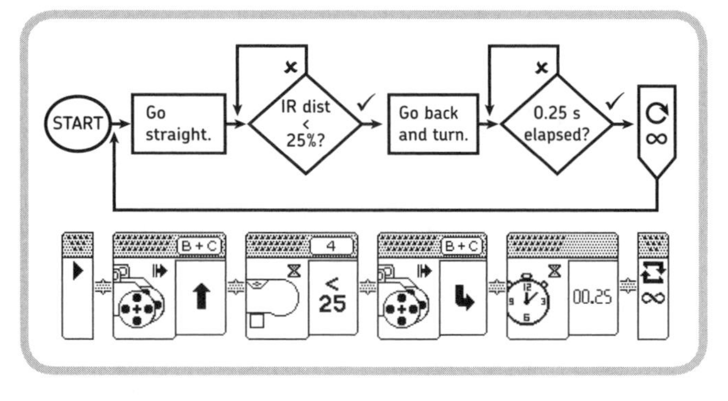

Figure 3-8: The flowchart of the obstacle avoidance program and its BP implementation

Now I'll guide you through building this program, step by step. You should start with a screen similar to the one on the right of Figure 3-6.

1. Press **Up** to access the **Block Palette**, which contains all the programming blocks you can place into your sequence. (You'll learn more in "The Block Palette" on page 53.)

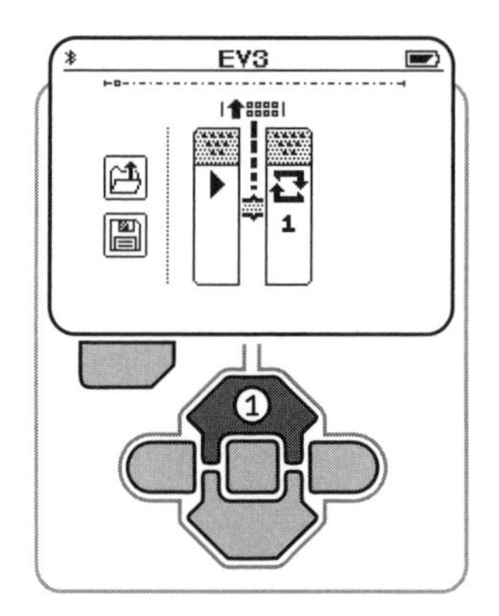

2. The Block Palette is now open. Select the **Move Action** block by pressing **Right**.

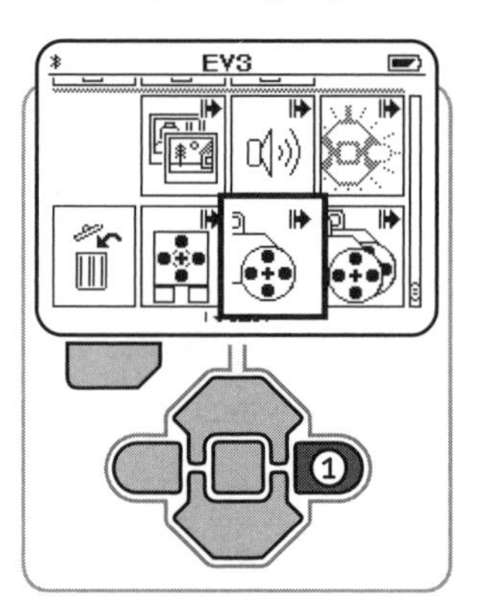

A QUICK GUIDE TO THE BRICK PROGRAM APP

adding a block

* Select a **Sequence Wire**.
* Press **Up** to access the Block Palette.
* Select a block using the arrow buttons.
* Add the selected block to the sequence by pressing **Enter**.

deleting a block

* Select the block you want to delete.
* Press **Up** to access the Block Palette.
* Select the trash bin icon.
* Press **Enter**.

replacing a block

* Select the block to replace.
* Press **Up** to access the Block Palette.
* Select a new block from the Block Palette.
* Press **Enter**.

editing the parameter of a block

* Select the block to edit.
* Press **Enter** to go into edit mode.
* Change the only editable parameter with the **Up** and **Down** buttons.
* Press **Enter** to confirm and exit edit mode.

3. Add the block to the program by pressing **Enter**.

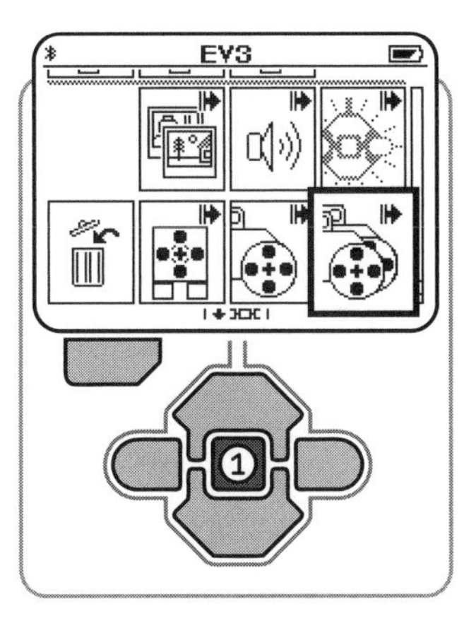

4. The parameter is already set to Forward. Press **Right** to select the next **Sequence Wire**. (The Sequence Wire connects the blocks of the sequence.)

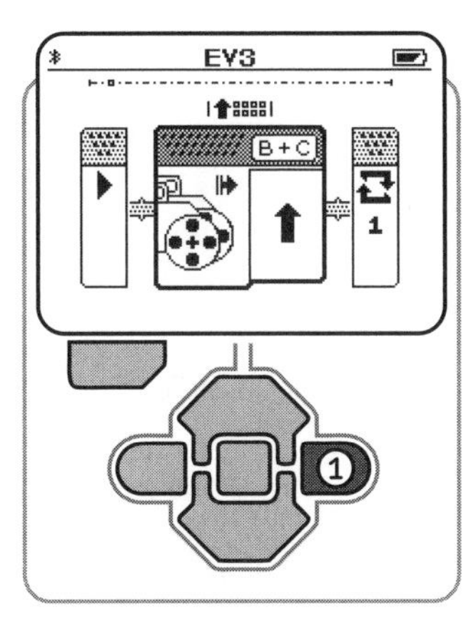

5. Press **Up** to open the **Block Palette**.

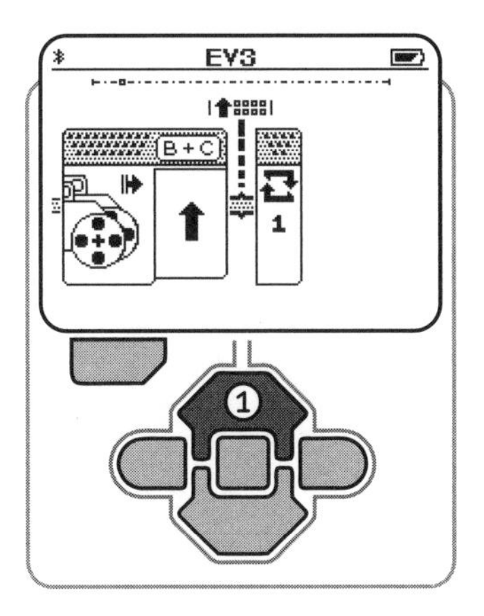

6. Press **Up** three times and **Left** once to select the **Wait IR Sensor** block. Then press **Enter** to add it to the program.

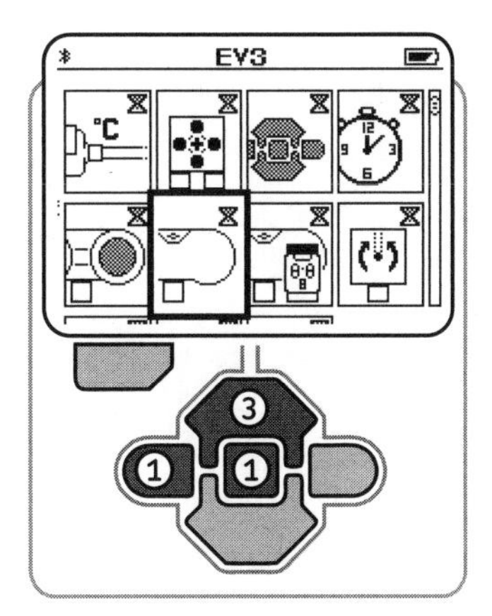

7. Press **Enter** to edit the distance threshold parameter.

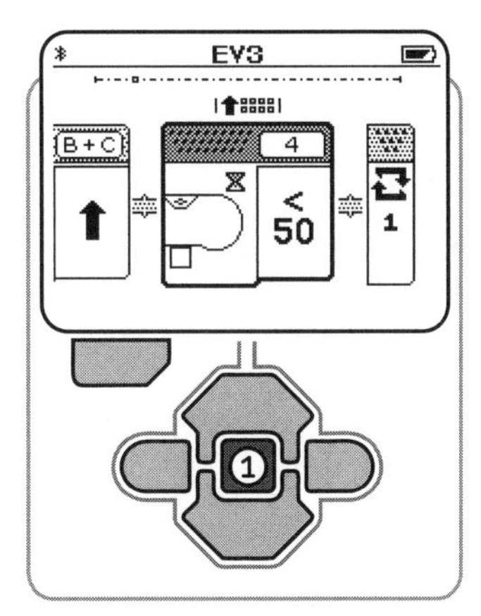

8. Press **Down** once to change the threshold to **<25**; press **Enter** to accept.

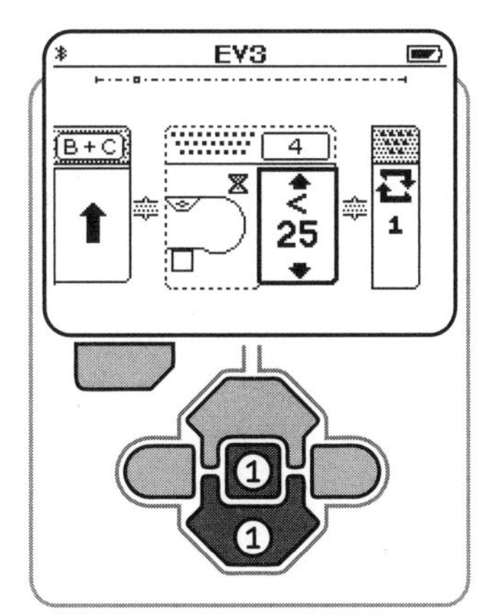

9. Now press **Right** to select the next **Sequence Wire**. Add another **Move** block, repeating steps 1 through 3.

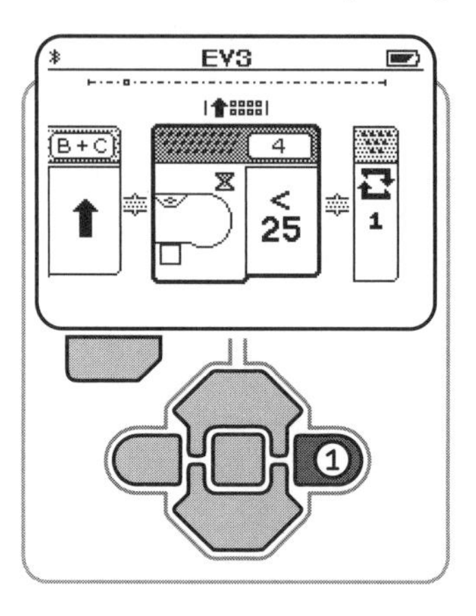

10. Press **Enter** to edit the Move block, pressing **Down** twice to make the robot go backward while turning right. Press **Enter** again to accept.

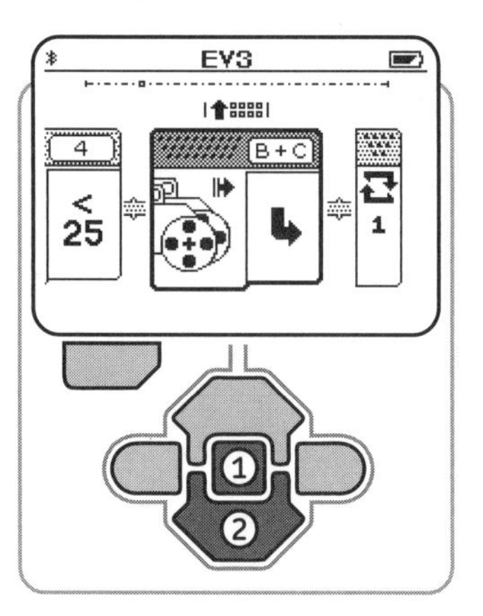

11. Add a **Wait Time** block and edit it to wait for **0.25** seconds. Then select the **Loop** block by pressing **Right** twice.

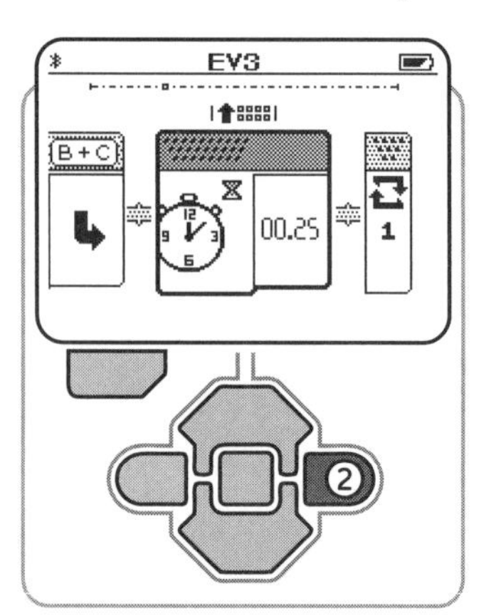

12. Press **Enter** to change the number of times you want the program to repeat.

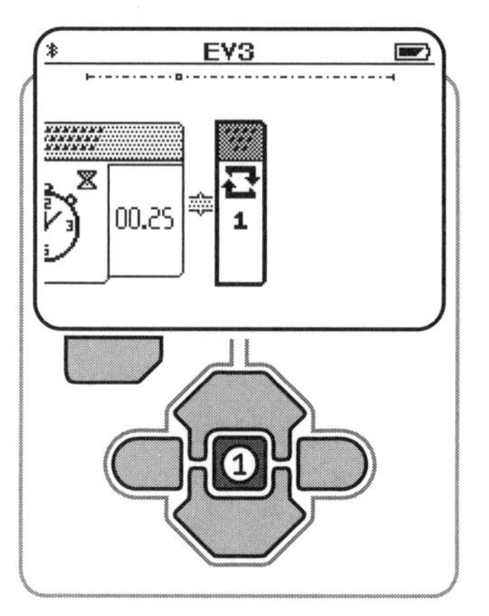

13. Press **Up** six times to set the Loop block to repeat the sequence forever.

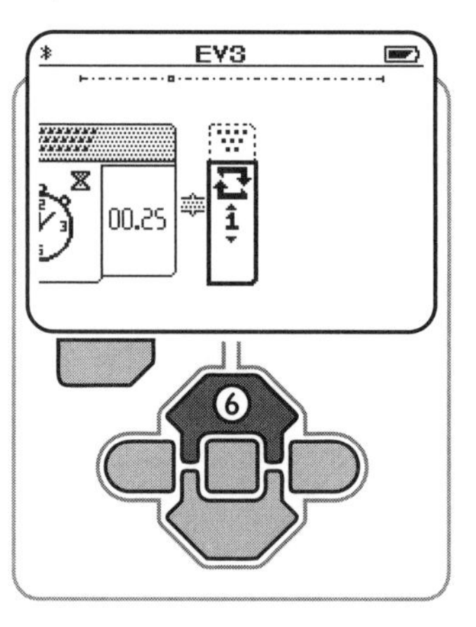

14. The infinity icon (∞) means that the program will repeat forever. Press **Enter** to accept.

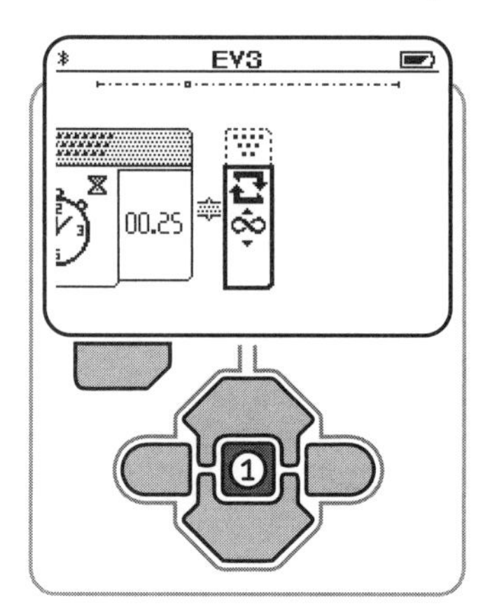

15. Go to the **Start** block by pressing **Left** 10 times or by simply pressing **Escape** once.

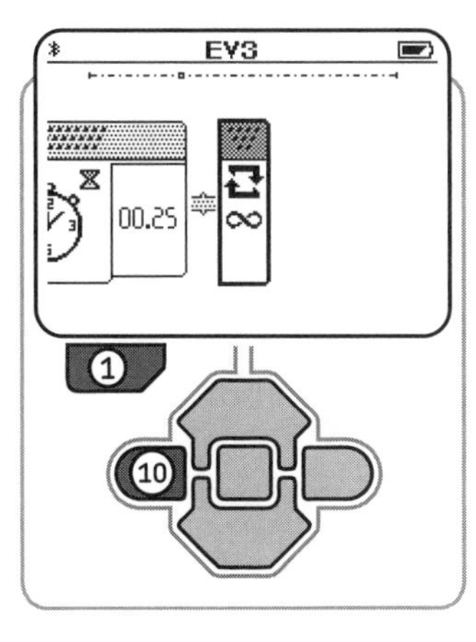

16. Press **Enter** to start the program.

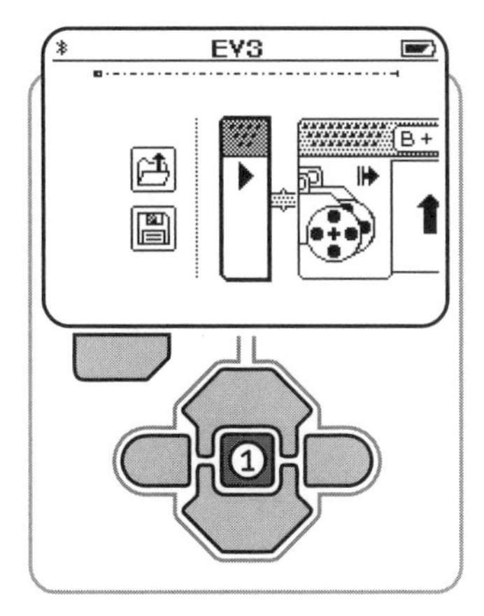

When you start the program, ROV3R will start traveling forward. Place your hand in front of the IR Sensor, and ROV3R will back up, turn a bit, and then start going forward again. To stop the program, press **Escape**. To save your program, press **Left** and then **Enter** to select the **Save** icon. The Brick Program Save Dialog should pop up. Now enter the name for your program like this:

- Highlight the letters on the onscreen keyboard using the navigation buttons.
- Insert the highlighted letter by pressing **Enter**.
- Delete a letter by highlighting the Backspace key (left-pointing arrow) and pressing **Enter**.
- Switch to capital letters by highlighting the Shift key (upward-pointing arrow) and pressing **Enter**.

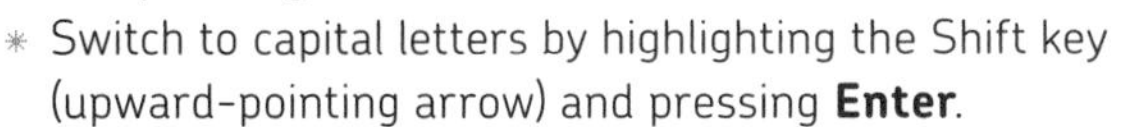

- Switch to the numbers and symbols keyboard by highlighting the 123 button and pressing **Enter**.
- Confirm the name and save by highlighting the Enter key (check mark) and pressing **Enter**.

Congratulations! You've just finished your first Brick Program! Now why not try to modify the program settings—for example, by changing the IR Sensor block threshold parameter or the Wait Time block parameter? Also, take some time to explore the Block Palette. In the next section, I'll describe in detail all the blocks available in the Block Palette.

the block palette

You can access the Block Palette of the Brick Program App by pressing the **Up** button while a block or a Sequence Wire is selected. To return to the programming sequence, press **Enter** (thus selecting a block) or **Escape**. The complete Block Palette is shown in Figure 3-9. Each block has only one customizable parameter.

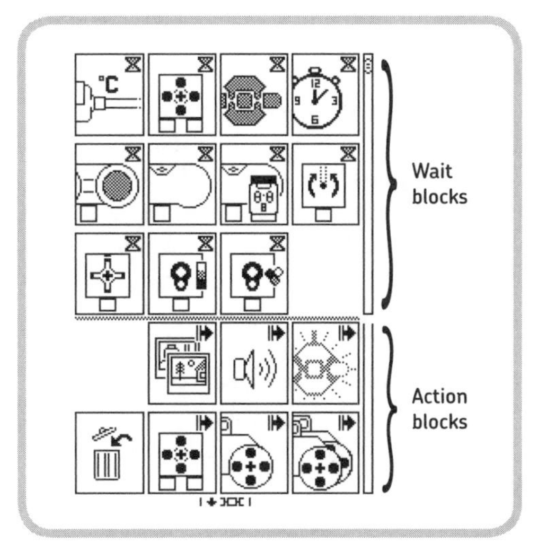

Figure 3-9: The complete Block Palette of the Brick Program App

For example, for the Move block, you can decide the direction the robot will move, but you can't customize the wheels' maximum speed or the output ports the motors are attached to. This simplifies the work it takes to program your robot, although it can be limiting.

The programming blocks can be divided into two groups: Action blocks and Wait blocks. In the following sections, I'll describe the blocks in detail. Each block is shown as it appears in the Brick Program App, and tables list the various icons and meanings of the blocks' parameters. The parameter icons with borders are the defaults.

the action blocks

The Action blocks allow you to move your robot, display images on the EV3 screen, turn the EV3 Brick light on and off, and play sounds. There are six Action blocks: Move, Large Motor, Medium Motor, Display, Sound, and the Brick Status Light.

the move block

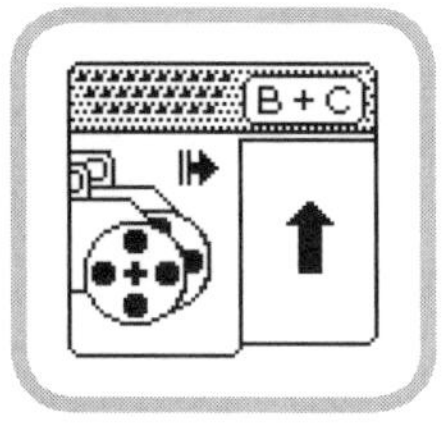

This block allows you to control a pair of Large Motors to drive a wheeled robot like ROV3R. Unlike a car with normal steering, ROV3R can go straight, steer, or spin in place if the two driving wheels turn at different speeds. If both wheels turn in the same direction at the same speed, the robot goes straight; if the wheels turn at different speeds, the robot will travel along a curved path; if the wheels turn in opposite directions, the robot will spin in place. A robot with two motors, each driving a wheel, is called a *differential drive robot*. Real-world examples of differential drive vehicles are tracked excavators and tanks.

Usually, wheeled robots don't drive in a straight line, because the motors don't turn at the same speeds. Using the Move block, the EV3 Brick keeps the two Large Motors' speeds synchronized to improve the robot's ability to drive straight. To see this in action, set the parameter to drive straight and then try to block a motor hub: You should see the other motor slowing down, waiting for the other to catch up.

NOTE Even if the motors' speeds are synchronized, a wheeled robot may still fail to travel straight due to uneven ground, small differences between the radii of the wheels, or other factors.

The parameter of this block is the direction of the movement, as shown below.

Go backward.

Spin right: The motors spin in opposite directions.

Back up left: The right wheel spins backward; the left wheel is still.

Steer right: The left wheel spins forward; the right wheel is still.

Go straight, keeping the motors' speeds synchronized.

Steer left: The right wheel spins forward; the left wheel is still.

Back up right: The left wheel spins backward; the right wheel is still.

Spin left: The motors spin in opposite directions.

Stop the motors.

NOTE If you directly attach wheels to the motor hubs without gears, the left wheel motor must be connected to port B, and the right wheel motor must be connected to port C. Otherwise the robot will turn right when the programming block is set to turn left, and vice versa.

the large motor block

This block controls a Large Motor attached to port D. Its parameter is the power and the direction of the motor, in steps of 25 percent of the full power, as shown below.

Forward at 100% power

Forward at 75% power

Forward at 50% power

Forward at 25% power

Stop and hold position

Backward at 25% power

Backward at 50% power

Backward at 75% power

Backward at 100% power

the medium motor block

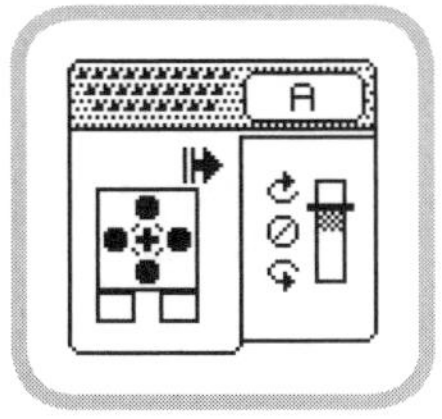

This block controls a Medium Motor attached to port A. Its parameter is the power and the direction of the motor, in steps of 25 percent of the full power.

The parameter icons for this block are the same as the ones for the Large Motor block.

NOTE You can safely attach a Medium Motor to port D or a Large Motor to port A. You can even connect two Large Motors to ports A and D and use them to propel a differential-drive robot. You can make the robot perform all kinds of maneuvers by setting different power levels for the motors. (Notice that the motors' speed will not be synchronized in that case.)

the display block

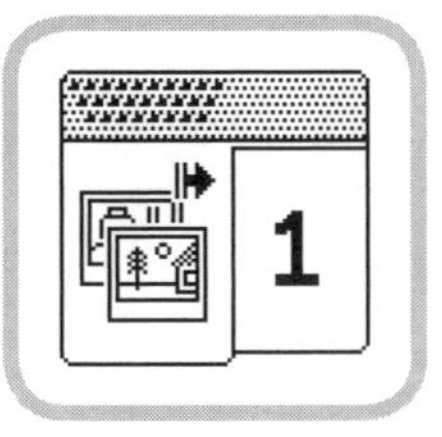

This block displays one of 12 available images or clears the display. You can select an image by changing the block parameter as listed in the table below. Using the Display block and the Sound block, you can make expressive robots.

Parameter	Image	Description
⊘		No image, reset the display
1		Neutral
2		Pinch right
3		Awake
4		Hurt
5		Accept
6		Decline
7		Question mark
8		Warning
9		Stop 1
10		Pirate
11		Boom
12		EV3 icon

the sound block

This block plays one of 12 available sounds. You can select a sound by changing the block parameter as listed below.

Parameter	Description
⊘	No sound, stop playing
1	Hello
2	Goodbye
3	Fanfare
4	Error alarm
5	Start
6	Stop
7	Object
8	Ouch
9	Blip 3 (electronic beeps)

10	Arm 1 (servo motor noise)
11	Snap (pneumatic noise)
12	Laser (laser "gunshot")

NOTE If you place the Sound block as the last block in a sequence that is executed just once, add a Wait Time block after it. (See "The Wait Time Block" on page 57.) Otherwise, the program will terminate before you can hear a sound played.

the brick status light block

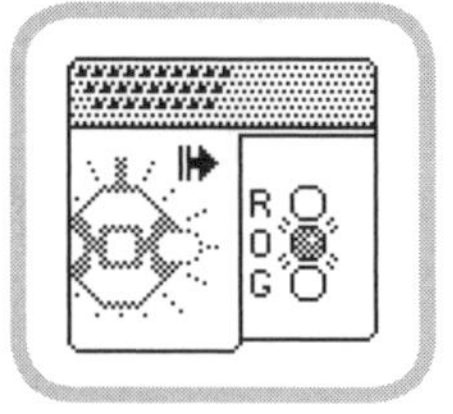

This block turns the status light surrounding the EV3 Brick buttons on and off. The chosen parameter allows you to change the light color and tell it to blink or not.

	Blinking red light
	Blinking orange light
	Blinking green light
	Steady red light
	Steady orange light
	Steady green light
	Turn light off

the wait blocks

The Wait blocks pause the program until a certain condition becomes true. They wait for a time period to pass or for a sensor reading to be equal to, greater than, or less than a specified value. The program can also wait for an EV3 Brick button to be pressed or for the built-in rotation sensor of the Servo Motor connected to port A to reach a certain value.

NOTE Even if the program execution is paused by a Wait block, the motors turned on by a previous Move block or Motor block in the sequence keep running.

Some Wait blocks present in the Block Palette are needed by sensors that are not included in the EV3 31313 set, like the Gyroscopic Sensor, the Ultrasonic Sensor, and the Temperature Sensor.

the wait touch sensor block

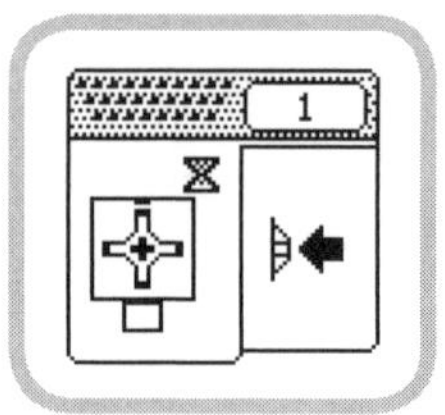

This block waits for the Touch Sensor connected to input port 1 to be pressed, released, or *bumped* (pressed and then released).

	Wait for the Touch Sensor to be pressed and released.
	Wait for the Touch Sensor to be released.
	Wait for the Touch Sensor to be pressed.

the wait reflected light sensor block

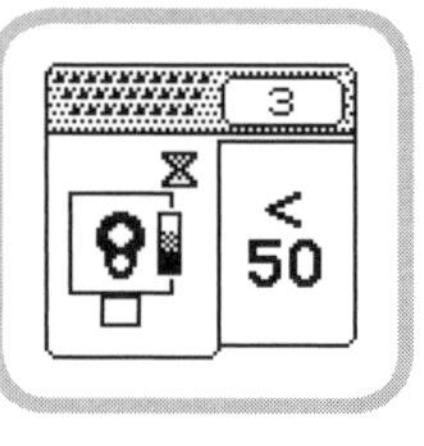

This block waits for the Color Sensor connected to input port 3, used in Reflected Light Intensity mode, to measure a value beyond a certain threshold expressed as a percentage. The sensor measures the light of the (red) LED reflected by surfaces. For greatest accuracy, the sensor must be held at a right angle, 5–10 mm above the surface it is measuring. Lighter surfaces will return higher values than darker surfaces.

≥ 90	≥ 35	< 90	< 35
≥ 75	≥ 25	< 75	< 25
≥ 65	≥ 10	< 65	< 10
≥ 50	< 100	< 50	< 5

the wait color sensor block

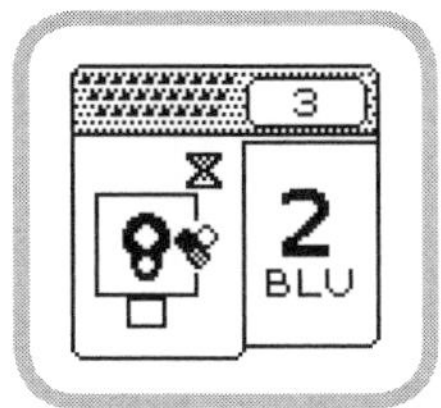

This block waits for the Color Sensor connected to input port 3 to detect the color that you have specified as the parameter. For greatest accuracy, the sensor must be held at a right angle, 5–10 mm above the surface it is measuring.

To detect colors, the Color Sensor flashes the built-in RGB LED, switching among all three colors (red, green, and blue) in a very fast loop and measuring the light returned by the object's surface. Depending on its color, the surface will return different reflection levels of red, green, and blue, which the sensor uses to estimate the color.

7 BRN	Wait until Color Sensor detects a brown object.
6 WHT	Wait until Color Sensor detects a white object.
5 RED	Wait until Color Sensor detects a red object.
4 YEL	Wait until Color Sensor detects a yellow object.
3 GRN	Wait until Color Sensor detects a green object.
2 BLU	Wait until Color Sensor detects a blue object.
1 BLK	Wait until Color Sensor detects a black object.
0 N/C	Wait until Color Sensor does not detect any object.

the wait brick buttons block

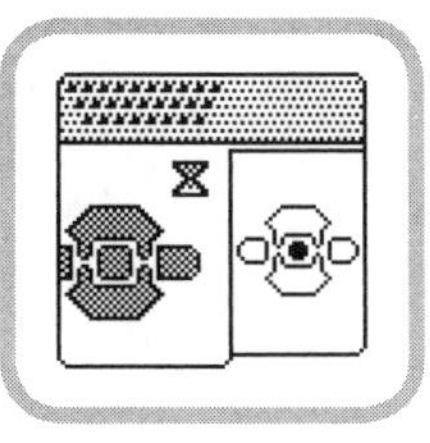

This block waits for an EV3 Brick button to be pressed.

The parameter is the button you specify the robot should wait for.

	Wait for the Right button to be pressed.
	Wait for the Left button to be pressed.
	Wait for the Down button to be pressed.
	Wait for the Up button to be pressed.
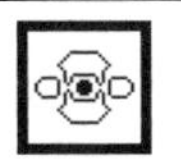	Wait for the Enter button to be pressed.

the wait motor rotation block

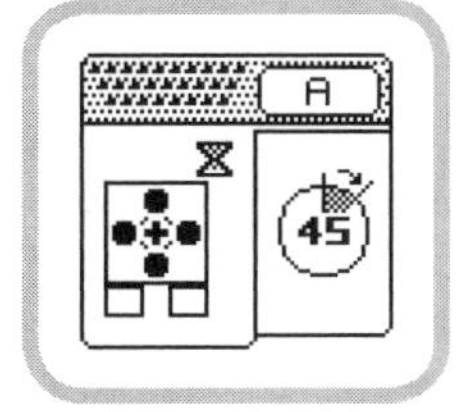

This block waits for the built-in rotation sensor of the Servo Motor attached to port A to measure a change in the shaft angle equal to the threshold specified as a parameter, expressed in degrees clockwise (CW) or counterclockwise (CCW). The block works in a relative way: It does not wait for the shaft to reach an absolute angle but, instead, waits for the shaft's angular position to change. You can combine Wait Motor Rotation blocks to specify other angles, like 45° + 10° = 55° CW.

360	360° CW	360	360° CCW
270	270° CW	270	270° CCW
180	180° CW	180	180° CCW
90	90° CW	90	90° CCW
45	45° CW	45	45° CCW
10	10° CW	10	10° CCW

the wait time block

This block waits for a certain number of seconds, as specified by the parameter. You can combine these blocks to wait for other intervals, like 0.50 + 0.25 = 0.75 seconds.

60.00	10.00	02.00	00.50
20.00	05.00	01.00	00.25

the wait infrared sensor block

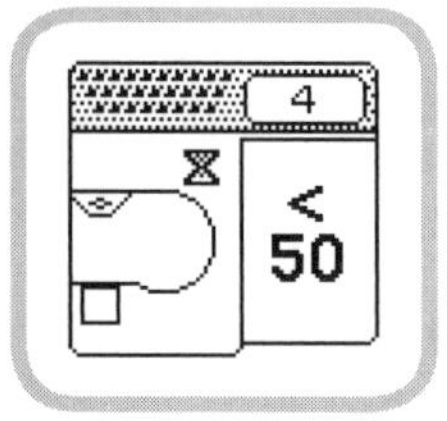

This block waits for the IR Sensor connected to input port 4 to measure the distance from an object lesser or greater than the specified value. The IR Sensor measures distances expressed as a percentage that does not correspond precisely to a distance (unlike the EV3 Ultrasonic Sensor included in the EV3 Education set 45544). The reading is affected by the color of the object the sensor is looking at. On a white surface, it reads 1% at about 2 cm distance, 50% at about 40 cm, and 100% above 90 cm.

< 100	
< 75	≥ 75
< 50	≥ 50
< 25	≥ 25
< 5	

the wait infrared remote block

This block waits for the IR Sensor connected to input port 4 to receive a command from the Remote IR Beacon on channel 1. (The remote has four channels to select from, so you can use up to four remote-controlled EV3 robots in the same room without interference.) You can specify the block parameter as the pressing of a Remote IR Beacon button.

	Wait for the IR Remote's central button to be pressed.
	Wait for the IR Remote's bottom-right button to be pressed.
	Wait for the IR Remote's top-right button to be pressed.
	Wait for the IR Remote's bottom-left button to be pressed.
	Wait for the IR Remote's top-left button to be pressed.
	Wait for all the IR Remote buttons to be released.

the loop block

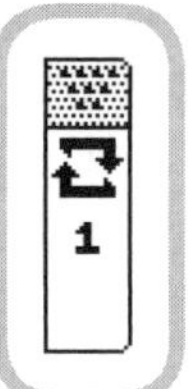

There's one more block that I haven't covered yet: the Loop block. This is the last block in every Brick Program sequence. You can't move or delete it, only change the number of times you want the program sequence to repeat. This block doesn't appear in the Block Palette.

∞	Repeat forever.
10	Repeat 10 times.
5	Repeat 5 times.
4	Repeat 4 times.
3	Repeat 3 times.
2	Repeat 2 times.
1	Don't repeat; execute just once.

EXPERIMENT 3-1

Make a traffic light program. Hint: Use the Brick Status Light block to change the color from green to yellow to red; use the Wait Time block to keep the light on for a good amount of time.

EXPERIMENT 3-2

Build ROV3R with Cleaning Tool and Touch Sensor Bumper as described in Chapter 2 (see Figure 3-10). Slightly modify the tutorial program from page 49 to make the ROV3R sweep the floor and avoid obstacles with the Touch Sensor Bumper. Hint: Just replace the Wait IR Sensor block with a Wait Touch Sensor block, and set the parameter appropriately.

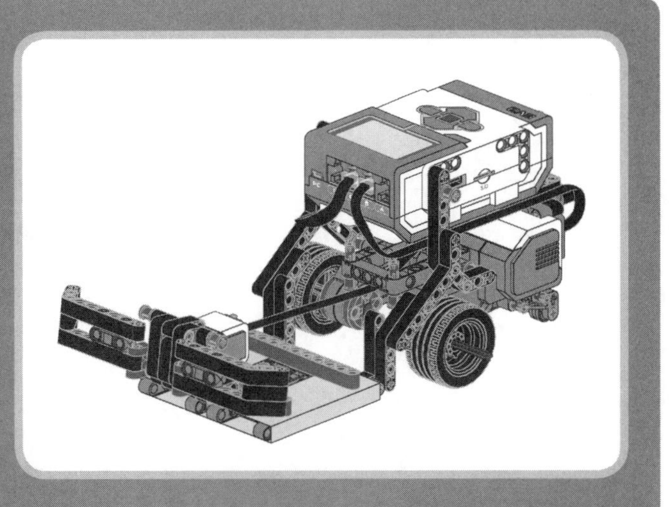

Figure 3-10: ROV3R with Cleaning Tool and Touch Sensor Bumper (page 38)

conclusion

In this chapter, you learned the basics of robot programming. In particular, you discovered how the Brick Program App allows you to program your EV3 robots without using the EV3 programming environment on a PC. Following a step-by-step tutorial, you made your first Brick Program for ROV3R. The final section described in detail the blocks included in the Block Palette of the Brick Program App. In the next chapter, you will learn more tips and tricks about programming with the Brick Program App and how to make ROV3R follow lines and walls!

advanced programming with the brick program app

In Chapter 3, you learned how to program a ROV3R that you built in Chapter 2 to travel across a room and avoid obstacles. You discovered that the EV3 Brick can be programmed without a PC, using the Brick Program App. In this chapter, you'll learn more about on-brick programming. You'll learn how to make ROV3R drive in a particular pattern, follow lines on the floor, or follow walls to explore a whole house!

ROV3R with touch sensor bumper

At the end of Chapter 3, I offered you a challenge ("Experiment 3-2" on page 59): to modify the program to make ROV3R sense obstacles with the Touch Sensor Bumper assembly instead of the IR Bumper. I give you the solution here. Figure 4-1 shows ROV3R equipped with the Touch Sensor Bumper, with and without the cleaning tool. (Chapter 2 has the building instructions for ROV3R and its modules.)

As you can see in Figure 4-2, the obstacle avoidance sequence is as follows: Go forward, wait for the Touch Sensor to be pressed, drive backward on a curved line, turn, and wait 0.25 seconds. The sequence is repeated forever in a loop. You can build this program using the Brick Program App. (See "Your First Brick Program" on page 49 for instructions.)

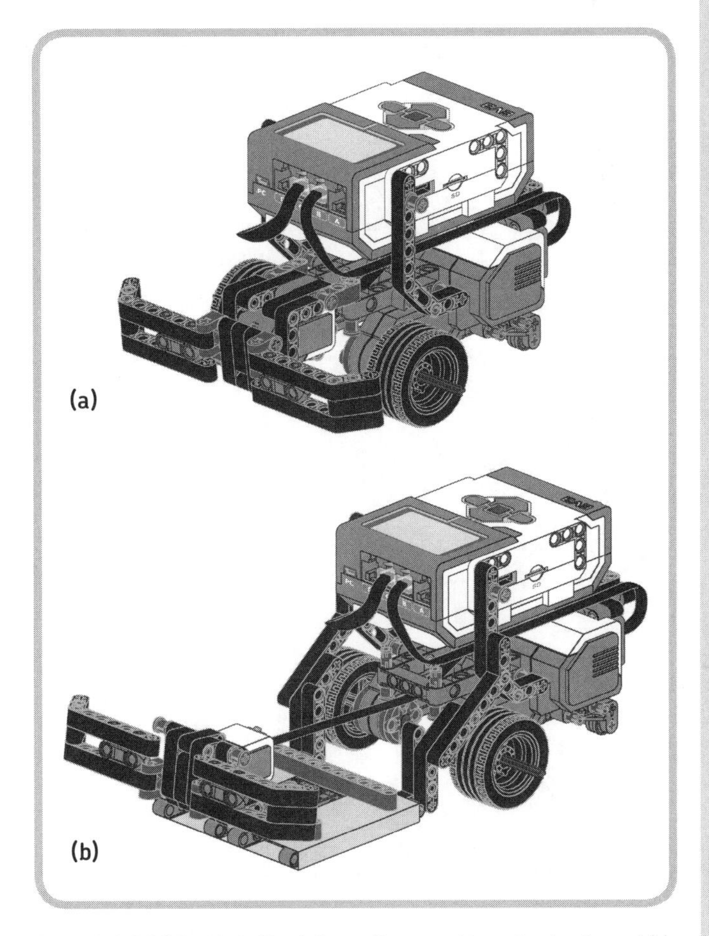

Figure 4-1: ROV3R with the Touch Sensor Bumper, without the cleaning tool (a) and with (b)

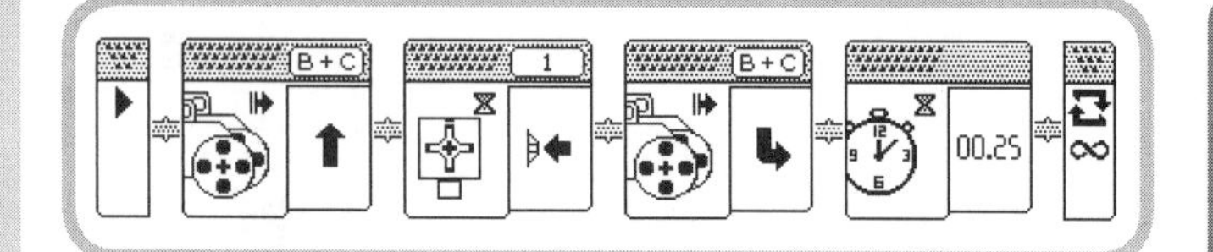

Figure 4-2: The Brick Program to avoid obstacles with the Touch Sensor Bumper

making ROV3R drive along geometric paths

Although you can't control precisely how many degrees the wheels on your robot will turn, you can—with the program shown in Figure 4-3—use the Wait blocks to adjust the wait time and make ROV3R drive along a square path. The robot should turn approximately 90 degrees at each bend.

To tweak the turn precisely, slightly increase the physical distance between the wheels by pulling them out along their axles. By keeping the motor on for the same amount of time (0.75 seconds) and increasing the distance between the wheels, you'll make the robot turn at a smaller angle because the wheels will have to travel along a bigger circumference.

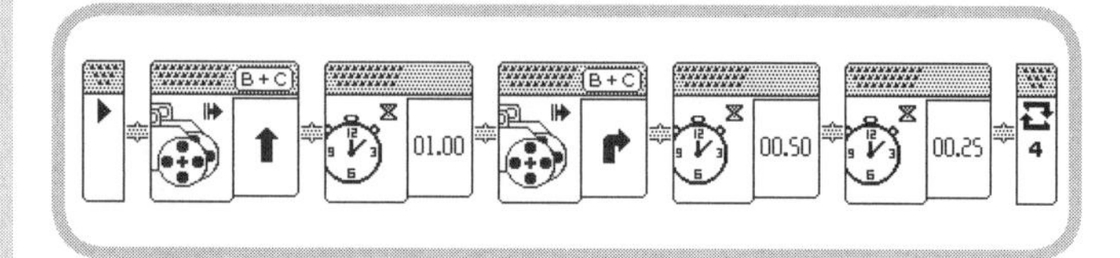

Figure 4-3: The Brick Program to drive ROV3R along a square path

EXPERIMENT 4-1

Which parameter would you change to increase the length of a side of the square? Which parameters would you change to make ROV3R drive in different patterns, like a triangle or a pentagon?

making ROV3R follow lines

One of the greatest challenges of robotics research is that of teaching a robot to navigate from one point to another. The easiest way to accomplish this is to have a robot travel along a predetermined path by following a line on the ground. This approach, which creates a *line-following robot*, is used even for real, goods-handling mobile robots in warehouses, to make them travel precisely from one point of the production line to another. In fact, the LEGO Group itself uses robots like these! (These robots usually follow painted lines on the ground by detecting them with cameras or follow metallic wires embedded in pavement by detecting them with magnetic sensors.)

ROV3R can follow the edge of a line on the floor by using a downward-pointing Color Sensor. The line to be followed must stand out with enough contrast for the Color Sensor to distinguish it from the surrounding floor. You can use either a dark line set against a light background or a light line on a dark background. Optimal colors are black and white; red lines on white (as on the EV3 paper test pad) may not work as well.

You could easily create paths to be followed by attaching some black tape (say, electrical tape) to a light surface or by printing thick black "paths" on white paper.

Figure 4-4 shows the Line-Following ROV3R (building instructions on page 28).

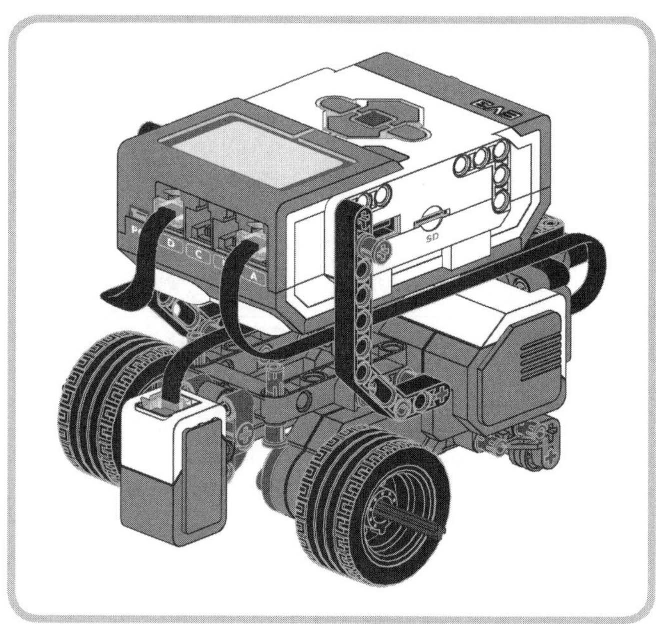

Figure 4-4: ROV3R equipped with the Color Sensor for line following. The motor cables should be attached to ports B and C or A and D, depending on the program.

Figure 4-5 shows how ROV3R follows a line. As the robot moves forward, it turns toward the dark line if the Color Sensor sees a light color (a) or toward the light ground if the Color Sensor sees a dark color (b). The result is a zigzag motion along the *edge* of the line (c).

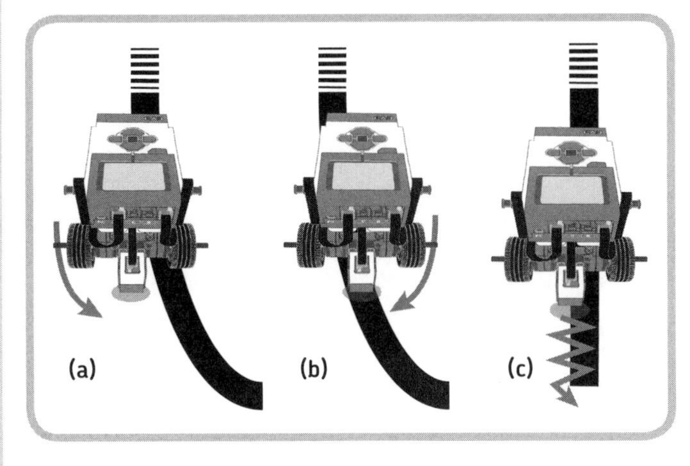

Figure 4-5: ROV3R using a simple line-following approach

using the brick program to follow lines

Now let's make ROV3R follow lines using just Action and Wait blocks from the Block Palette of the Brick Program App. Remember from Chapter 3 that when using the Brick Program App, you can't choose to have your robot perform different actions depending on the sensor reading. So how can you get it to react to different sensor readings in a fast loop? Simple! You set the robot to turn right *until* (not *if*) it sees the line edge, and then you switch it to turn left *until* it sees the light ground, and so on, in a loop. You can use Wait blocks to detect the change in the sensor reading and Action blocks to make the robot steer. The program repeats forever, using the infinite repetitions setting in the final Loop block as shown in Figure 4-6 and described below. Four blocks do all the work!

NOTE For this program, the right motor should be attached to port C and the left motor to port B.

* The first Move block makes the robot steer right.
* The Wait block waits until the Color Sensor (in Reflected Light Intensity mode) reads a value less than 10 percent (a dark color). When it does, the program continues.
* The second Move block makes the robot steer left.
* The second Wait block waits until the Color Sensor reads a value equal to or greater than 25 percent (a lighter color). Since the Loop block is set to *forever* (∞), when the second Wait block lets the program continue, the sequence starts again from the first Move block.

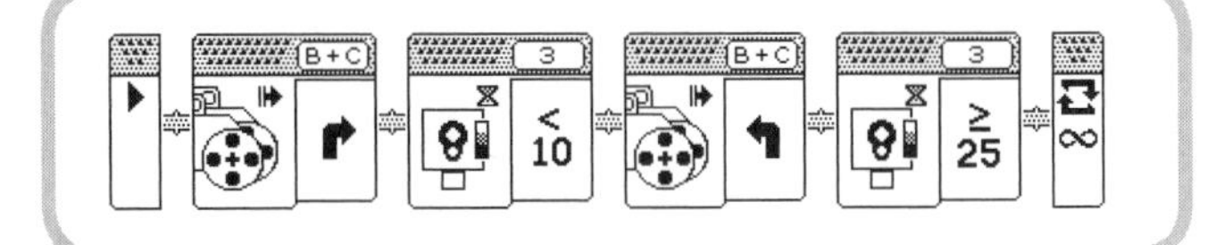

Figure 4-6: The Brick Program to follow dark lines

If the robot is not behaving correctly (if it is missing the line or traveling in circle, for example), try to fine-tune the program by changing the parameter thresholds of the Wait blocks. For example, try changing the threshold for the darker color from <10 to <25 or <5. Or you could change the threshold related to the lighter color from ≥25 to ≥35 or ≥10.

improving the motion

You might notice that the robot's motion is nervous and jerky and that its abrupt change in direction causes some skidding. This is because the Move blocks drive the motors at high power and make the robot turn by stopping one wheel while driving the other.

To improve this behavior, switch the right motor to port D and the left motor to port A, and replace the Move blocks with Action blocks that control motors A and D separately: The Large Motor block will control the motor attached to port D, and the Medium Motor block will control the motor attached to port A. (These blocks can drive either the Large or Medium Motors and allow you to control the power separately.) By modifying the power levels, you'll be able to smooth out the robot's motion. You can see the improved Brick Program for line following in Figure 4-7.

As you can see, instead of using a single Move block to steer right, we use two Motor blocks. The left motor (port A) is set to turn faster than the right one (port D), which makes the robot proceed forward while steering slightly to the right. Similarly, we replace the second Move block that steers left with two Motor blocks that drive the right motor (D) faster than the left motor (A). The Wait blocks remain the same as in the previous program. The resulting motion is smoother because the motors are driven at different speeds and they never stop, as they did when commanded by the Move blocks in Figure 4-6.

EXPERIMENT 4-2

Try setting the Color Sensor's threshold to <10 and ≥10. How does the robot's performance change?

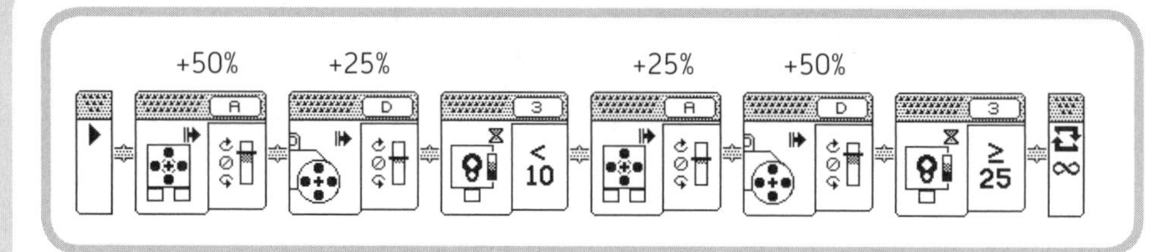

Figure 4-7: The improved Brick Program to follow dark lines

making ROV3R follow walls

Let's make ROV3R explore a space and return to its starting point. How? By making it follow walls!

As you can see in Figure 4-8, the robot can explore any environment (your room, your house, your school) by trying to keep a constant distance from walls or any other objects (such as furniture, shoes, cats, and so on) that it sees with the IR Sensor.

Figure 4-8: ROV3R can explore a space and return to its starting point if its path is not too cluttered.

The method for wall following is similar to the method used for line following, as shown in Figure 4-9. The robot turns toward the wall until the measured distance drops below a certain threshold (a), at which point it turns away from the wall until the measured distance rises above the threshold (b). The resulting movement is a wiggling path at an average constant distance from the wall (c). As long as the robot keeps a good distance from the wall, it can deal with corners and edges without getting stuck (d).

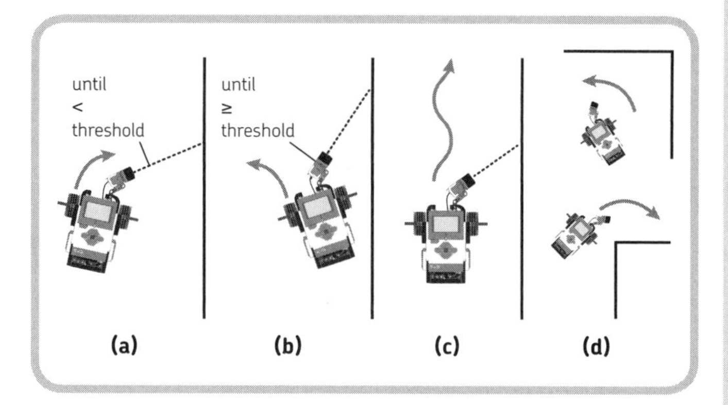

Figure 4-9: ROV3R using a simple wall-following strategy

Build ROV3R with the IR Sensor mounted as a wall-following sensor, as shown in Figure 4-10 (see "Wall-Following ROV3R" on page 32). The IR Sensor placed diagonally on the right side of your robot will see objects ahead of it. For the program, just replace the Wait Reflected Light Sensor blocks used in the line-following program (Figure 4-6) with Wait IR Sensor blocks to produce a program that looks like the one in Figure 4-11.

improving the motion

As in the line-following program shown in Figure 4-6, this wall-following program uses Move blocks to drive the motors attached to ports B and C, and the resulting movement is quite abrupt. To smooth out ROV3R's path, try the program shown in Figure 4-12. As in the program in Figure 4-7, you can use separate blocks to set the motors on ports A and D to run at slower speeds and thereby avoid stopping one wheel in order to turn.

If you lower the thresholds to <25 and ≥25, the robot will follow the wall more closely and will try to explore narrow passages, but it could get stuck when passing near edges (convex corners) or going around thin walls. If you increase the thresholds to <75 and ≥75, it will stay farther from walls and objects, which will smooth out its travel around corners, but it could end up traveling through the middle of a room and may skip narrow passages.

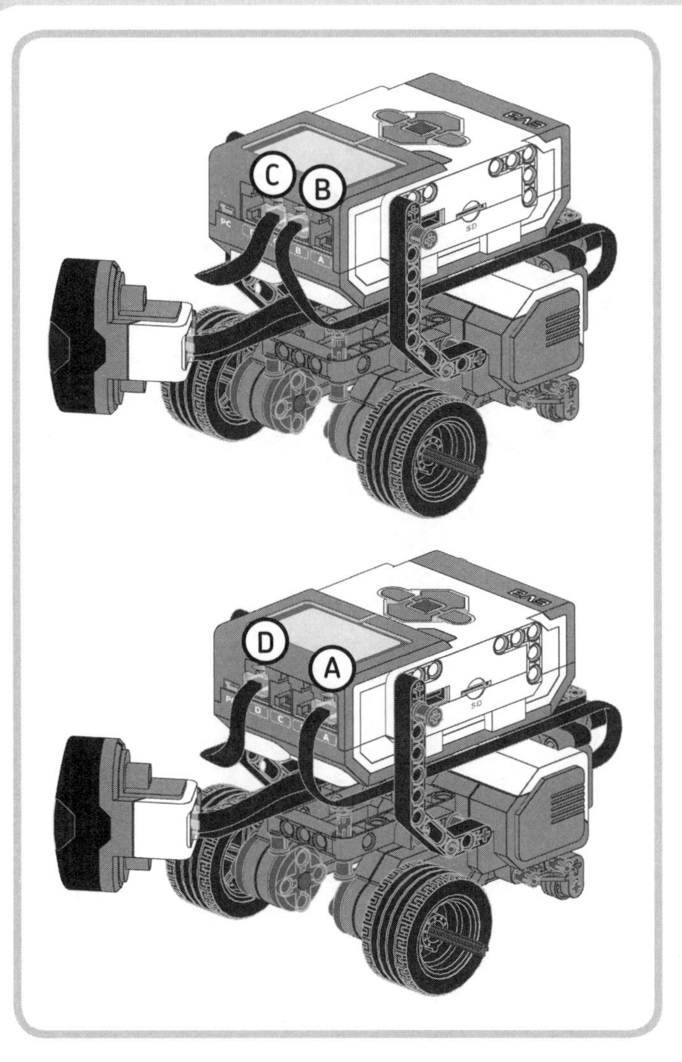

Figure 4-10: ROV3R equipped with the IR Sensor assembly for wall following. The motor cables should be attached to ports B and C or A and D, depending on your program.

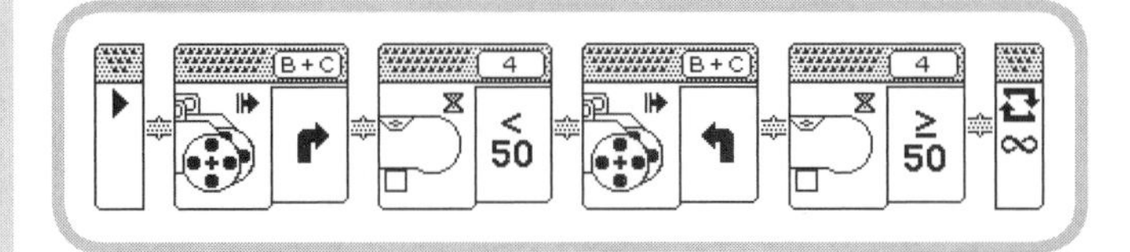

Figure 4-11: The wall-following program. The motor cables should be attached to ports B and C.

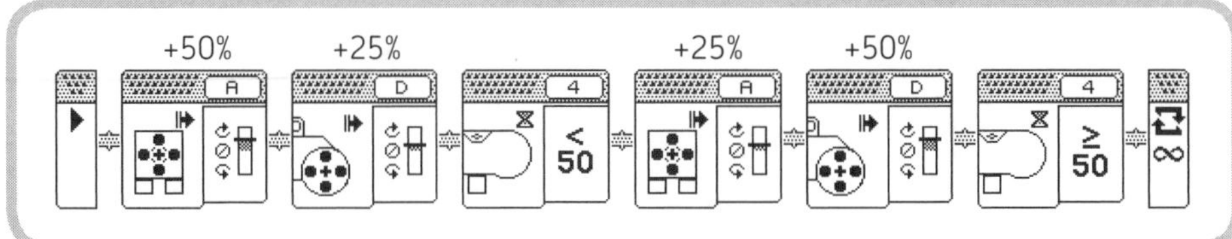

Figure 4-12: Alternative wall-following program. The motor cables should be attached to ports A and D.

EXPERIMENT 4-3

Build a frame to hold a video camera facing forward. Then switch recording on and let ROV3R explore. When your robot returns, you'll have a video of the trip. If you use a smartphone with video chat software (for example, Skype), you can even transmit the live video of ROV3R exploring its environment!

conclusion

In this chapter, you learned how to make ROV3R drive in patterns and follow lines and walls. A wall-following robot can explore any environment autonomously, transmit video back to the base, and even help get you out of trouble. What trouble? Read on!

I MADE ROV3R NAVIGATE AVOIDING OBSTACLES, FOLLOWING LINES AND WALLS...WHAT'S THE USE OF WALL FOLLOWING?
DID YOU EVER GET LOST IN A MAZE? WELL, YOU CAN GET OUT BY JUST FOLLOWING A WALL, ALWAYS KEEPING IT ON THE SAME SIDE!
SOB!
BUT BACK THEN, I DIDN'T KNOW...
WILL I EVER NEED IT?
POF!
YOU NEVER KNOW!
04X1

PROGRAMMING THE EV3 IS REALLY SIMPLE! I DON'T EVEN NEED A COMPUTER!
EASY, BOY! THERE'S STILL MUCH TO DO. GET YOUR LAPTOP!
NGH!
HEY, WHAT'S UP WITH YOU?
PAF!
I FORGOT MY LAPTOP AT HOME!
04X2

EV3 programming

In this chapter, we'll cover programming with the EV3 Software. I think you'll find that the EV3 Software allows you to program your robots in an easy and intuitive way. It also gives you the tools to document your projects with comments, pictures, and videos.

The official *EV3 User Guide*, which gives you a basic understanding of the EV3 Brick interface and software, is provided in digital format (PDF) with the EV3 Software. Since I think it is very useful to have a printed manual and to make sure we're all starting in the same place, in this chapter I'll give you an overview of the EV3 Software's essential features.

I'll also show you how to move easily from the limited programming with the Brick Program App to proper, complete EV3 programming.

EV3 software setup

Before you begin, make sure that your Windows or Macintosh computer meets the minimum system requirements listed on the back of the EV3 set box. The EV3 Software does not come on CD, so you'll have to download the installer from the Downloads section of the LEGO MINDSTORMS official website (*http://LEGO.com/mindstorms/*). Once the download is finished, double-click the installer file to begin installation, and then follow the onscreen instructions. In addition to the EV3 Software, the installer will also install the drivers that let your computer communicate with the EV3 Brick.

EV3 software overview

Once the software has been installed, double-click its icon to open the LEGO MINDSTORMS EV3 Home Edition software. The main parts of the software that we'll work with are the Lobby, the Programming Interface, and the Project Properties. I'll describe each in detail in the following sections.

the lobby

Every time you launch the software, you should see (and unfortunately hear) the Lobby. This is a welcome screen that lets you access the main working features quickly, as you can see in Figure 5-1.

1. **Menu bar:** This includes File, Edit, Tools, and Help items. You'll use the menu bar mostly while programming. (See "The Tools Menu" on page 72.)
2. **Lobby tab:** Click here to return to the Lobby at any time.
3. **Add Project tab:** Click the tab with a plus sign (+) to create a new project.
4. **Missions:** Here you can learn more about the five official models that come with the set. This area is interactive: Click the robots to access more information and to get started with their tutorials.
5. **Open Recent:** Click here to quickly open your recent projects.
6. **Quick Start, News, and More Robots tabs:** The Quick Start tab gives you access to the official video tutorials, the guide, and documentation. The News and More Robots tabs give you access to online content from the LEGO MINDSTORMS website, including 12 bonus models, among which is the EL3CTRIC GUITAR I designed.
7. **User Guide:** The *EV3 User Guide* (in PDF format) is an essential manual, containing information about the EV3 system and how to use the EV3 Brick and connect it to your computer. You'll also find a brief introduction to the EV3 Software and the complete list of the LEGO elements included in the set.
8. **EV3 Help:** This contains details about the EV3 Software's features, tools, and programming blocks.

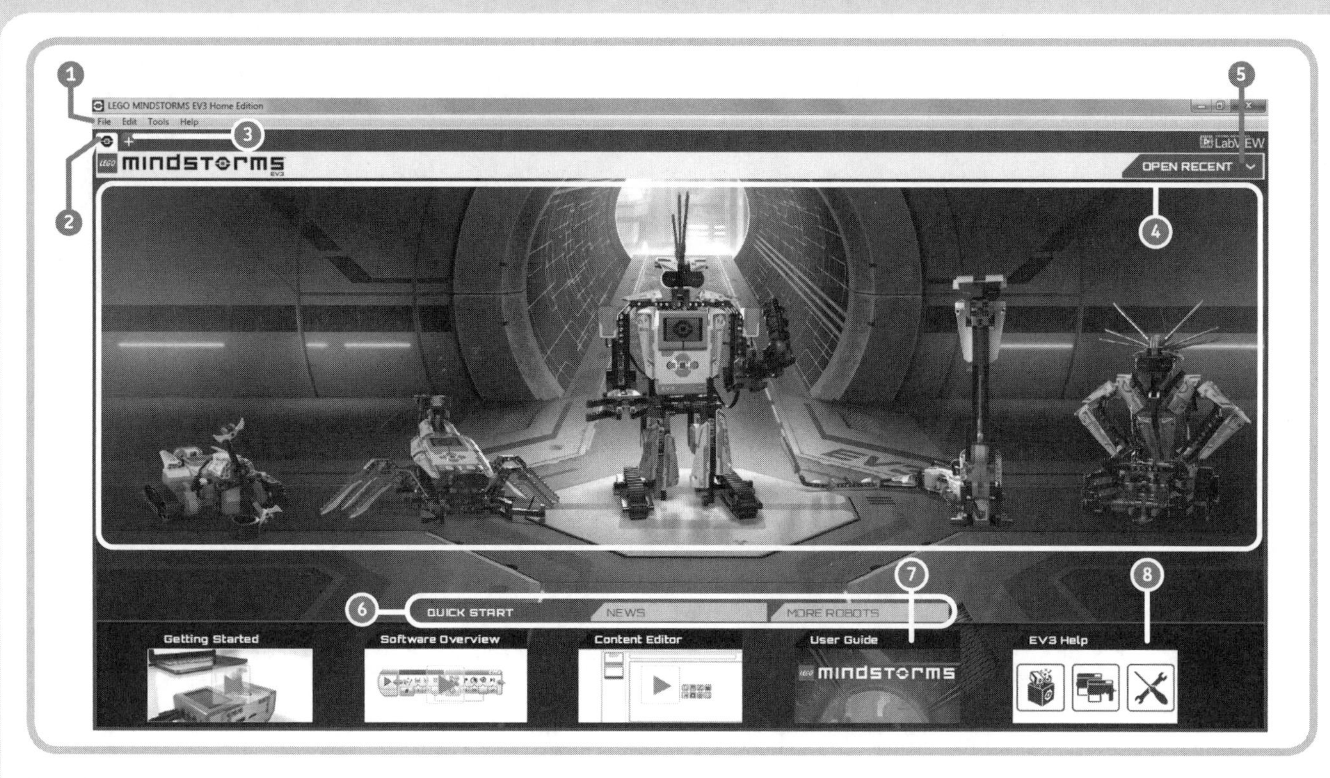

Figure 5-1: The Lobby welcomes you every time you open the EV3 Software.

NOTE Be sure to consult the *EV3 User Guide* and EV3 Help for detailed documentation. I've chosen not to duplicate their information in this book.

To get started, create a new project by clicking the **Add Project** tab [see Figure 5-1(3)]. This takes you to the Programming Interface, and an empty program called *Program* is created automatically.

the programming interface

The Programming Interface shown in Figure 5-2(a) is where you build the programs for your robots. You'll find descriptions of its various controls below.

1. **Programming Canvas**: This is where you build your programs by adding programming blocks.
2. **Content Editor**: This is like a workbook built into your project where you can document your projects with descriptions, videos, images, and even building instructions! Hide it by clicking the **EV3** icon to the right of the Edit icon (a pencil icon). When collapsed, this tab shows a book icon.
3. **Programming Toolbar**: Use this to see all open documents, switch between the Selection and the Pan tools, add comments to your programs, save a project, undo/redo, zoom out/in, and change the program view to the original magnification.
4. **Programming Palettes**: These contain all the blocks you need to program your robots. You'll learn more about these in "The Programming Palettes" on page 73.
5. **Hardware Page**: Use this to manage the connection to your EV3 Brick, see sensor and motor readings in real time, browse the EV3 Brick's memory, and more. Even when collapsed, this still shows the **Controller** (Figure 5-3), allowing you to *Download* (1), *Download and Run* (2), and *Download and Run Selected* blocks (3). When a program is running, the (2) button changes to *Stop*. The Download command sends all project data (including other programs, images, and sounds) to the EV3 Brick before running the program. Using Download and Run sends all the project data and runs the program you are currently working on. Using Download and Run Selected is quicker because it allows you to download and run only the selected chunk of a program. When a project gets large, downloads can become time consuming, especially if you just need to test a couple of blocks!
6. **Lobby tab**: Click here to return to the Lobby.
7. **Project Properties**: Click here to see Project Properties (see "Project Properties" on page 75).
8. **Add Project tab**: Click here to create a new project.
9. **Add Program tab**: Click here to add a new program to your project.

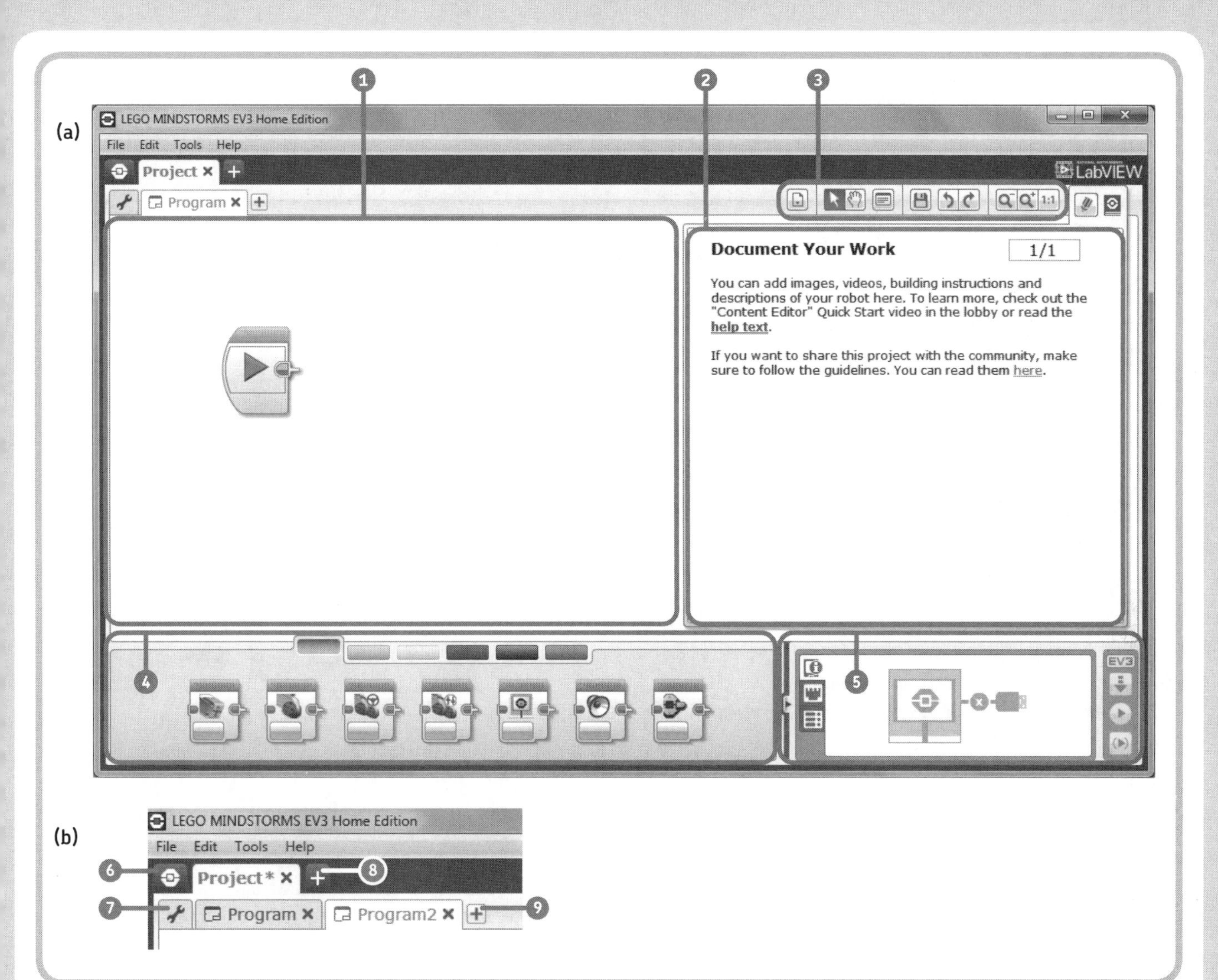

Figure 5-2: The EV3 Programming Interface with the Content Editor window expanded on the right (a); a zoomed-in view of the Project and Program tabs, the Lobby, and the Project Properties tab (b)

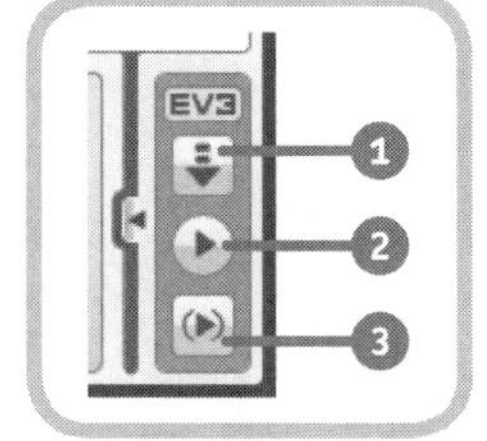

Figure 5-3: The Controller is shown when the Hardware Page is collapsed. The buttons are Download (1), Download and Run (2), and Download and Run Selected (3).

COMPILING PROGRAMS

To be understood by a computer, programs must be compiled. A *compiler* is software that translates human-readable programming code into binary digits. When you click Download and Run, all the compiling happens behind the scenes.

the hardware page

The Hardware Page is located at the bottom right of the Programming Interface. When no EV3 Brick is connected, most of the icons and controls are grayed out. The Hardware Page has three tabs:

* **Available Bricks tab** (Figure 5-4): From here, you can search for EV3 Bricks that are Bluetooth enabled or that are connected to your computer via USB, Bluetooth, or Wi-Fi.

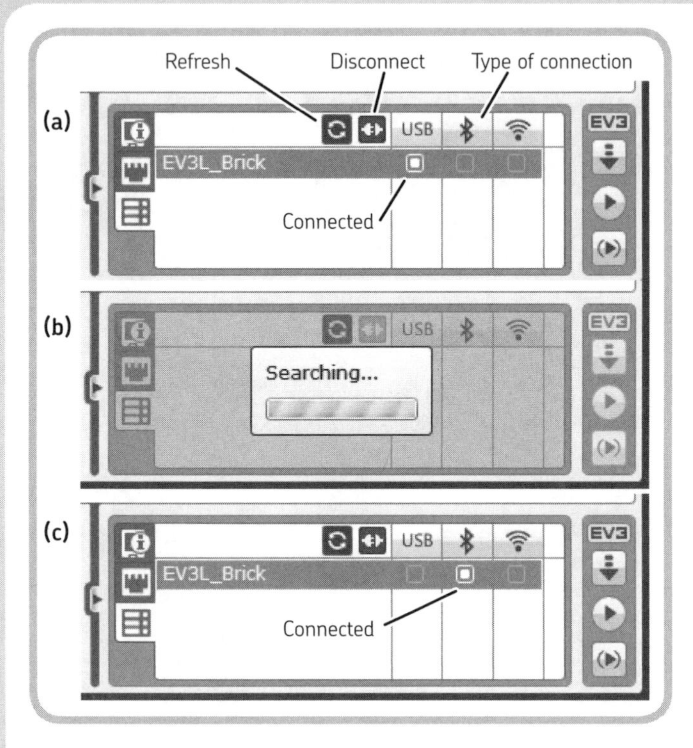

Figure 5-4: The Available Bricks tab of the Hardware Page: EV3 Brick connected via USB (a), searching for EV3 Bricks available on Bluetooth (b), EV3 Brick connected over Bluetooth (c)

* **Brick Information tab** (Figure 5-5): When the EV3 Brick is connected, use this tab to check the battery level, the amount of memory available, and the firmware version of the EV3 Brick. You can change the name of the EV3 Brick, set up the wireless network, and browse files in the EV3 Brick's memory.

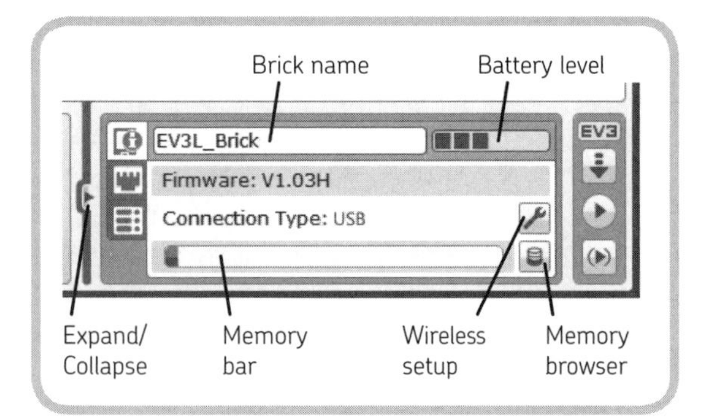

Figure 5-5: The Brick Information tab of the Hardware Page

* **Port View tab** (Figure 5-6): When the EV3 Brick is connected, use this tab to check the readings of all the sensors and motors attached to it in real time. This feature is really useful when you need to set thresholds in your programs for Wait blocks or when you need to measure how much a motor shaft should rotate. When you connect a sensor or a motor to an input or an output port on the EV3 Brick, the EV3 Brick recognizes the type of device automatically, thanks to the Auto-ID feature. You can change the mode of the sensors (for example, the Color, Reflected Light Intensity, and Ambient Light modes for the Color Sensor) by clicking the related icons. You can also reset the Motor Rotation sensor values by clicking the port names. When Daisy-Chain mode is enabled (see "Project Properties" on page 75), the Port View tab shows the sensors of all the daisy-chained EV3 Bricks [Figure 5-6(b)].

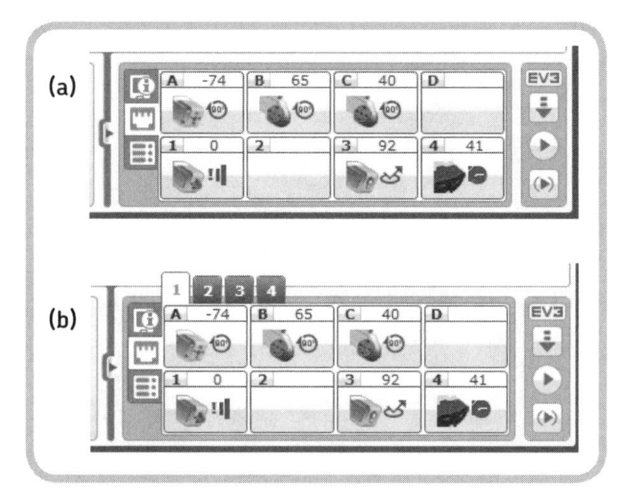

Figure 5-6: The Port View tab of the Hardware Page without (a) and with (b) Daisy-Chain mode enabled.

the tools menu

The Tools menu is located in the menu bar [see Figure 5-1(1)]. You'll find many useful tools here:

* **Sound Editor**: This allows you to record or import a sound and save it to your project, thus enabling your robot to play custom sounds using the Sound block.
* **Image Editor**: This allows you to create, import, or edit an image and save it to your project. Images can be shown on the EV3 Brick screen using the Display block.
* **My Block Builder**: This allows you to create custom blocks (called *My Blocks*) that contain small subprograms. Grouping blocks into a single My Block is useful for creating a small sequence that can be used as a module in many parts of a project or to make your programs look tidier. My Blocks can

also have inputs or outputs for data. (You'll use this tool a lot, so I'll describe it in detail in Chapter 10.)

* **Firmware Update**: This allows you to update the firmware of your EV3 Brick. The firmware makes the EV3 Brick work, and it should be updated if the LEGO Group releases new versions to fix bugs or add new features.
* **Wireless Setup**: This allows you to configure the Wi-Fi network to connect the EV3 Brick using the Wi-Fi Dongle. (You can access the same tool from the Brick Information tab of the Hardware Page.)
* **Block Import**: This allows you to import new programming blocks made by the LEGO Group or third-party developers (for example, to program your robot to work with a new sensor).
* **Memory Browser**: This allows you to manage the files stored in the EV3 Brick's memory. (You can access the same tool from the Brick Information tab of the Hardware Page.)
* **Download as App**: Using this tool, you can download a program as an app, making it appear in the Brick Apps menu together with Port View, Motor Control, IR Control, and Brick Program Apps.
* **Import Brick Program**: This tool will be our launchpad to proceed smoothly from Brick programming to actual EV3 programming! In fact, it allows you to import a Brick Program saved in the EV3 Brick's memory to the EV3 Programming Canvas so that, instead of starting from scratch, you can improve an existing program by using the full EV3 Software programming capabilities.

the programming palettes

The Programming Palettes include all the programming blocks you need to create programs for your robots. You may recall from Chapter 3 that the basic structures that make every computer program work are sequences of actions, choices, and loops. Programs have inputs and outputs, and they can store, retrieve, and transform data. Programming blocks are grouped in the Palettes according to their function, so they're easy to find and use.

Each palette has a different color, and all the programming blocks belonging to the same palette have headers of the same color. For example, all the Action blocks have green headers, and all the Flow blocks—like loops and switches—have orange headers.

The programming blocks will be described as soon as you use them in the following chapters. To find information about a particular block, look it up by name in the index.

the action blocks

The Action blocks (Figure 5-7) control the output of your robots' programs. They rotate motors, display text and images on the EV3 Brick screen, play sounds, and light up the EV3 Brick Status Light.

the flow control blocks

The Flow Control blocks (Figure 5-8) control the flow of the program. Every program sequence begins with a Start block. (You'll learn in Chapter 12 how to get multiple sequences of blocks to run in parallel by placing more than one Start block in a program.) As in the Brick Program App, here you find a Wait block and a Loop block. There is also a Switch block that changes the program flow according to conditions that you specify. The Loop Interrupt block stops the execution of a Loop block even if some blocks inside the loop are still running; the program will continue with the blocks after the Loop. You can set the Loop Interrupt block to interrupt a specific Loop by name, as every Loop block has a name label on top.

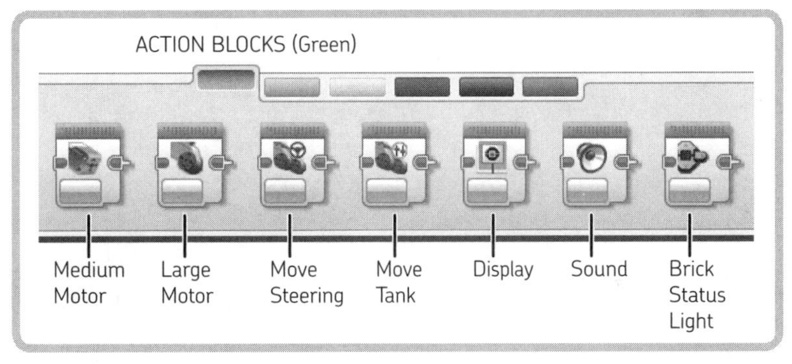

Figure 5-7: The Action blocks palette

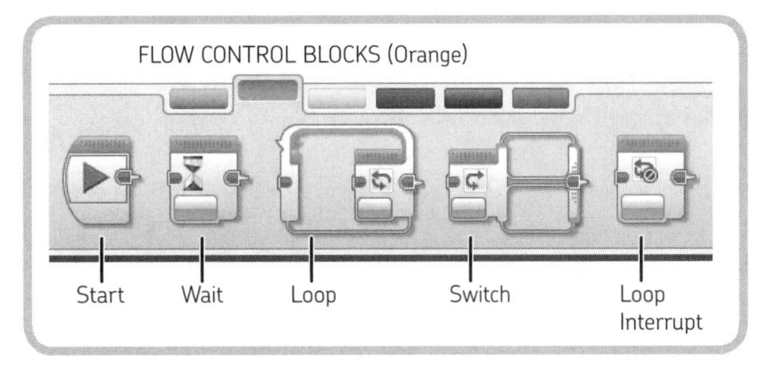

Figure 5-8: The Flow Control blocks palette

the sensor blocks

The Sensor blocks (Figure 5-9) allow you to read the inputs for your program. In addition to the blocks that read the Touch, Color, and Infrared Sensors, you'll find blocks to read the EV3 Brick buttons and Motor Rotation sensors and a block to read time intervals from the EV3's internal timer. We will learn about these blocks when programming the robots in Chapters 10, 12, 14, and 16.

You can download additional Sensor blocks from the Downloads section of the LEGO MINDSTORMS official website (*http://LEGO.com/mindstorms/*) and from third-party sensor producers' websites, like *http://www.hitechnic.com/* and *http://www.mindsensors.com/*.

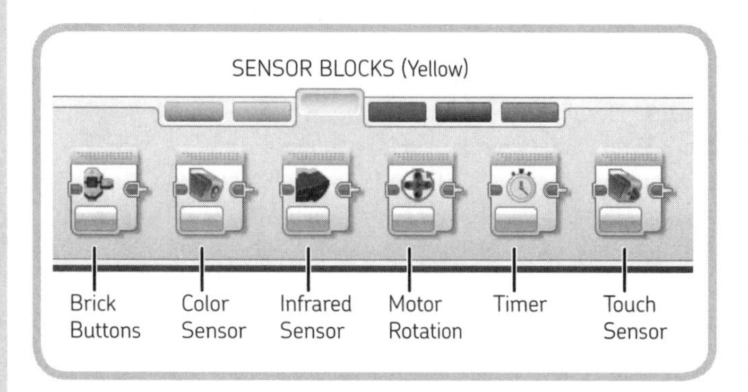

Figure 5-9: The Sensor blocks palette

the data operations blocks

Use the Data Operations blocks (Figure 5-10) to write and read variables and arrays, manipulate data with Math and Logic Operations blocks, compare values, combine strings of text, and generate random numbers. (Each of these will be explained in the following chapters.)

the advanced blocks

The Advanced blocks (Figure 5-11) let you manage files and Bluetooth connections, send Bluetooth messages, keep the EV3 Brick awake (it turns off automatically according to the sleep setting), invert a motor's direction, read raw sensor values (for third-party sensors), drive the motors without internal speed regulation, and stop the program.

My Blocks

The last palette on the right contains the My Blocks you create or import into a project. Initially, it's empty. You'll learn how to create My Blocks in Chapter 10.

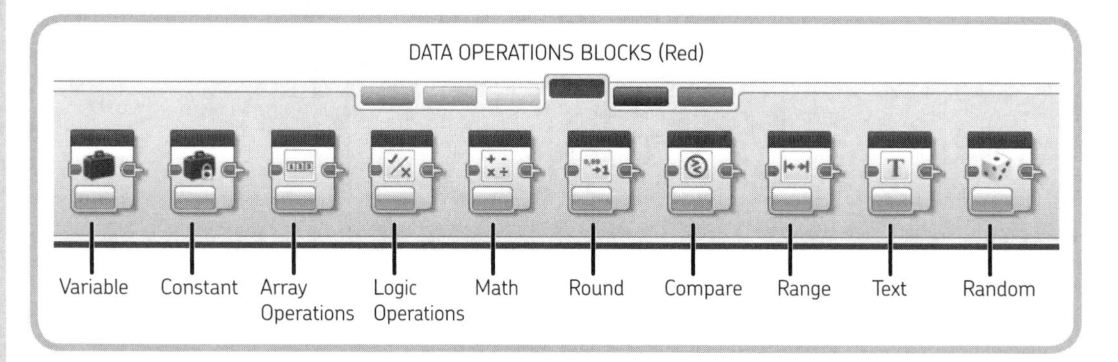

Figure 5-10: The Data Operations blocks palette

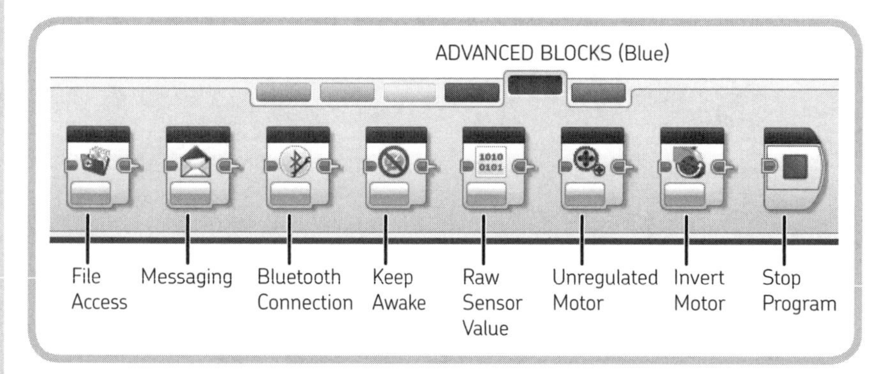

Figure 5-11: The Advanced blocks palette

project properties

An EV3 project file is actually an archive that contains all the programs, custom blocks, sound files, image files, videos, and documentation for your robot. When you're working on a project, you can access Project Properties (Figure 5-12) by clicking the tab with the wrench icon [Figure 5-2(7)]. Project Properties features the following:

1. **Project Description**: Here you can document your project by giving it a name and adding a main picture, a video, and a description.
2. **Share Project button**: Click this button to share your project with other LEGO MINDSTORMS users.
3. **Daisy-Chain Mode**: Check this box to enable the software to program a single EV3 Brick to control up to three slave EV3 Bricks connected together. (Each *slave brick* is connected to its master using a USB cable that goes from the master's USB host port to its miniUSB port.)
4. **Project Content**: Here is a list of all assets included in the project, grouped in categories—Programs, Images, Sounds, My Blocks, and Variables. From here, you can manage the project's files and variables.

connecting the EV3 brick to your computer

Let's connect the EV3 Brick to your computer using the Available Bricks tab in the Hardware Page (refer to Figure 5-4). The EV3 Brick must be on for the computer to detect it.

To connect via USB, just plug the USB-to-miniUSB cable into the miniUSB port of the EV3 Brick and the other end into any available USB port on your computer [Figure 5-4(a)].

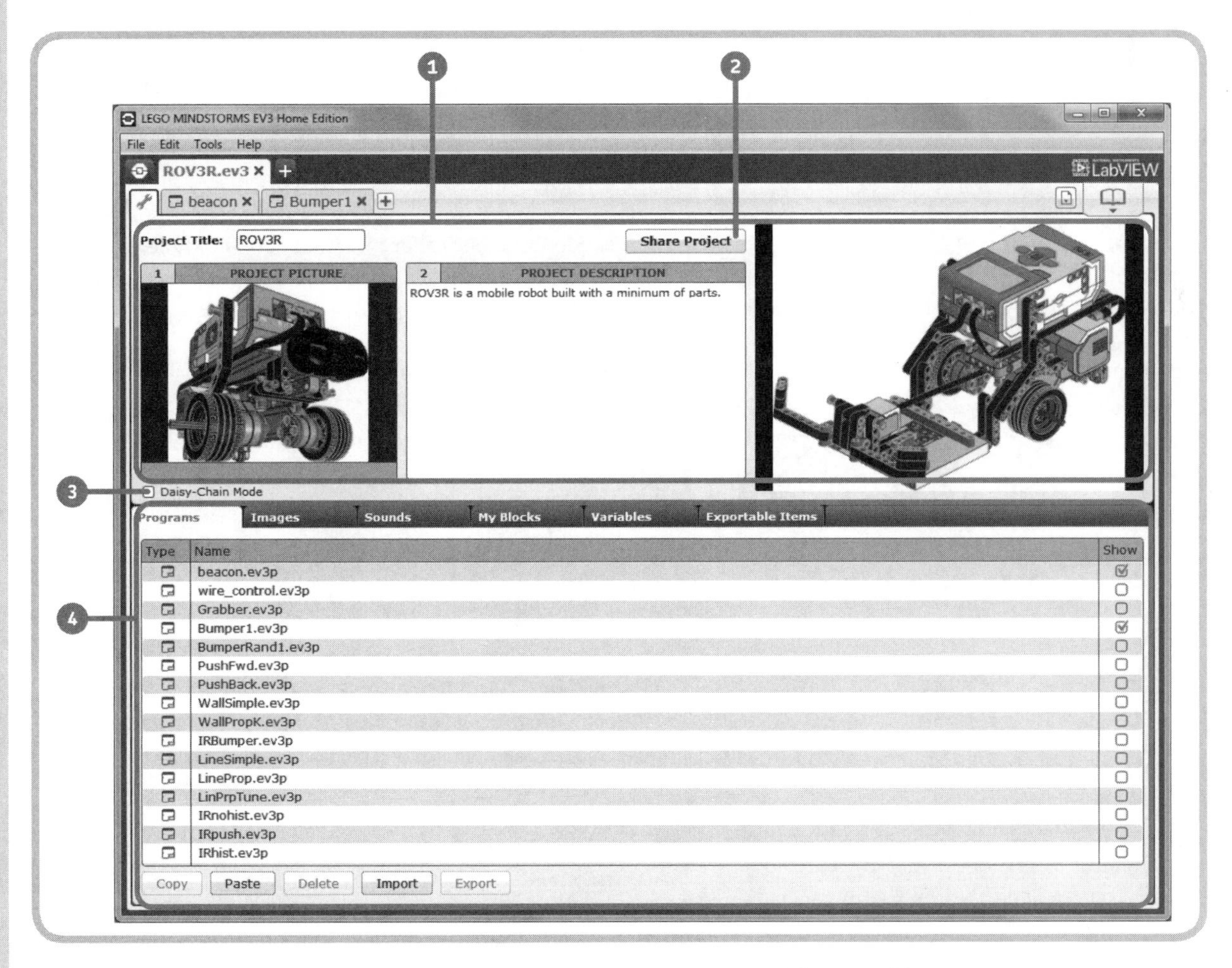

Figure 5-12: Project Properties is available at any time by clicking the wrench icon at the top-left portion of the screen.

To connect using Bluetooth, enable Bluetooth on the EV3 Brick (see the *EV3 User Guide* for details) and make sure that your computer is Bluetooth enabled. Click the **Refresh** button with a two-arrow icon [Figure 5-4(b)], and once the EV3 Brick is found and paired (you'll be asked to enter a passkey on the EV3 Brick and in the software), you can connect to it by clicking the square corresponding to Bluetooth [Figure 5-4(c)]. The entire Bluetooth scanning, pairing, and connection process is handled by the EV3 Software, not by your operating system.

To connect using Wi-Fi, you must insert a Wi-Fi USB Dongle (to be purchased separately) into the EV3 USB host port located on the side of the EV3 Brick.

NOTE If the EV3 Software fails to find, pair, or otherwise connect to the EV3 Brick via Bluetooth, try to disable and reenable Bluetooth from the menu on the EV3 Brick. Alternatively, close and reopen the EV3 Software.

importing a brick program

Now that you have a complete overview of the EV3 Software, it's time to learn how to use the Programming Palettes' blocks. We'll start by importing the Brick Program you built in Chapter 4, which makes ROV3R travel along a square path (duplicated in Figure 5-13). If you didn't save the program earlier, build it again now and save it on the EV3 Brick with the name *SQUARE*. (You'll find instructions on how to make a Brick Program in Chapter 3.)

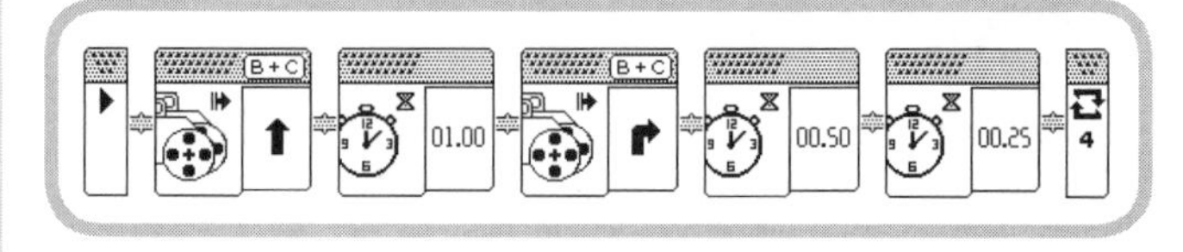

Figure 5-13: The Brick Program for ROV3R to drive in a square

To import your Brick Program, create a new project by clicking the **Add Project** (+) tab as shown in Figure 5-1(3). Then, connect the EV3 Brick using USB or Bluetooth. Now select **Tools ▸ Import Brick Program**, and you should see a dialog similar to Figure 5-14.

If the EV3 Brick is connected, the dialog should show a list of the Brick Program files stored in the EV3 Brick's memory. Select your *SQUARE* program and click **Import**. You should see that program imported into your project, as shown in Figure 5-15.

Import Brick Program
Select Program to Import:
Program Name
BUMPER
Demo
SQUARE
Status: Idle
Import
Close

Figure 5-14: The Import Brick Program dialog

analyzing the imported brick program

Now, before we begin modifying this program, let's analyze it. Right after the Start block, you have a Loop block, which includes all the other programming blocks.

At the right side of the Loop block is a button with a hash mark (#) on it. This is the Mode Selector, set to **Count** mode. The Loop block input beside the Mode Selector lets you specify how many times you want the sequence inside the loop to repeat—**4** times in this case, as in the original program. The Loop block supports many modes, so you can also repeat the inner sequence of blocks for a certain period of time, for a certain number of times, forever, until a certain logic condition becomes true, or until one of the sensors reads a specified value.

Inside the Loop block are Move Steering blocks and Wait blocks. The Move Steering blocks can be set to many modes: *Off*, *On*, *On for Seconds*, *On for Degrees*, and *On for Rotations*. Here the Move Steering blocks are in **On** mode: The blocks turn on the motors attached to the specified Motor Ports, and then the program continues.

In On mode, the block shows two inputs, Steering and Power, which let you control the steering direction and the power applied to the motors of a differential drive robot (like ROV3R), respectively. The first Move Steering block has the **Steering** input set to **0** and the **Power** input set to **70**, which makes the robot travel straight with motors at 70 percent power. The second Move Steering block has the **Steering** input set to **45** and the **Power** input set to **50**, which makes the robot turn by almost stopping the right motor and driving the left motor at 50 percent power.

The Steering input accepts values from –100 to 100. A value of 0 makes the robot go straight, a positive value (>0) makes the robot turn to the right, and a negative value (<0) makes it turn left. The farther the Steering value

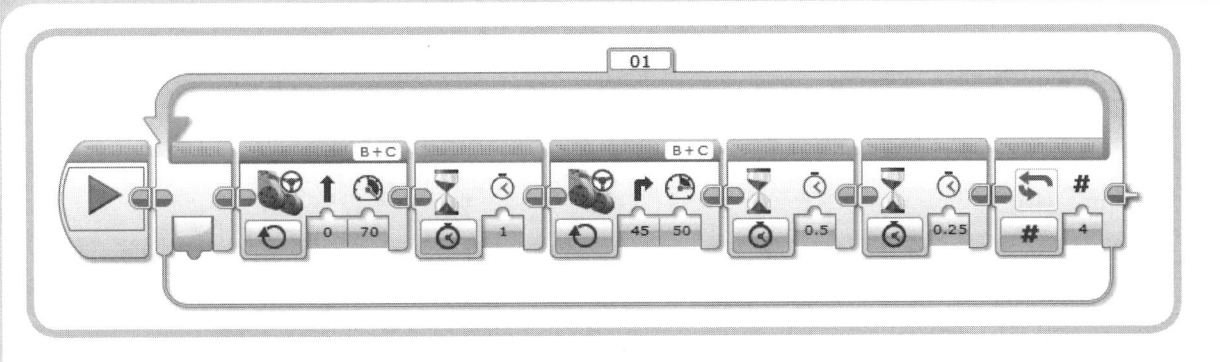

Figure 5-15: The imported Brick Program in EV3 language form

is from 0, the tighter the curve will be. (A value of 100 or –100 will make the robot spin in place.)

The Wait blocks are in **Time** mode, so their inputs let you specify how many seconds they should pause the program flow before the program continues. The time is expressed in seconds, but you can specify values as fractions of a second, like 0.25 seconds or 0.5 seconds.

Save the project now with the name *myROV3R*, and test the program on ROV3R by pressing the **Download and Run** button in the Hardware Page Controller [Figure 5-3(2)].

The program should work just like the Brick Program from Chapter 3.

NOTE To download and run a program, you can also click the green arrow on the Start block or use the keyboard shortcut CTRL-R (⌘-R on Mac).

GET RID OF THAT BLOCK!

To remove blocks from a program, select them and press DEL on the keyboard. Alternatively, drag them away from the sequence and into the Programming Palette area. The remaining blocks should move together to fill the space left by the deleted block.

NOTE Blocks that have multiple modes might change their appearance as the modes change, showing different inputs. The EV3 graphical language is designed so that you can check the configuration of the blocks at a glance. When you place the mouse over the blocks or buttons, corresponding hints will pop up to guide you. (For details, see EV3 Help from the menu: Help ▸ Show EV3 Help.)

editing the imported brick program

Let's modify the program shown in Figure 5-15. Select the **Loop** block and then select **Edit ▸ Copy** (keyboard shortcut CTRL-C, or ⌘-C on Mac OS) to copy the loop and its contents. Click the **Add Program** button [Figure 5-2(9)]; then paste the Loop block using **Edit ▸ Paste** (CTRL-V or ⌘-V), and drag it to snap it to the Start block of the new empty program. (When blocks are not connected to the Start block, they appear faded out.) Rather than use the Wait blocks, let's set the Move blocks to **On for Seconds** mode to produce the same robot behavior (that is, drive along a square path). In this mode, the *Seconds* and *Brake at End* inputs are also shown. In each Move block's **Seconds** input, enter a time interval that's the same as that of the Wait blocks right after it, and then delete the Wait block. Leave the **Brake at End** input set to **True** so that the motors stop when the motion is complete. In this mode, the Move Steering blocks pause the program flow until the duration has elapsed. You can see the complete program in Figure 5-16.

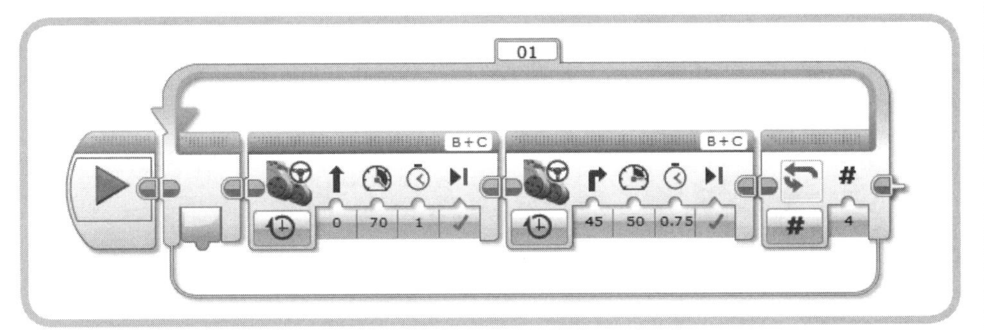

Figure 5-16: The modified program for ROV3R to drive in a square

Now try changing the Power and the Steering parameters. When you click an input, you should see a *slider* that allows you to change the value quickly. You can also enter a new value from the keyboard.

going for precision

In Chapter 4, we tweaked the precision of the 90-degree turn by increasing the distance between the ROV3R's wheels. Now that we have full control of the motors, let's make the robot turn a precise number of degrees by changing the software rather than the hardware. In order to make ROV3R travel and turn by a precise number of degrees, change the Move blocks' mode to **On for Degrees**. Their inputs should change again, as the Seconds input is replaced by the Degrees input. Now tweak the **Degrees** parameter of the second Move Steering block to make the robot turn by 90 degrees, as shown in Figure 5-17. You'll see why this works in just a bit.

NOTE The Degrees input of the second Move Steering block controls the rotation of the faster of the two motors, not the number of degrees the robot turns. The change in the robot's direction also depends on the wheel radius and the distance between the wheels.

How can you set the Degrees parameter of the first Move Steering block to make the robot travel a precise distance? For that matter, how can you set the Degrees parameter of the second Move Steering block to make the robot change its heading by precisely 90 degrees, without proceeding by trial and error? The easiest way is to use the Port View App on the EV3 Brick or the Hardware Page Port View tab in the EV3 Software to measure the degrees of the Motor Rotation sensors as you move the robot by hand. But first you'll need to reset the rotation count of these motors.

On the EV3 Brick, go the **Apps** tab (third from left) and open the **Port View App**. Use the arrow buttons to select the motor port you want to view (Figure 5-18). To reset the rotation reading, just unplug the motor and plug it in again or close the Port View App and open it again. In the EV3 Software, open the **Port View** tab in the Hardware Page and reset each motor rotation count by pressing the corresponding port letter to reset the port.

traveling a precise distance

To tune the parameter for the first block, move the robot 20 cm by hand, using a ruler for reference. The Port View should show about 520 degrees for each motor. That's the measure of how much each wheel has turned to travel 20 cm forward; enter this value in the **Degrees** input of the first Move block. The actual value might differ due to wheel slippage and other mechanical accidents, so try the Move Steering block by selecting it and clicking **Download and Run Selected** [Figure 5-3(3)]. With a ruler, measure the distance traveled, and adjust the Degrees parameter as necessary.

DIGGING DEEPER: COMPUTING THE DEGREES PARAMETER TO DRIVE PRECISELY

Here's how to compute the Degrees parameter—using a little math! Measure the wheel radius (for the ROV3R's tires, R = 21.6 mm) using a ruler, or just halve the dimension that you see embossed on the tire (43.2 mm). To travel X millimeters, you should use the Move Steering block with the **Steering** input set to **0** (the wheels spin forward at the same speed) and the **Degrees** input set to $X / R \times 57.3$. (The 57.3, an approximation of the constant $180 / \pi$, is needed to convert radians to degrees.) In the example above, to travel by 20 cm (200 mm), the formula is $200 / 21.6 \times 57.3 \approx$ 530 degrees. You'll probably need to tweak the calculated value a bit to allow for uncertainties in your measurements or even wheel slippage.

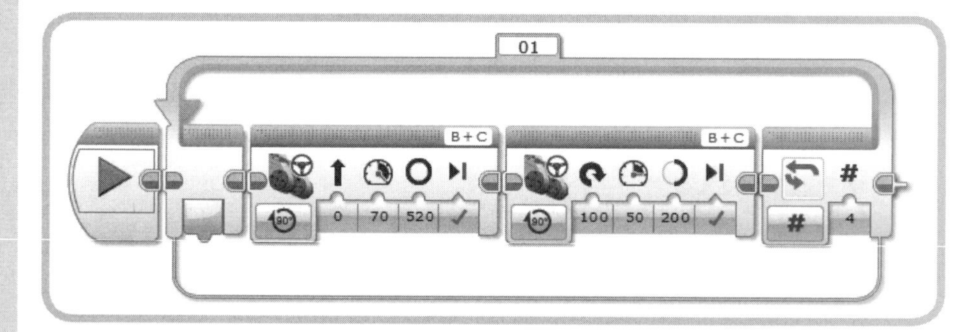

Figure 5-17: The program to drive ROV3R in a square by turning the motors a precise number of degrees

Figure 5-18: The Port View App on the EV3 Brick allows you to read all the sensors' values in real time.

NOTE **The distance traveled by a wheel attached to a motor is proportional to the amount of rotation of the motor shaft (in degrees or rotations) and to the wheel radius.**

turning a precise number of degrees

To make the robot turn by a precise number of degrees, you can apply a similar method to tweak the settings of the second Move Steering block shown in Figure 5-17. To do so, open the **Port View** in the EV3 Software or on the EV3 Brick and reset the motor rotation count. While pivoting the robot in place on its right wheel (which must not rotate), measure the number of degrees the left motor needs to turn in order to change the ROV3R's heading by 90 degrees clockwise. Use the number of degrees read in the Port View as the value to set the **Degrees** input of the second Move Steering block, with the **Steering** parameter set to **50** (to keep one wheel still). The Move Steering block will run until the faster motor (in this case, the left one) has turned by the number of degrees specified by the Degrees input.

Now try executing the Move Steering block alone by selecting it and clicking **Download and Run Selected** [Figure 5-3(3)]. The EV3 Brick executes only the Move Steering block that is responsible for making the robot turn. Check the angle turned by the robot, and adjust the **Degrees** parameter as necessary to achieve the 90-degree heading change.

Instead of pivoting on a wheel, you can also have ROV3R change its heading by spinning in place around its center. To do so, set the **Steering** input to **100** and the **Degrees** input to ___? I'll leave it to you to discover the value to fill in here! (You'll find the answer in the box below.)

experimenting with action blocks

The best way to learn the features of the Action blocks is to experiment with them. Using Figure 5-19 as your guide, add a new program to your project by clicking the **Add Program** (+) tab (1). Next, drag and drop a **Display** block to the sequence (2). Hold down the left mouse button while dragging the block; when the block is in almost in place, a gray shadow should appear. Release the mouse, and the block should snap into place automatically.

By default, the new Display block is in **Image** mode. To change the image it displays, click the **File Selection** field (3) and choose from the available images. You can also change the block's inputs (4) and switch between available modes: Text Pixels, Text Grid, Shapes Line, Shapes Circle, Shapes Rectangle, Shapes Point, Image, and Reset Screen (this last mode simply resets the screen to the Info screen when a program is running).

Now, following the procedure outlined above, add programming blocks from the Action blocks palette, from left to right, to produce the program in Figure 5-20. Because the page has limited space, I put some blocks below the others and then connected them using a Sequence Wire, which defines the program flow, like the arrows that connect the blocks in the flowcharts in Chapter 3. This is called *snaking*. Laying out the program in this way can save space horizontally and separate groups of blocks for better readability.

DIGGING DEEPER: COMPUTING THE DEGREES PARAMETER TO STEER PRECISELY

In addition to the trial-and-error method discussed above, we can use math to steer ROV3R precisely. To do so, measure the wheel radius R and the distance L from the center of the tire tread to the center of the robot (for ROV3R, $R = 21.6$ mm and $L = 50$ mm, respectively). To change the robot heading by a precise angle T, you can use the Move Steering block with the Steering input set to 100 or –100 (the wheels spin at the same speed in opposite directions) and the Degrees input set to $(T \times L) / R$.

$T \times L$ is the length of the arc of the circumference of radius L, corresponding to the central angle T in degrees. To determine how many degrees a wheel should rotate in order to travel on that arc, divide the length by the radius R of the wheel.

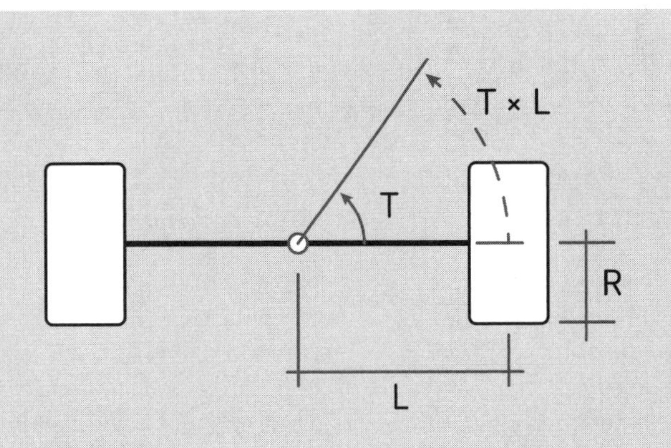

In the example of Figure 5-17, to turn the robot 90 degrees, the equation is $(90 \times 50) / 21.6 \approx 208$ degrees. Tweak the calculated value to allow for uncertainties in measurement, especially due to the fact that it's hard to know the exact point at which the flat tires contact the ground.

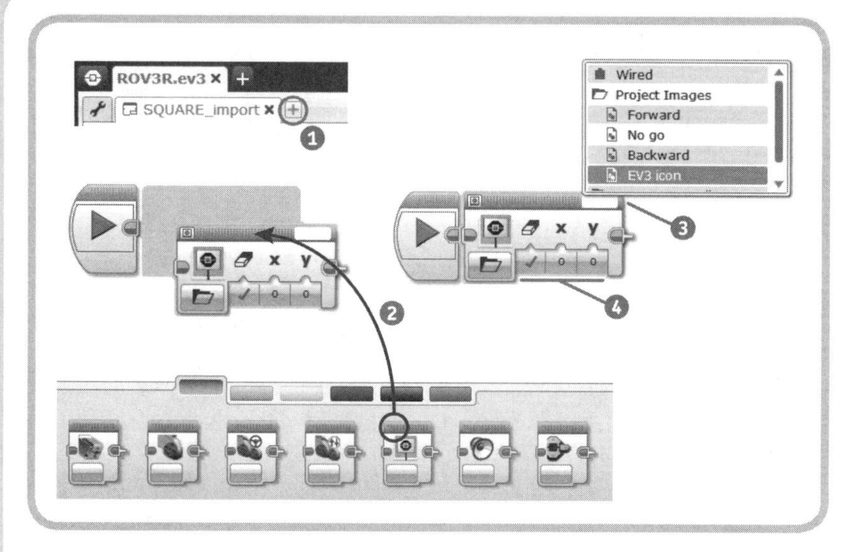

Figure 5-19: Adding and configuring blocks: the first steps in building a new program

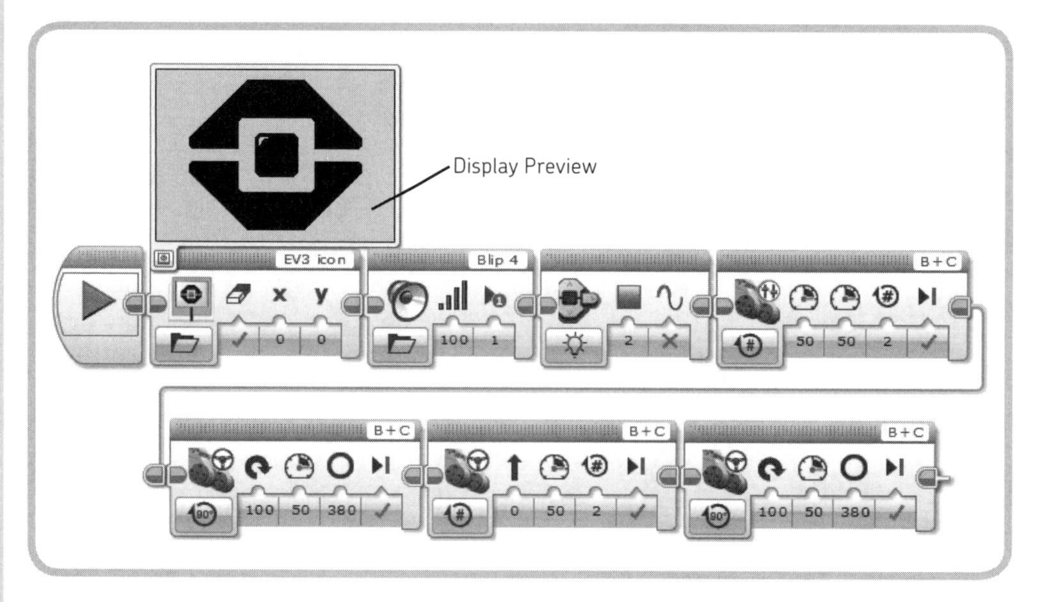

Figure 5-20: A program for experimenting with all the Action blocks

To connect blocks with a Sequence Wire, click the Sequence Exit Plug of the last block of the first row and drag it to the Sequence Entry Plug of the first block of the second row, as shown in Figure 5-21(a, b, c). The mouse cursor should change into a spool while performing this action. You can move the straight pieces of wire by clicking and dragging them [Figure 5-21(d)].

To improve the readability of your program, click the Sequence Exit Plugs of the blocks connected side by side to make a short, straight Sequence Wire appear between them, as shown in Figure 5-21(e). Click the Block Exit Plug again to collapse the wire and move the blocks close together again. Click the end plug of a Sequence Wire to disconnect the blocks.

NOTE To fit more blocks on the Programming Canvas, change the zoom factor by using the mouse scroll wheel while holding down the CTRL (⌘) key or by using the Zoom control buttons in the Programming Toolbar [Figure 5-2(a)(3)]. To move the Canvas, you can pan by dragging the mouse in an empty area while holding down the ALT key, or you can use the Pan tool in the Programming toolbar, next to the Select tool. When a program is large, the Programming Canvas will show small triangular arrows near its edges. Click these arrows to shift the program in the corresponding direction so you can view it in its entirety.

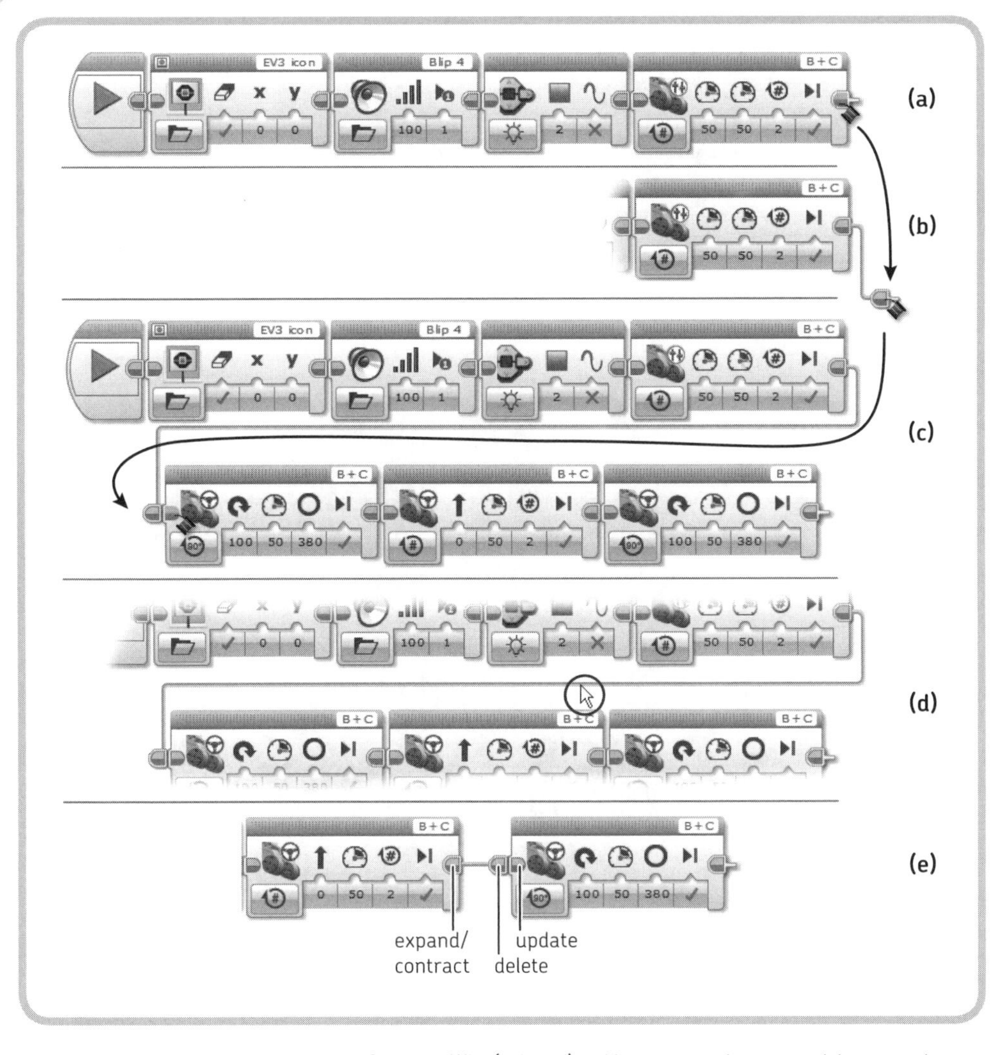

Figure 5-21: How to connect blocks with a Sequence Wire (a, b, c, d) and how to expand, contract, delete, or update a Sequence Wire (e)

In the program in Figure 5-20, the first Move block is a Move Tank block, not a Move Steering block. These two blocks have slightly different properties: The Move Tank block allows you to configure the power of the two driving motors separately, but for the Move Steering block, you must set the power and the amount of steering for both motors together. The Move Steering block computes the power of each motor for you.

Notice that the Display block can display a preview of what will appear on the EV3 Brick screen. To show this preview, click the **Display Preview** button in the top-left corner of the Display block. (In Figure 5-20, a preview of the EV3 icon is shown.)

What does this collection of Action blocks do? To find out, run the program and try changing the parameters of the various blocks. For example, change the EV3 Brick Status Light color, change the displayed image, or change the sound file played by the Sound block.

controlling the program flow

The Flow blocks in the Palette with the orange tab can control program flow by pausing the program, repeating sequences, or choosing to execute different actions depending on a condition. To better understand these blocks, let's import the wall-following program from Chapter 4 (Figure 4-11 on page 65) using the Import Brick Program tool. The imported program is shown in Figure 5-22.

The Loop block is configured to repeat the inner sequence forever (as marked by the infinity symbol). The Wait blocks are in **Infrared Sensor Compare Proximity** mode. Their **Compare** input is set to **Less Than (4)** and **Greater Than**

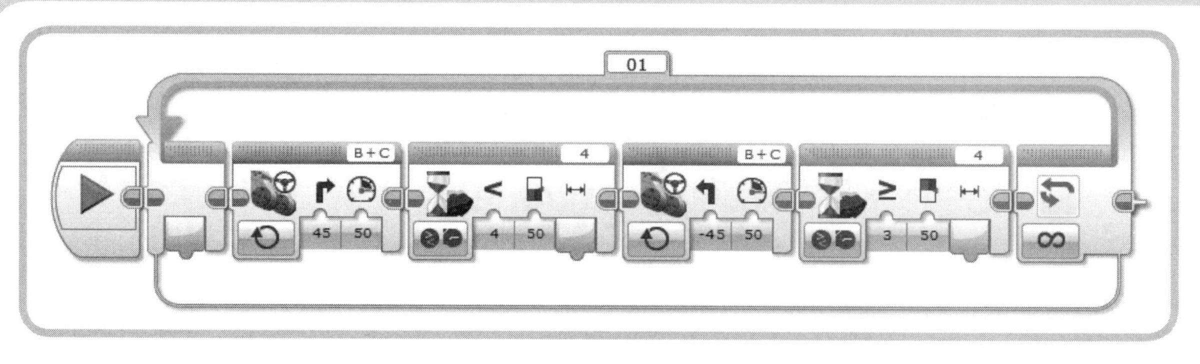

Figure 5-22: The imported wall-following program

or Equal To (3), while their **Threshold** inputs are set to **50**. Configured this way, the Wait blocks compare the IR Sensor proximity value against the specified threshold, and when the test succeeds (proximity < 50 or proximity ≥ 50), they let the program continue.

the switch block

In Chapter 4, when we first considered the challenges of building a wall-following robot, we determined that the algorithm should check a condition and act accordingly. To allow an EV3 program to choose between two different actions, we use a Switch block. Let's see how that works.

Build the program shown in Figure 5-23 as follows:

1. Place a **Loop** block (from the Flow Control blocks palette) and leave it set to its default mode, **Unlimited** (repeat forever).
2. Place a **Switch** block inside the loop.
3. Change the Switch block's mode to **Infrared Sensor Compare Proximity**. Set the **Compare** input to **Less Than (4)** and the **Threshold** input to **45**.
4. Drag a Move Steering block into the True case of the Switch block at the top, indicated by a check mark.
5. Change the Move Steering block's mode to **On** and set **Steering** to **25** and **Power** to **40**.
6. Add another Move Steering block inside the False case of the Switch block at the bottom, indicated by an *x*.
7. Change this second Move Steering block's mode to **On**, and set **Steering** to **–25** and **Power** to **40**.

> **EXPERIMENT 5-1**
>
> Starting from the imported program shown in Figure 5-22, try smoothing out the robot's jerky movement by changing the Move Steering blocks' **Steering** input to 25 and –25, respectively, and changing the Wait block's **Threshold** inputs to something less than 50, like 45 or 40. Keep experimenting to see how the robot's wall-following behavior changes. How would you make the robot follow a wall more closely? How would you make it travel faster?

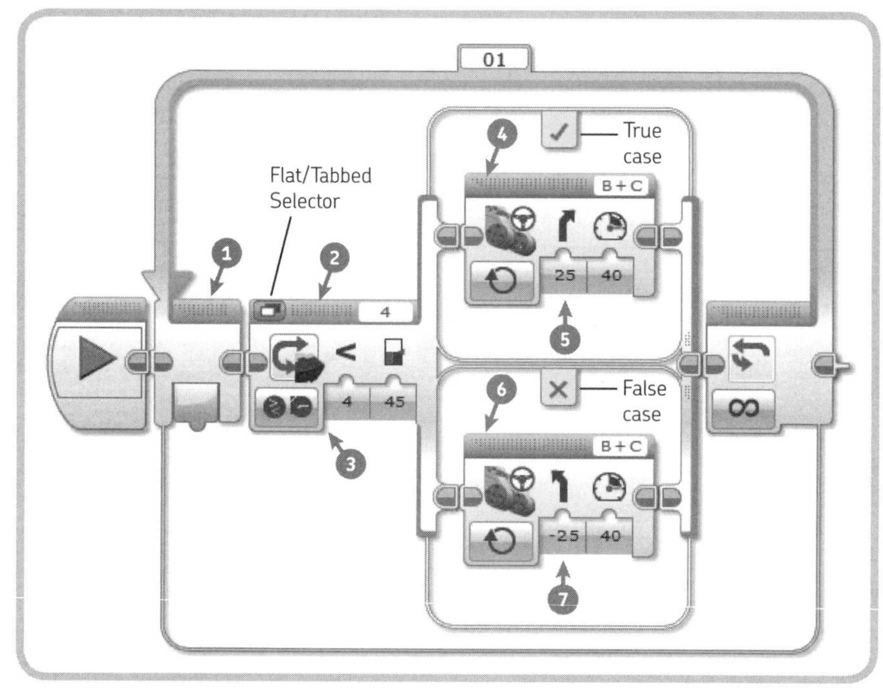

Figure 5-23: The wall-following algorithm implemented with a Switch block

EXPERIMENT 5-2

Import the Brick Programs that you made in Chapters 3 and 4 into the EV3 Software. In the imported line-following program, notice how the Wait blocks are configured to wait for certain Color Sensor readings (in Compare Reflected Light Intensity mode). Try to create a line-following program using a Loop block and a Switch block, as you did for the wall-following program in Figure 5-23.

The Switch block is shown in *Flat view* by default, which means that every case is visible. To change to Tabbed view, click the **Flat/Tabbed Selector** button (shown in Figure 5-23). *Tabbed view* takes up less space, but you'll see just one case at a time. If only one case contains programming blocks, you can keep it in Tabbed view, showing only the case filled with blocks. Otherwise, Flat view should be fine.

To fit all the blocks inside them, resize the Loop blocks and the Switch blocks by dragging their Resize Handles, as shown in Figure 5-24. In the Switch block, each case can be resized independently. The Resize Handles are shown when the block (or case) is selected.

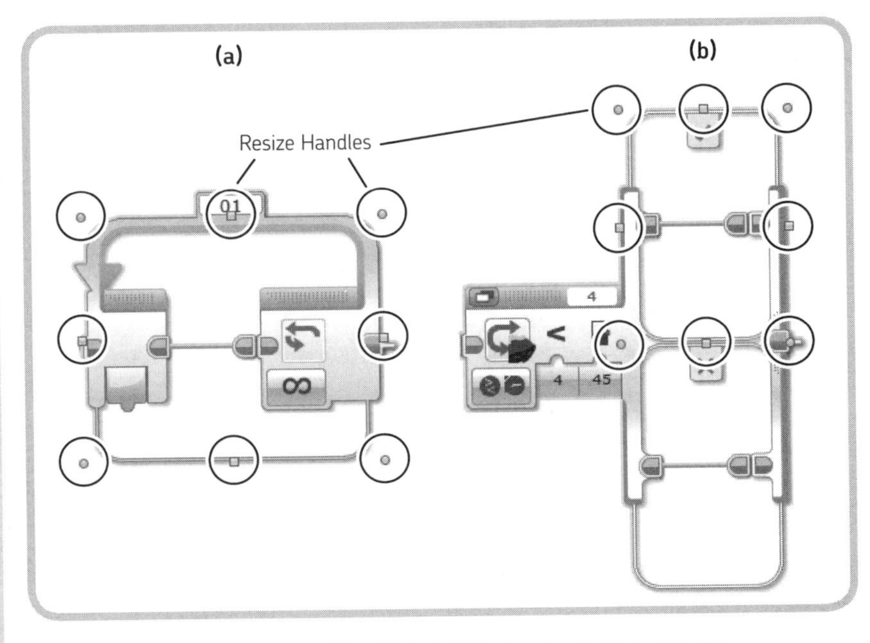

Figure 5-24: How to resize the Loop (a) and Switch (b) blocks using the Resize Handles

conclusion

This chapter has given you a complete overview of the EV3 Software features, tools, and work areas. You learned how the programming blocks are organized in the Programming Palettes and, by importing Brick Programs, you took your first steps toward proper EV3 programming. You also learned how to use Action and Flow Control blocks. In the next chapter, you'll discover all the features of the Remote IR Beacon, as well as some new programming concepts.

WHIIIRRRRRRR
BUMP!
I HOPE YOU DON'T WANT TO USE THE EV3 AS NOTHING BUT A REMOTE-CONTROLLED TOY!
NO, BUT IT'S FUN! WHY DID THEY PUT THIS THING IN THE BOX THEN?
THAT IS A REMOTE. BUT IT HAS ANOTHER, MUCH MORE INTERESTING USE! IT CAN BE USED AS A BEACON THAT THE ROBOT TRACKS TO FIND ITS RELATIVE POSITION, JUST LIKE SHIP CAPTAINS USE LIGHTHOUSES TO NAVIGATE NEAR THE COAST.
IF YOU WEAR THE BEACON, ROV3R WILL FOLLOW YOU LIKE A PUPPY! IF YOU THROW IT, ROV3R WILL GO AFTER IT.
WHIIIRRRR
05X1

6

experimenting with the EV3 infrared components

In this chapter, you'll learn about the Remote Infrared (IR) Beacon and the Infrared (IR) Sensor. In addition to measuring the proximity of objects, the IR Sensor can detect infrared signals from the Remote IR Beacon, allowing you to send commands to your robot just as you send commands to a television with a remote control. The IR Sensor can also estimate its distance and orientation with respect to the Remote IR Beacon; this cool feature will allow you do fun things with your robots like play tag, chase prey, or locate and reach a mission base.

remote IR beacon

The Remote IR Beacon is powered by two AAA batteries. It has holes on each side, which make it easy to incorporate into a LEGO Technic model. As you can see in Figure 6-1, it has four small buttons (labeled 1, 2, 3, and 4), a large button (9), and a red Channel Selector switch (12). The Channel Selector lets you choose among four channels, so you can use up to four Remote IR Beacons at once. A number engraved in red plastic in the small circular window shows the current channel. (As long as each remote is set to a different channel, its signal won't interfere with other remotes' signals.)

The large button (9) turns on Beacon Mode. When in Beacon Mode, the device transmits a continuous signal until any button is pressed or until one hour has elapsed. The IR Sensor can estimate the proximity and the heading to a beacon set in Beacon Mode. This feature allows a robot to follow a moving beacon or find its distance and heading relative to a fixed beacon.

The four small buttons (1, 2, 3, and 4) send commands (the numbers in the following list) to the IR Sensor using two infrared light-emitting diodes (LEDs) at the front of the remote. (The plastic that houses these LEDs is a dark blue filter that lets only infrared light pass through it.)

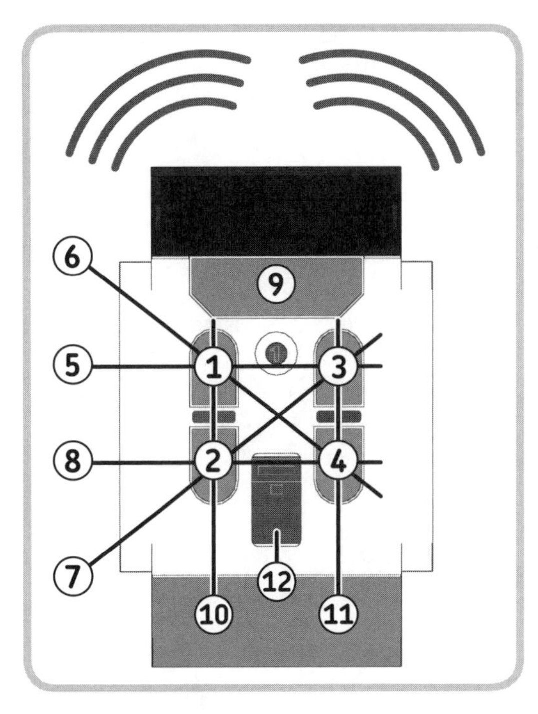

Figure 6-1: The Remote IR Beacon

0	No button is pressed and Beacon Mode is off.
1	Button 1
2	Button 2
3	Button 3
4	Button 4
5	Buttons 1 and 3
6	Buttons 1 and 4
7	Buttons 2 and 3
8	Buttons 2 and 4
9	Beacon Mode is on.
10	Buttons 1 and 2
11	Buttons 3 and 4

using the remote IR beacon as a remote

Let's see how we can use the Remote IR Beacon as a simple remote control for your robot using only the IR Control App (and no programming—yet). In the EV3 Brick menu, go to the **Apps Tab** (third from the left) and open the **IR Control App**. You should see a screen like Figure 6-2(a).

There are two modes to choose from, as shown in Figure 6-2. In the mode shown in Figure 6-2(a), you can control the motors using Remote IR Beacon channels 1 and 2; in the mode shown in Figure 6-2(b), you can control the motors using channels 3 and 4. To switch between modes, press the **Enter** button on the EV3 Brick.

In the first mode, with the Remote Channel Selector (labeled 12 in Figure 6-1) on channel 1, you can control a motor connected to port B with buttons 1 (forward) and 2 (backward), and you can control a motor connected to port C with buttons 3 (forward) and 4 (backward). While in the same mode, you can control motors connected to ports A and D with another remote on channel 2. The second mode works similarly but receives commands from remotes set on channels 3 or 4.

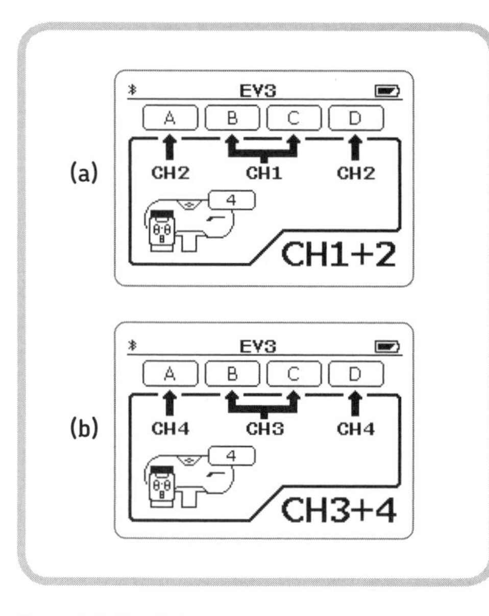

*Figure 6-2: The IR Control App. To switch between controlling channels 1 and 2 (a) and channels 3 and 4 (b), press the **Enter** button on the EV3 Brick.*

The IR Control App makes it easy to remotely control a wheeled robot like ROV3R. You'll also find it useful when building and testing a motor-powered mechanism, since you can test the mechanism by turning the motor forward and backward without having to build a test program.

To test out the remote control, build ROV3R in any of the versions from Chapter 2 and connect the motors to ports B and C. Now start the IR Control App on the EV3 Brick, take the Remote IR Beacon, and select channel 1. You should be able to use the small buttons on the remote to drive ROV3R around. Table 6-1 shows how you would control ROV3R.

table 6-1: controls for a differential drive robot such as ROV3R

Buttons pressed	Motion
1 & 3	Drive forward.
2 & 4	Drive backward.
1 & 4	Spin right.
2 & 3	Spin left.
1	Turn right by pivoting on the right wheel.
2	Turn left by going backward and pivoting on the right wheel.
3	Turn left by pivoting on the left wheel.
4	Turn right by going backward and pivoting on the left wheel.

NOTE If the controls aren't working at first, make sure that the IR Sensor is connected to port 4 and that the IR Control App is in the right mode, receiving commands from channel 1 (or channel 2, if you connected the motors to ports A and D).

See? Now you can use the Remote IR Beacon as a remote control for ROV3R, without having to create a program for it! This same setup works for real-world vehicles, such as tanks or tracked vehicles like excavators. The human driver controls these vehicles by moving two levers, and each lever controls the motor that drives the track on the corresponding side. This is just like pressing the Remote IR Beacon's buttons 1 and 2 or 3 and 4. With the Remote IR Beacon and some extra programming, you can also control vehicles with different steering, as you'll see in Chapter 12 with the SUP3R CAR.

using sensor blocks and data wires

A robot uses the data provided by its sensors to perceive the world around itself. In Chapters 3, 4, and 5, we compared sensor readings against thresholds to trigger Wait blocks or Switch blocks. To directly access sensor readings, we can use the Sensor blocks (found in the Programming Palette with the yellow tab).

Each Sensor block has several modes that serve different functions. In Measure mode, Sensor blocks provide measurements as numeric values to other blocks. In Compare mode, they compare measured values against a threshold to provide logic values (see "Understanding Data Types" on page 89 for a discussion of these). Some Sensor blocks (like the Motor Rotation block and the Timer block) also have a Reset mode, which resets their measured values to 0.

The IR Sensor block has a Proximity mode (which measures the distance to the nearest object), a Remote mode (which receives commands from the Remote IR Beacon), and a Beacon mode (which estimates the robot's proximity and heading in relation to the Remote IR Beacon).

Let's get some data from the IR Sensor block. Build ROV3R with Front IR Sensor (page 31), and create the program shown in Figure 6-3 by adding blocks as described in Chapter 5. The IR Sensor block is set in **Measure Proximity** mode, while the Move Steering block is in **On** mode. The key idea of this simple program is to take the Proximity value provided by the IR Sensor block and send it to the Move block for use as a Steering input. ROV3R should steer according to the distance the sensor measures to an object. This simple program makes ROV3R spin in place until you place your hand in front of the IR Sensor; then it will follow your hand, going straight until you remove your hand.

To send the sensor output to the Steering input, we need to set up a Data Wire. (Data Wires carry values from one block to another.) To create the Data Wire, use your mouse to click and hold the **Sensor block output** and drag the wire—left to right—to the **Steering input** (Figure 6-4). When you place the mouse cursor on an output, its shape should change into a wire spool. When you click an output, a wire plug appears. When you drag the wire near a block's inputs, all inputs that can accept that type of data should be highlighted in blue. Place the plug on your desired input, and release the left mouse button.

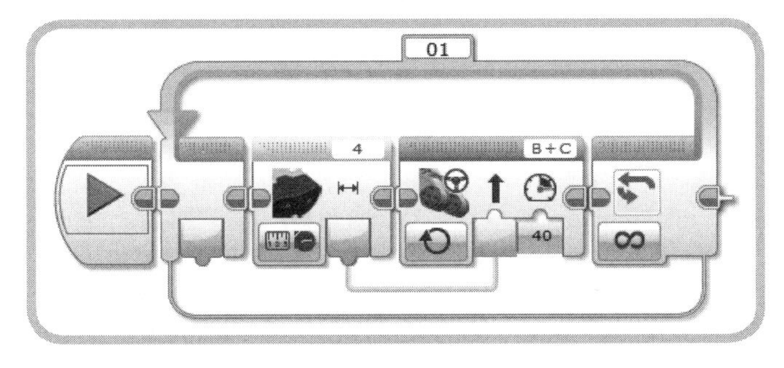

Figure 6-3: This program makes the robot steer according to the distance measured by the IR Sensor.

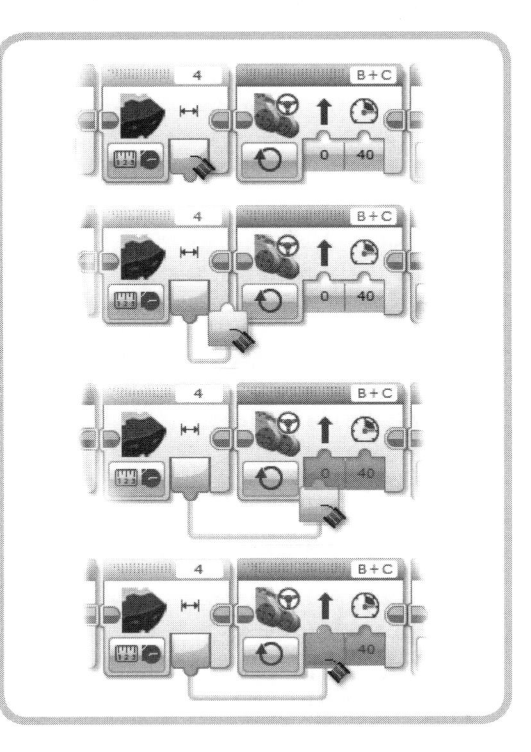

Figure 6-4: To add a Data Wire, click a block output. A plug appears on the end of the wire. Drag the plug to another block's input. The wire automatically follows the plug on the screen.

UNTANGLING DATA WIRES

To delete a Data Wire, click and drag its end slightly away from the input (the reverse of what we did in the last two steps shown in Figure 6-4). To move a Data Wire, just click and drag it. To make the EV3 software rearrange and compact the wire (in case things are getting messy), double-click the wire.

Remember that the block that provides the output value must precede the block that receives the value in its input and that the blocks are executed in sequence from left to right: The output block where the wire begins must be to the left of the input block where it ends. However, a Data Wire can skip over many blocks and connect distant blocks.

Download and run the program in Figure 6-3. What happens? ROV3R should spin until you put your hand near the sensor. Once your hand is near, ROV3R should go toward your hand and follow it as you slowly pull it away.

How does this all work? When the IR Sensor measures a large distance (when no object is near), its Proximity output sends a high value, around 80 to 90 percent. When you place your hand in front of the sensor, its Proximity output drops closer to 0 percent. The Proximity value is carried by the Data Wire into the Steering input of the Move Steering block, which accepts values from –100 to 100 (percentage of steering). When the value is high, the robot spins; when it's low, the robot will go almost straight.

EXPERIMENT 6-1

What does the program in Figure 6-3 do if you connect the Data Wire to the Power input, setting the Steering input to 0?

EV3 software features for debugging programs

In Chapter 5, you learned that the Port View tab in the Hardware Page lets you see the values of sensors even while a program is running. But the EV3 Software allows you to do even more!

Since the Data Wires carry data, you can display the sensor's current readings on the wire. Place the mouse pointer over a Data Wire, and a small window pops up displaying the current value, as shown in Figure 6-5. The number in the pop-up window changes continuously because the blocks are inside a loop that runs forever, very quickly. Note that this feature works only if you run the program from the EV3 Software using the Controller (Figure 5-3 on page 71). It won't work if you run the program from the EV3 Brick menu, even if the Brick is connected to the EV3 Software.

Notice in Figure 6-5 that the header of each block contains diagonal stripes (they are animated in the software). These animated stripes indicate the blocks currently being executed. Both this *Execution Highlight* feature and the real-time Data Wire pop-up display really help with *debugging* programs (that is, finding and fixing errors, or *bugs* in programming jargon).

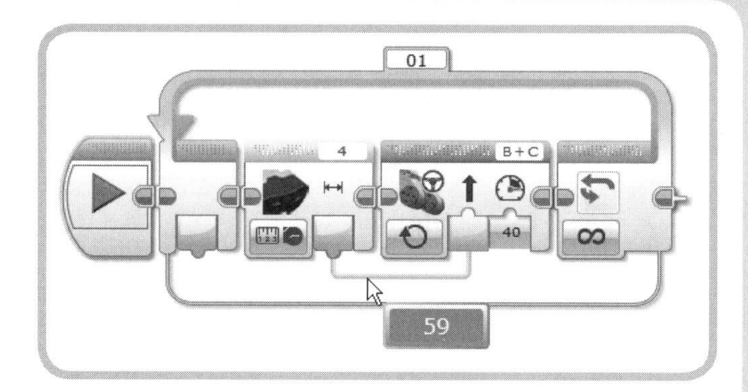

Figure 6-5: Place the mouse pointer over a Data Wire to display a pop-up window with the wire's current value. The blocks currently being executed are highlighted with animated diagonal stripes.

For example, you can use these features to see whether a Data Wire is carrying the expected values or whether the program is stuck somewhere due to a stalled Wait block.

NOTE The origin of the terms *bug* and *debugging* is curious and controversial. It's rumored that the term *bug* originated back in the 1950s, when computers were as big as walk-in closets. Back then, real bugs (moths and roaches) sometimes snuck into those huge relay-based computers, causing electrical and mechanical problems. Engineers had to literally remove the bugs to get the computer to work correctly!

displaying data nicely with the text block

Let's add some blocks to the program in Figure 6-5 that will display the IR Sensor values on the EV3 Brick screen as the robot follows a hand. Using Figure 6-6 as reference, add a Text block (Data Operations palette, red header) and a Display block to the program.

Text blocks have only one mode, Merge mode, which combines strings of text provided by its inputs **a**, **b**, and **c**. A *string* is just a bit of text with any combination of letters, numbers, spaces, or symbols: `!”#$%&’()*+,-./:;<=>?@[\]^_°{|}~`. To enter characters into the Text block, you can either type text into one of its input fields or connect a Data Wire from another block to one of its inputs.

If we had connected the IR Sensor output directly to the Display Block Text input, the numeric values would have converted to text automatically. However, when many numeric values are displayed, what they mean is not always clear. You can use the Text block to generate more meaningful strings

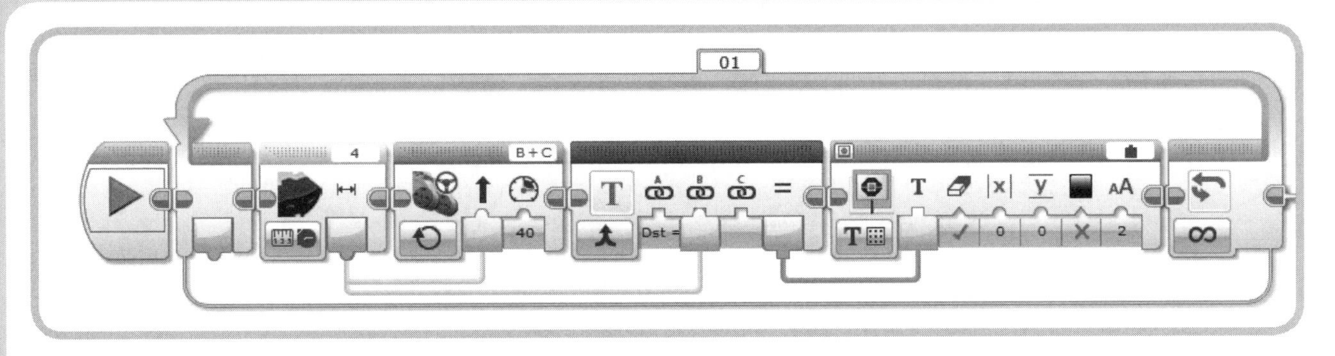

Figure 6-6: You can display meaningful messages on the EV3 Brick screen using the Text block. This program is similar to the one in Figure 6-3.

by wrapping text around numeric values. For example, you could create a string like *Proximity is 20%*, where the number 20 comes from a Data Wire, or the Text block could report *Distance = 40*, where 40 is a sensor reading.

Let's give this a try. Enter **Dst =** (with a space after the equal sign) into **Text field A**. Set the Display block to **Text Grid** mode. Then select **Wired** instead of static text by clicking the **Text field** in the header (as shown in Figure 6-7). The Display block should show a new Text input, where you will provide the variable text data to be displayed. Set the Clear Screen input to **True** (as indicated by the check mark under the eraser icon) so that the block will clear the screen every time it is executed.

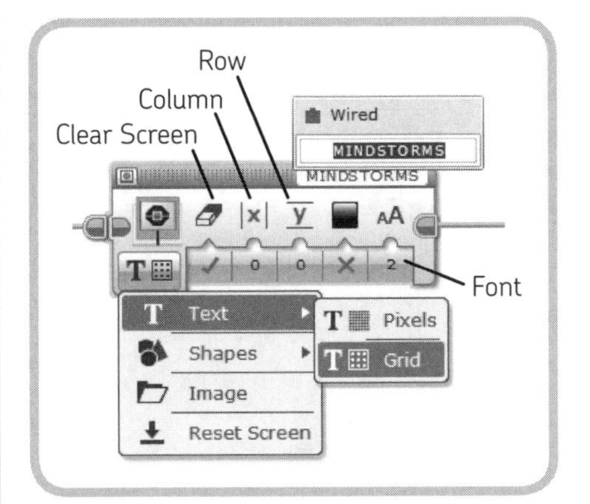

Figure 6-7: Configure the Display block to show text on a grid, with the text input coming from a wired input.

Text Grid mode allows you to display the text aligned to a grid of rows (Y input) and columns (X input). The dimension of a cell grid is one character (Normal font = 0; Bold font = 1). A character set in Large font (2) is two rows and two columns wide.

Now drag a new Data Wire from the IR Sensor output to the Text block's second input, and drag another Data Wire from the Text block's output to the Text input of the Display block. (You'll now have two Data Wires coming from one output.) Download and run the program. The robot's behavior should be the same as before, but onscreen you should see a nice big report of the IR Sensor's distance readings.

WARNING You can have many Data Wires coming out of a single output, but you cannot connect multiple Data Wires to a single input.

understanding data types

In the programs above, you used either numbers or text as data. There's a third type of data: *logic values* (true or false), which are often used to express the result of a comparison. To help differentiate the data types, Numeric, Text, Logic, Numeric Array, and Logic Array inputs have different plug shapes, and the corresponding Data Wires have different colors. These are shown in Figure 6-8.

The plug shapes and colors are as follows:

* Numeric inputs/outputs have a rounded shape, and the wire is yellow.
* Logic inputs/outputs have a triangular shape, and the wire is aqua.
* Text inputs/outputs have a square shape, and the wire is orange.
* Numeric Array inputs/outputs have a double rounded shape, and the wire is thick and yellow.
* Logic Array inputs/outputs have a double triangular shape, and the wire is thick and aqua.

NOTE You will learn about arrays in Chapter 13.

data type conversion

The EV3 Software will not allow you to attach a Data Wire to the wrong type of input. For example, you can't connect a

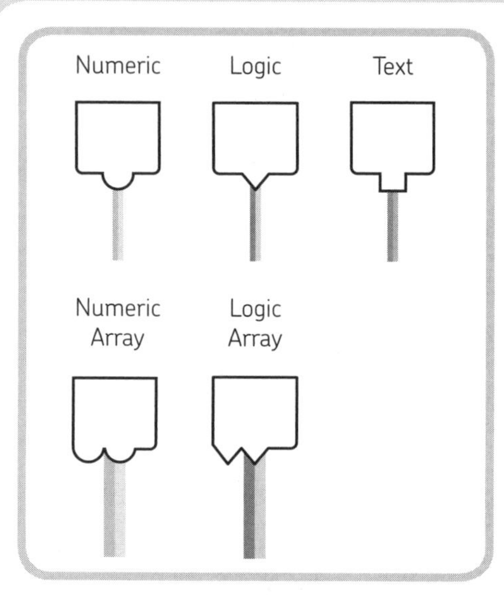

Figure 6-8: Different data types are distinguished by their input/output plug shape and wire color.

Text output to a Numeric input with a Data Wire. However, you can convert some types of data from one to another other automatically simply by connecting an output of one type to an input of another type.

As a visual guide, a data type can be converted if its plug can fit into another. For example, because the triangular shape can fit into the round or square shape, we know that a logic value can be converted into a number or text. Likewise, because the square shape does not fit into the round or triangular shape, we know that text can't be converted to a number or logic value. Table 6-2 lists all possible conversions.

table 6-2: the automatic type conversions

From	To	Resulting data
Numeric	Text	A number becomes a text string. For example, 3.1415 would become the text string "3.1415".
Logic	Numeric	The logic value True becomes 1; the logic value False becomes 0.
Logic	Text	The logic value True becomes the text string "1"; the logic value False becomes the text string "0".
Logic	Logic Array	The logic value becomes the first and only element of the resulting logic array.
Numeric	Numeric Array	The numeric value becomes the first and only element of the resulting numeric array.

As you can see, numeric and logic values can be converted into text (shown in quotation marks), but text cannot be directly converted into a numeric or logic value. Also, a logic value can be converted to a numeric value—but not the other way around. Computers (like the EV3 Brick) represent all kinds of data using only the binary digits 0 and 1, which are equivalent to the truth states False and True, respectively. By convention, we convert the True logic value to 1 and False to 0, but the EV3 Software doesn't know how to directly convert any numeric value to a logic value. In fact, there could be several options. For example, should any nonzero number be converted to True? Or should any number less than a certain threshold be converted to False?

With some programming effort, we can overcome the limits of direct data conversion: You'll learn how to convert numeric values into logic values in Chapter 7 and how to convert text into a numeric value in Chapter 14.

DIGGING DEEPER: DECIMAL NUMBERS

The Numeric type represents numbers that can be positive or negative and can have digits after the decimal point; that means the EV3 Brick supports floating-point arithmetic operations, such as dividing 3 by 10, the result of which is 0.3. If you did the same operation on a system that supports only integers (numbers without digits after the decimal point), the result would be 0. Pretty different, isn't it?

following the remote IR beacon

By using the IR Sensor in Measure Beacon mode, you can make ROV3R follow the Remote IR Beacon. ROV3R will drive toward the remote as long as it can detect it. The Remote IR Beacon can be thought of as a kind of landmark the robot can use to determine its relative position and orientation.

You can reuse this concept in many creative ways. For example, you could build a robot that plays tag by attaching the Remote IR Beacon to your belt and having the robot chase you! Or you could put the beacon in the corner of a room and have the robot return to it, even after a long exploration, like a home base.

WARNING **The Remote IR Beacon must be in Beacon Mode to be detected by the IR Sensor in Measure Beacon mode! To enable Beacon Mode, press button 9 on the Remote IR Beacon. To disable Beacon Mode, press any other button.**

The key to creating programs like these lies in using the Proximity and Heading outputs of the IR Sensor block (in Measure Beacon mode) as Power and Steering inputs for a Move Steering block. For example, in the program shown in Figure 6-9, the robot will steer toward the beacon with steering action that is proportional to the measured heading: The greater the heading, the greater the steering action. When the heading is near 0 (frontal), the robot will drive straight toward the beacon.

Likewise, the robot's speed will be proportional to its distance from the beacon. When the beacon is far away, the robot will travel fast; when it's close by, the robot will slow to a stop. If the beacon is not detected, the robot will glide to a stop. (This program would be clearer if the Switch block was in Flat View, but I had to set it in Tabbed View to pass Data Wires into it.)

NOTE **Data Wires can pass through a Switch block when the Switch block is in Tabbed View. As soon as you drag a Data Wire through the border of a Loop block or a Tabbed Switch block, a tunnel appears. To pass values through Switch blocks in Flat View, you need to use variables. (You'll learn how to use variables in Chapter 12.)**

Here's how the program works. First of all, the entire sequence is set to repeat forever by using a Loop block in Unlimited mode.

The IR Sensor block in Measure Beacon mode has its **Channel** input set to **1**. It has three outputs:

* The first output, *Heading*, provides the sensor's heading to the Remote IR Beacon. The values range from a low of –25 (indicating that the beacon is directly to the left of the beacon) to a high of 25 (which indicates the beacon is directly to the right). A value of 0 says that the beacon is directly ahead of the sensor.
* The second output provides the *Proximity* of the Remote IR Beacon. Its values range from 0 (the nearest position) to 100 (the farthest).

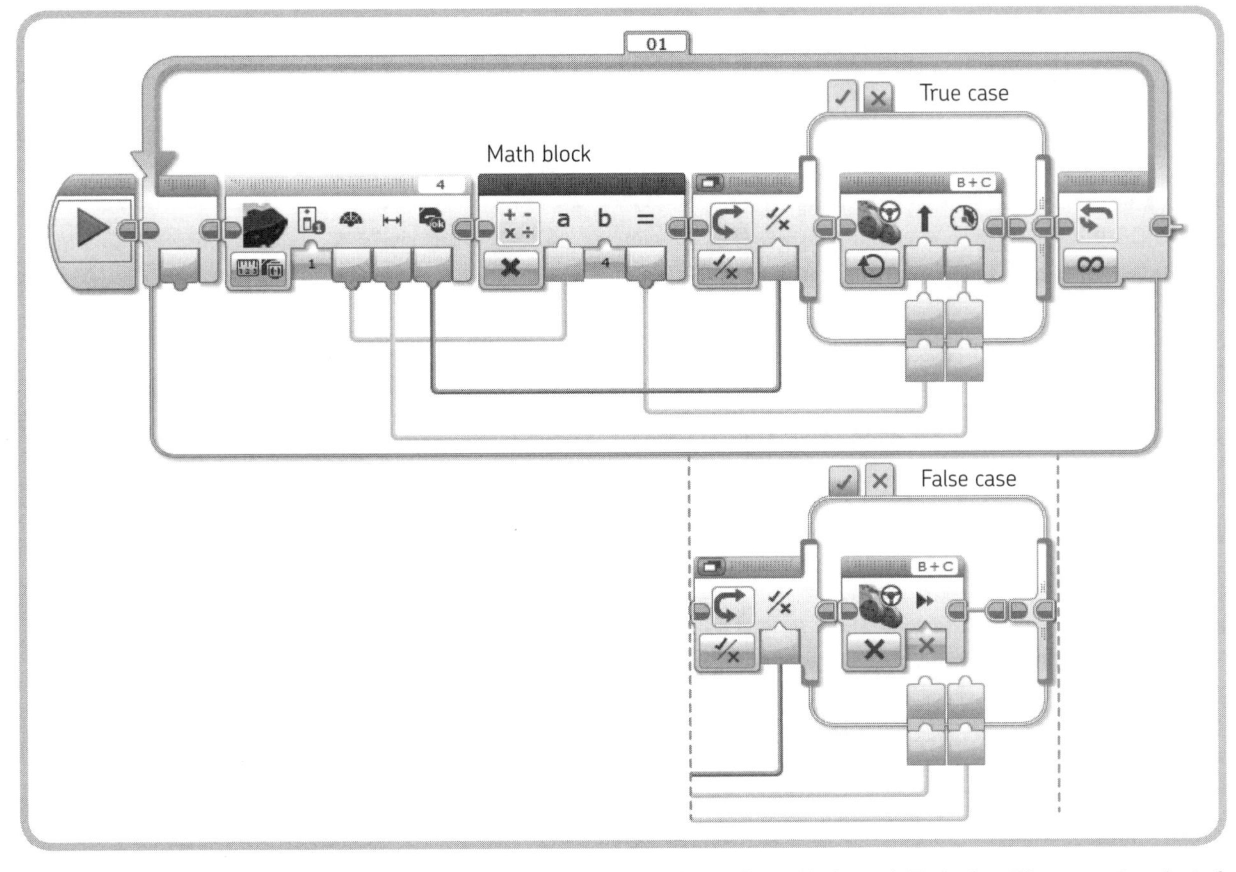

Figure 6-9: The beacon-following program. Both the True and False cases of the Tabbed Switch block contain blocks. Data Wires can go in and out of a switch only in Tabbed View.

* The third output provides the *Detected* state. The value is False when Beacon Mode is off or the signal is not detected, and it is True when the Beacon Mode signal is detected.

We need to take these three outputs and send them to other blocks in the program. First, we'll send the IR Sensor block Heading output to the Steering input in the Move Steering block. However, we can't use the raw data, because we need a number from –100 to 100 for the Steering input and the Heading output only ranges from –25 to 25. To solve this problem, we use a Math block in **Multiply** mode to multiply the value coming from the IR Sensor block Heading output by **4** before sending that value to the Steering input.

Next, we want to send the Proximity output to the Power input of the Move Steering block. Since the ranges of the values are the same, we use a Data Wire to connect them directly.

DIGGING DEEPER: ROBOT LOCALIZATION

A robot that is aware of its position and can navigate through an environment is one of the most interesting and challenging topics in mobile robotics research. Why not take up the challenge with EV3?

For the purposes of this experiment, you could use up to four Remote IR Beacons, set at known coordinates and on different channels, to build a robust localization system for your EV3 robot. This would give you up to four proximity and heading measurements from the IR Sensor. However, heading measurements to beacons are very inaccurate, so use them only as a very rough reference. On the other hand, proximity measurements should be reliable, as long as the IR Sensor's line of sight to the beacon is unobstructed and the beacon is more or less straight ahead of the sensor.

For best performance, avoid building things around the IR Sensor that could shield the beacon's signal. You want to maximize the sensor's field of view (especially sideways).

By merging the *eteroceptive* (external) measurements that the robot takes of its environment with its *proprioceptive* (internal) measurements of the distance it traveled and changes in orientation (using the rotation sensors in the Servo Motors, for example), you can create a robot with the ability to precisely determine its location and navigate. This technique requires some complex math, but the LEGO EV3 has enough computational power to handle it.

We use the Detected output of the IR Sensor block to choose between two cases of a Tabbed Switch block. When the Detected state is True, a Move Steering block in On mode is executed. Since the Move Steering block's Power and Steering inputs are connected to Data Wires coming from the IR Sensor block, the robot's speed and direction will change according to its distance and heading relative to the Remote IR Beacon.

When the Detected state is False, a Move Steering block in Off mode is executed, and the robot will stop moving.

The Move Steering block has the *Brake at End* input set to False, so the robot will glide to a stop when the beacon disappears or when Beacon Mode is turned off.

using the basic operations of the math block

The Math block is a Data Operations block. Depending on its mode, it performs mathematical operations on numeric inputs, producing the result as an output. The Math block can handle both integers and numbers with significant digits after the decimal point. The available operations are listed in Table 6-3.

table 6-3: basic operations of the math block

Mode	Inputs	Output
Add	a, b	$a + b$
Subtract	a, b	$a - b$
Multiply	a, b	$a \times b$
Divide	a, b	a / b
Absolute Value	a	a if $a \geq 0$, $-a$ if $a < 0$ (Result is always positive.)
Square Root	a	$\sqrt{a}$
Exponent	a, n	a^n

You'll learn about the Advanced mode of the Math block in Chapter 7. In Advanced mode, you can enter a formula in the Block Equation field that involves up to four operands.

EXPERIMENT 6-2

Expand the program in Figure 6-9 to display the IR Beacon's Proximity and Heading outputs on the EV3 Brick screen. You'll need Display blocks and Text blocks. To display more lines of text without having them overlap, you'll have to set different values for the Row input of the Display blocks (the input with the red Y icon). Experiment with various ways to display the messages using the fonts (Normal, Bold, and Large) and colors. And because you will probably use more than one Display block, remember to set the Clear Screen input to False in any Display block that follows. Otherwise, that block will clear the text displayed by the preceding one.

When a Display block with the Clear Screen input set to False shows a text string on a row that previously contained a longer string, the old string will not be cleared completely. Any characters from the old string that are beyond the length of the new string will remain displayed, and the screen will look messy. To avoid that without completely clearing the display, just enter a series of spaces in the Text block's **c** input field when building the text string. For example, if one of the Text blocks has "DIST = " as input **a** and a Data Wire carrying the data from the Proximity output of the IR Sensor block as input **b**, enter a few blank spaces as input **c**.

EXPERIMENT 6-3

Create a program that plays a tone whose frequency is proportional to the Proximity output of the IR Sensor in Proximity mode. You'll need a Sound block in Tone mode, with the Proximity output wired to a Math block whose result is wired to the Sound block's Frequency input. Try a frequency range from 300 to 3000 Hz. For example, use the formula

```
Frequency = 10 * Proximity + 330
```

WARNING If you divide a number by zero, the result will be an error shown as *Inf* (infinite) on the Display block, with a sign depending on the dividend. If you use this value for the Rotations input of a Move block, the motors will turn forever. If you compute the square root of a negative number, the output will be an error displayed as ---- on the Display block. Errors displayed as ---- are interpreted as zero when they are used as inputs. (Errors are still sent through Data Wires connected to the output.)

conclusion

In this chapter, you learned all the features of the Infrared devices included in the EV3 set, the Remote IR Beacon and the IR Sensor. You learned how the Remote IR Beacon can be used with the IR Sensor to build a remote-controlled robot, and you discovered how to make your robot follow the beacon. You also learned how to read sensor data and how to transmit data between blocks using Data Wires. You read about the different types of data that you can use in your programs, how to pass them from one block to another, and how to display them on the EV3 screen. Finally, you saw how to use the Math block to perform simple operations. Along the way, you learned how to drive a tracked excavator!

ALREADY DONE? VERY GOOD! I'VE GOT SOMETHING ELSE FOR YOU.
WHAT NOW?
UHM, NOTHING SPECIAL. YOU SHOULD GO AND REBOOT THE **EMP*** SYSTEM IN THE DEAD SECTOR.
OUTSIDE THE DOOR YOU'LL FIND INSTRUCTIONS, AND INSIDE YOU'LL FIND THE MAINFRAME.
***EMP**: ELECTRO-MAGNETIC PULSE
WHY ME? CAN'T SOMEONE ELSE DO THAT?
DO YOU WANT TO BE MY APPRENTICE OR NOT?
YES, DOC!
GOOD! THEN DON'T ARGUE. TAKE ROV3R AND HIS ACCESSORIES WITH YOU, AND GET TO WORK!
UMPH!
06X1

the math behind the magic!

"I'm afraid we need to use . . . math!"
—Prof. Farnsworth in Bender's body,
Futurama, season 6, episode 10

In Chapter 6, you learned how to use Data Wires and how to work with the Math block. In this chapter, you'll learn about the Compare block and the Advanced mode of the Math block. Then we'll develop new programs for the Wall-Following ROV3R.

dealing with measurement noise

Build the Wall-Following ROV3R (page 32), turn on the EV3 Brick, and open the Port View App. Go to the Input 4 tab, and set the IR Sensor into **IR-PROX** mode. You will find that the IR Sensor's proximity readings vary from one sampling to the next—the last digit changes frequently—even if the sensor's distance from an object doesn't change. This is normal: All sensor output is affected by measurement errors, called *measurement noise*.

Just as ambient noise may interfere with your ability to enjoy music, measurement noise interferes with the sensor's ability to measure. We call the interference *noise* because it's random. One good way to reduce, or filter, measurement noise is to take multiple measurements of the distance to one object and average them; in other words, add up the measurements and divide the sum by the total number of readings.

For example, imagine you are measuring the length of a corridor with a measuring tape. You repeat the procedure a few times, trying to do everything the same way (you stretch the tape with the same hand, you align the end of the tape at the same wall edge, and so on). Astonishingly, your measurements differ from each other: for example, 5.10, 5.12, 5.09, 5.11, 5.08. To average the measurements, divide the sum of that list of numbers by the length of the list; that is, $A = (5.10 + 5.12 + 5.09 + 5.11 + 5.08) / 5 = 5.10$.

This result is the *arithmetic mean*. Calculating the mean is an effective method for reducing measurement noise because these random repeated errors tend to be distributed evenly around zero; that is, positive and negative errors are equally likely to occur. We can implement this filtering technique in a wall-following program, as shown in Figure 7-1.

I'll describe each programming block in more detail in the following sections, but let's start with a high-level overview. In this program, three IR Sensor blocks provide three Proximity measurements that are fed into a Math block in Advanced mode. This block averages the measurements using the following formula (the output is called `Result`):

```
Result = (a+b+c)/3
```

One cool thing about using the Math block in Advanced mode is that you can enter a mathematical formula into the Formula field, so you won't need three separate Math blocks to compute the average.

As you can see in Figure 7-1, the result of this calculation is carried by a Data Wire into input **a** of a Compare block, which compares it against a fixed threshold (40) as specified in input **b**. The logic output of the Compare block, which is carried by a Data Wire into the input of a Switch block (shown in Flat View), is used to choose between the logic cases True or False. These cases, in turn, make the robot steer left or right, adjusting its trajectory to keep the wall at the correct distance.

Build, download, and run this program on your ROV3R to test it. You should find that ROV3R, by averaging the sensor readings, becomes less sensitive to random sensor noise and thus moves more smoothly.

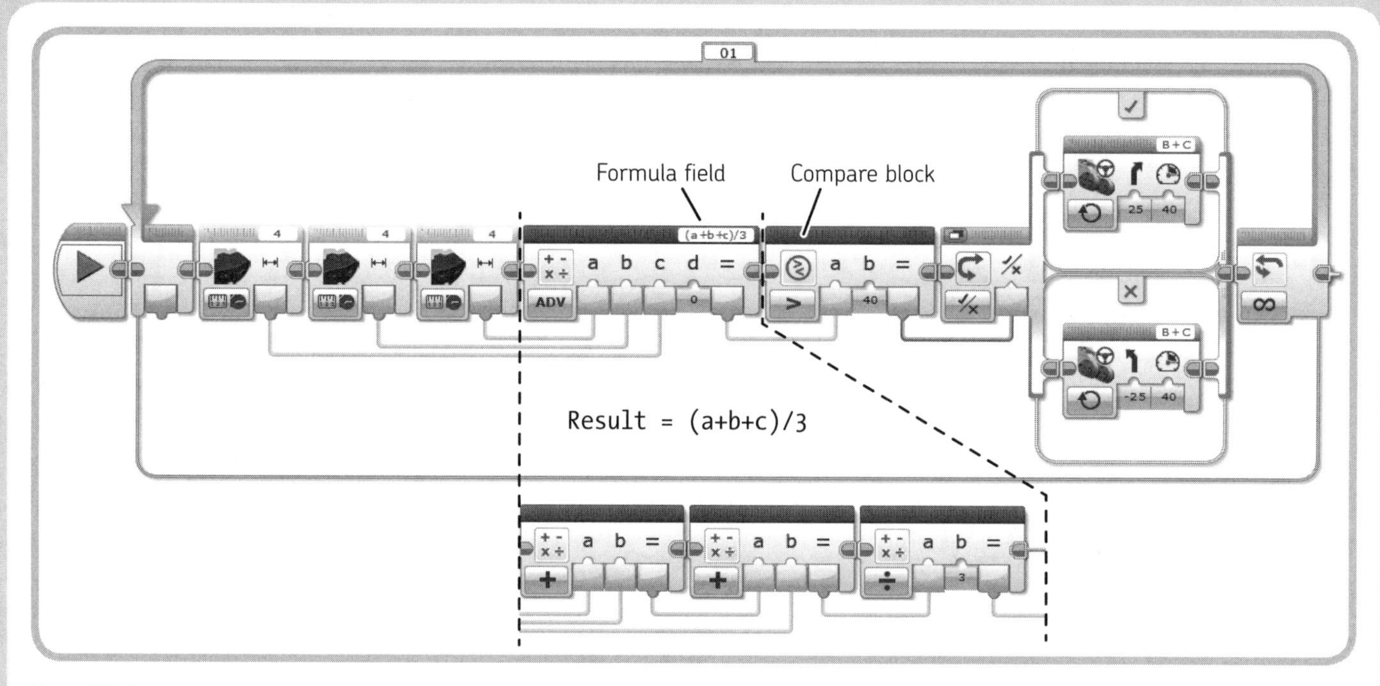

Figure 7-1: This wall-following program uses an average of three IR Sensor readings.

the math block in advanced mode

We saw above that by setting a Math block in Advanced mode, we could use just one Math block to compute several math operations by entering a formula into its Formula field. In Advanced mode, the Math block has four numeric inputs: **a**, **b**, **c**, and **d**. You can use these inputs as operands in either lower- or uppercase. The EV3 Software will give the operands the actual numeric values from the inputs and then compute the result according to any valid formula in the Formula field. In addition to using the Math block in Advanced mode for common operations, you can access several built-in functions by clicking the Formula field. Table 7-1 lists these functions by name (as shown in the block's function list), showing the symbols added to the formula and an example of each operation.

the round block

Some functions in Table 7-1 are equivalent to using the Round block (Data Operations palette, red header). The Floor function is equivalent to using the Round block in Round Down mode, the Ceil (ceiling) function is equivalent to using the Round block in Round Up mode, and Round is equivalent to using the Round block in To Nearest mode. The Round block also has the Truncate mode, which truncates the value, keeping only the number of digits after the decimal point specified by the *Number of Decimals* parameter. For example, truncating 3.2 to 0 decimals returns 3, and truncating –3.34 to 1 decimal returns –3.3.

table 7-1: the functions available from the math block in advanced mode

Name	Symbol	Example
Add	+	$a + b$
Subtract	–	$a - b$
Multiply	*	$a * b$
Divide	/	a / b
Modulo	%	*a* % *b* computes the remainder of the integer division. For example, 4 % 2 = 0, 10 % 3 = 1, 4 % 7 = 4.
Exponent	^	*a*^*b* computes *a* to the power of *b*.
Negate	–	–*a* (adds a minus sign, like Subtract)
Floor	floor(	floor(*a*) brings the real number *a* to the greatest integer smaller than *a*. For example, floor(3.2) = 3, floor(3.5) = 3, floor(–3.2) = –4, floor(–3.5) = –4.
Ceil	ceil(	ceil(*a*) brings the real number *a* to the smallest integer greater than *a*. For example, ceil(3.2) = 4, ceil(3.5) = 4, ceil(–3.2)= –3, ceil(–3.5) = –3.
Round	round(	round(*a*) brings the real number *a* to its nearest integer, with a half unit as the decision point. For example, round(3.2) = 3, round(3.4999) = 3, round(3.5) = 4, round(–3.2) = –3, round(–3.4999) = –3, round(–3.5)= –4.
Absolute	abs(	abs(*a*) returns *a* if $a \geq 0$ and –*a* if $a < 0$. The result is always positive.
Log	log(	log(*a*) computes the base 10 logarithm of *a*.
Ln	ln(	ln(*a*) computes the natural logarithm of *a*.
Sin	sin(	sin(*a*) computes the sine of angle *a* (where *a* is given in degrees).
Cos	cos(	cos(*a*) computes the cosine of angle *a* (where *a* is given in degrees).
Tan	tan(	tan(*a*) computes the tangent of angle *a* (where *a* is given in degrees).
Asin	asin(	asin(*a*) computes the arcsine of *a* (result in degrees).
Acos	acos(	acos(*a*) computes the arccosine of *a* (result in degrees).
Atan	atan(	atan(*a*) computes the arctangent of *a* (result in degrees).
Square Root	sqrt(	sqrt(*a*) computes the square root of *a*.

DIGGING DEEPER: HANDLING ERRORS FROM MATH BLOCKS

Some operations will produce errors as output when the inputs are not legal. All error values propagate through Data Wires and will affect the results of other Math blocks. You can track down which Math block produced an invalid value by connecting a Data Wire from a Math block output to a Display block in Text mode.

One potential error situation occurs when you divide by zero. The result will be an error value displayed as *Inf*, or infinite, with a sign that depends on the sign of the quotient. Also, the result of log(0) and ln(0) is *-Inf.* If you use this value for a Rotations input on a Move block, the motors will run forever.

As another example, if you compute the square root of a negative number or you make a syntax error in a formula (by misspelling a function name or using an unbalanced parenthesis), the output will be an error displayed as ---- on the Display block. Such errors will be interpreted as zero by other inputs. Also, be careful with the results of functions that are usually implemented as approximate. For example, the result of tan(90) should be *Inf*, but it turns out to be a huge negative number (-22,877,332).

the compare block

The Compare block lets you compare two numbers using its inputs **a** and **b** to see if they're equal or if one is greater than the other. The result of the comparison is a logic value that can be True or False. Table 7-2 lists the six comparisons you can perform by changing the mode.

table 7-2: the comparisons available on the compare block

Mode	Result
Equal To	True if $a = b$; False otherwise
Not Equal To	True if $a \neq b$; False otherwise
Greater Than	True if $a > b$; False otherwise
Less Than	True if $a < b$; False otherwise
Greater Than or Equal To	True if $a \geq b$; False otherwise
Less Than or Equal To	True if $a \leq b$; False otherwise

converting numeric values to logic values

Recall from Chapter 6 that a numeric value can't be automatically converted into a logic value. However, you can perform that computation with the Compare block (Figure 7-2) by setting it to Not Equal To mode and setting input **b** to zero. In this way, any number provided to input **a** other than zero will be converted to True. Any number equal to zero will be converted to False.

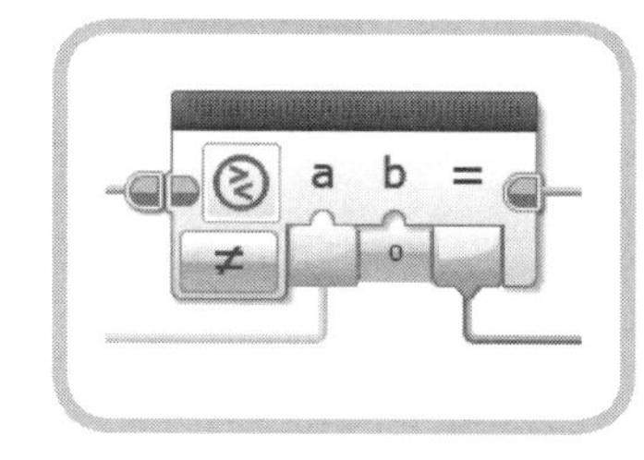

Figure 7-2: Converting a numeric value to a logic value using a Compare block

NOTE Treating any nonzero number as True is just one of many options, but it's exactly how a C or Java compiler would perform the conversion.

embedded compare blocks

We used the functionality of Compare blocks in previous chapters without seeing these blocks themselves. In fact, some programming blocks embed the comparison functionality. Let's see how the Switch and Wait blocks use the code of the Compare blocks.

In Figure 7-3(a), the Switch block selects which case to execute by comparing the sensor value against a threshold. This is equivalent to a Sensor block in Compare mode that provides a logic value to a Switch block, as shown in (b), or to a Sensor block and a Compare block connected to a Switch block, as in (c). Similarly, a Wait block (d) is equivalent to a Loop block that checks for the exit condition with a comparison (e). The Loops in (f) and (g) function equivalently to (d), with (g) showing the Compare block explicitly.

the constant block

The Constant block simply provides a constant output as specified in its top Text Field. Depending on its mode, a Constant block can provide all kinds of data types, numbers, logic values, strings of text, and arrays. (See "Understanding Data Types" on page 89.)

improving our wall-following program

Now to improve our wall-following program to make it smoother and more intelligent than ever! Our current wall-following program (from Figure 5-23 on page 82) switches rigidly between two steering directions if the proximity read by the IR Sensor is greater or less than a threshold. The result is a jerky motion, even after we try to reduce the jerkiness with steering adjustments.

To improve our wall-following program, we'll use the same concept used by the program in Figure 6-3 on page 87: When the robot moves forward, the IR Sensor's Proximity output will control the amount of steering of a Move Steering block. In particular, we'll make the Steering control **U** proportional (by a constant gain **K**) to the difference **E** (error) between

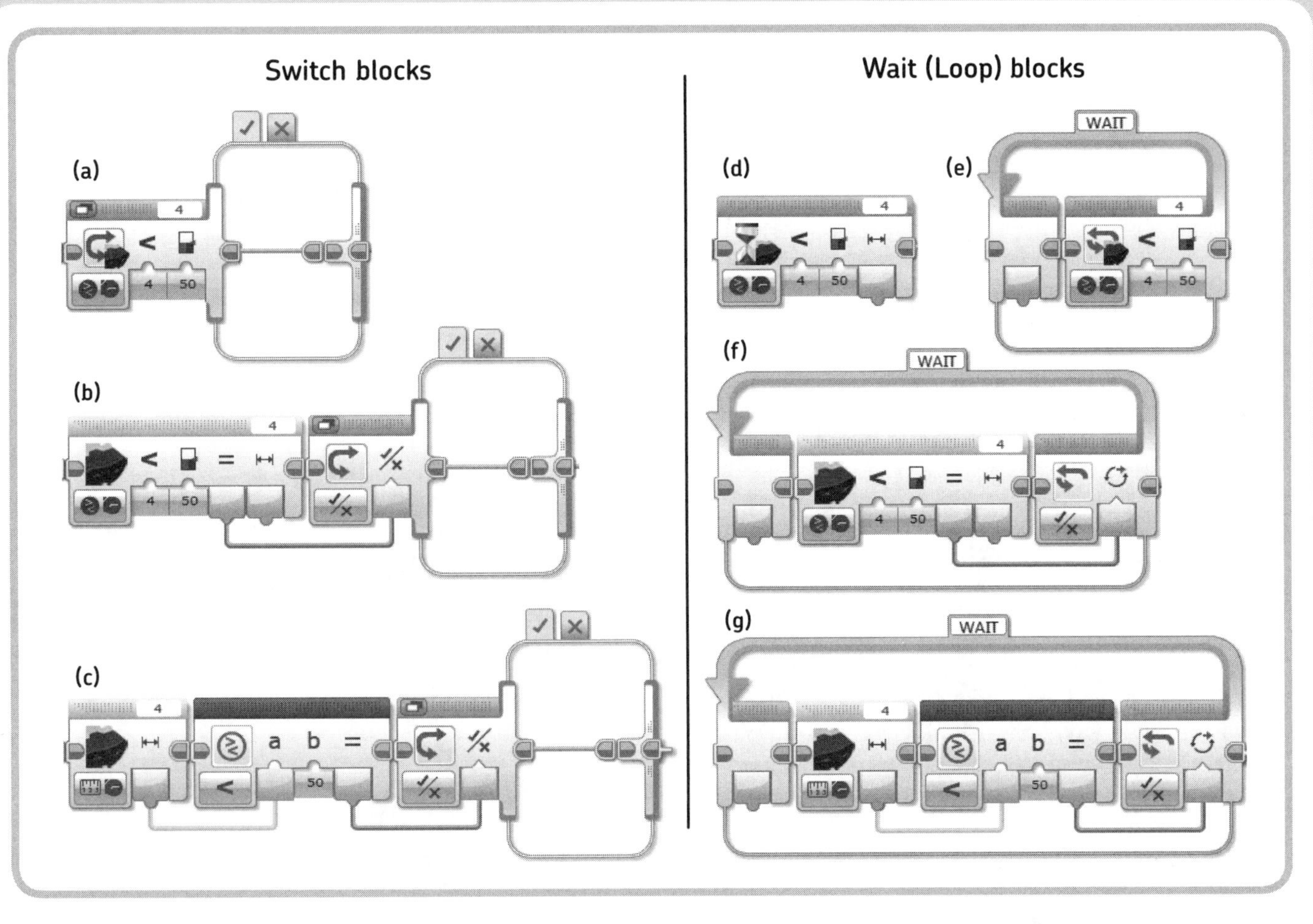

Figure 7-3: Equivalent programming examples using Compare blocks

the desired distance **R** from the wall and the actual distance measured by the sensor **Y**. (I'll explain what gain is and why it matters shortly.) This formula will do the trick:

```
U = - K * E = - K * (R - Y) = - gain * (reference - measure)
```

NOTE We need the minus sign in front of the gain to make the robot react properly: Steer toward a wall when too far away from it and away from the wall when too close. We could rewrite the formula as `U = K * (Y - R)`, but traditionally the error is defined as $E = R - Y$ and not vice versa.

The greater the difference between distance values, the stronger the steering input will be! For example, when the robot is at about the desired distance from a wall, the difference **E** will be small, as will the Steering control **U**: The robot will move almost straight ahead, making only small adjustments to its trajectory. However, when the robot is very near a wall, the difference **E** will be bigger, and the Steering control will have a stronger effect, moving the robot away from the wall by making the robot spin in place.

Notice, in the program shown in Figure 7-4, that the IR Sensor's Proximity value is fed into Input **c** of a Math block in Advanced mode. The two Constant blocks are there to show the value of the desired distance from the wall, *R* (50), and the gain, *K* (2). The *gain* controls the strength of the robot's reaction: The larger the gain, the more "nervous" the behavior of the robot; the smaller the gain, the weaker the steering intervention will be, but the robot may not be able to handle corners. Adjusting the gain involves a trade-off between smoothness and reaction time. A gain of 2 works well in this program.

This wall-following method has some limitations; for example, the robot can only follow walls on its right. However, you can modify the sensor assembly and the program to allow the robot to follow walls on its left. Also, this relatively minimal program will fail if the robot is too far from the wall; in that case, it will spin in place, with no chance of finding the wall again.

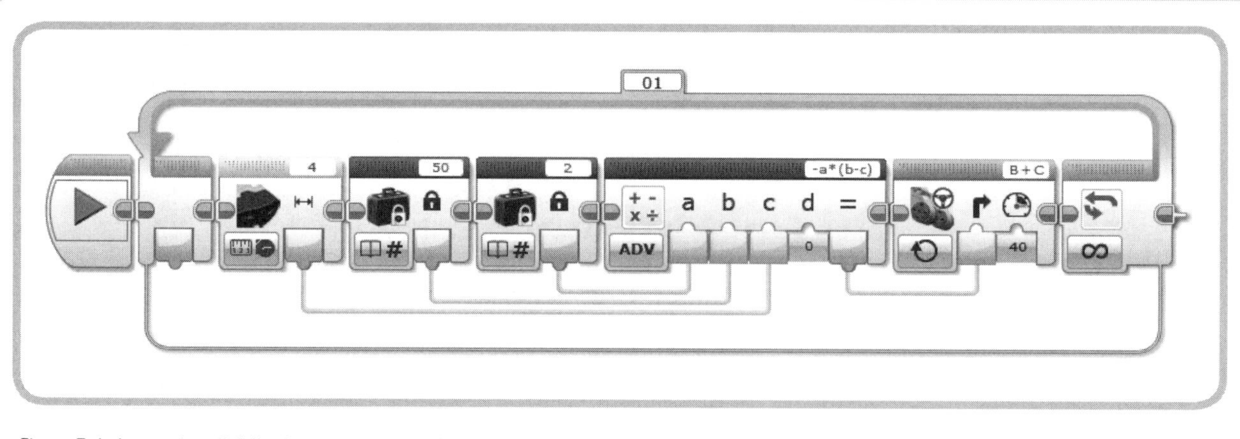

Figure 7-4: A smooth wall-following program using Steering that is proportional to the robot's distance from the wall

DIGGING DEEPER: FEEDBACK CONTROLLERS

The wall-following programs are examples of feedback controllers. A *controller* is a device (a program) that monitors and changes the behavior of a system (your robot). In the case of wall following, we want to keep the robot at a constant distance from a wall. The input channel that provides the feedback to the controller is the IR Sensor, and the output comes from the Servo Motors that drive the robot. Our first Brick Programs for line and wall following in Chapter 4, and the EV3 programs derived from them in Chapter 5, implemented a so-called *Bang-Bang controller* (or On-Off controller), because the control command (the steering amount) switched abruptly between two states. The controller discussed in this chapter is a *proportional feedback controller*, where the control command U is proportional (based on gain K) to the error E between the desired distance R and the actual distance Y. The formula is $U = -K * (R - Y)$.

conclusion

In this chapter, you learned how to enter formulas into a Math block set in Advanced mode. In particular, you learned how to compute the arithmetic mean of a set of numbers using a single block. You met the Compare block and realized that it was an old acquaintance, as it is embedded in Switch blocks and Wait blocks. Finally, you saw how to build a proportional controller as part of a smooth and sensitive wall-following program.

EXPERIMENT 7-1

Try changing the gain and the proximity reference in the program shown in Figure 7-4 to see how the robot's behavior changes.

EXPERIMENT 7-2

Modify the wall-following program shown in Figure 7-4 so that you can set the reference distance from the wall before starting the loop that controls the robot. When you press **Enter** on the EV3 Brick, you should be able to set the distance and start the robot; when you press **Enter** again, the robot should stop and return to setup mode. (Hint: Put the wall-following loop inside an external loop.)

EXPERIMENT 7-3

As noted earlier, the program in Figure 7-4 fails if the robot is too far from a wall. Once you've learned all of the programming in this book (including the Timer block, which I'll discuss in Chapter 10), develop a strategy to overcome this limitation. For example, the program could detect that the robot is spinning in place by checking whether the distance from a wall is too great for a long period of time; then it could make the robot go straight for a while, hopefully getting closer to a wall.

0x0000DEAD
IT'S PITCH BLACK!
DOC DIDN'T TELL ME ON PURPOSE! I COULD HAVE BROUGHT A FLASHLIGHT!
SBAM!
OUCH!
07X1

...TO REACH THE MAINFRAME, FOLLOW THE DARK LINE ON THE FLOOR... TO ENABLE THE EMP SYSTEM, TYPE EMP ENABLE ON THE COMMAND CONSOLE...THE LIGHTING SYSTEM AND DOORS ARE AUTOMATIC, TRIGGERED BY INFRARED SENSORS...
YEAH, TOO BAD THOSE SENSORS NEAR THE ENTRANCE ARE OUT OF ORDER! I'M GOING TO CRACK MY HEAD ON SOMETHING!
WAIT! THE EV3 BRICK EMITS LIGHT, AND ROV3R CAN FOLLOW THAT BLACK LINE ON THE FLOOR!
WHIIRRRRR
IT WORKED!
YESSSS!
EMP SYSTEM ENABLED!
07X2

LEGO recipes

In this chapter, you'll learn the basics of LEGO geometry, like how to build sturdy structures, make functional gear trains, and transmit and transform motion. You'll also see some building ideas for working with EV3 motors.

the angular beams unveiled

As you know, the basic unit in the LEGO Technic system is the fundamental LEGO unit, and every piece is designed to respect this basic unit. The angular beams are no exception. However, their geometry isn't always simple to understand. The 2×4 and 3×5 angular beams, the T beam, and the double angular beams are easy enough because they have right angles (90°) and half-right angles (45°). But what about the other angular beams? Why are they bent at odd angles?

For example, Figure 8-1 shows a 3×7 angular beam. The triangle overlaying the beam has a right angle (90°) at its base, and its sides measure 3M, 4M, and 5M in whole LEGO units. (I'm measuring the lengths from center to center of the holes: six holes should correspond to 5M because you subtract half a unit from each end.)

The angle at the base indicated with α (alpha) results from the LEGO Group having designed this beam to fit in the geometry of the LEGO Technic system. You'll see the same triangle in the 4×6 and 4×4 angular beams.

The Pythagorean Theorem (see Figure 8-2) says that in any right triangle, the area of the square of the hypotenuse (the side opposite the right angle) is equal to the sum of the areas of the squares of the two sides that meet at the right angle.

For example, in Figure 8-1, the legs of the triangle measure 3 and 4. Their squares are 3 × 3 = 9 and 4 × 4 = 16. The sum of their squares is 9 + 16 = 25, and therefore the hypotenuse is equal to 5, the square root of 25. (The story is that the ancient Egyptians knew this trick long before Pythagoras and that they used knotted ropes to make right triangles in order to set land boundaries.)

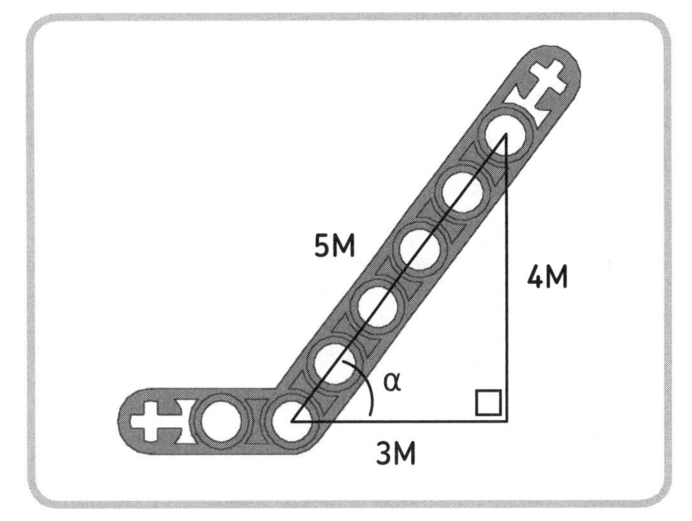

Figure 8-1: Geometry of the LEGO angular beam

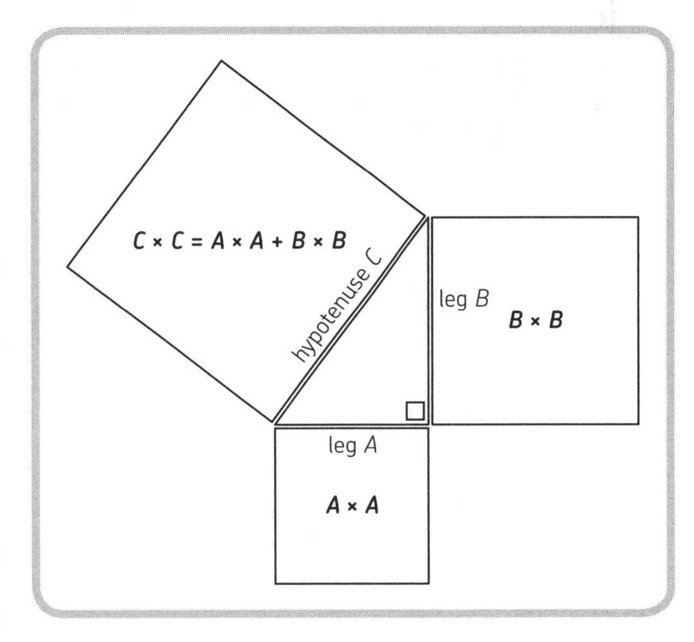

Figure 8-2: Graphical representation of the Pythagorean Theorem

DIGGING DEEPER: ANGULAR BEAMS MYSTERY SOLVED!

How about measuring angle α? Since the 3-4-5 triangle (as it's known) in Figure 8-1 is a right triangle, the angle $\alpha = \arctan(4/3) \approx 53.13°$. You may have seen this number before, perhaps in the name of the angular beams in LDraw computer-aided design (CAD) elements. Now you know where it comes from!

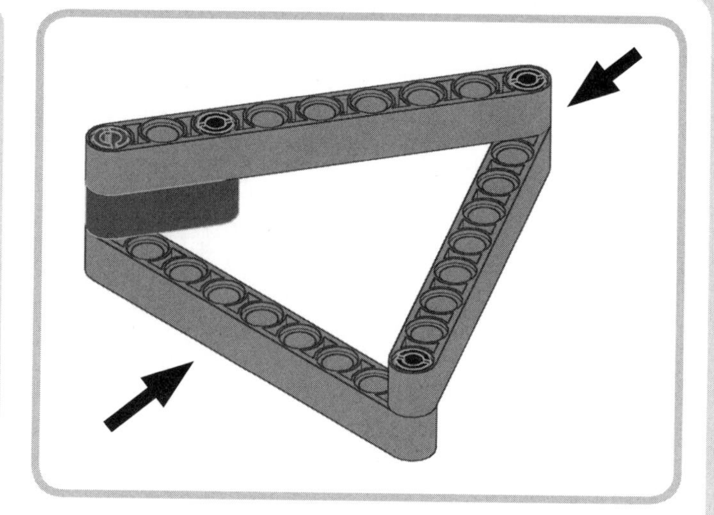

Figure 8-4: A triangular structure can resist applied force without being deformed.

triangles vs. rectangles

There's a difference between building rectangular structures and triangular structures. As you can see in Figure 8-3, it's easy to squash a parallelogram by applying force to it. On the other hand, a triangular structure, like the one in Figure 8-4, can resist applied forces without getting squished. That's why bridges and many support structures are built out of a lattice of triangles, and it's why you should try to use triangular structures when building LEGO models that need to stand up to applied force.

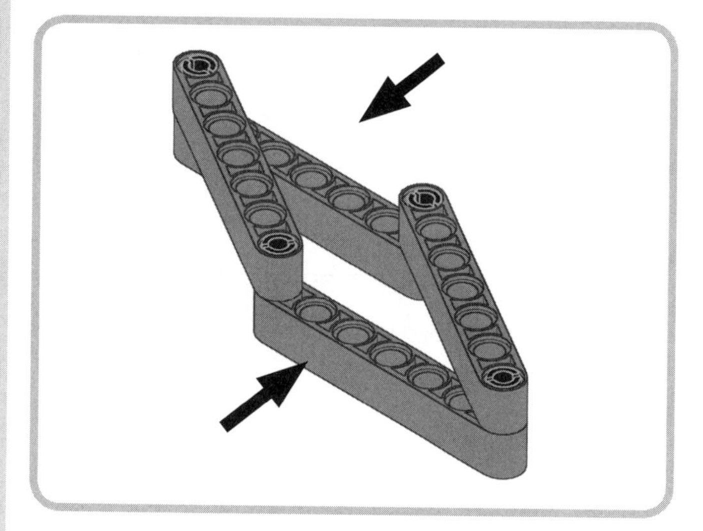

Figure 8-3: You can easily squash a parallelogram by applying force to it.

Is it possible to build strong rectangular structures? If you need to build a rectangular frame that is as resistant to pressure as a triangular one, just add a diagonal beam, as shown in Figure 8-5. By doing this, you effectively create two triangles. (Notice that the diagonal beam is like the hypotenuse in the triangle in Figure 8-1.)

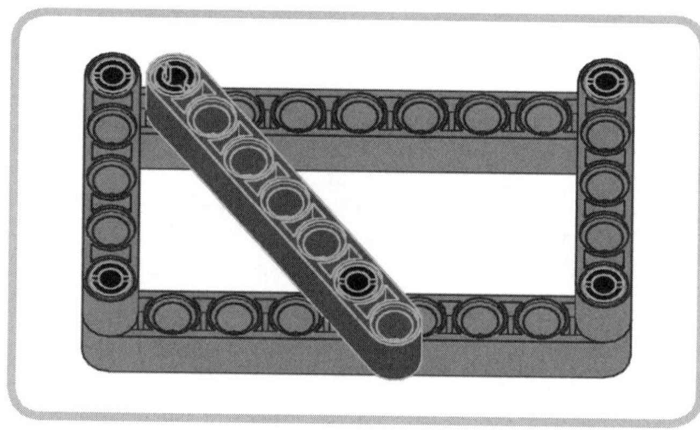

Figure 8-5: Adding a diagonal beam to a rectangular structure makes it resistant to applied force; with the beam, it's essentially two triangles.

Now for some examples of assemblies that show how to use angular beams and triangular structures to build solid assemblies. Have a look!

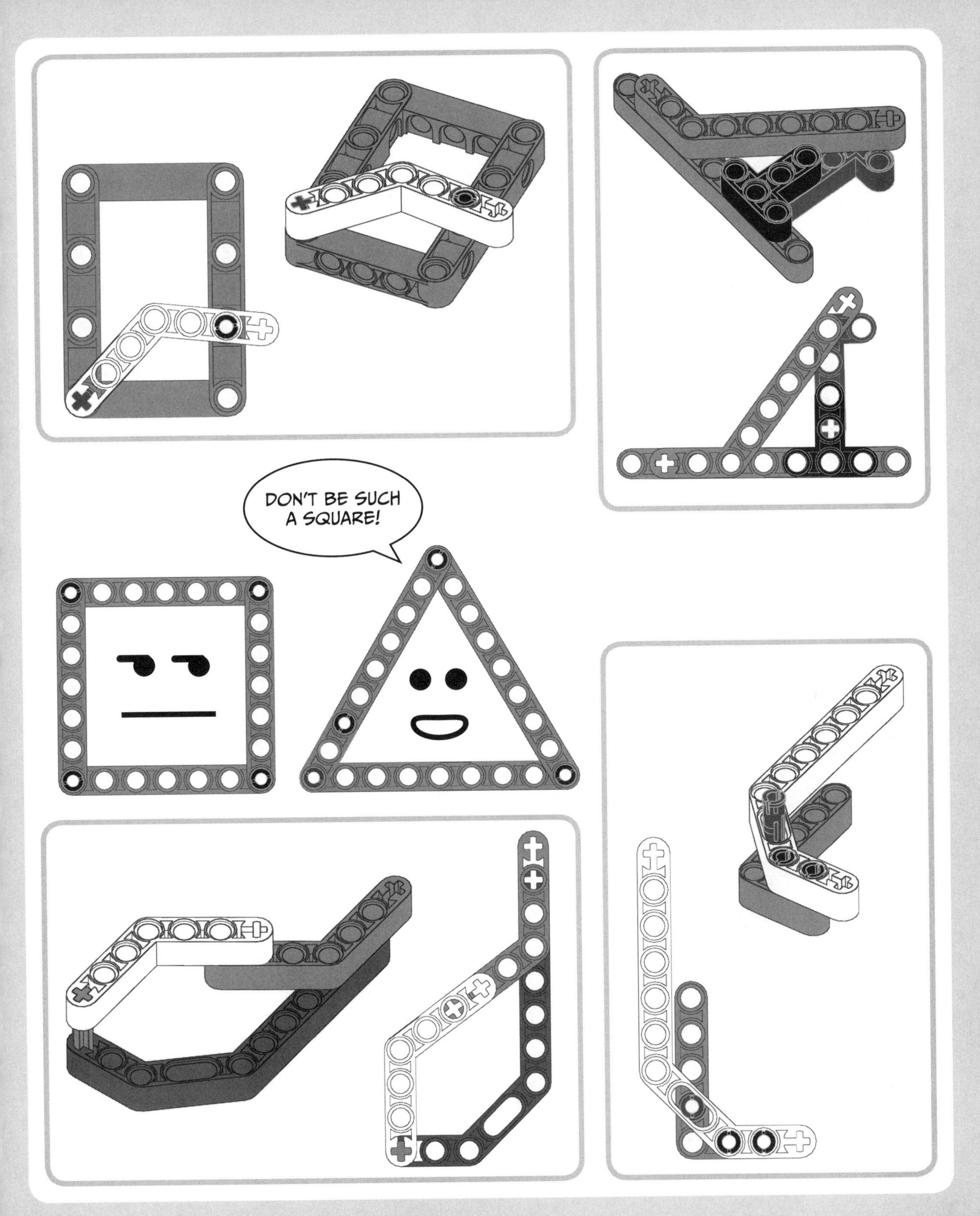
DON'T BE SUCH A SQUARE!

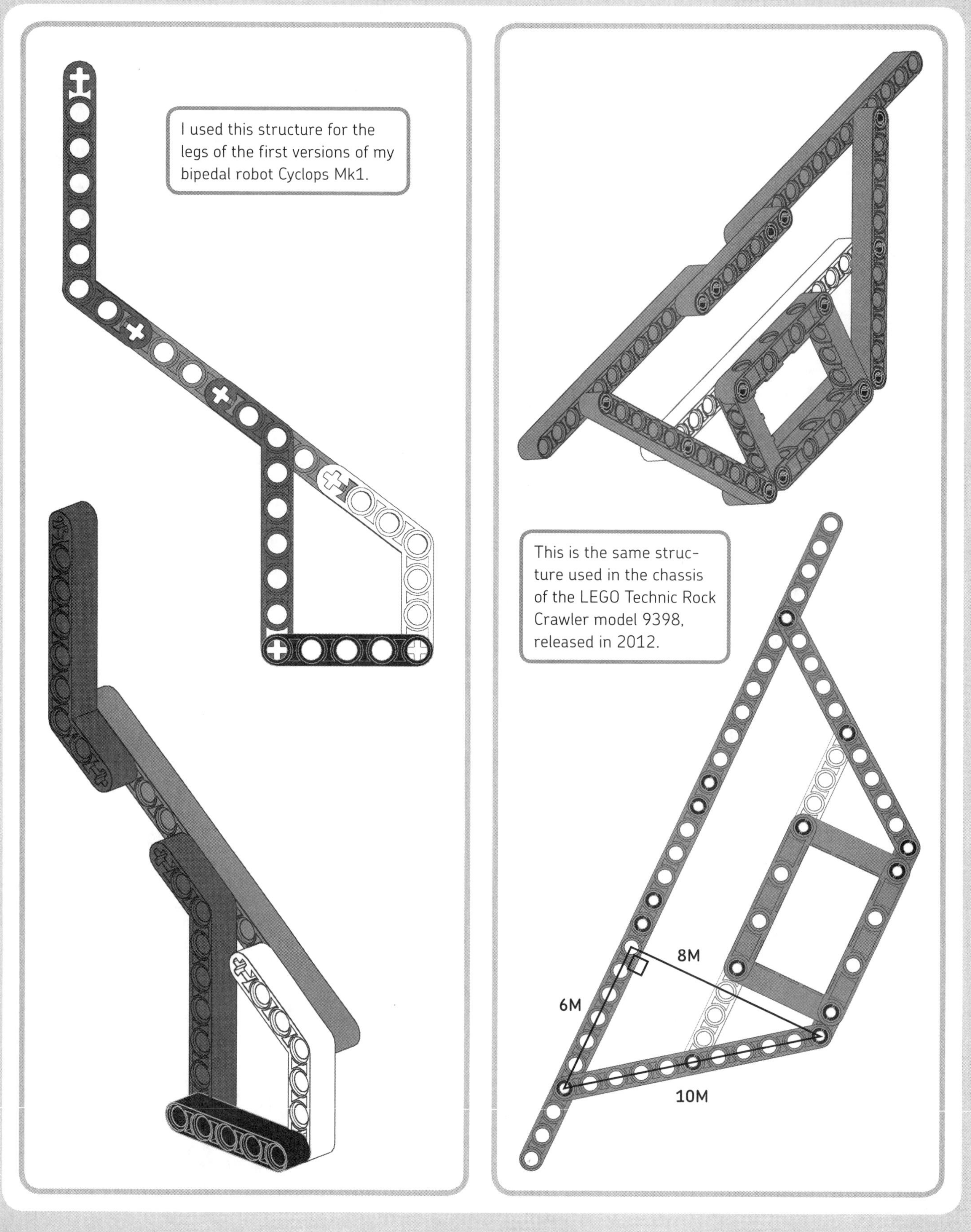
I used this structure for the legs of the first versions of my bipedal robot Cyclops Mk1.
This is the same structure used in the chassis of the LEGO Technic Rock Crawler model 9398, released in 2012.
8M
6M
10M

extending beams

When you need to extend the length of a beam, use the black pins with friction or the blue long pins with friction to connect two or more beams. Figure 8-6 shows various ways to extend beams and strengthen structures.

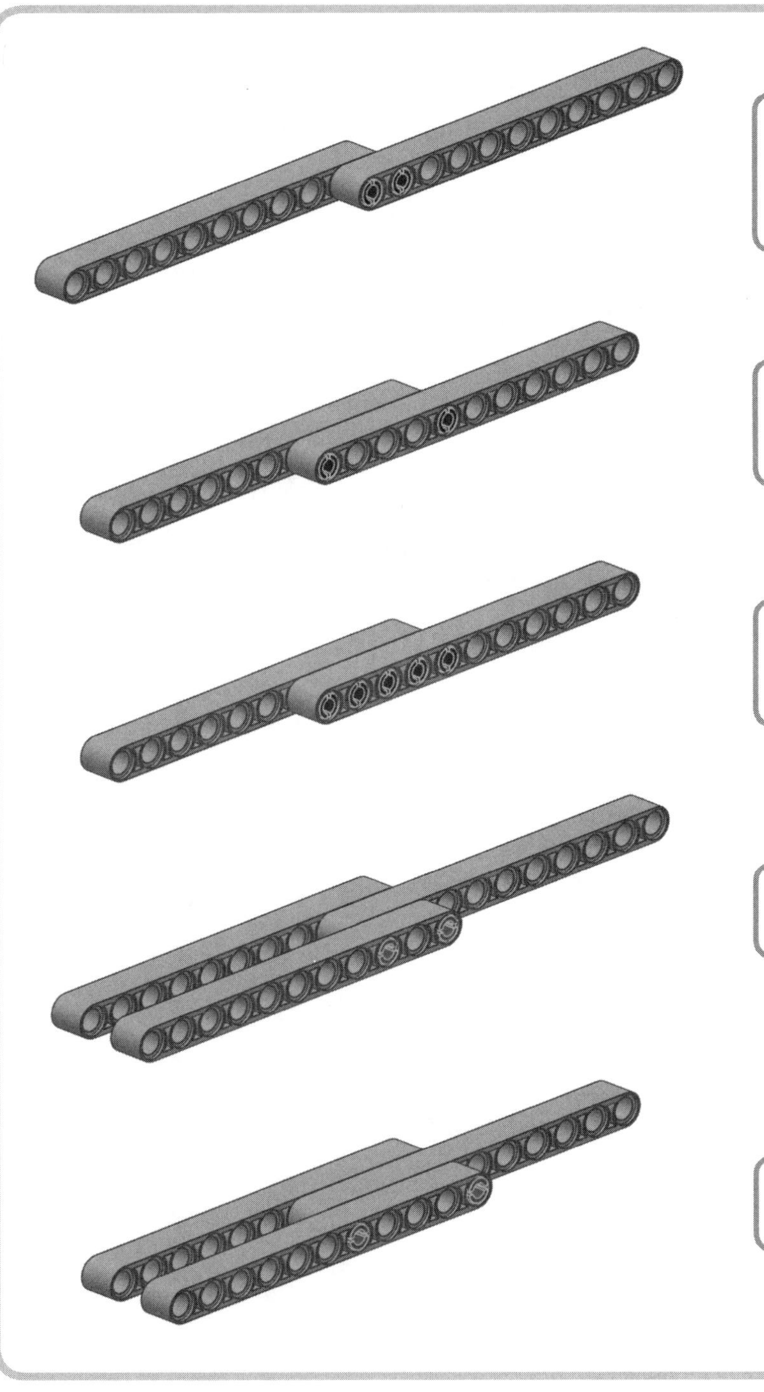

These two beams are connected with black pins and with an overlap of just two holes. Notice that the assembly is straight but still flexible.

Increasing the overlap of the beams to five holes makes the resulting structure more rigid.

Adding more pins can help hold the beams together better, but they're usually not needed.

You can use long blue pins to strengthen the assembly by making it thicker.

Increasing the amount of overlap makes the structure more rigid.

Figure 8-6: Extending and thickening beams with friction pins

bracing

LEGO assemblies are good at resisting compression and shearing forces, but they're designed to come apart easily when subjected to pulling forces. (Otherwise, you couldn't disassemble them!) Therefore, when subjected to stress and strain, assemblies can come apart because the beams pull away.

The solution to this problem is *bracing*, using one or more beams to brace the parts that would otherwise come apart. (You saw this technique in action when you built ROV3R in Chapter 2.) You'll gain a better understanding of this concept as you build the two assemblies shown next, and I'll point out bracing when it's used in the various robots throughout this book.

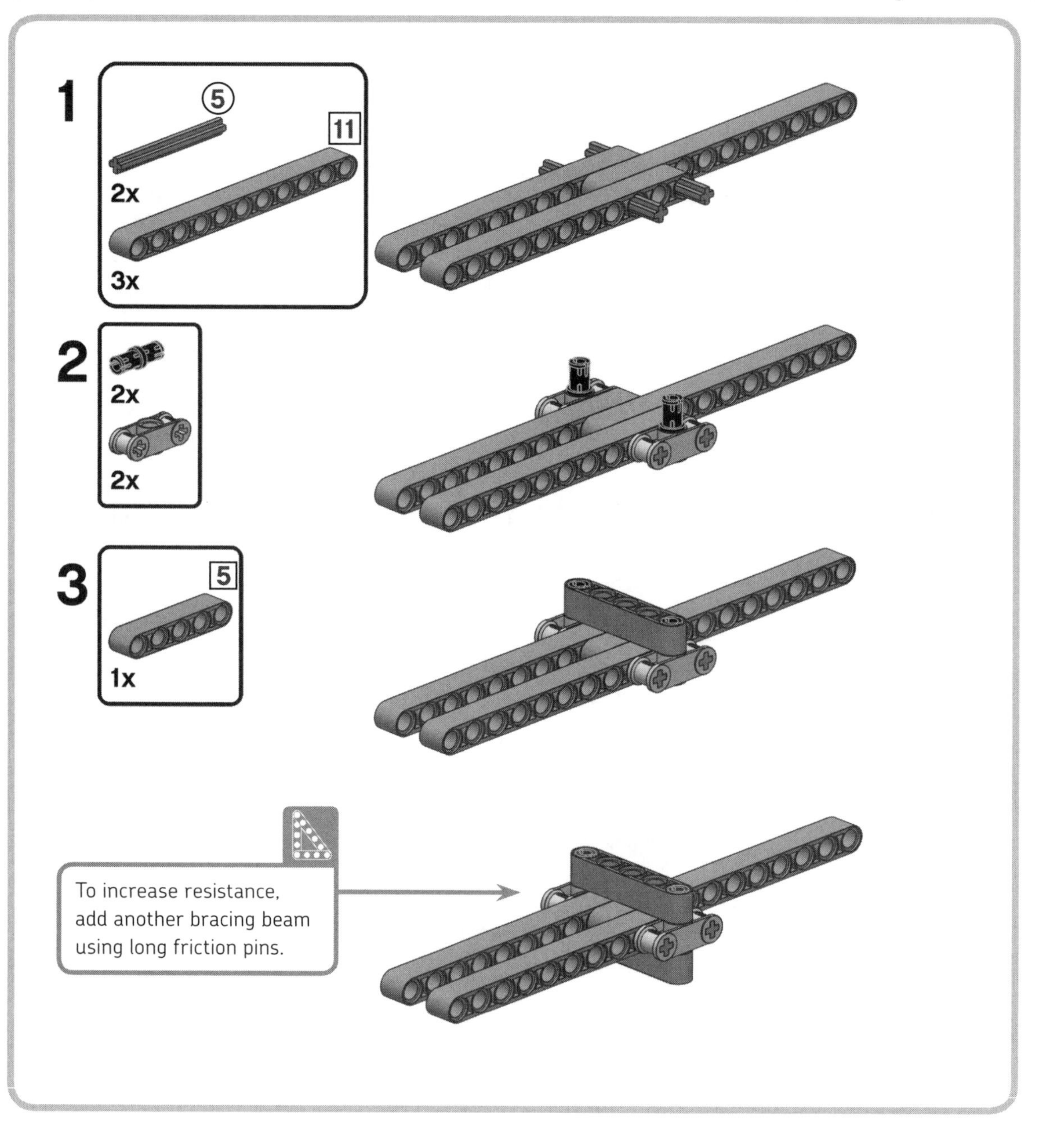

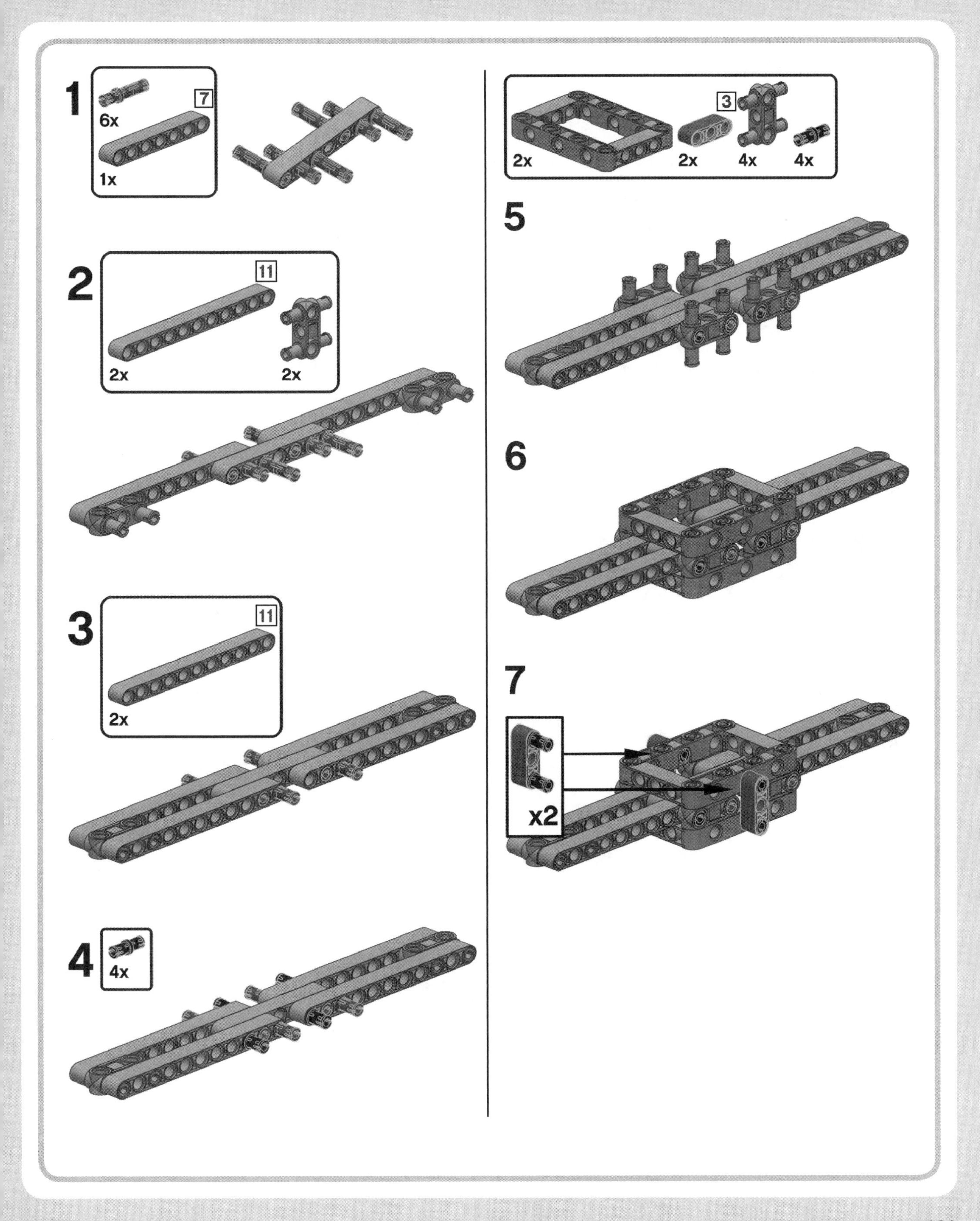
1
6x
7
1x
2
11
2x
2x
3
11
2x
4
4x
3
2x
2x
4x
4x
5
6
7
x2

cross blocks

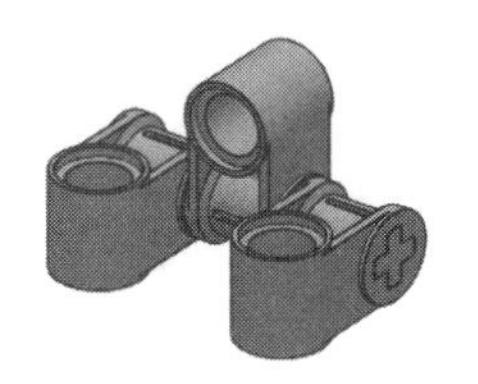

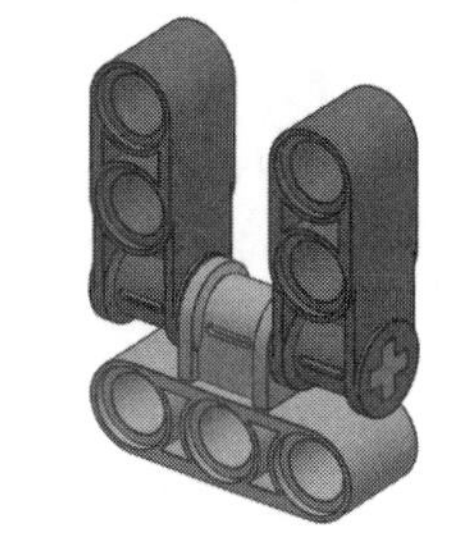

Cross blocks allow you to expand your LEGO structures in three dimensions. Remember, the LEGO Technic studless way of building requires you to think in 3-D. You're not stacking bricks—think through your design from the inside out!

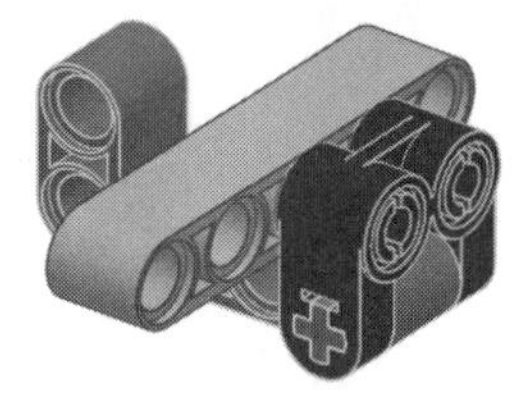

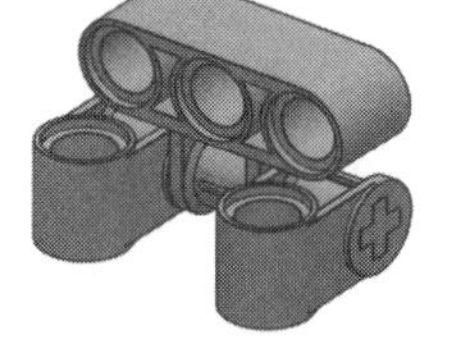

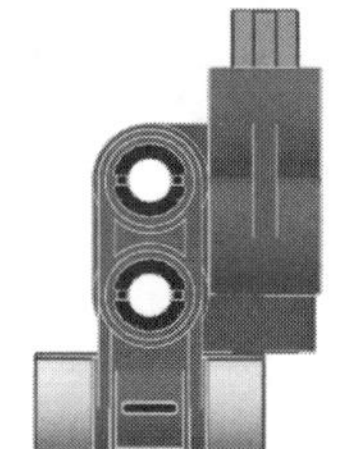
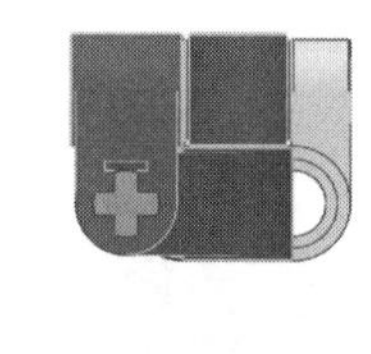

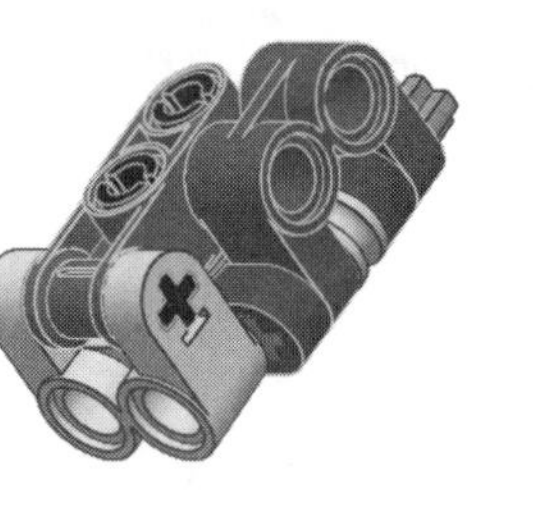

Use cross blocks to create half-module offsets.

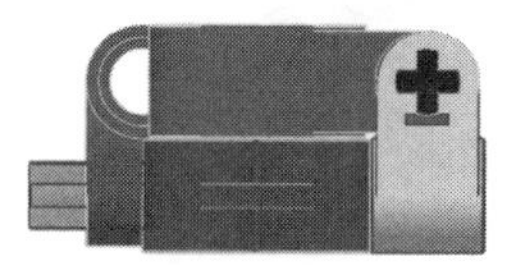

gears revisited

We first met gears in Chapter 1 (Figure 1-13 on page 12). Now we'll discover their secrets! Here are some basic things to remember about gears:

* You measure a gear by counting its teeth.
* You combine gears according to radius and thickness.
* Gears have a cross hole in the middle so that they can fit on axles and transmit rotation from one axle to another.

Build the assembly shown at the top of Figure 8-7: a 12z double-bevel gear meshing with a 36z double-bevel gear. If you rotate the 12z gear three times, how many times will the 36z gear rotate? If you thought "just once," you're right! When the smaller gear is the *input gear* (the driving gear), both the number of rotations and the speed of the larger *output gear* (the driven gear) decrease.

How does this relate to the number of teeth? The relationship between gears is expressed in terms of the *ratio* between the numbers of their teeth. In this example, the ratio is 12:36 = 1:3. Thus, the 36z gear goes one-third as fast as the 12z gear.

Figure 8-7: Gears change the speed and the torque of rotating axles.

The advantage of this gear combination is that if you turn the 12z input gear, the 36z output gear has three times more torque than the input gear. *Torque* is a twisting force applied to an object that tends to make it rotate. Put another way, torque relates to rotating objects as force relates to pushing objects. Aside from changing how fast the driven axle turns, gears change the amount of torque transmitted to it.

Now try turning the 36z gear with the crank for one complete turn. The smaller gear will perform three turns, and its torque will be one-third as much as the larger gear's torque.

To sum things up, if the input gear is smaller than the output gear, the speed of the output gear will be decreased and its torque increased. On the other hand, if the input gear is the bigger one, the speed of the output gear will be increased but its torque decreased. The illustrations below the gears in Figure 8-7 show how gears change the speed and torque of axles.

getting gears to mesh together well

How do you make two gears mesh together well so that you won't have to hear that awful noise when they disengage and their teeth slip? This is a question that nags the novice LEGO builder, but you'll find its answer in this section. Table 8-1 lists the radii (plural of *radius*) of the gears, expressed in LEGO units. (To verify these values, you can measure the radius of gear wheels by using a beam as a reference. There's no need to memorize them, but they are helpful in understanding how to make gears mesh correctly.)

table 8-1: the radius of each gear

Name	Radius (in LEGO units)
8z gear	0.5
12z double-bevel gear	0.75
16z gear	1
4z knob wheel	1
20z double-bevel gear	1.25
24z gear	1.5
Small turntable (28z)	1.75
36z double-bevel gear	2.25
40z gear	2.5
Large turntable (56z)	3.5

NOTE Table 8-1 lists some gears and turntables (shown in italic) that are not included in the 31313 set.

There are two kinds of gear combinations: perfect and imperfect.

* You have a perfect gear combination when the sum of the radii of the gears equals a whole number of LEGO units. Examples of perfect combinations are 8z with 24z (0.5 + 1.5 = 3M) and 12z with 20z (0.75 + 1.25 = 3M).
* You have an imperfect gear combination when the sum of the radii of the gears is not equal to a whole number of LEGO units.

Don't think that you need to avoid imperfect combinations! On the contrary, they can be very useful. Just know that LEGO didn't plan for you to use imperfect combinations of gears, so the teeth may mesh a bit loosely, or you may find it hard to build a frame to hold the gears correctly. But bending (and breaking) LEGO design rules is my passion, so give imperfect gear combinations a chance!

Of course, before bending the rules, you need to know what the rules are. Let's consider all the gears in the EV3 set as well as some extra gears not included in the EV3 set (like the 8z, 16z, and 40z gears), while leaving the worm gear and the knob wheel gear aside for now. As you can see in Table 8-2, there are 28 possible combinations of these gears. (The bottom half of the table is greyed out because it's symmetrical: From the point of view of the radii, combining 12 and 20 is the same as combining 20 and 12.) Here's how to read this table:

* A checkmark (✓) means that the combination is perfect and building a frame to hold the gears will be easy.
* An x (**×**) means that the combination is imperfect and building a frame to hold the gears may be tricky, or the combination may not be strong enough to be usable.
* A question mark (**?**) means that I could not find any usable or reliable way to make that combination. See if you can find the missing solution!

table 8-2: possible combinations of LEGO gears

	8	12	16	20	24	36	40
8	✓	?	×	×	✓	×	✓
12		×	×	✓	×	✓	×
16			✓	×	×	×	×
20				×	×	×	?
24					✓	?	✓
36						×	?
40							✓

Now that you know how to choose gears that fit together, I'll show you some examples of how to build with gears. Think of this as your LEGO gear recipe book!

assembling gears

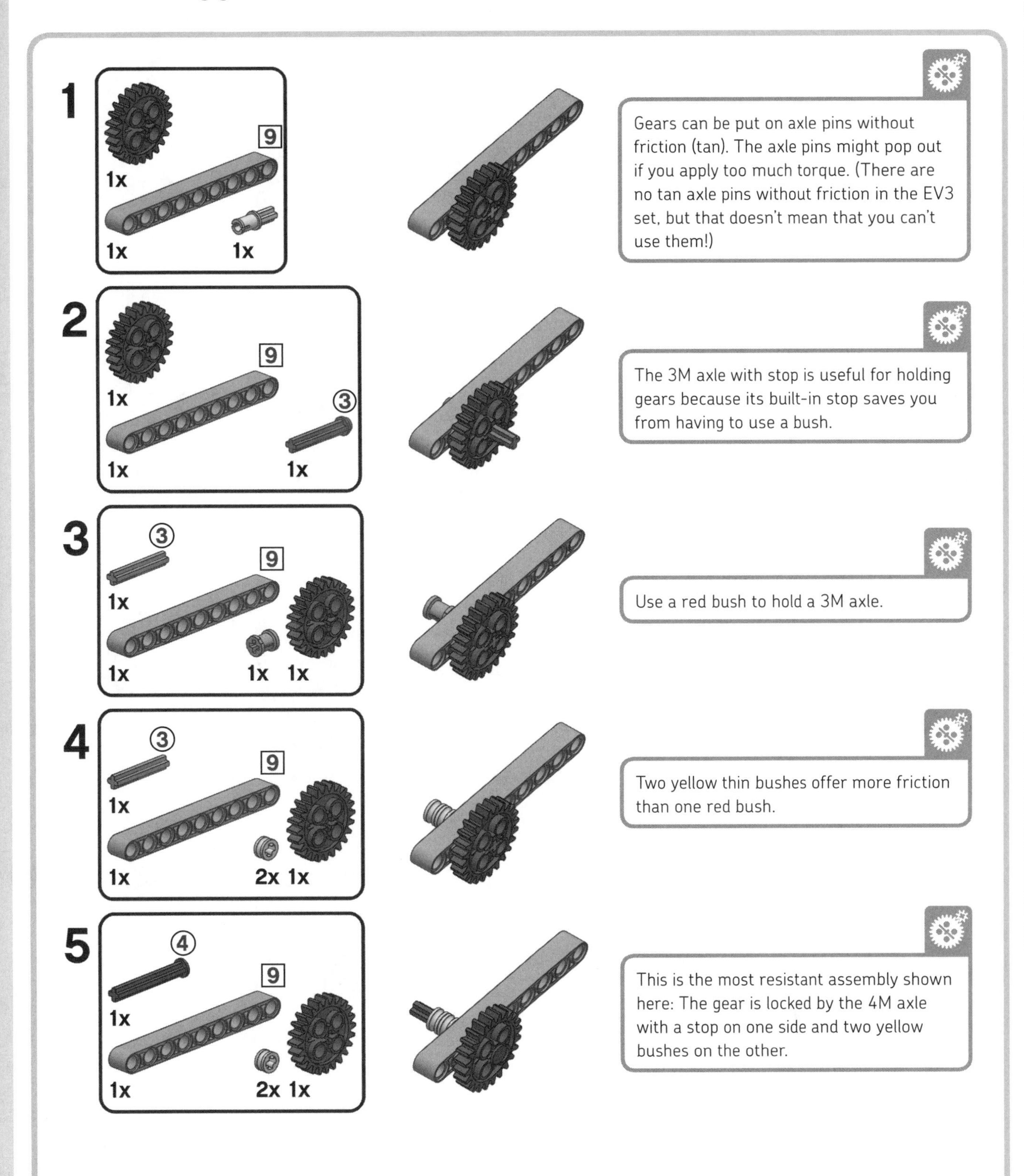

Gears can be put on axle pins without friction (tan). The axle pins might pop out if you apply too much torque. (There are no tan axle pins without friction in the EV3 set, but that doesn't mean that you can't use them!)

The 3M axle with stop is useful for holding gears because its built-in stop saves you from having to use a bush.

Use a red bush to hold a 3M axle.

Two yellow thin bushes offer more friction than one red bush.

This is the most resistant assembly shown here: The gear is locked by the 4M axle with a stop on one side and two yellow bushes on the other.

gear combinations

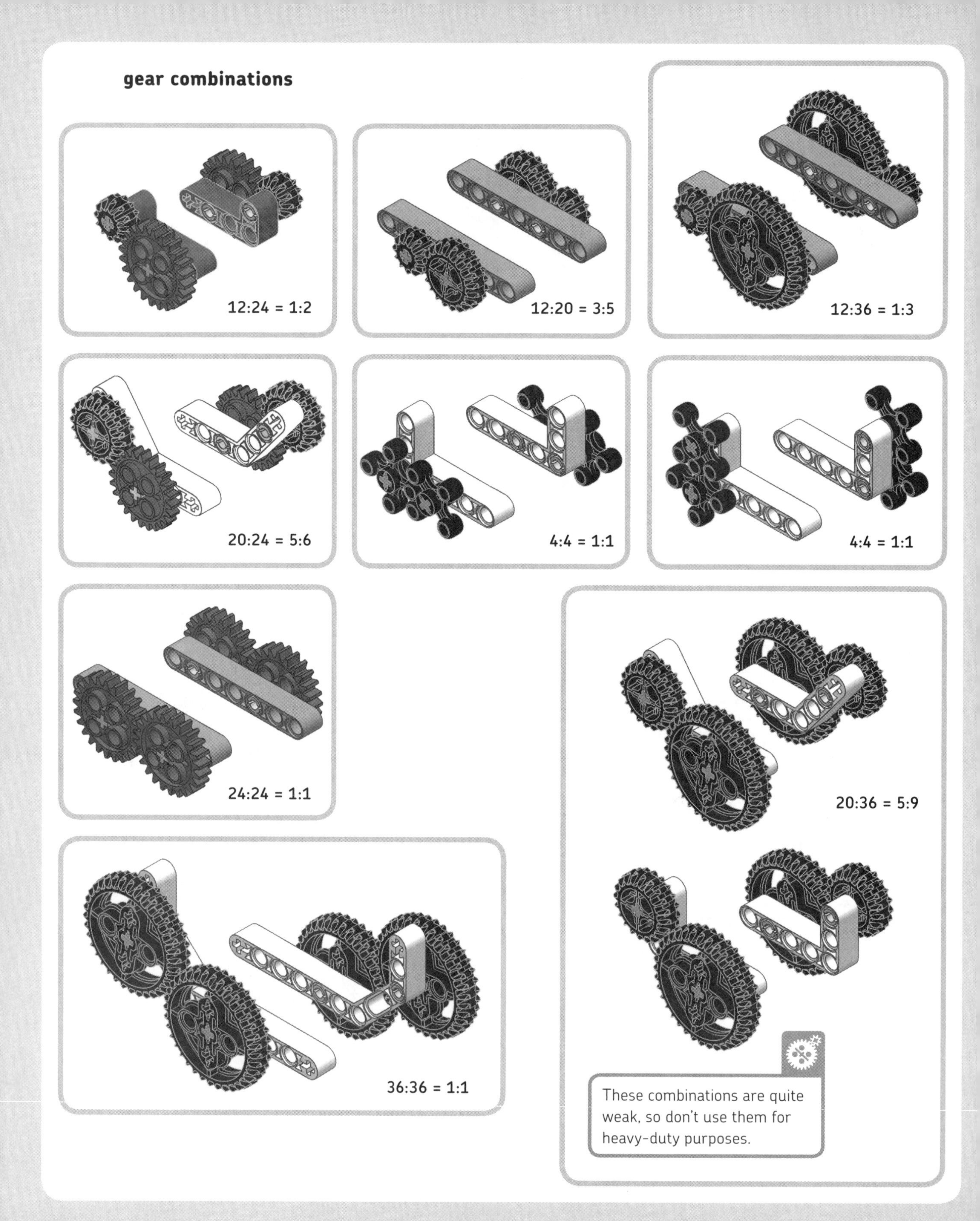

These combinations are quite weak, so don't use them for heavy-duty purposes.

90-degree-coupled gears

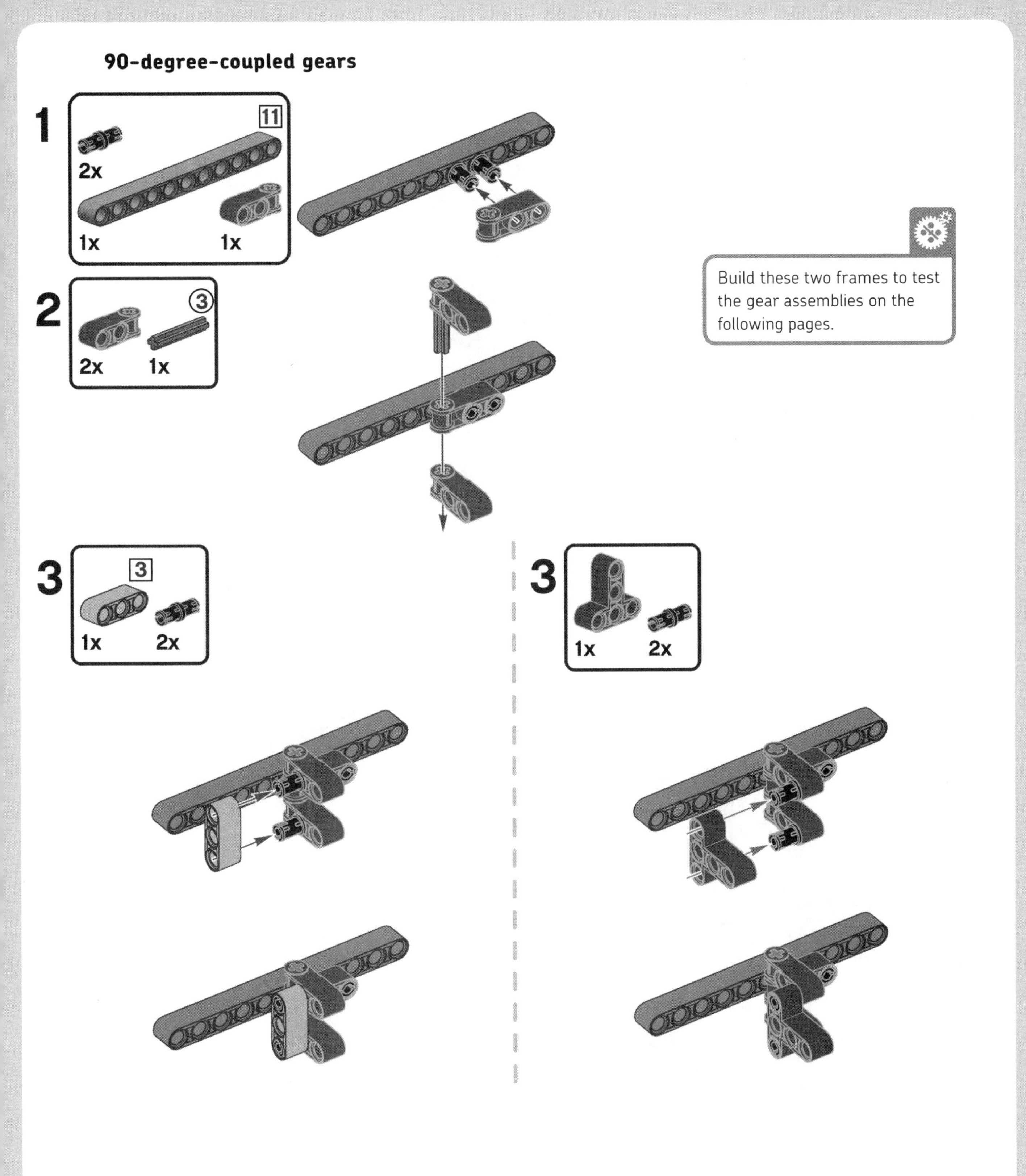

Build these two frames to test the gear assemblies on the following pages.

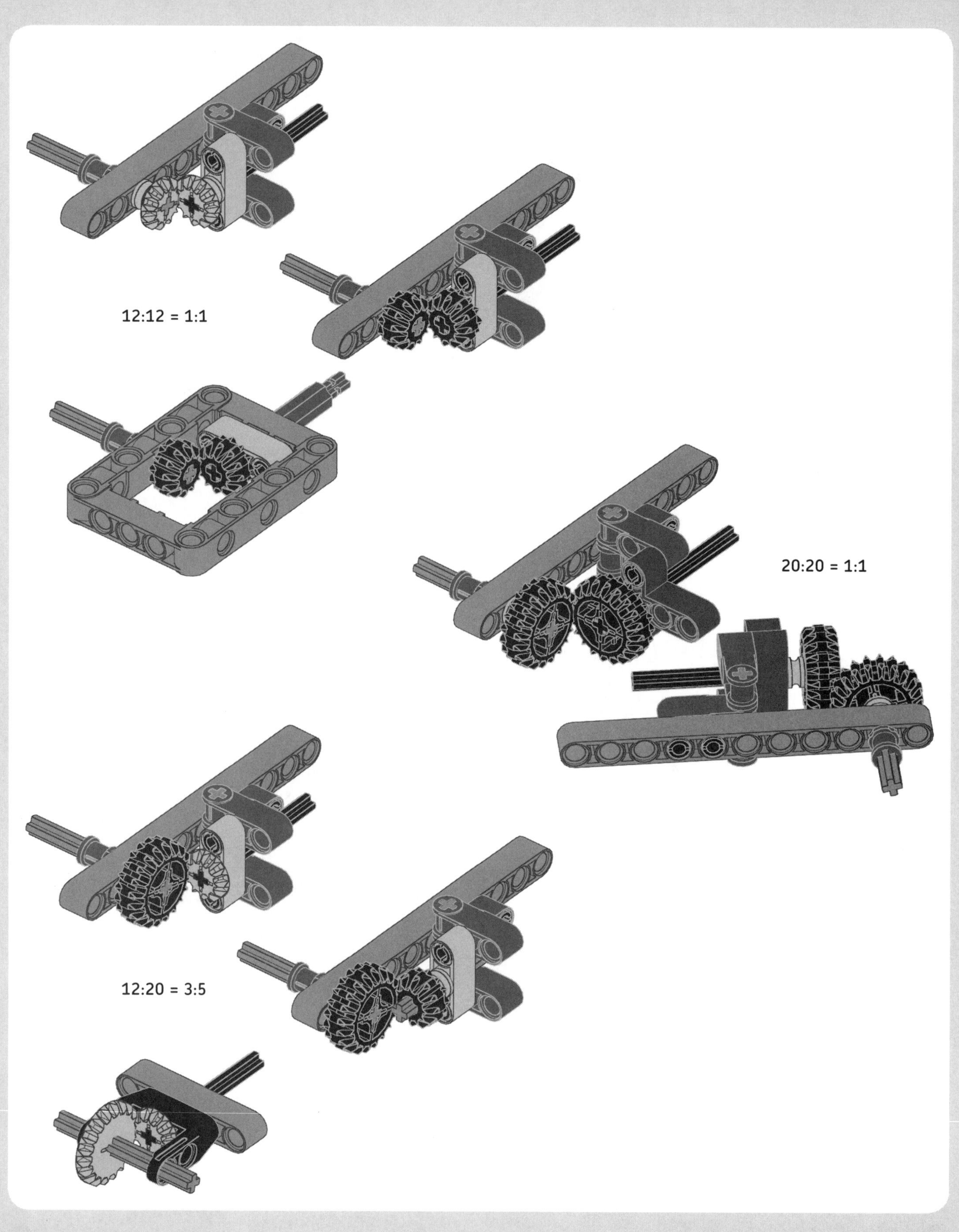
12:12 = 1:1
20:20 = 1:1
12:20 = 3:5

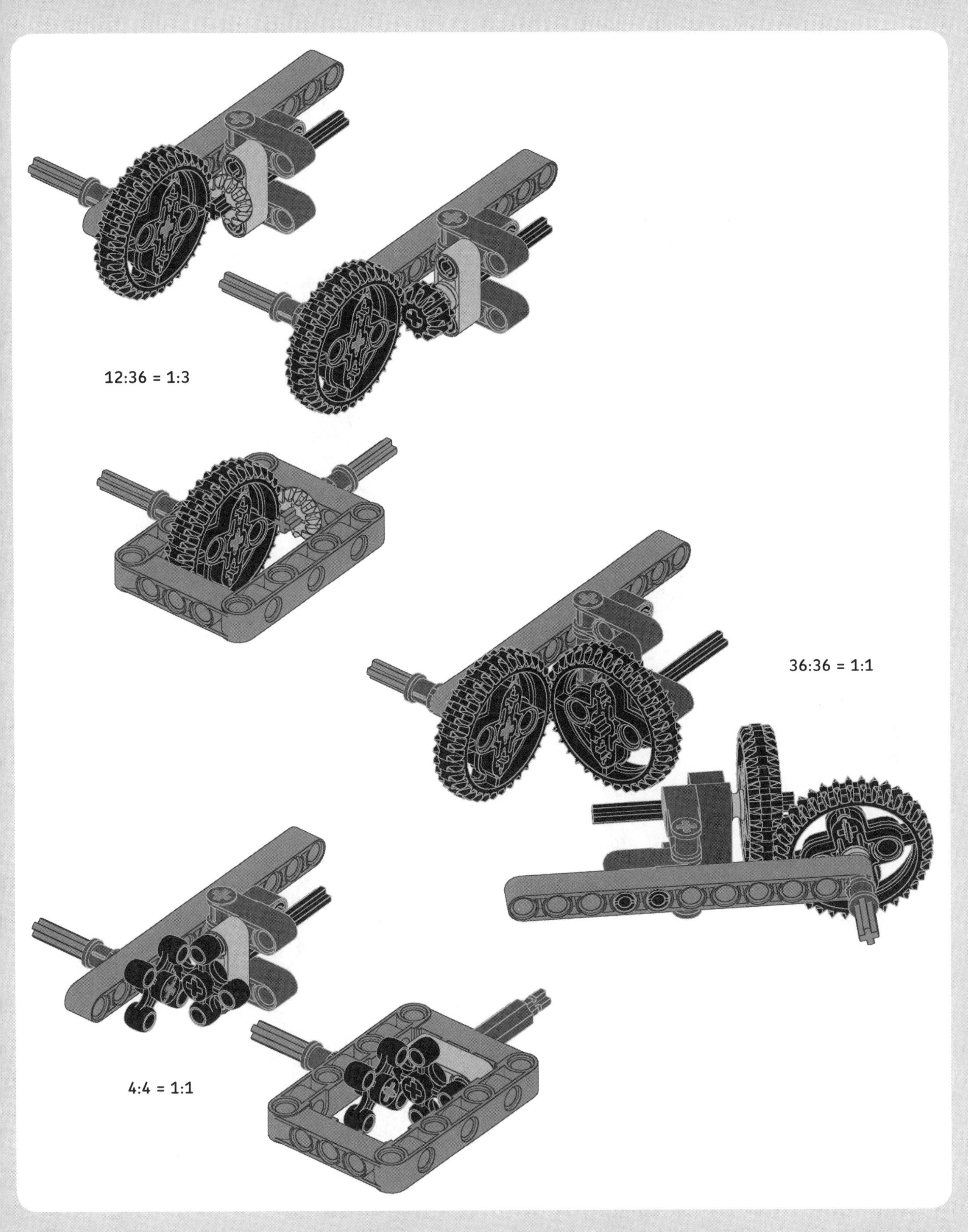
12:36 = 1:3
36:36 = 1:1
4:4 = 1:1

gear trains

You can combine multiple gears to build *gear trains*, essentially a combination of cascading gear couples. You'll find gear trains useful when you need to get higher gear ratios or transmit rotation over greater distances. An *idler gear* is one that is inserted between two or more gears to change the direction of rotation of the output axle without affecting the ratio of the gear train. The ratio depends only on the number of teeth of the input and output gears, as you can see in Figures 8-8 and 8-9. In Figure 8-10 you can see a gear train with a 1:5 ratio.

NOTE In general, if the gear train has an odd number of idler gears, the first and last gears turn in the same direction.

Idler gear
(12:20) × (20:12) = 12:12 = 1:1
Idler gear
(20:12) × (10:20) = 20:20 = 1:1

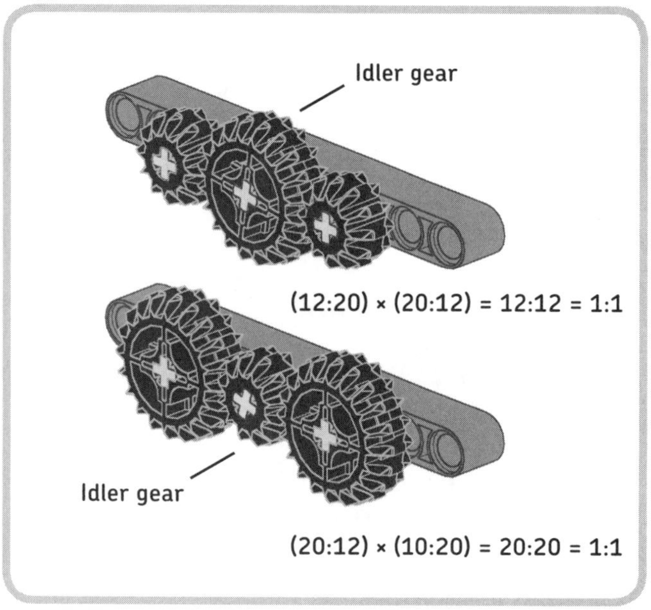

Figure 8-8: Gear train resulting in a 1:1 ratio

(12:36) × (36:12) = 12:12 = 1:1
(36:12) × (12:36) = 36:36 = 1:1

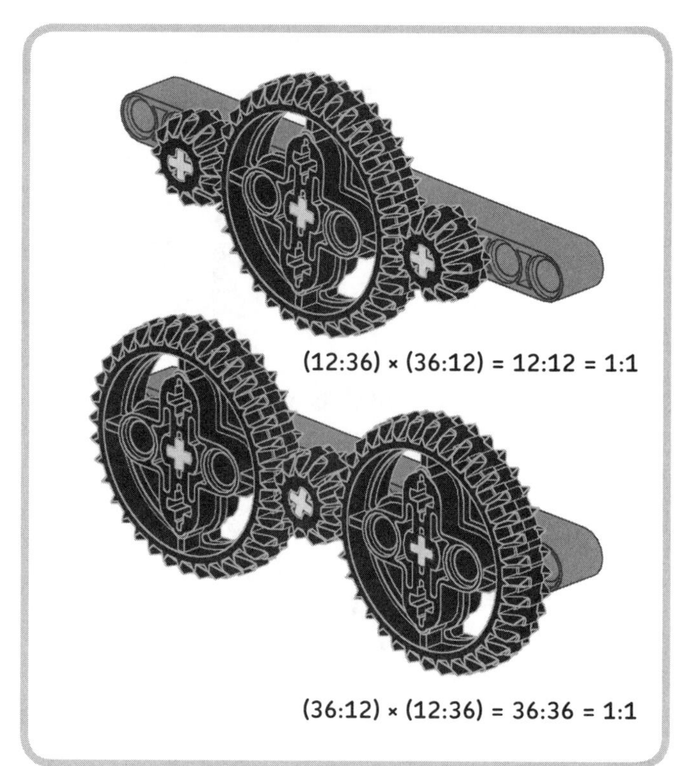

Figure 8-9: Gear train resulting in a 1:1 ratio

(12:36) × (12:20) = 1:5

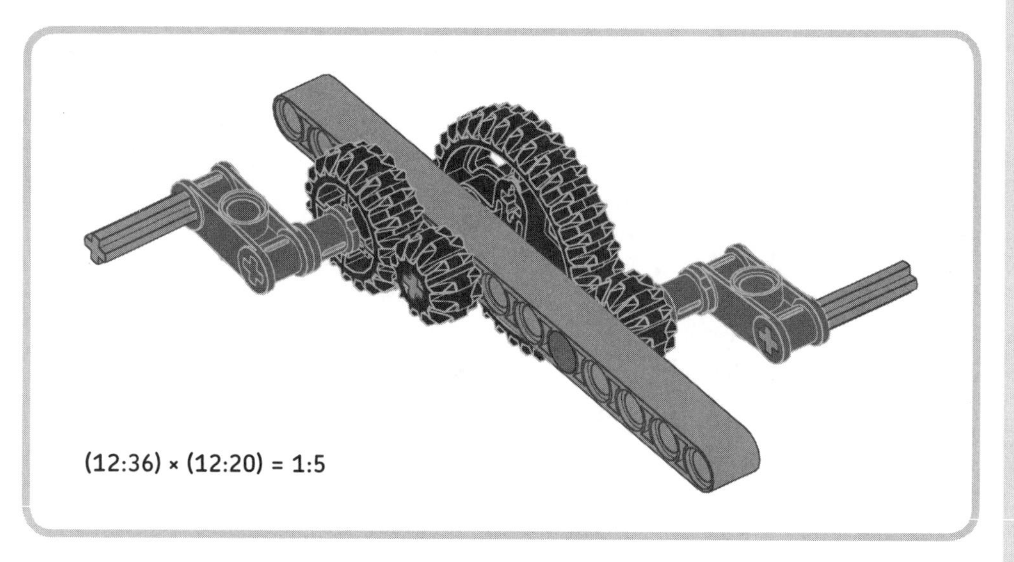

Figure 8-10: Gear train resulting in a 1:5 ratio

the worm gear

The worm gear looks like a kind of screw. When computing the ratio of a gear train that uses the worm gear, think of the worm gear as having just one tooth.

The worm gear is *self-locking*. This means that while turning, the worm gear makes the other gear turn, but you can't make the worm gear move by turning the other gear. (See Figure 8-11.)

In the next pages, you'll see two robust frames that you can use to hold worm gear–based assemblies. Notice how bracing prevents the worm gear from being pushed away when resisting torque is applied to the output axle.

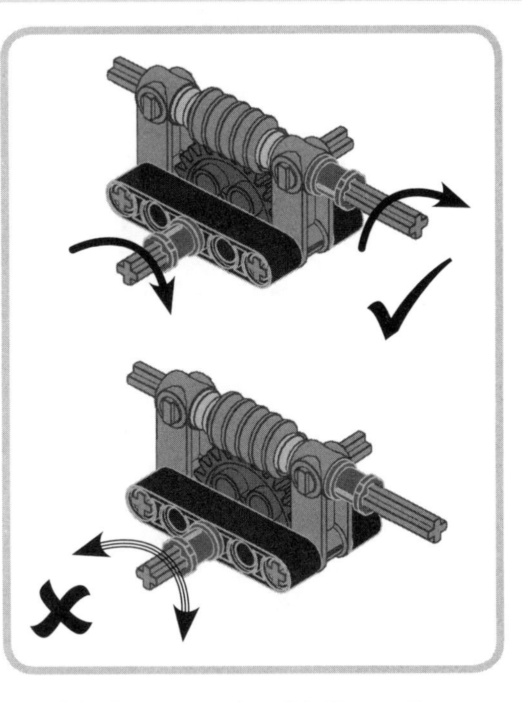

Figure 8-11: The worm gear is a self-locking gear. You cannot make it turn by driving the other gear.

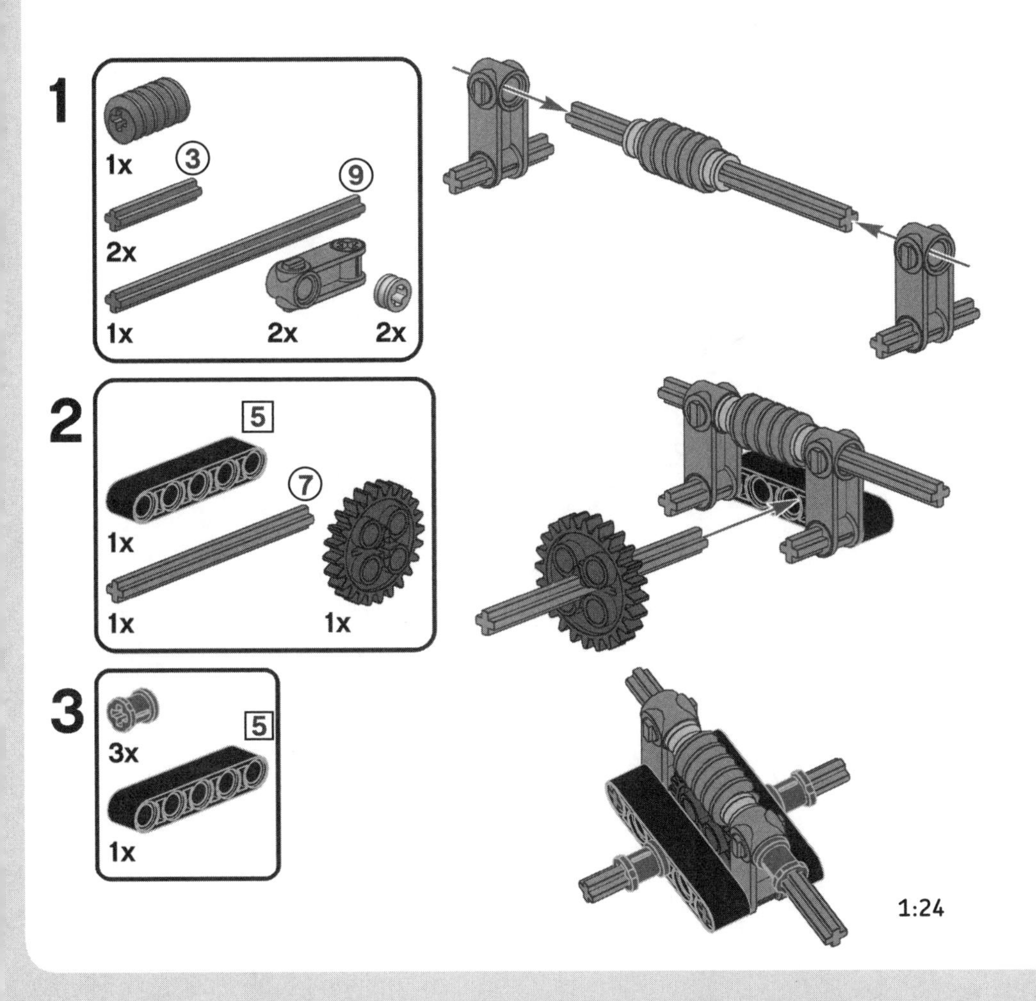

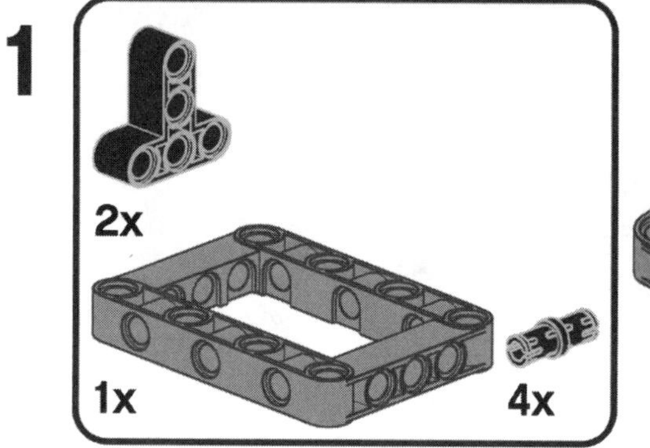

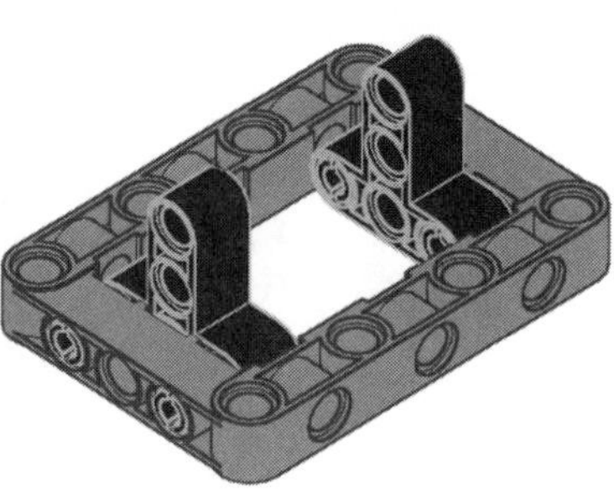

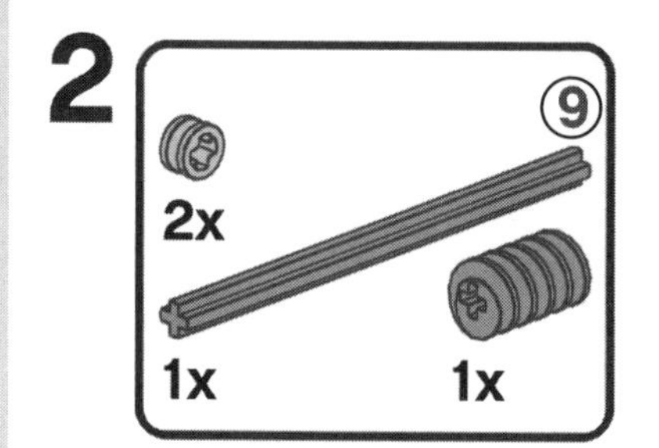

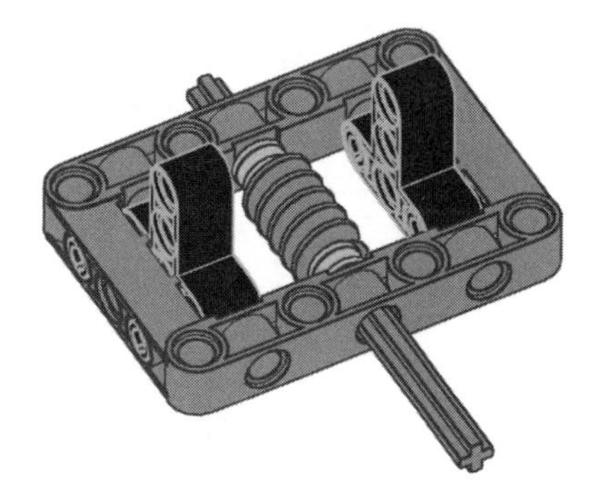

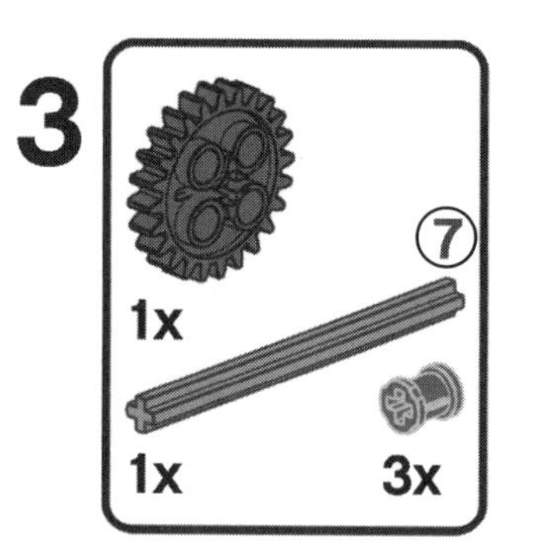

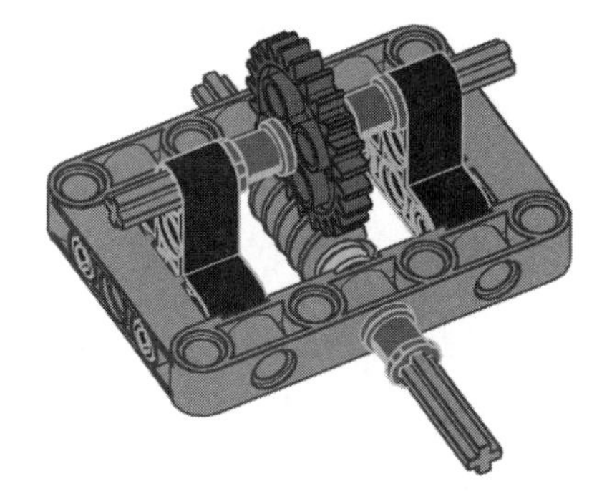

1:24

motion transformation

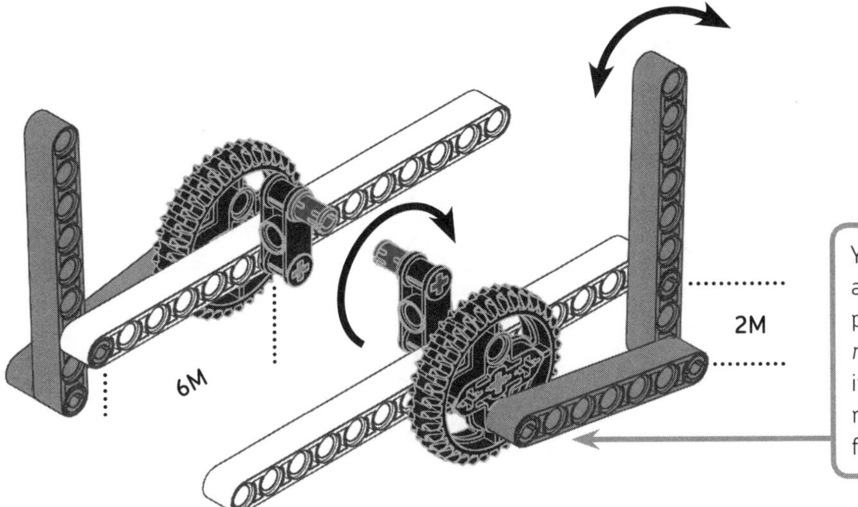

You can use a gear as a crank to make an eccentric mechanism by putting a pin in the off-center holes. An *eccentric mechanism* transforms circular motion into *reciprocating motion*—that is, a repetitive up-and-down or back-and-forth motion.

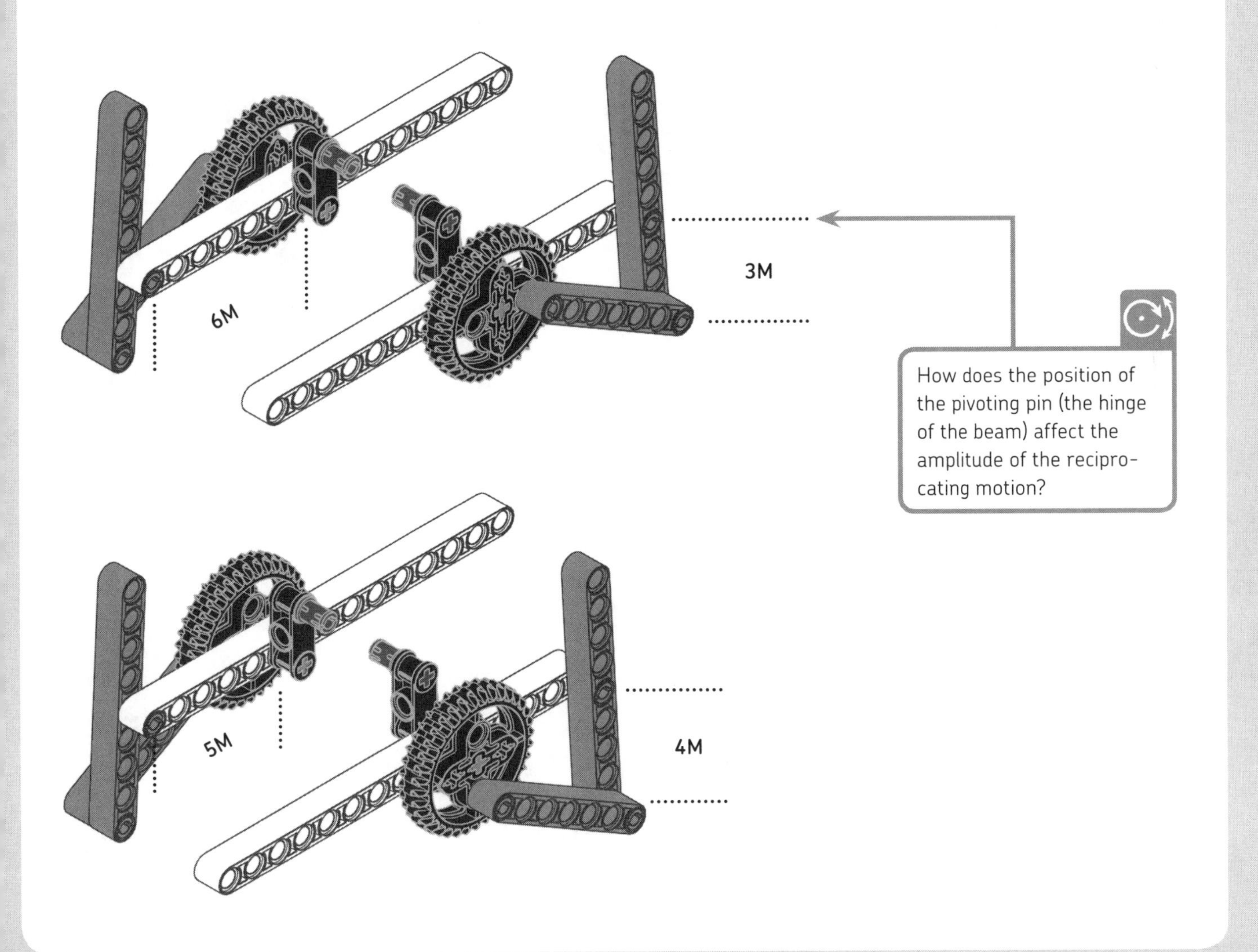

How does the position of the pivoting pin (the hinge of the beam) affect the amplitude of the reciprocating motion?

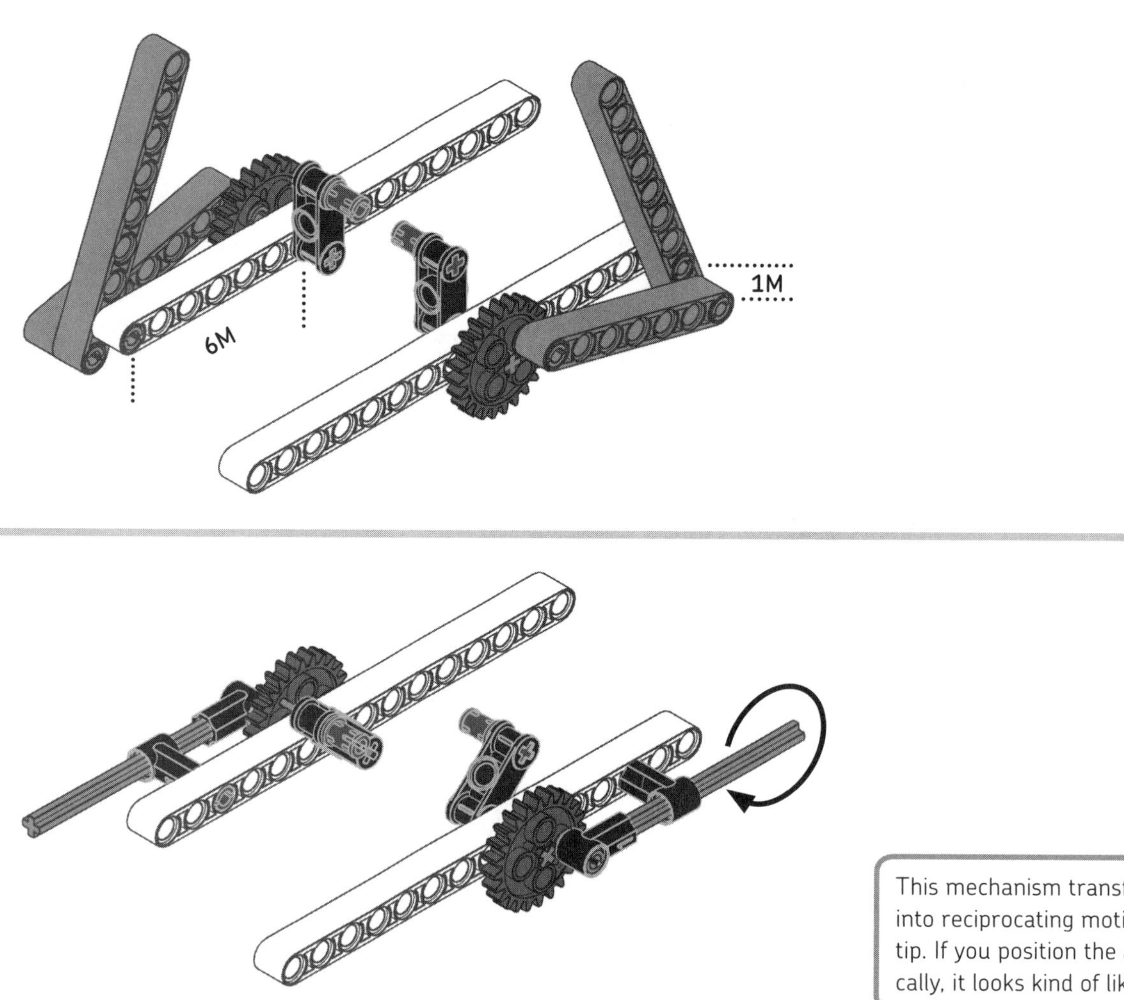

This mechanism transforms rotation into reciprocating motion at the axle's tip. If you position the assembly vertically, it looks kind of like a leg.

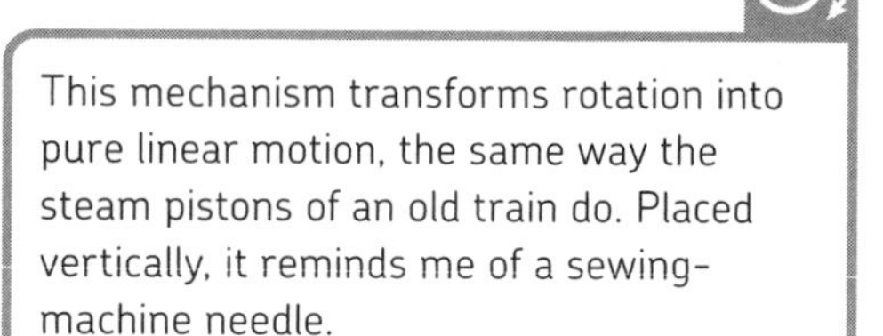

This mechanism transforms rotation into pure linear motion, the same way the steam pistons of an old train do. Placed vertically, it reminds me of a sewing-machine needle.

building ideas for the motors

The EV3 set contains two Large Motors and a Medium Motor. Since these motors are the core of your robots, this section lists some useful modules and assemblies to build motors inside your robots.

medium motor with front output #1

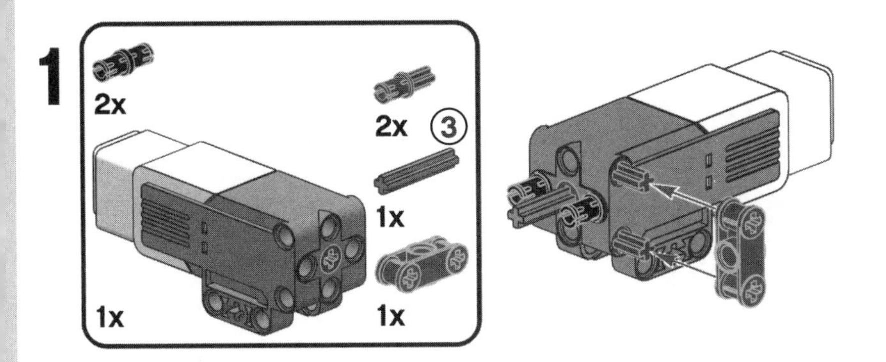

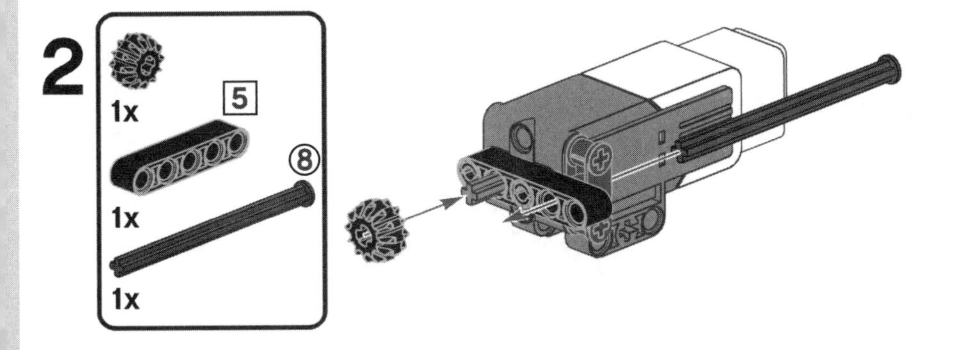

medium motor with front output #2

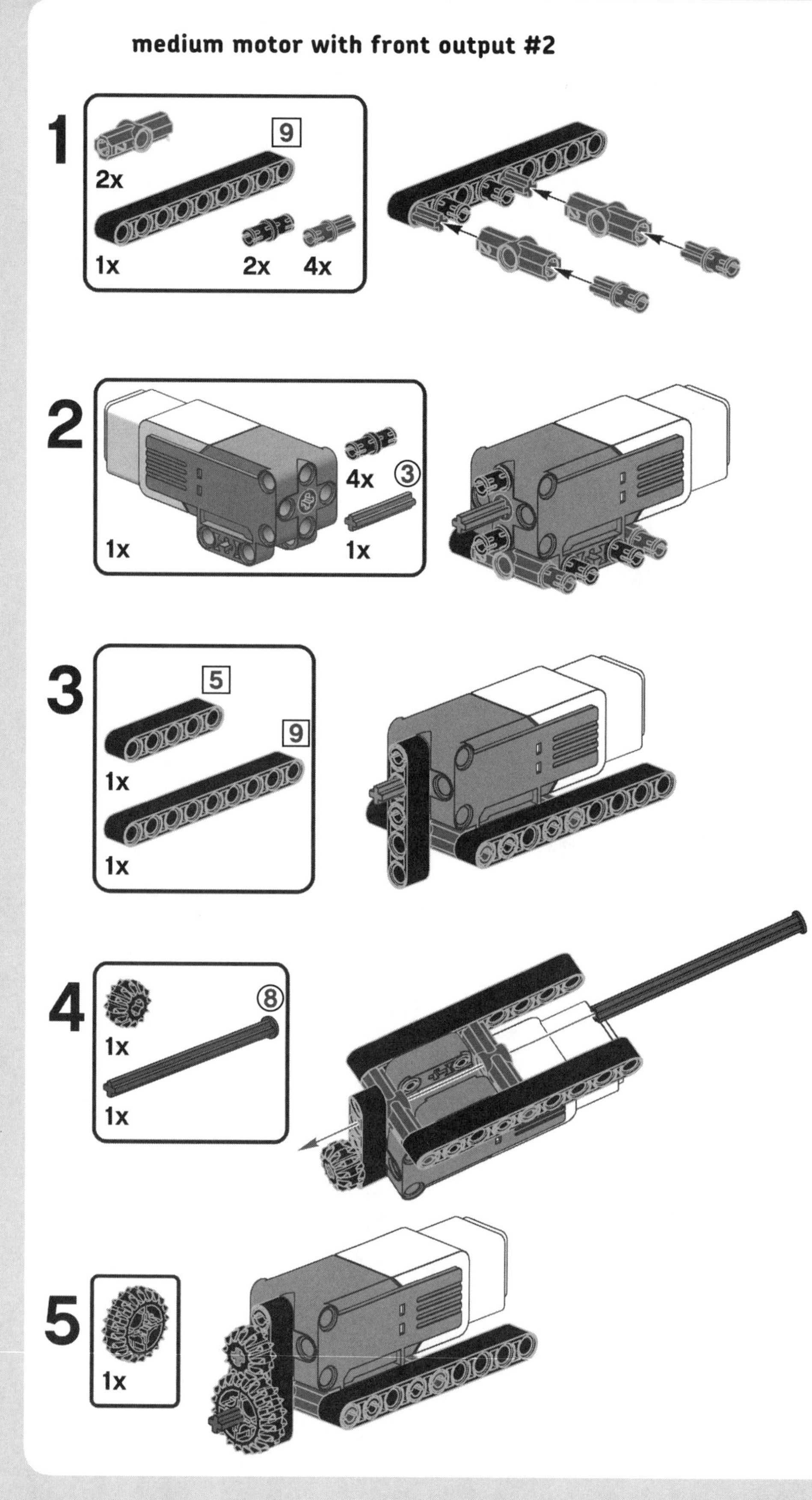

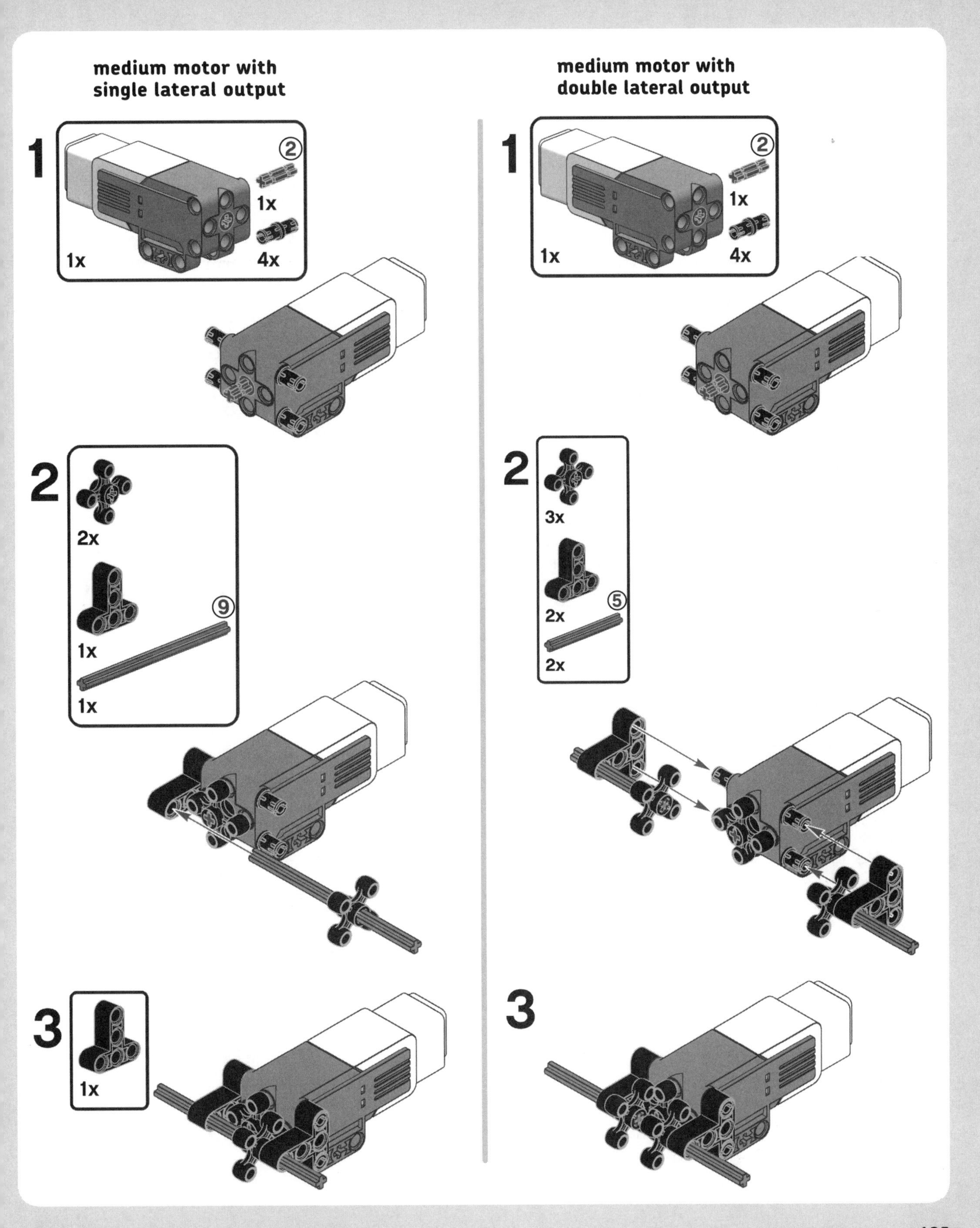
medium motor with
single lateral output
1
2
1x
4x
1x
2
2x
1x
9
1x
3
1x
medium motor with
double lateral output
1
2
1x
4x
1x
2
3x
2x
5
2x
3

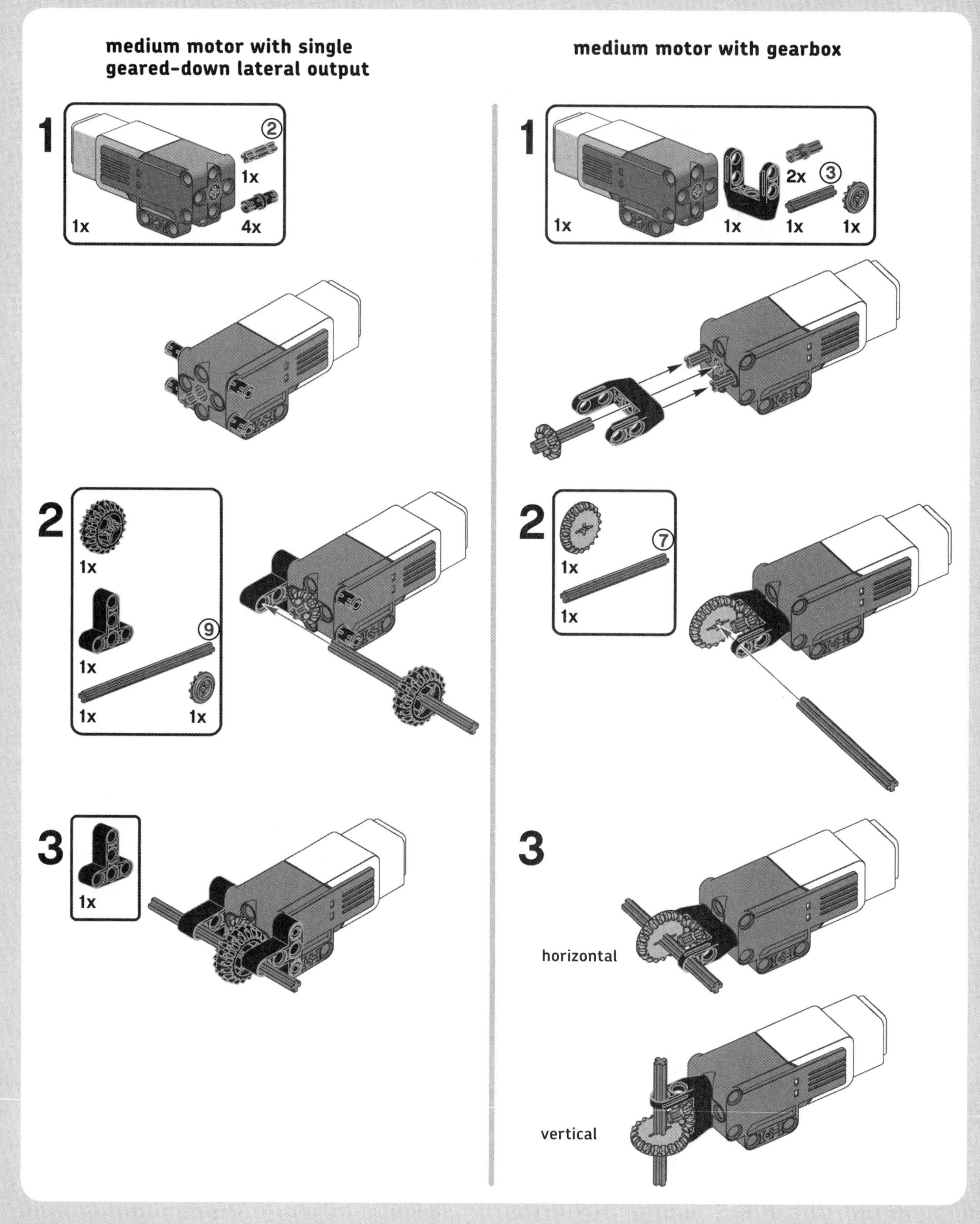
medium motor with single geared-down lateral output
1
1x
1x
4x
2
2
1x
1x
1x
1x
9
3
1x
medium motor with gearbox
1
1x
1x
2x
1x
1x
3
2
1x
1x
7
3
horizontal
vertical

medium motor with multiple outputs

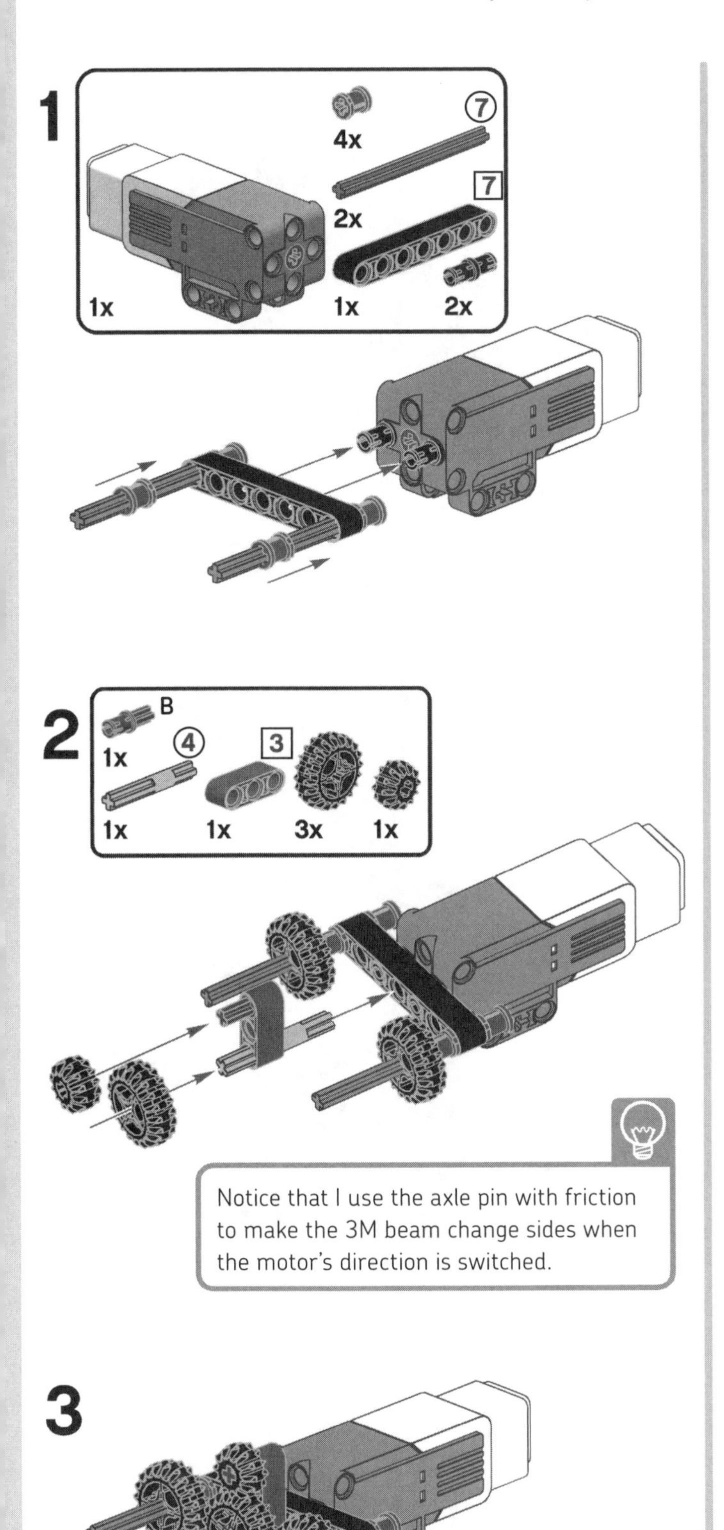

Notice that I use the axle pin with friction to make the 3M beam change sides when the motor's direction is switched.

By switching the motor direction, you can drive two different axles.

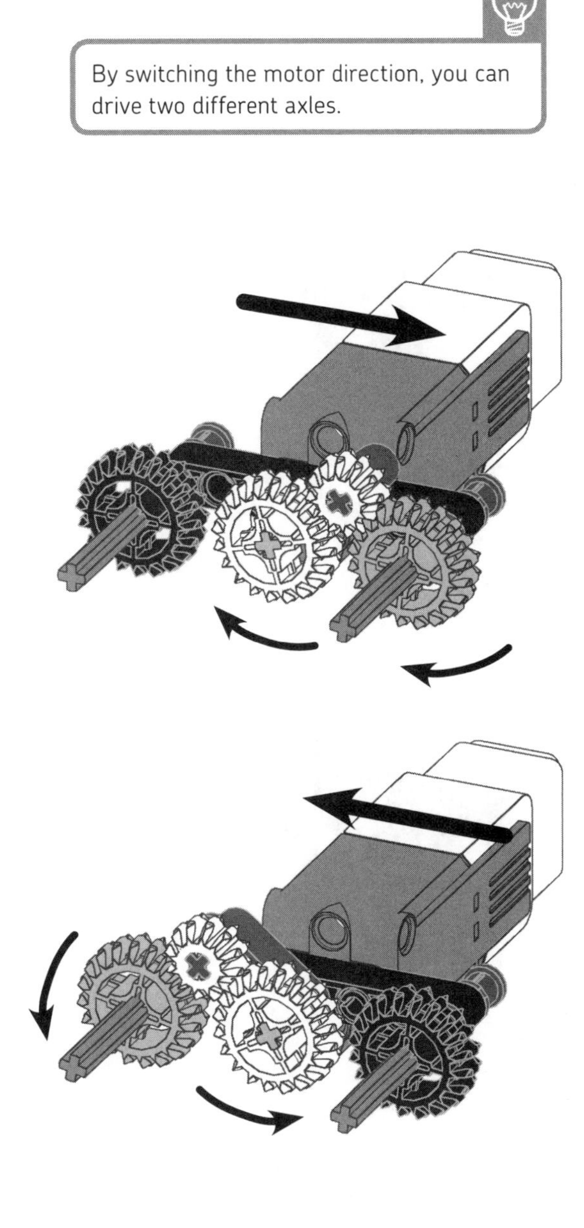

large motor with horizontal output

large motor gearing options

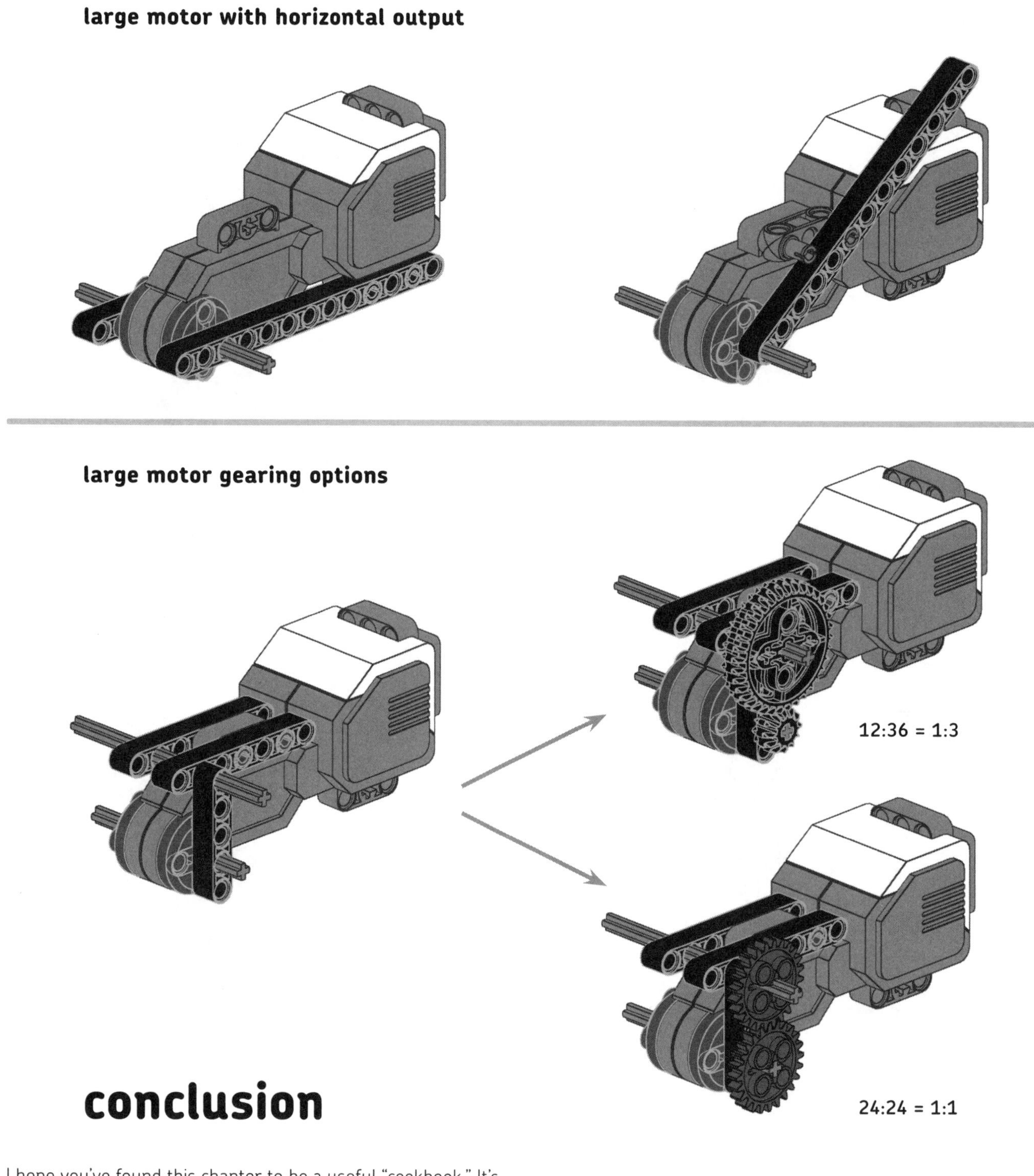

conclusion

I hope you've found this chapter to be a useful "cookbook." It's given you some basic knowledge about LEGO geometry and shown you how to build sturdy structures and functional gear trains. You've also seen some ideas for modules to transmit motion and transform motion, as well as some sample Servo Motor assemblies to include in your amazing creations!

SO, HOW WAS IT IN THE ***DEAD*** SECTOR? IS THE ***EMP*** SYSTEM ENGAGED?
YES, DOC! BUT YOU COULD HAVE TOLD ME THAT IT WAS ***PITCH BLACK!***
SURE, BUT THAT WOULDN'T BE HALF AS FUN! WAS ROV3R USEFUL TO YOU?
YES, I COULDN'T HAVE MADE IT WITHOUT ITS HELP!
I HAVE A NEW MISSION FOR YOU. GO AND CATCH MY WATCHGOOZ3!
SOUNDS... OMINOUS...
DON'T WORRY, IT'S HARMLESS. THE ROBOT JUST NEEDS A TUNE-UP. TURN IT OFF AND BRING IT HERE.
I HOPE THE BOY HAS GUTS!
08X1

building WATCHGOOZ3

Goose on the loose! In this chapter, you'll build WATCHGOOZ3 (Figure 9-1), a robot goose that can patrol and guard a room while moving on only two legs. This robot is inspired by a creation of Bruno Zarokian.

The mechanics of this robot are designed so that you can program the robot to walk, turn, and avoid obstacles with just the Brick Program App (see Chapter 10). No computer required! In this chapter I'll describe the key design ideas that allow us to simplify the software so that the limited Brick Program App is sufficient to program the robot. In addition to the building instructions, you'll find building techniques for working with structures, gears, and motor assemblies. You'll learn how to make sturdy, braced structures that hold gears; how to gear down a motor to increase its torque; and how to build parallelogram linkages.

how does WATCHGOOZ3 walk?

A creature that walks on two legs is called a *biped*. Humans are bipeds, and so are kangaroos, some primates, dinosaurs like *Tyrannosaurus rex* and velociraptors, and birds like ostriches (all the time) and geese (when not flying).

For the biped to maintain *static equilibrium* while walking on two legs (that is, to avoid falling), the projection of its *center of mass (COM)* must always lie inside the *support area* (the feet). Imagine the center of mass as the only spot where all of an object's mass is concentrated. An imaginary vertical line, always forming a 90-degree angle with the walking surface (like a string with a weight hung at its end), connects the center of mass to its *projection* on the ground. When a biped lifts a foot [Figure 9-2(c)], the support area decreases.

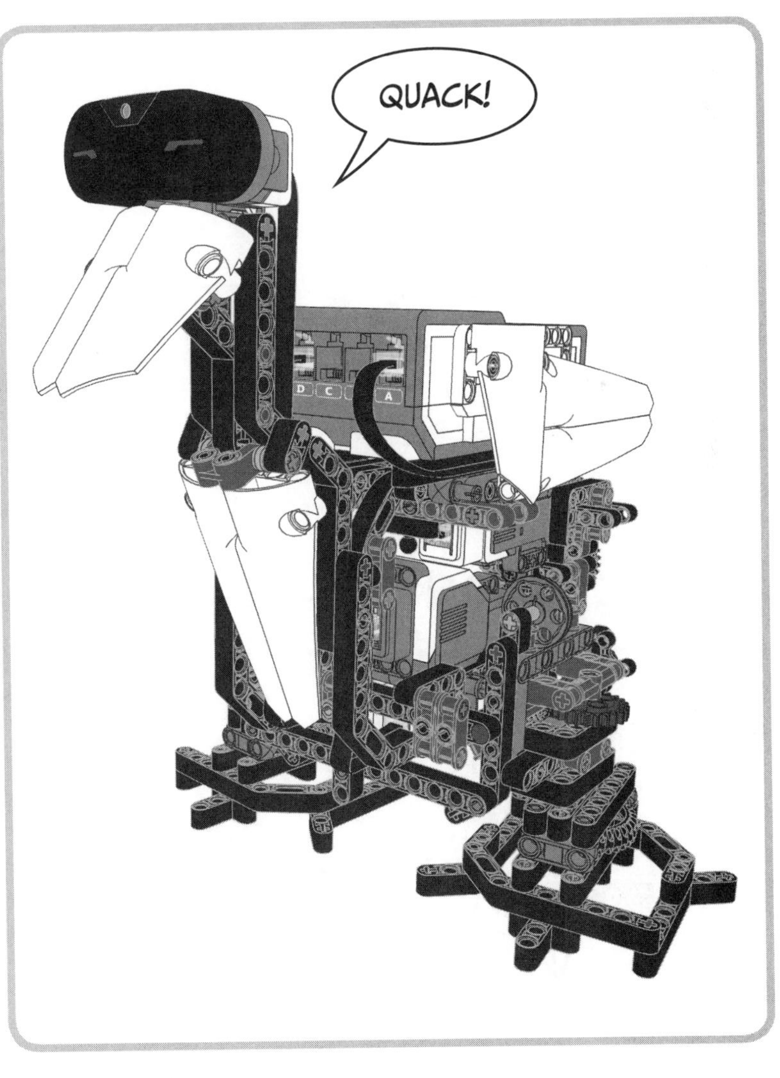

Figure 9-1: WATCHGOOZ3

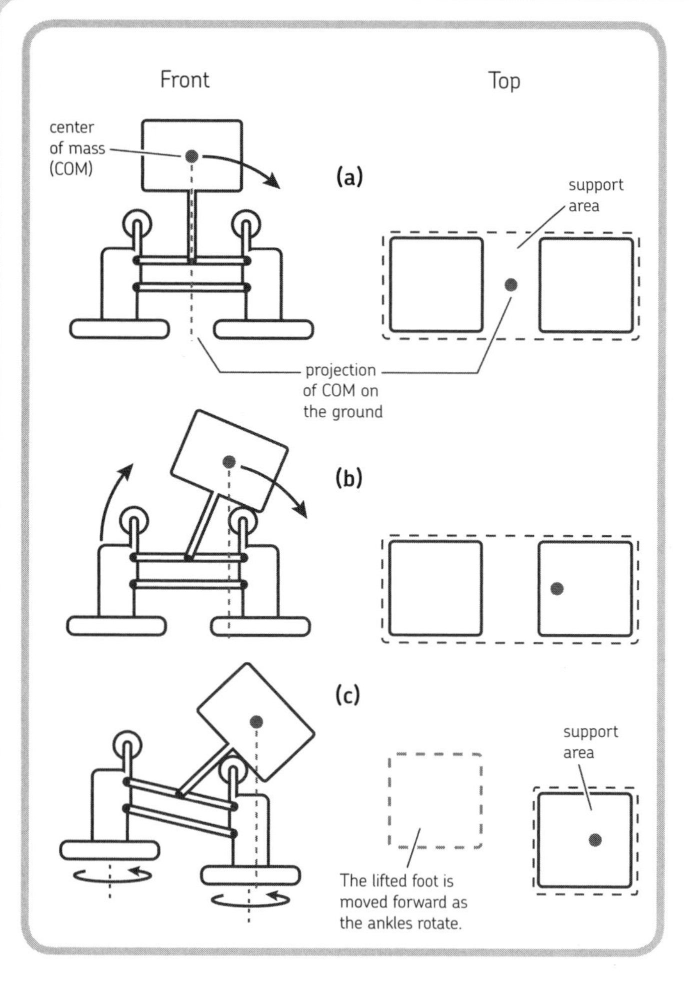

Figure 9-2: The workings of a weight-shifting biped robot

If the projection of the center of mass is not within the support area of the foot that's still on the ground, the biped falls over.

To move the center of mass inside the support area of its standing foot, a biped will naturally lean to one side. Figure 9-2 shows how WATCHGOOZ3 maintains static equilibrium while walking. As you can see, it shifts the weight of the EV3 Brick to the foot that remains on the ground (a); then the swinging frame touches a rolling wheel (b), unloading the other foot and lifting it with a parallel linkage. The robot takes steps by turning its feet, pivoting on the foot on the ground (c). WATCHGOOZ3 is thus a *weight-shifting biped robot*.

Because WATCHGOOZ3 walks with alternating weight-shifting and stepping actions, it has a kind of swinging gait. When the robot sees an object, it avoids it by turning the foot on the ground more than usual. In fact, the foot on the ground keeps turning until the object is no longer in its sight, at which point the robot shifts its weight to the other side so it can continue walking straight.

right leg assembly

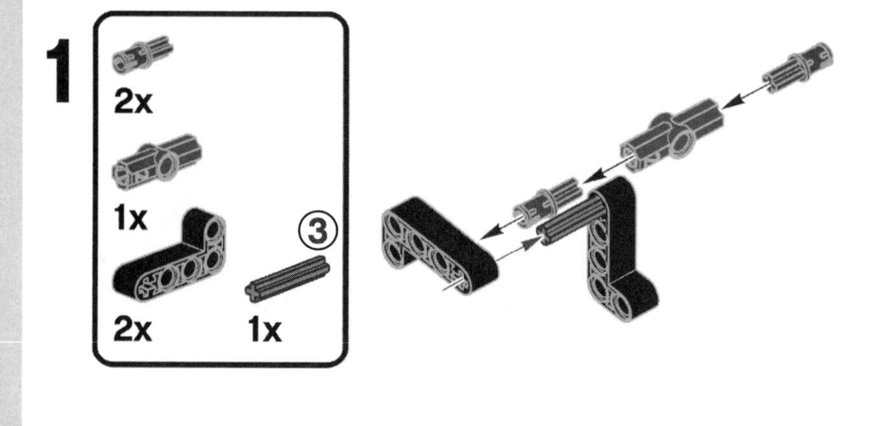

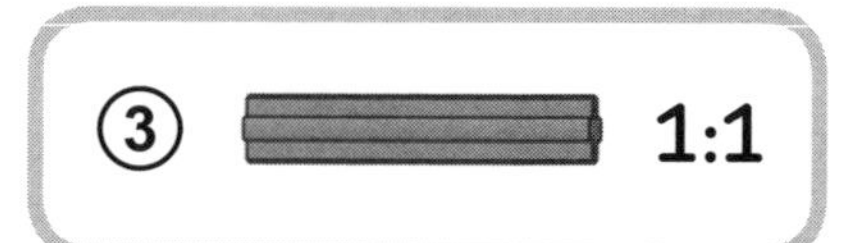

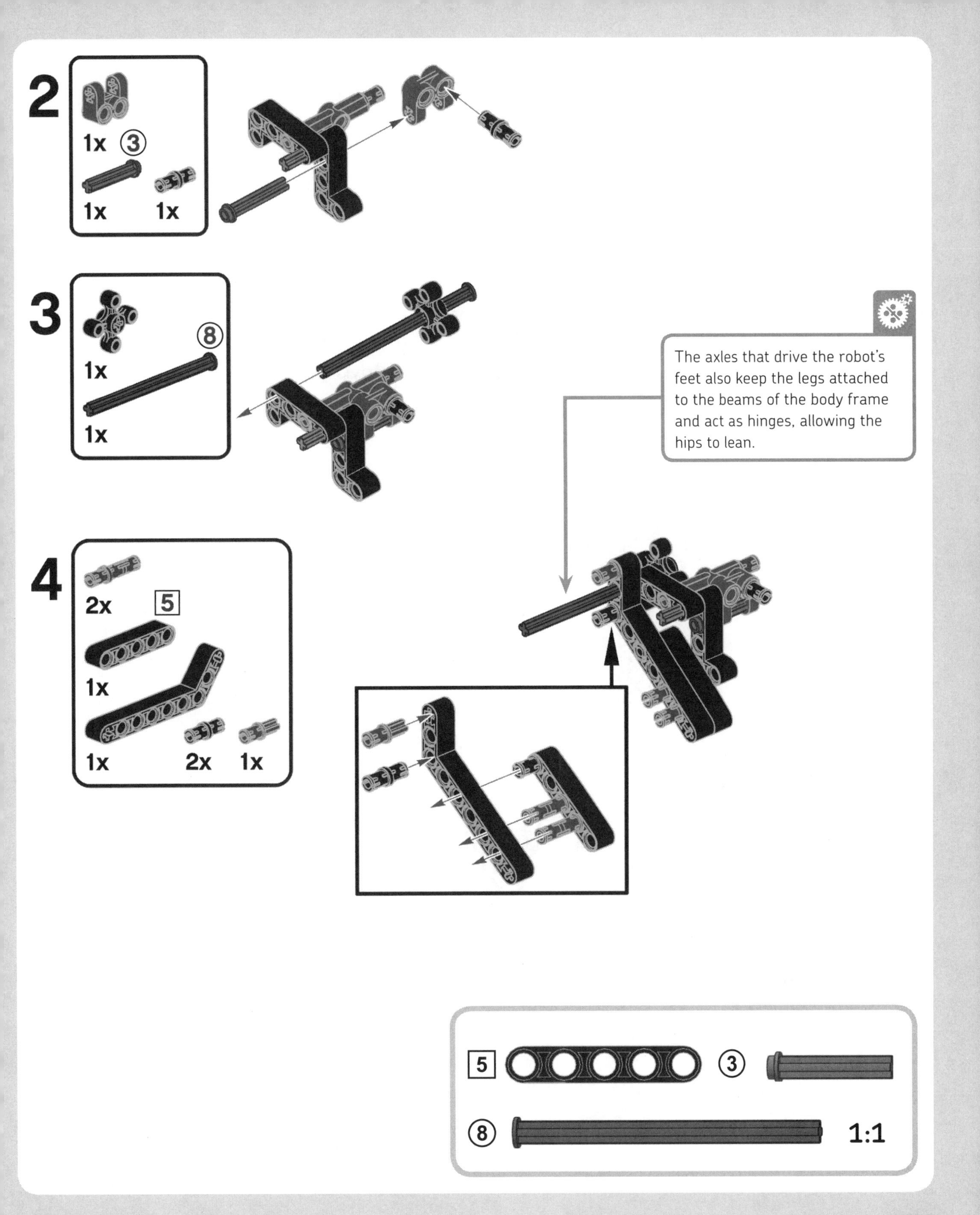
2
1x 3
1x 1x
3
8
1x
1x
The axles that drive the robot's feet also keep the legs attached to the beams of the body frame and act as hinges, allowing the hips to lean.
4
2x 5
1x
1x 2x 1x
5
3
8
1:1

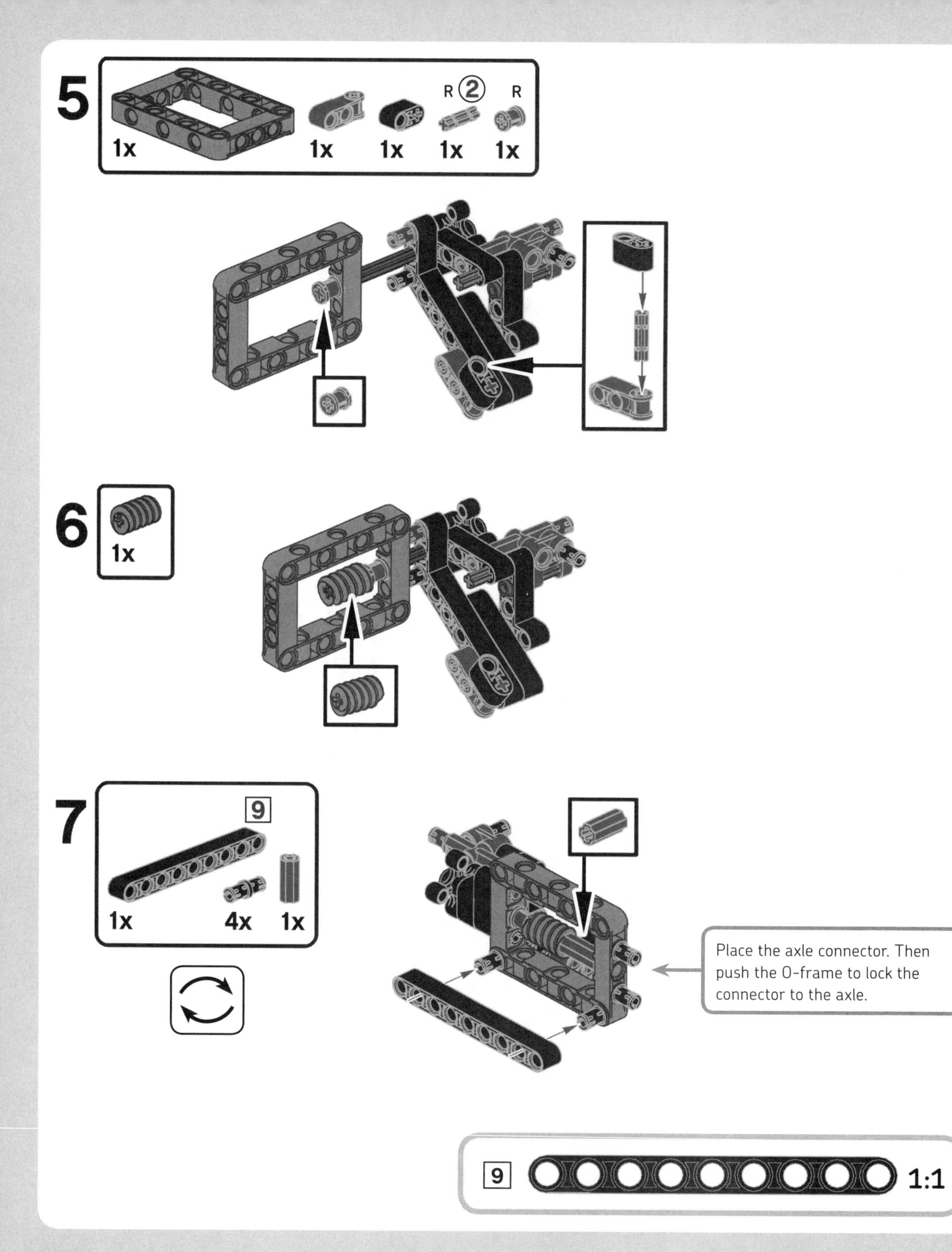
5
R
2
R
1x
1x
1x
1x
1x
6
1x
7
9
1x
4x
1x
Place the axle connector. Then push the O-frame to lock the connector to the axle.
9
1:1

8

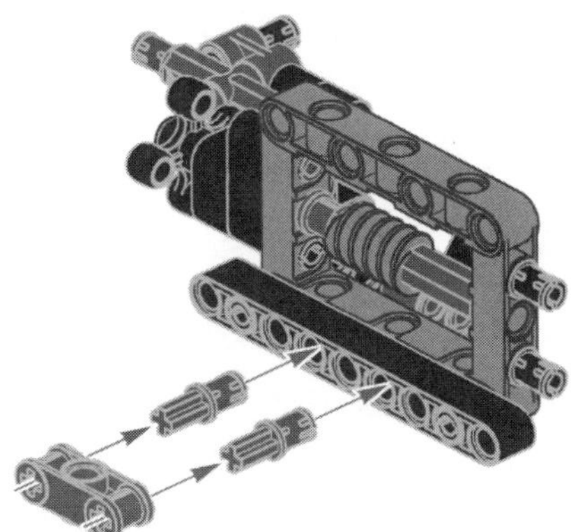

9

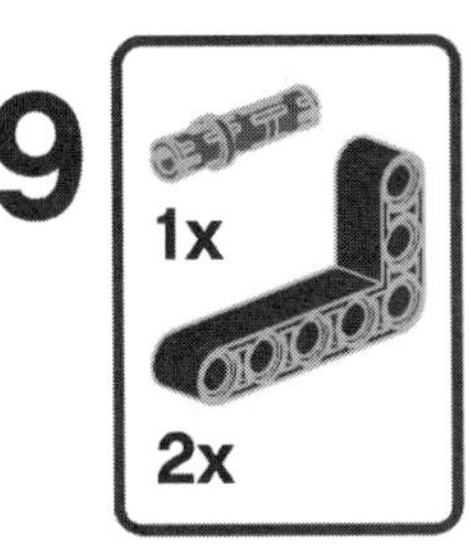

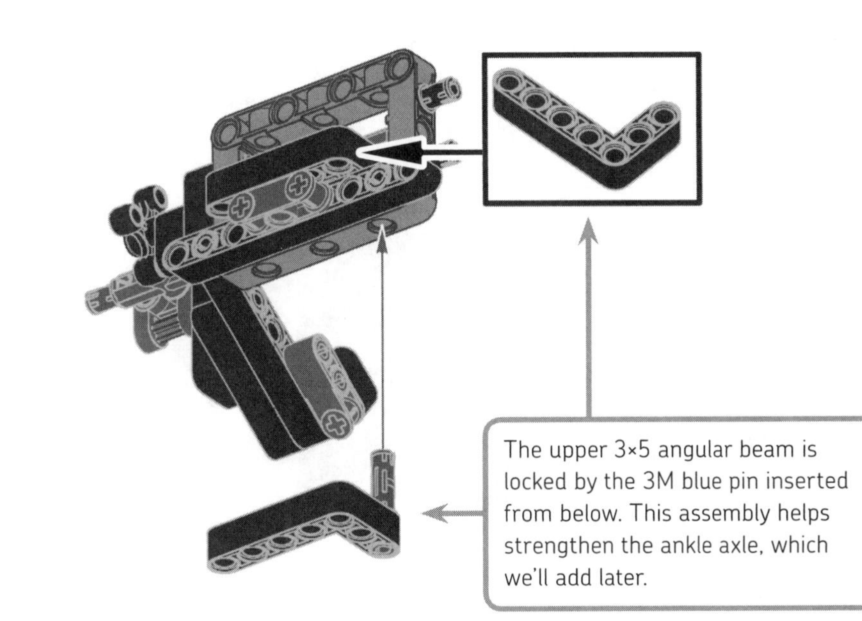

The upper 3×5 angular beam is locked by the 3M blue pin inserted from below. This assembly helps strengthen the ankle axle, which we'll add later.

10

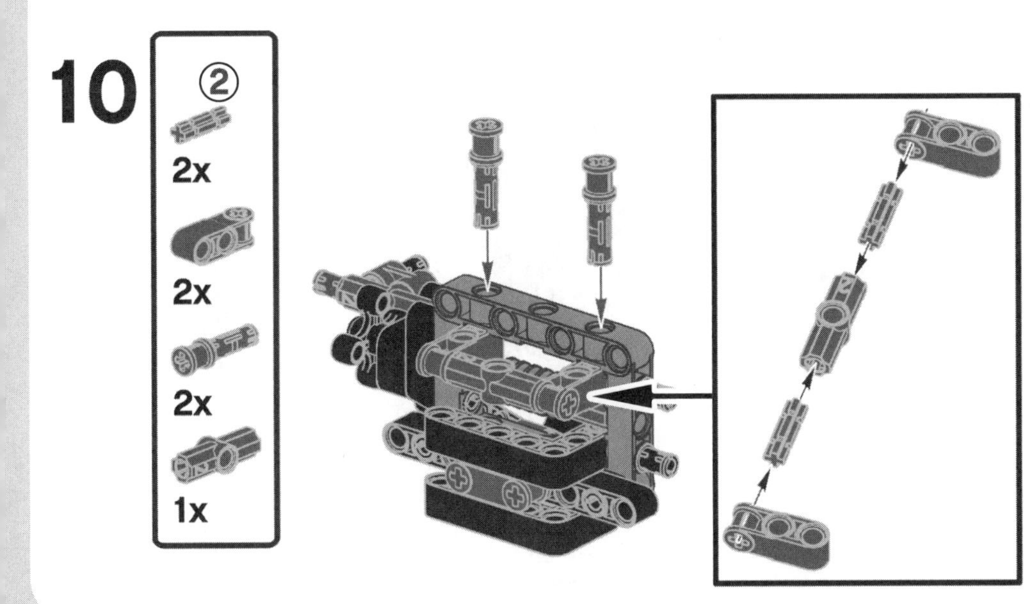

11

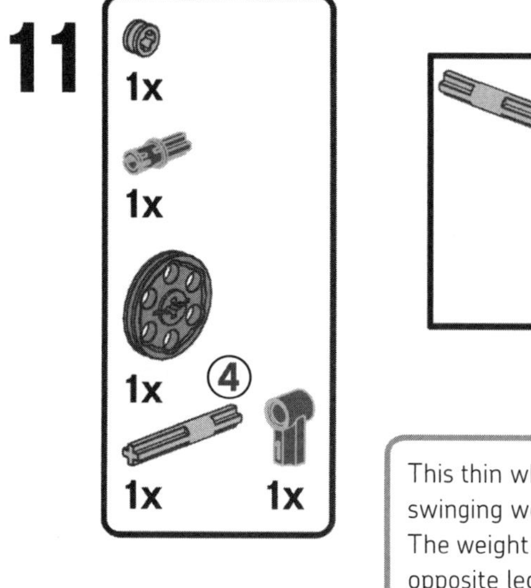

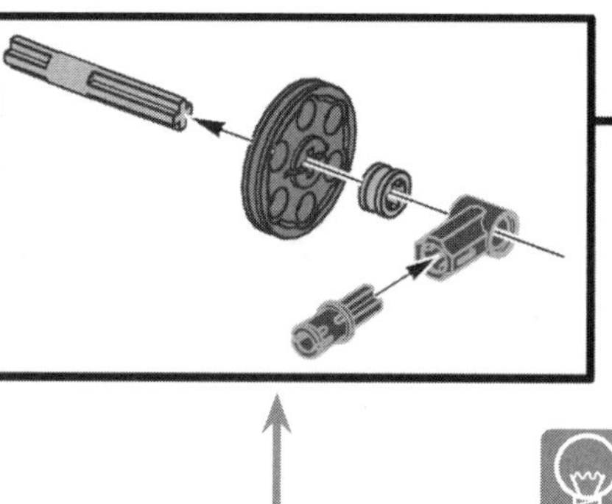

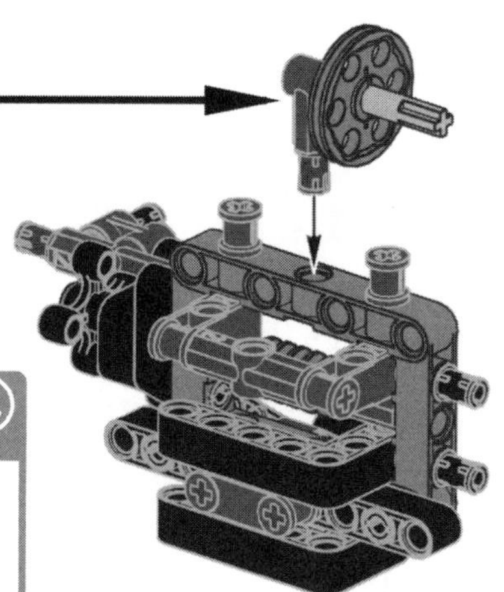

This thin wheel will support the robot's swinging weight when it leans to the side. The weight of the EV3 Brick will lift the opposite leg from the ground.

12

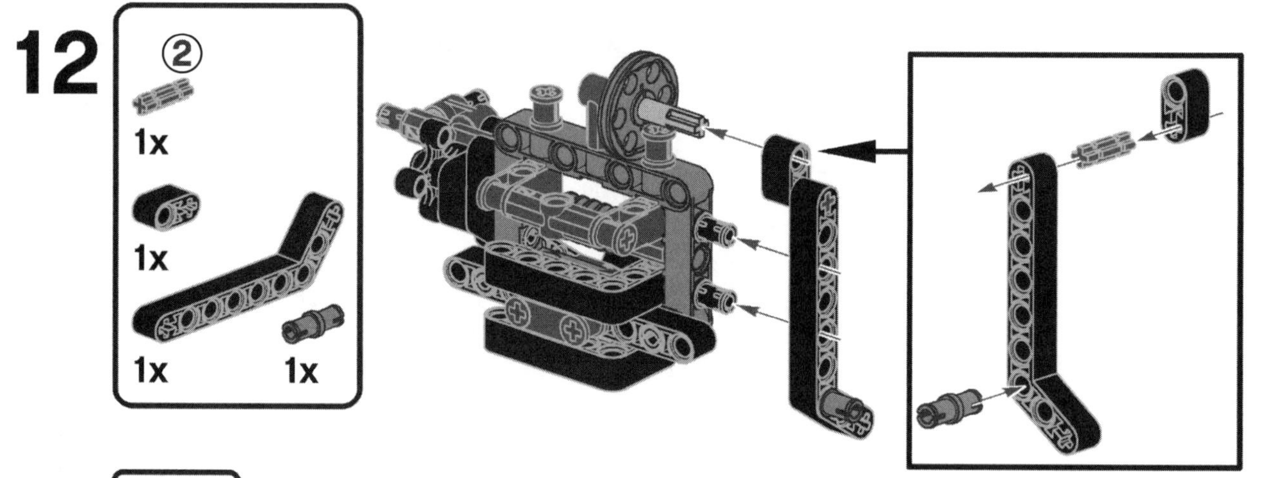

13

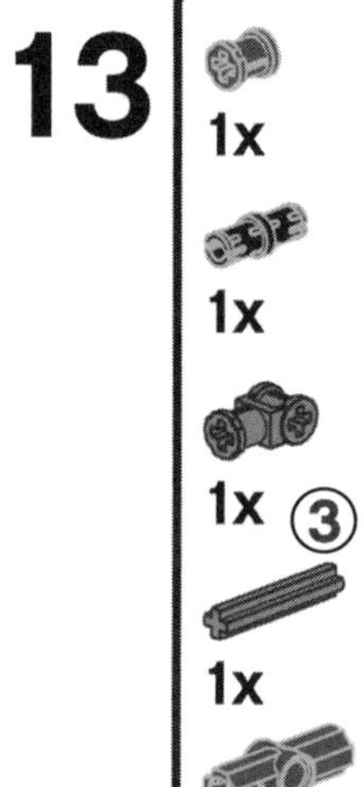

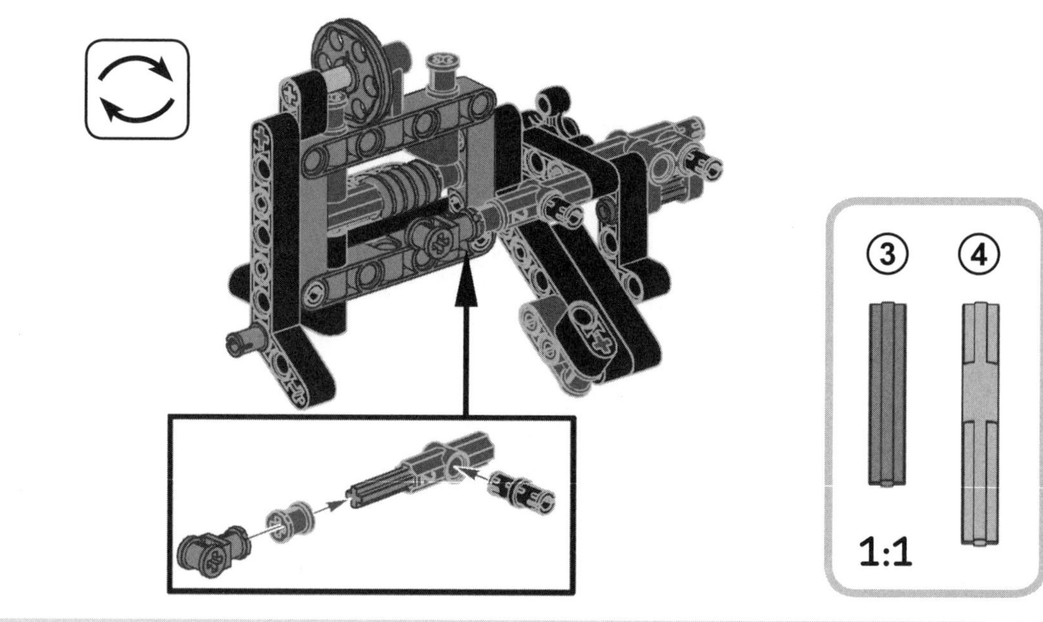

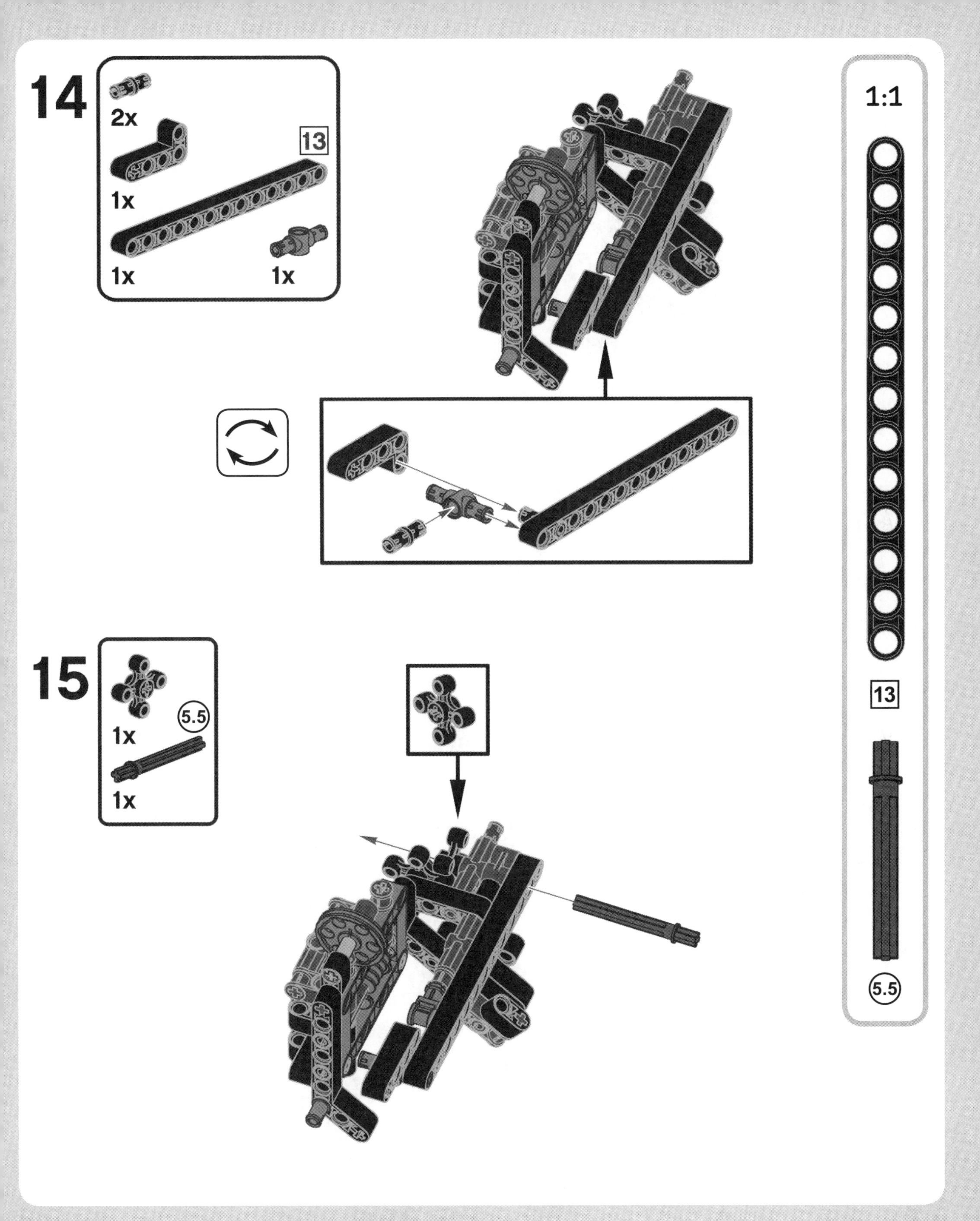
14
2x
13
1x
1x
1x
15
1x
5.5
1x
1:1
13
5.5

16

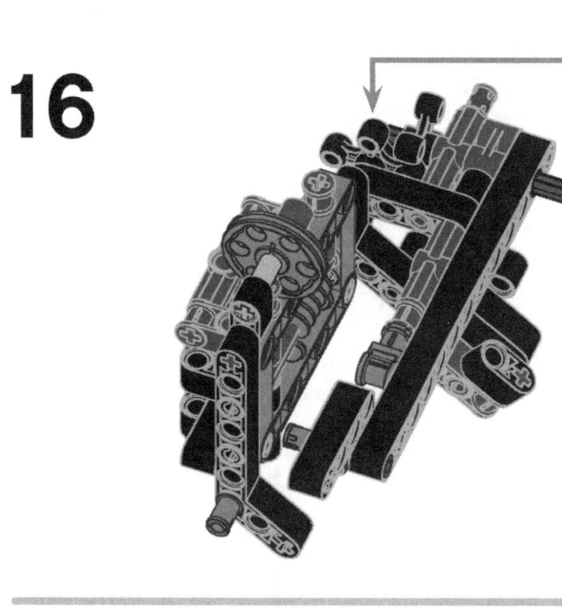

The knob wheels are ideal for transmitting the motion to perpendicular axles (axles that form a right angle).

left leg assembly

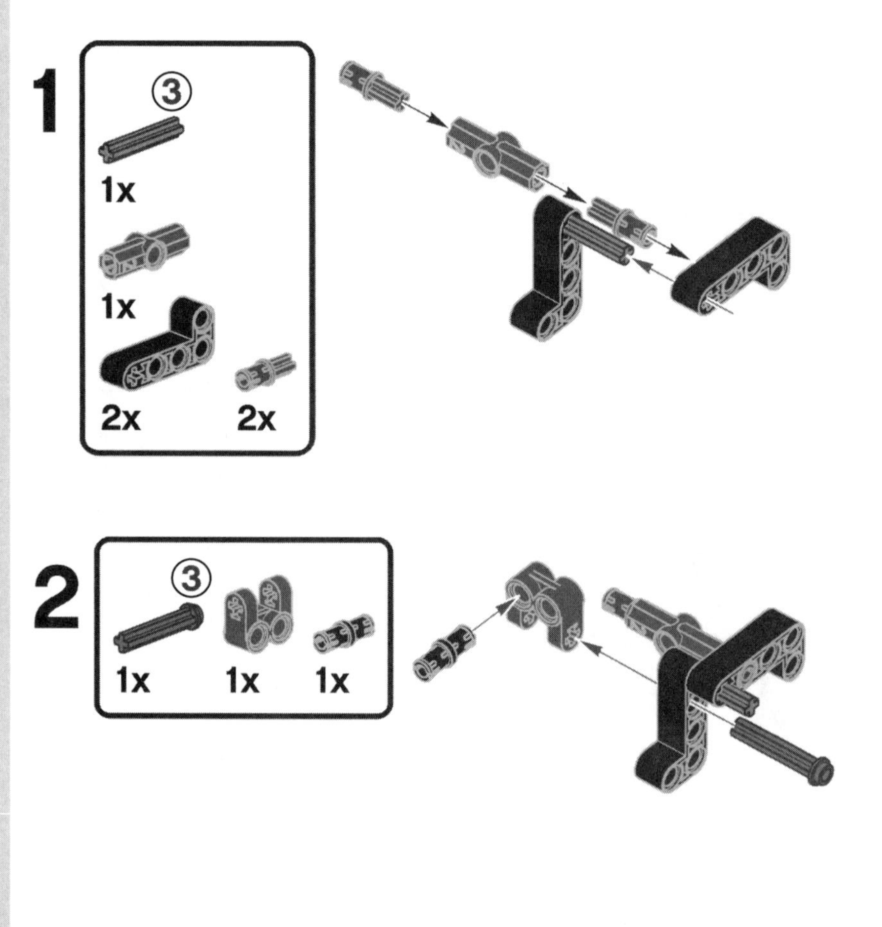

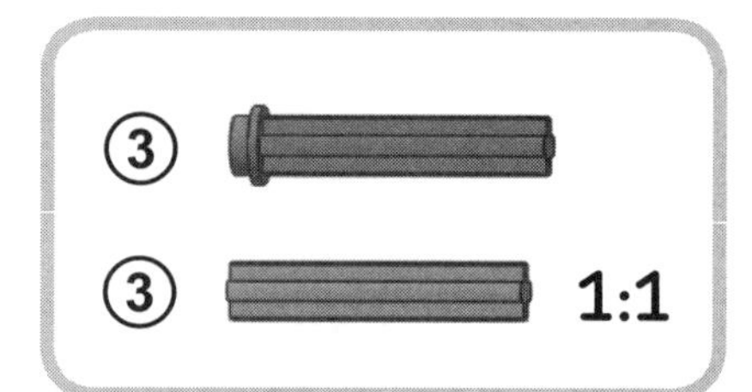

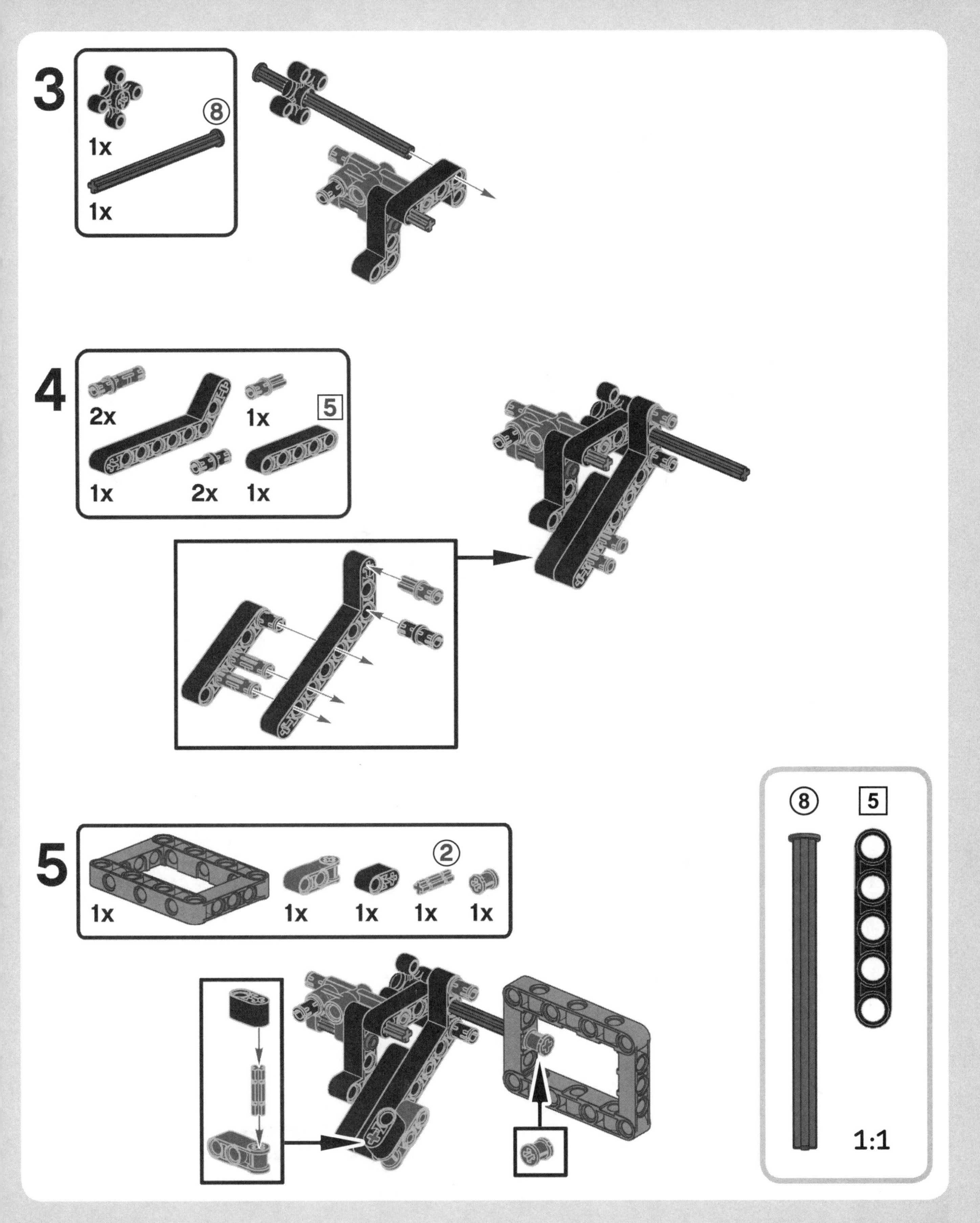
3
1x
8
1x
4
2x
1x
5
1x
2x
1x
5
1x
1x
1x
2
1x
1x
8
5
1:1

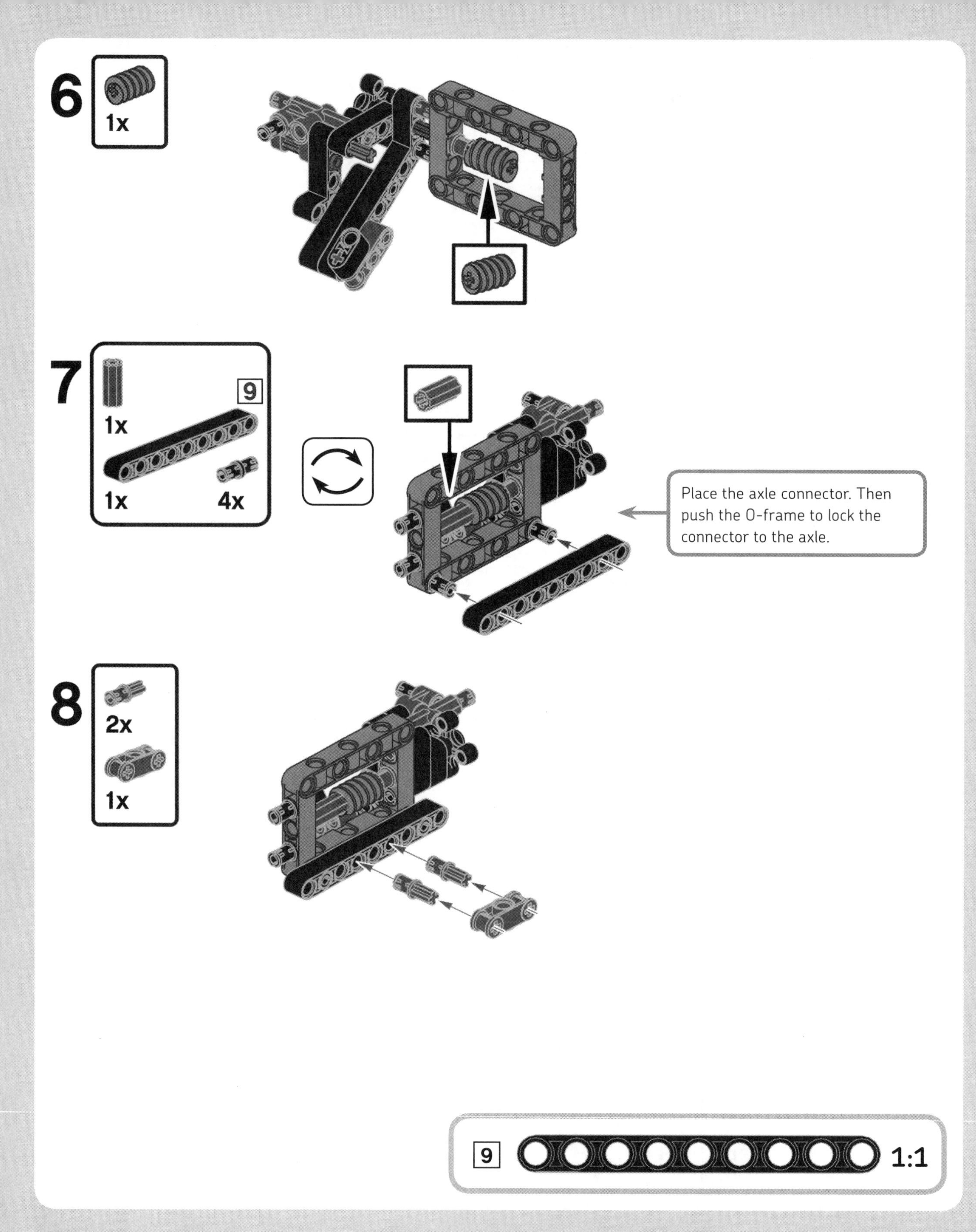
6
1x
7
1x
9
1x
4x
Place the axle connector. Then push the O-frame to lock the connector to the axle.
8
2x
1x
9
1:1

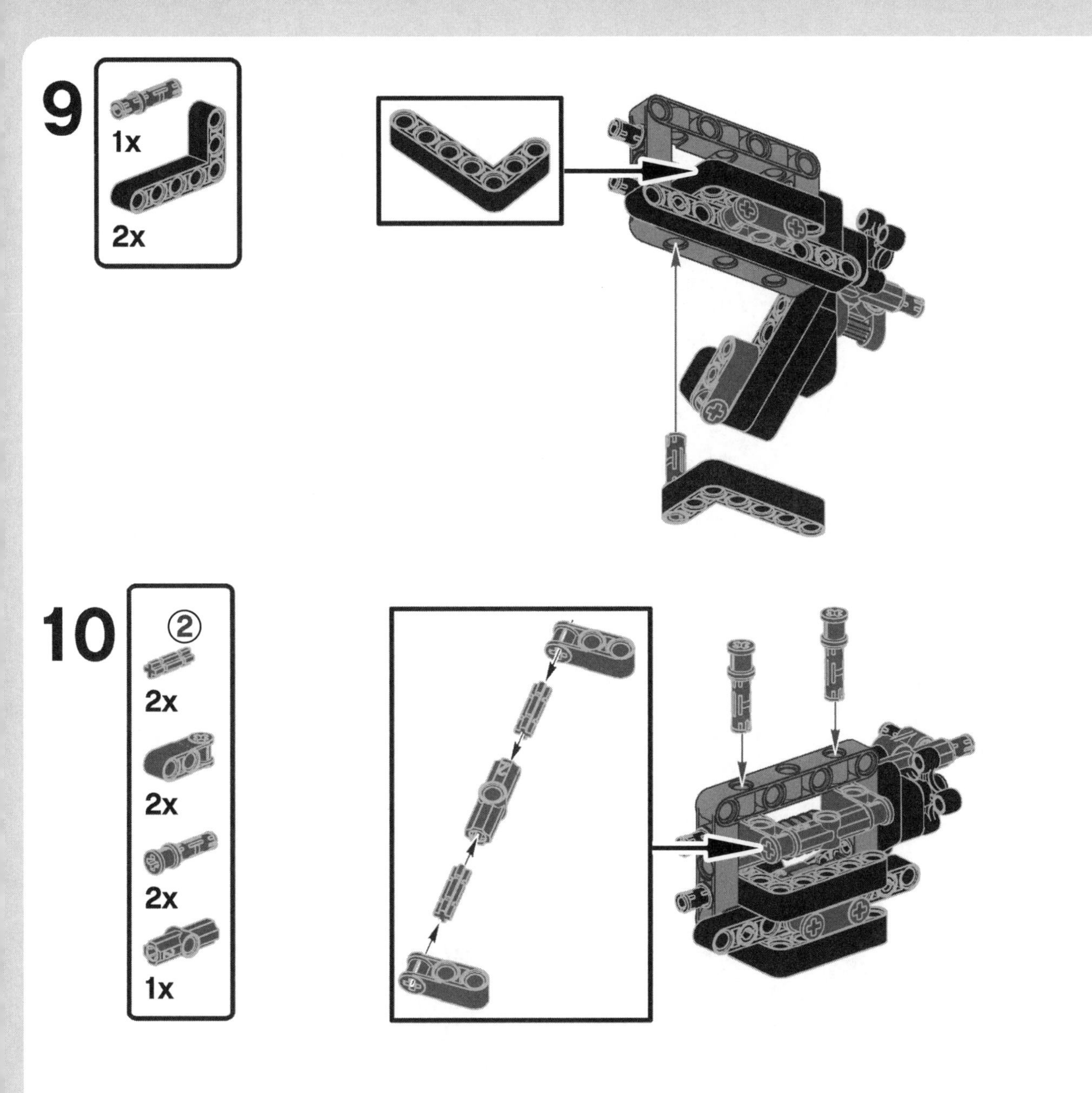
9
1x
2x
10
②
2x
2x
2x
1x

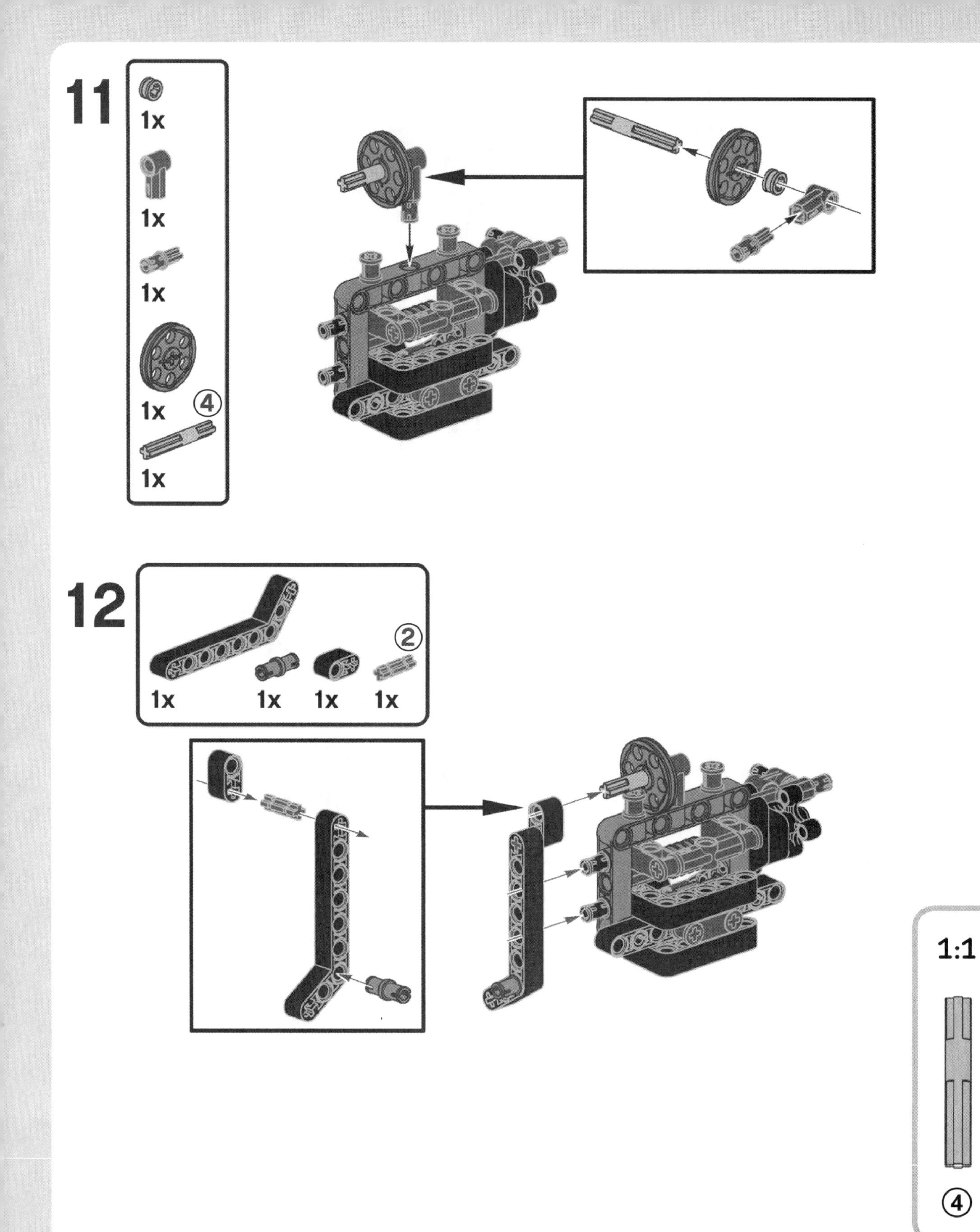
11
1x
1x
1x
1x
4
1x
12
1x
1x
1x
2
1x
1:1
4

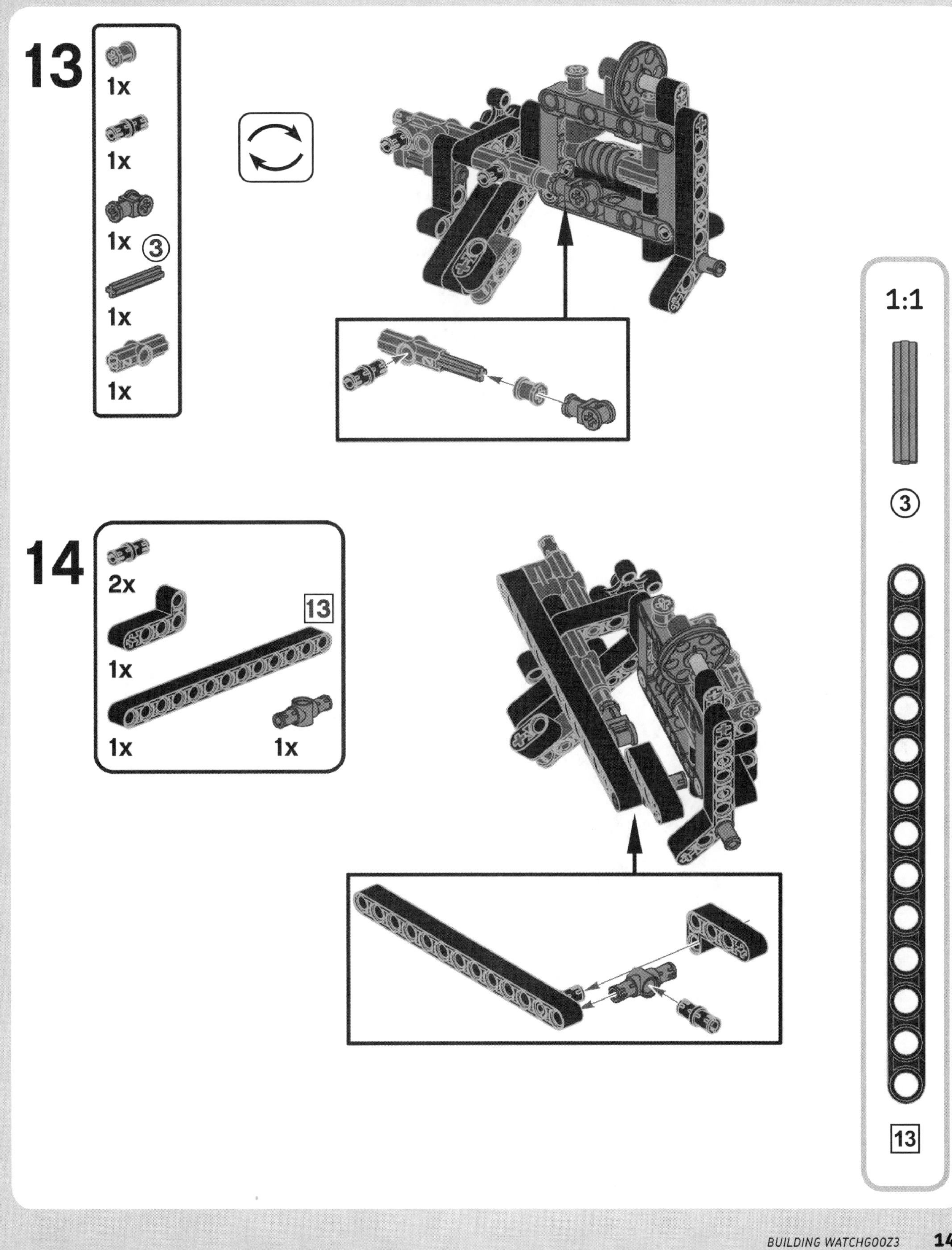
13
1x
1x
1x
3
1x
1x
14
2x
13
1x
1x
1x
1:1
3
13

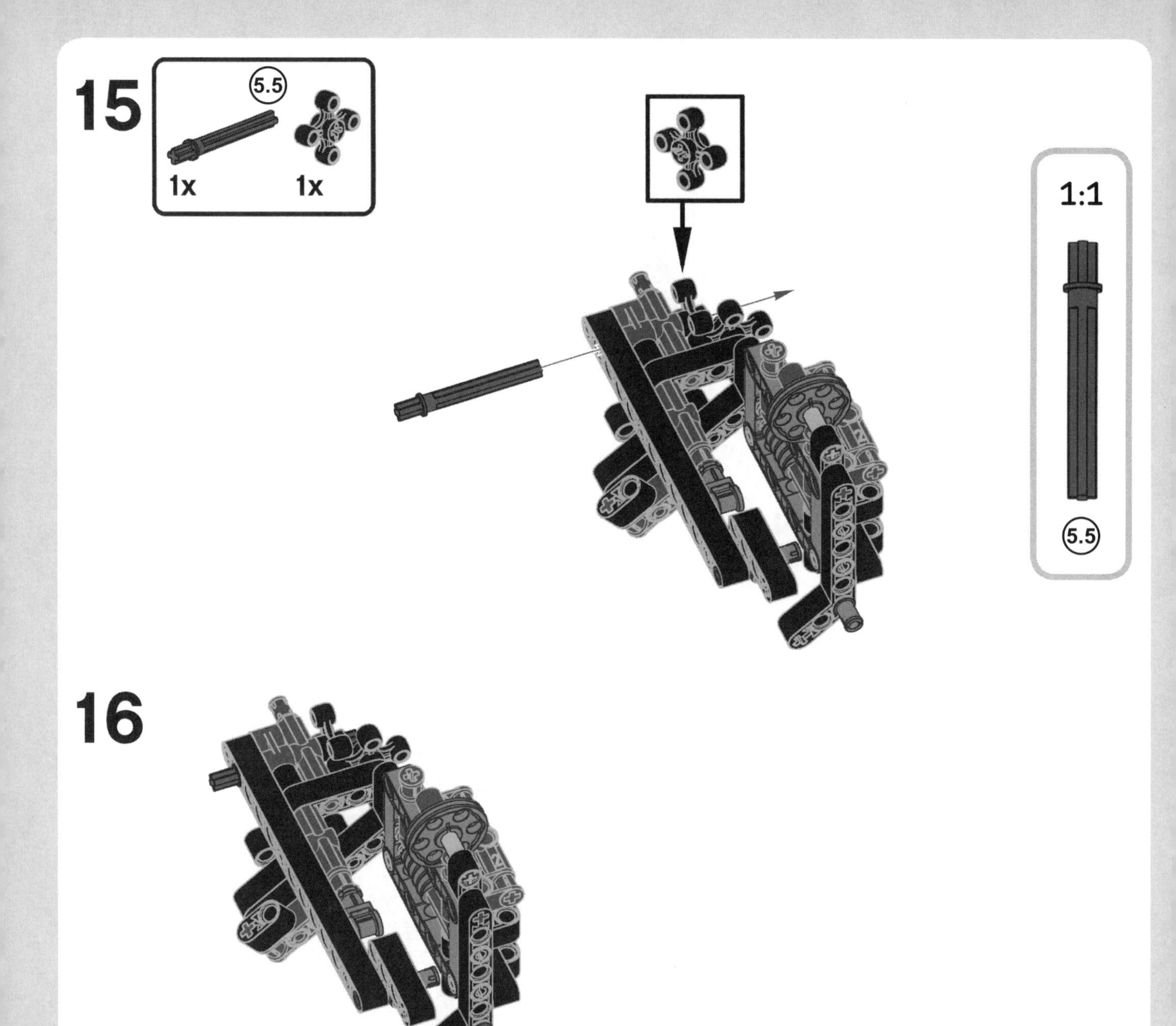
15
5.5
1x
1x
1:1
5.5
16

main assembly

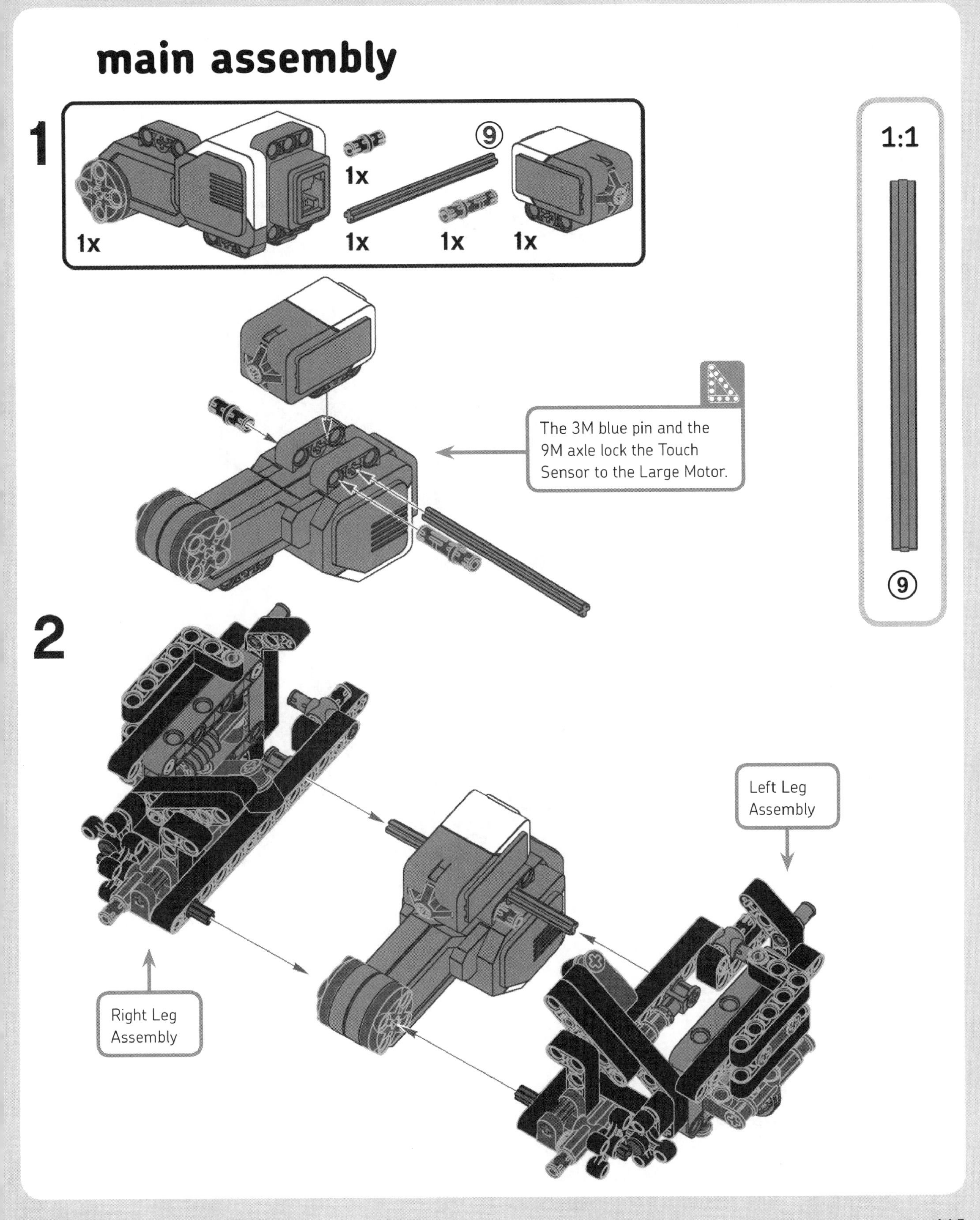

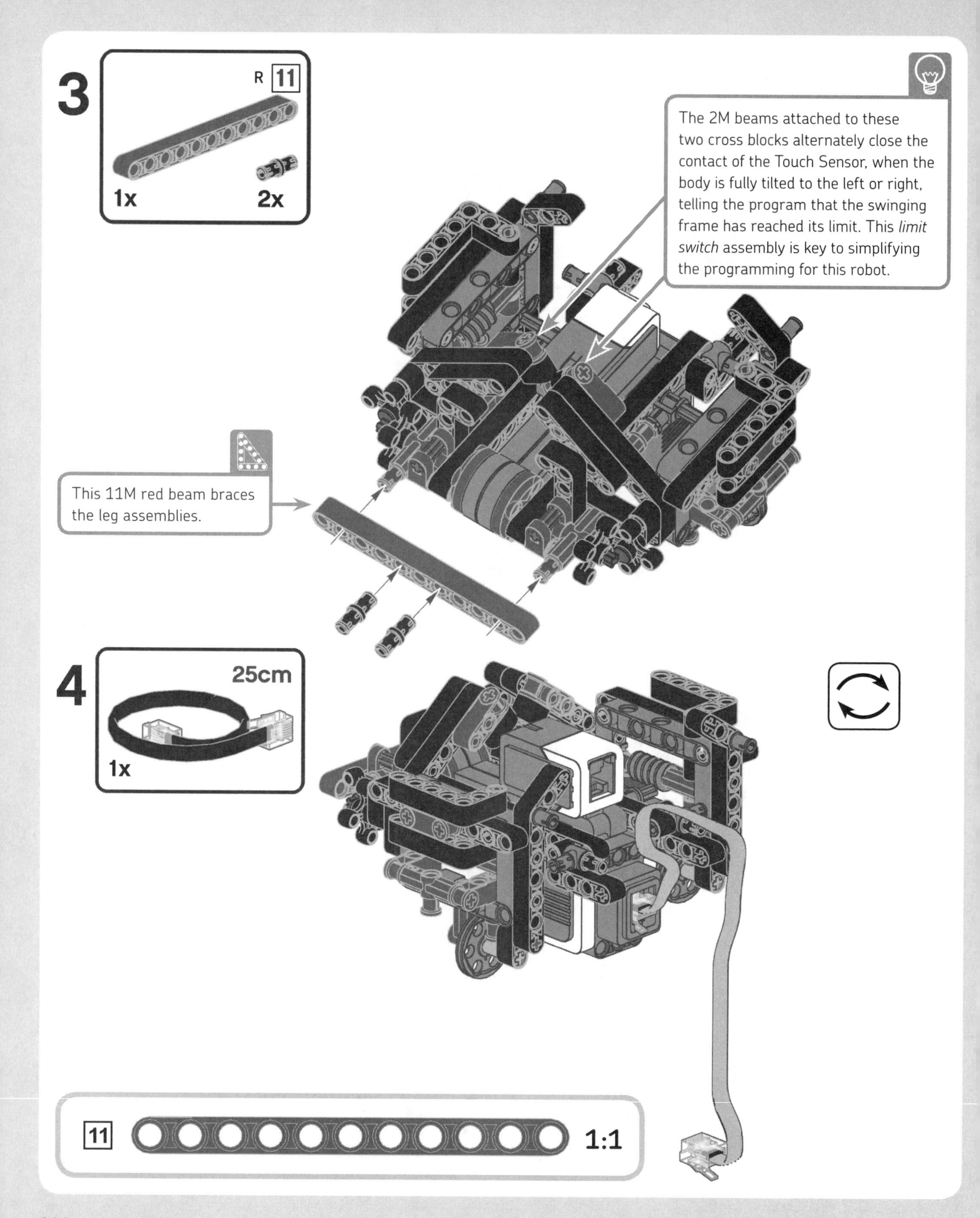
3
R 11
1x
2x
The 2M beams attached to these two cross blocks alternately close the contact of the Touch Sensor, when the body is fully tilted to the left or right, telling the program that the swinging frame has reached its limit. This *limit switch* assembly is key to simplifying the programming for this robot.
This 11M red beam braces the leg assemblies.
4
25cm
1x
11
1:1

5

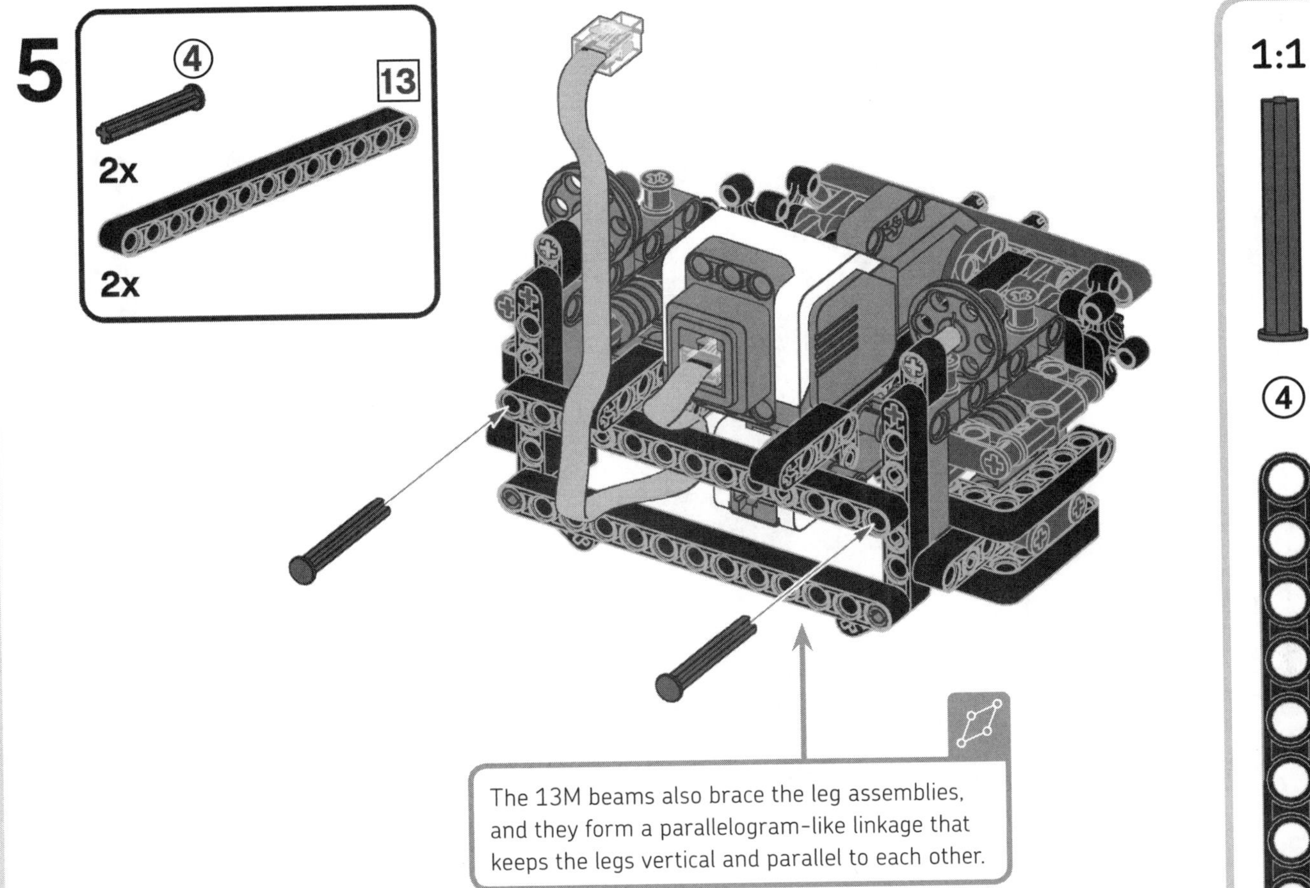

The 13M beams also brace the leg assemblies, and they form a parallelogram-like linkage that keeps the legs vertical and parallel to each other.

left foot assembly

1

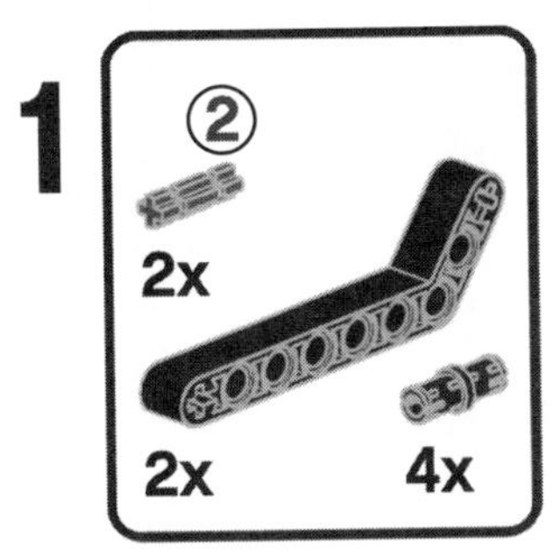

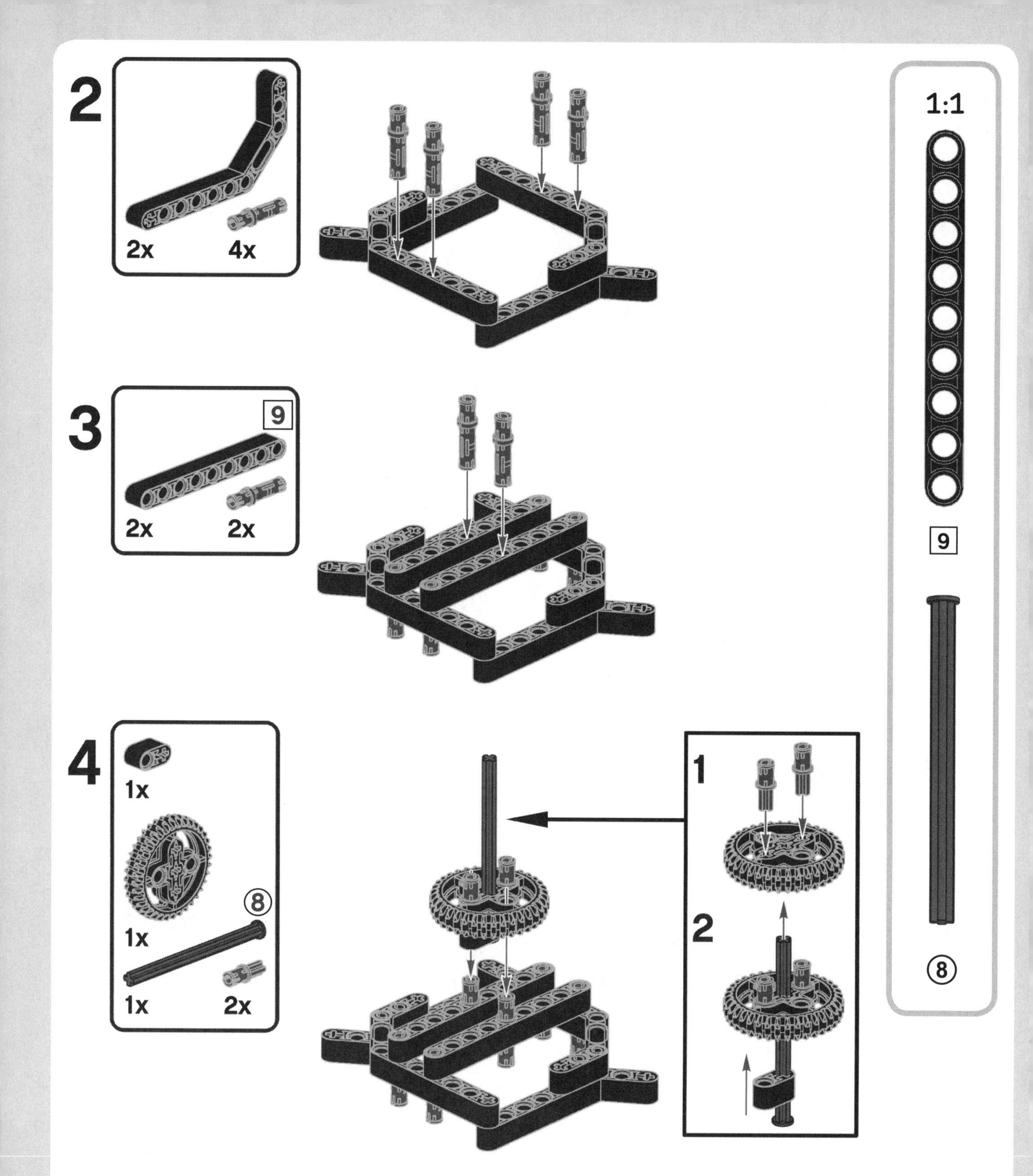
2
2x
4x
3
9
2x
2x
4
1x
1x
8
1x
2x
1
2
1:1
9
8

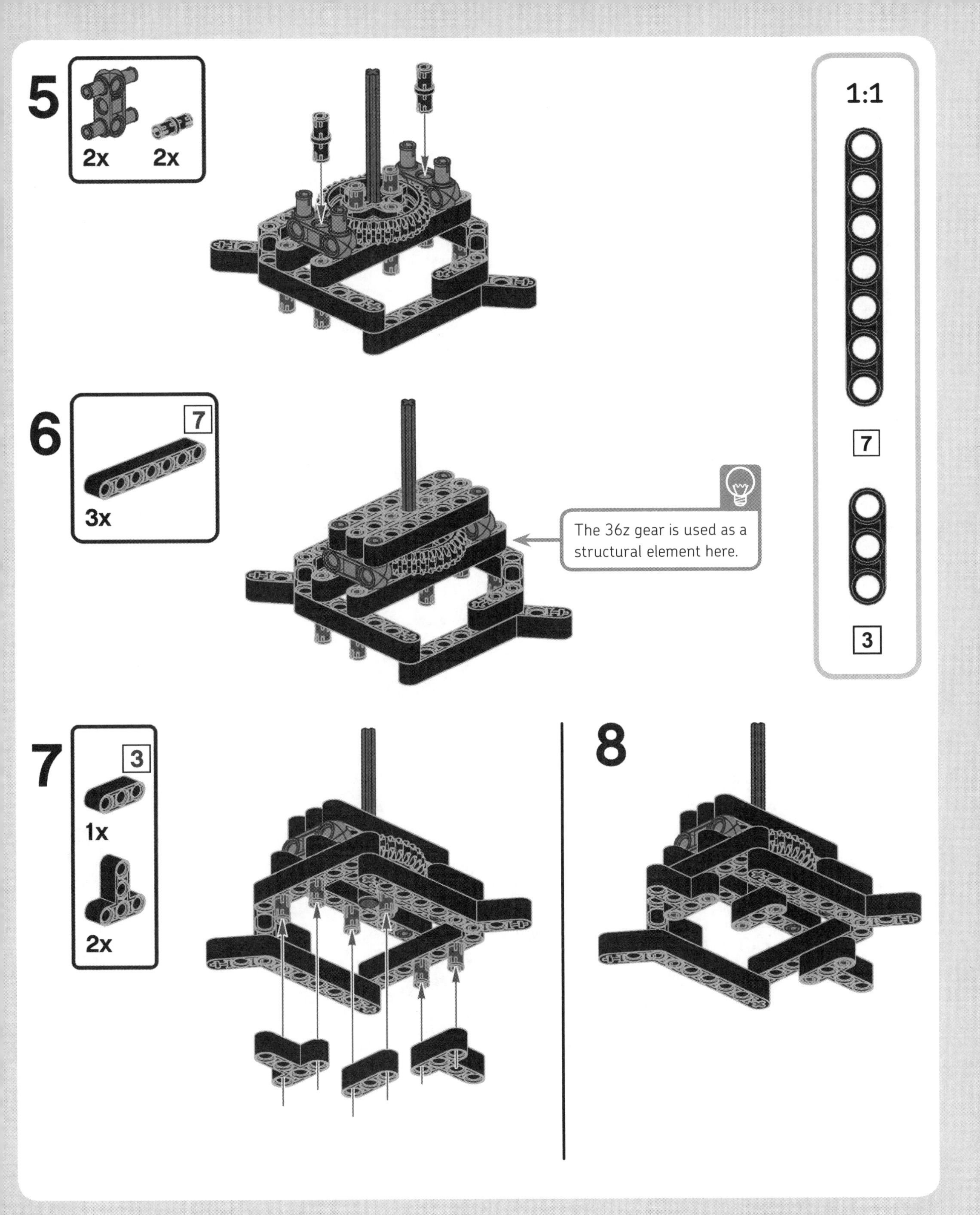
5
2x
2x
1:1
6
7
3x
The 36z gear is used as a structural element here.
7
3
7
3
1x
2x
8

right foot assembly

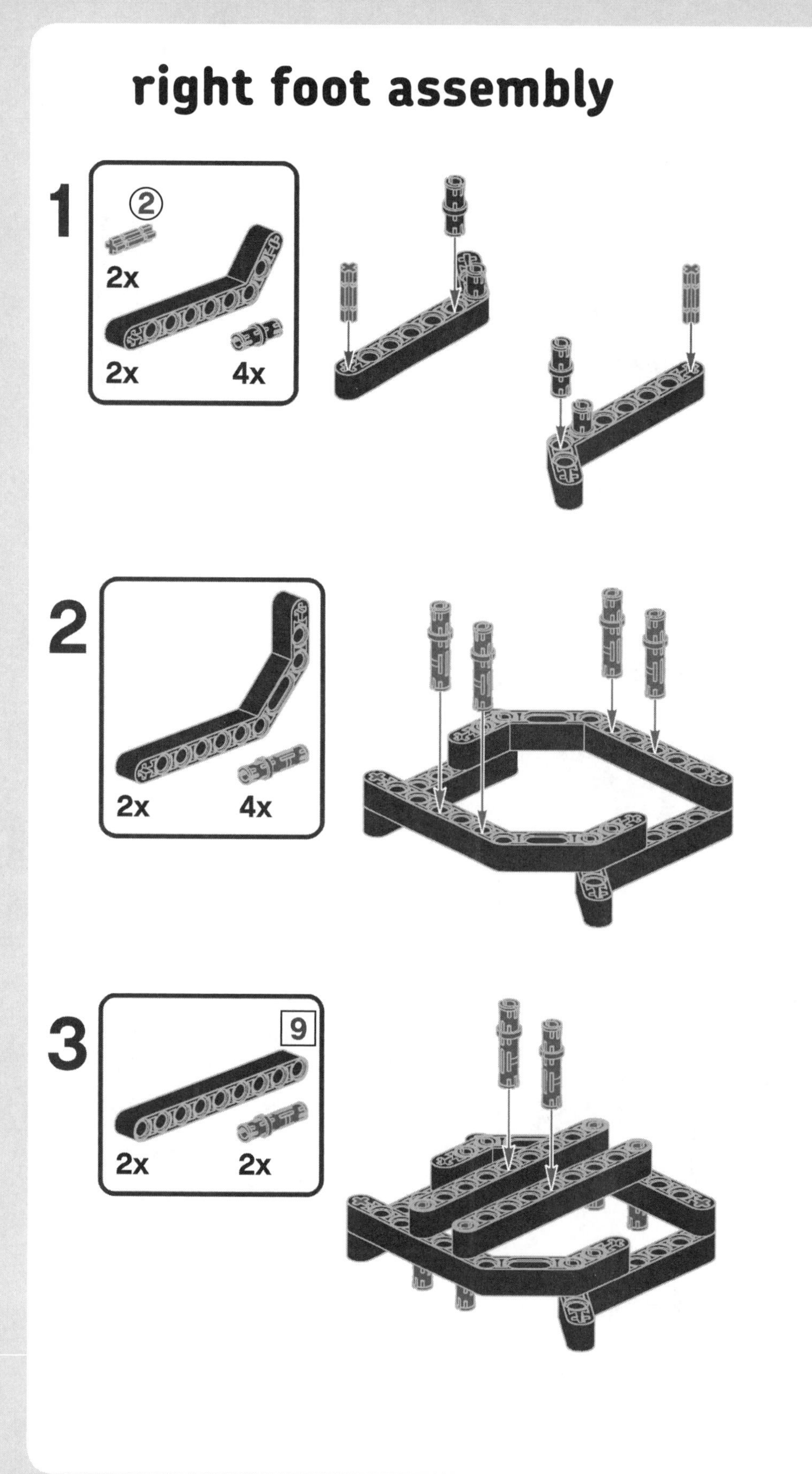

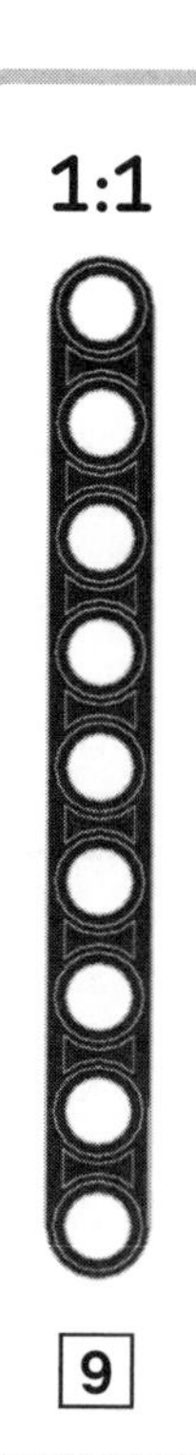

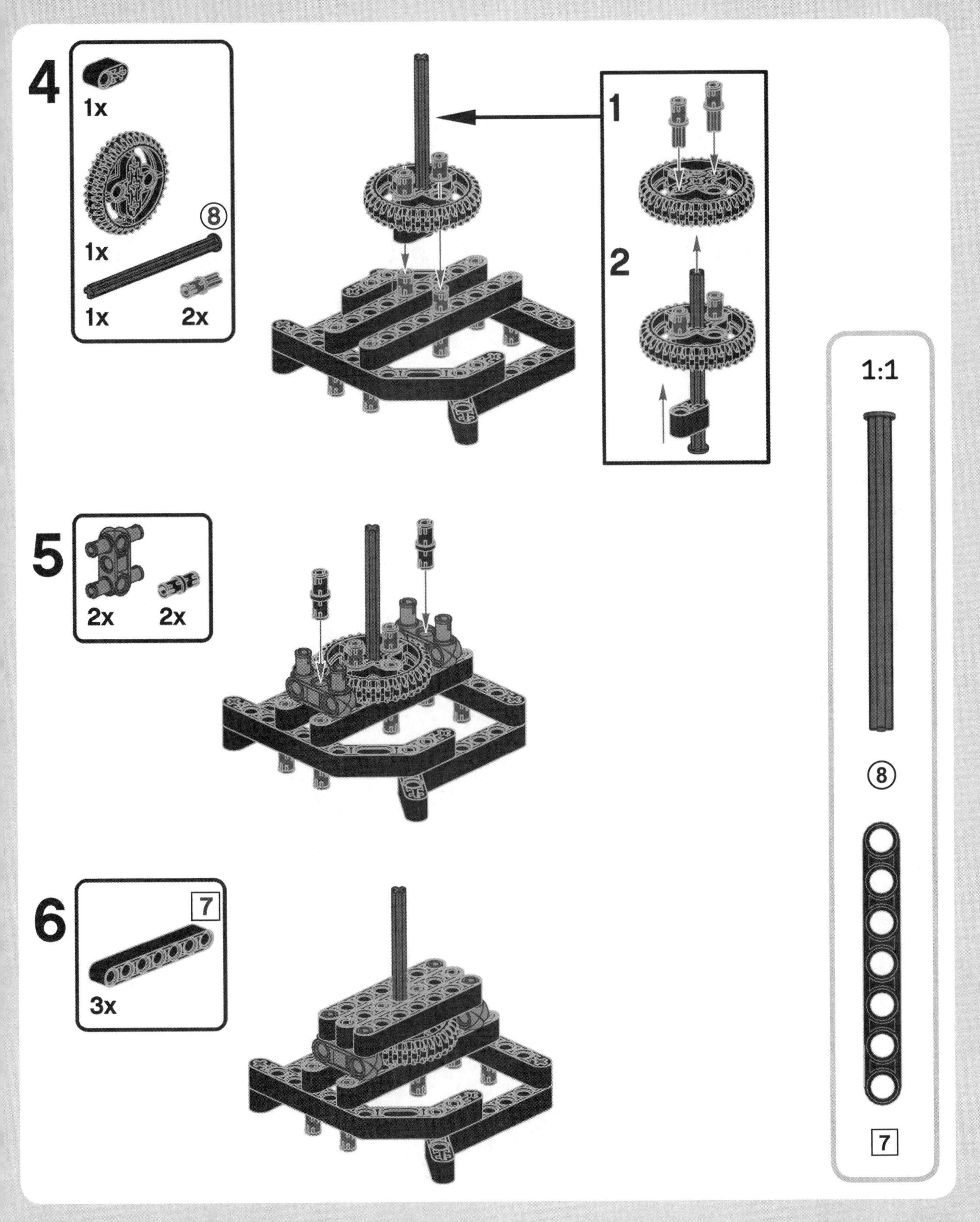
4
1x
1x
8
1x
2x
1
2
5
2x
2x
6
7
3x
1:1
8
7

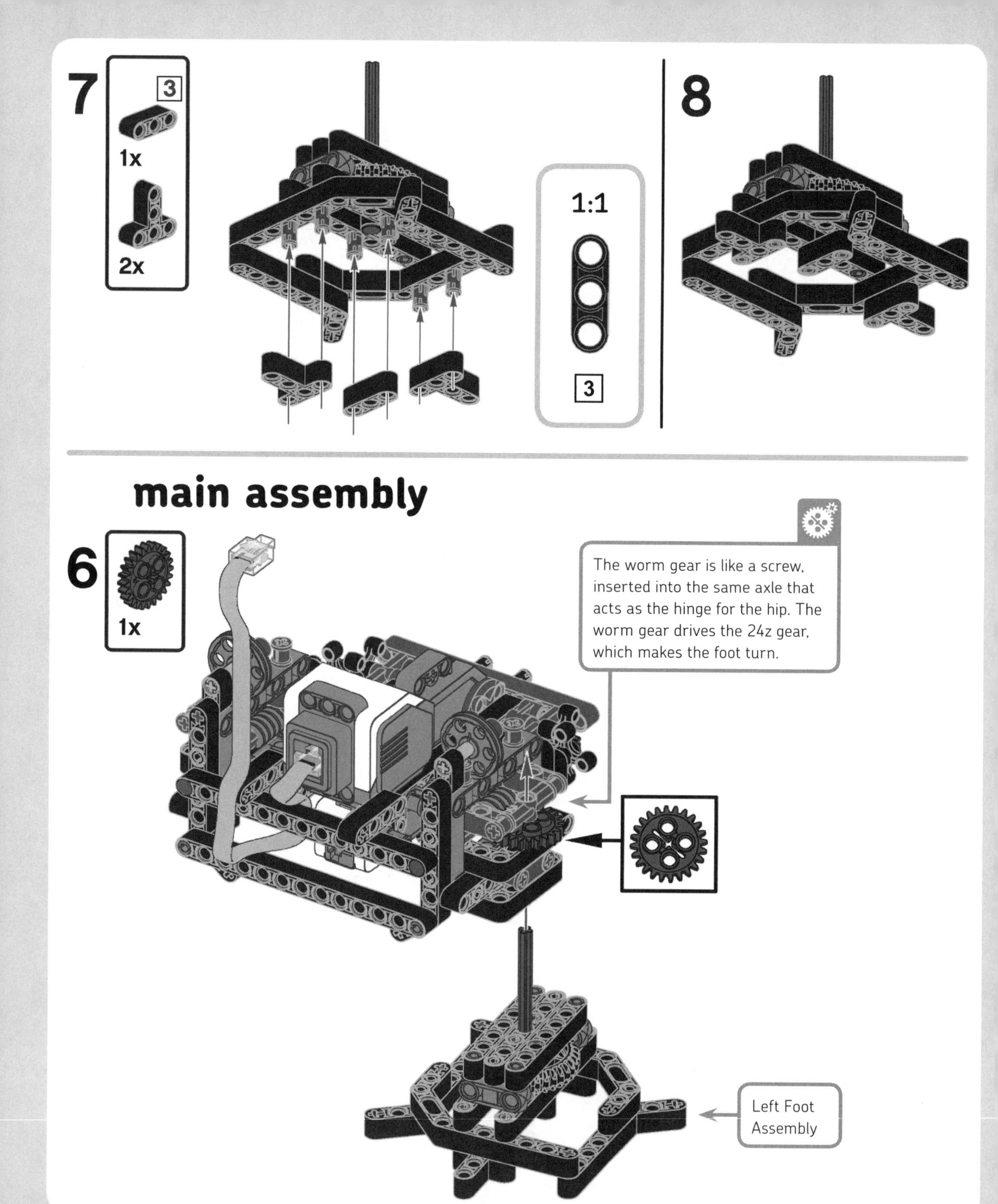
7
3
1x
2x
1:1
3
8
main assembly
6
1x
The worm gear is like a screw, inserted into the same axle that acts as the hinge for the hip. The worm gear drives the 24z gear, which makes the foot turn.
Left Foot Assembly

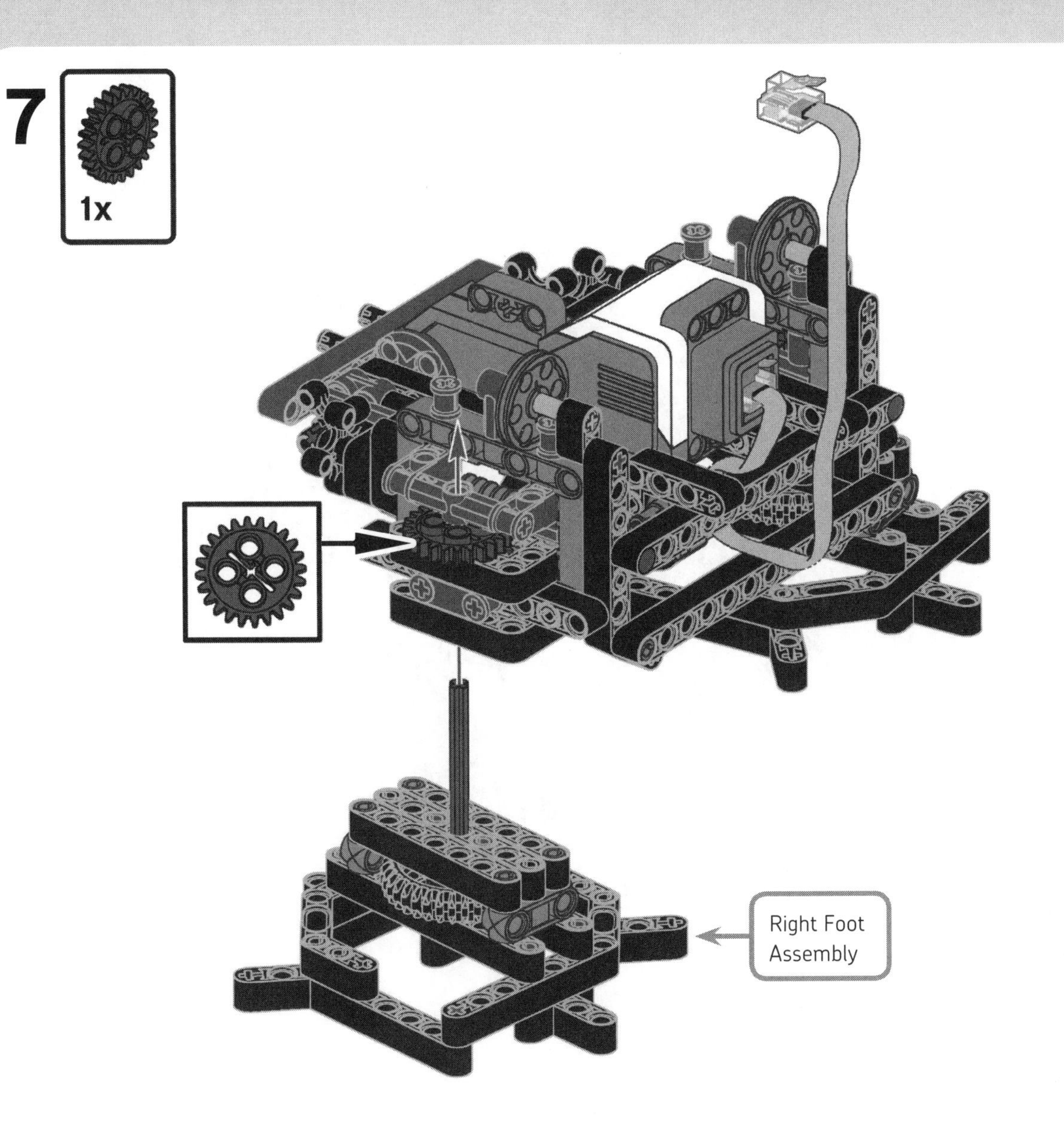
7
1x
Right Foot
Assembly

8

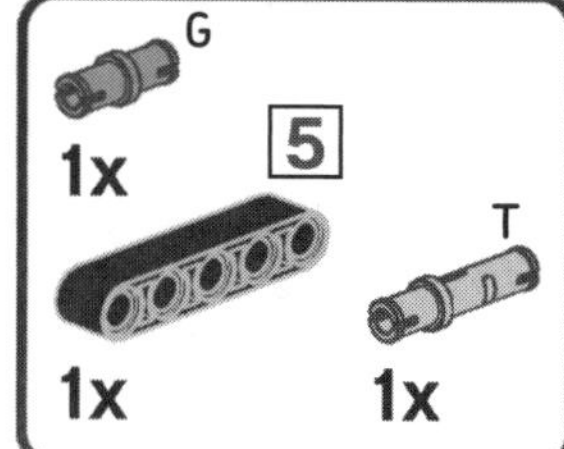

Try turning the knob wheel to see how movement is transferred from the Large Motor shaft to the feet while allowing the hips to swing freely.

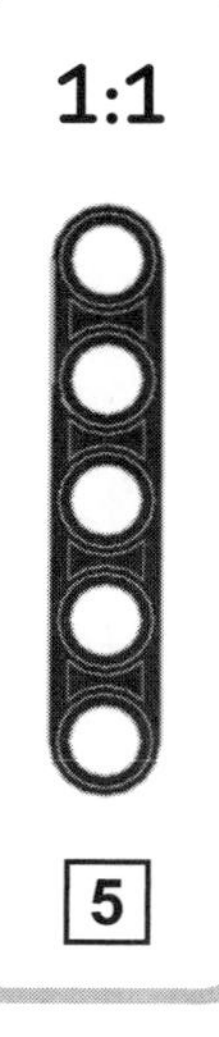

back bracket assembly

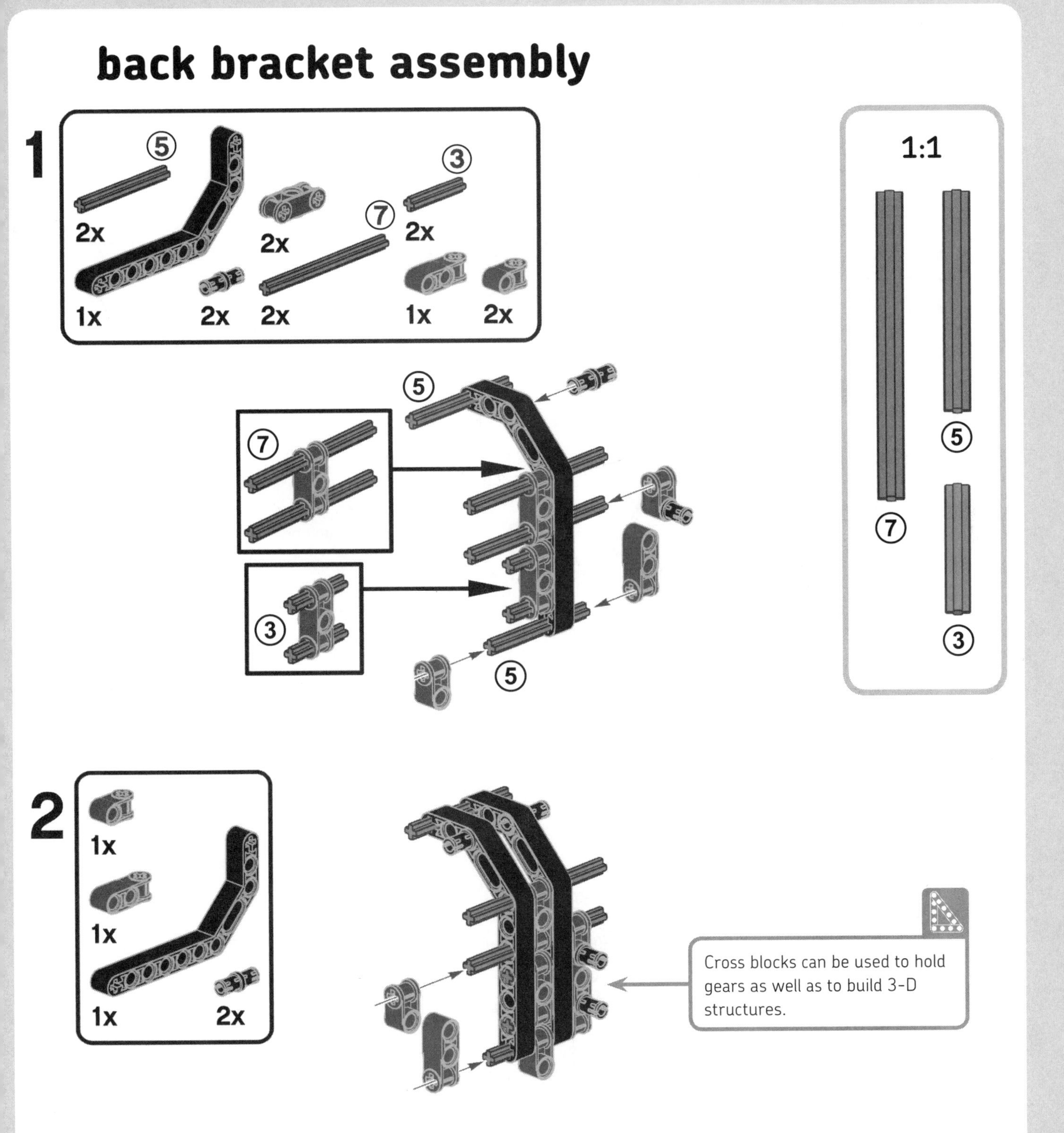

Cross blocks can be used to hold gears as well as to build 3-D structures.

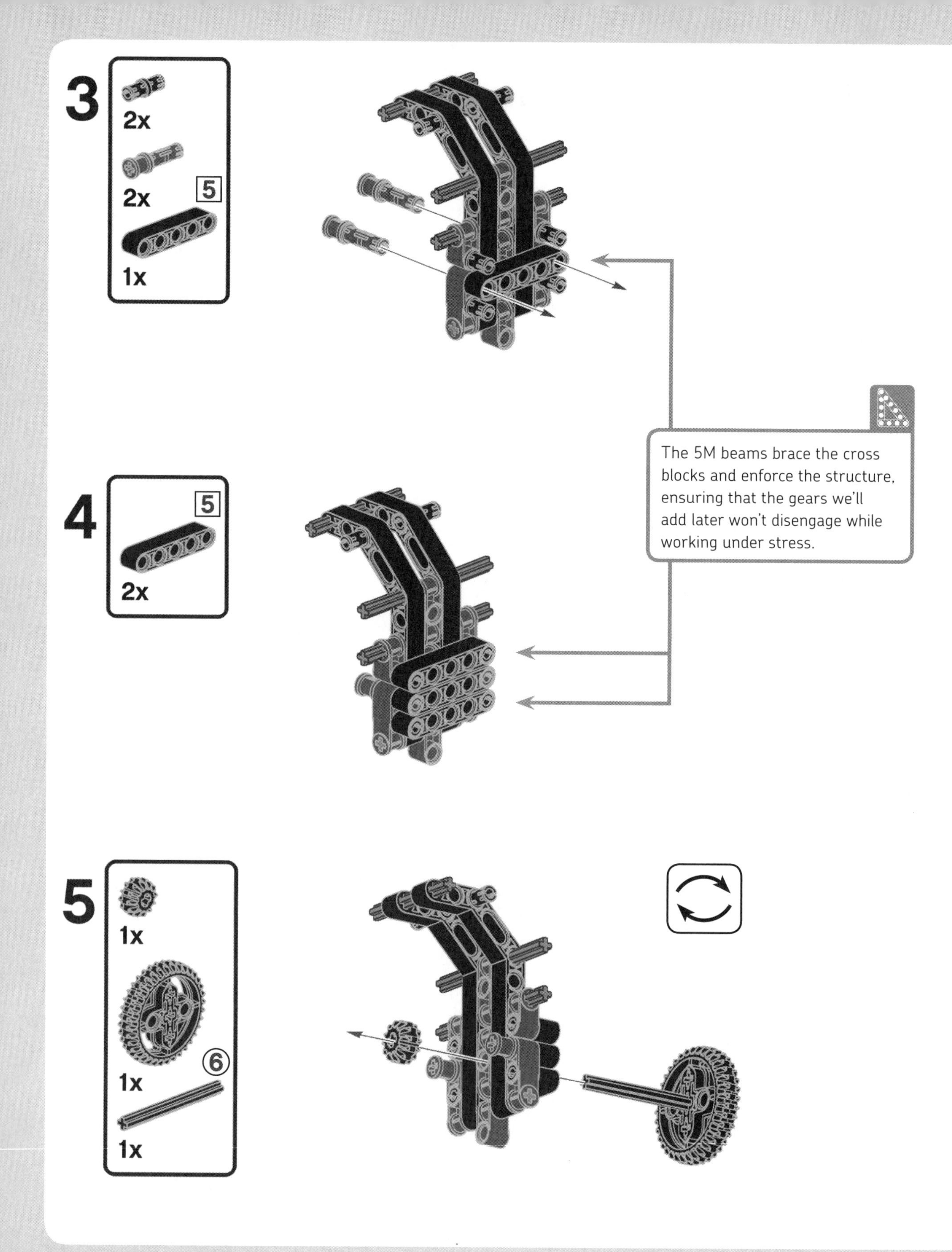
3
2x
2x
5
1x
The 5M beams brace the cross blocks and enforce the structure, ensuring that the gears we'll add later won't disengage while working under stress.
4
5
2x
5
1x
1x
6
1x

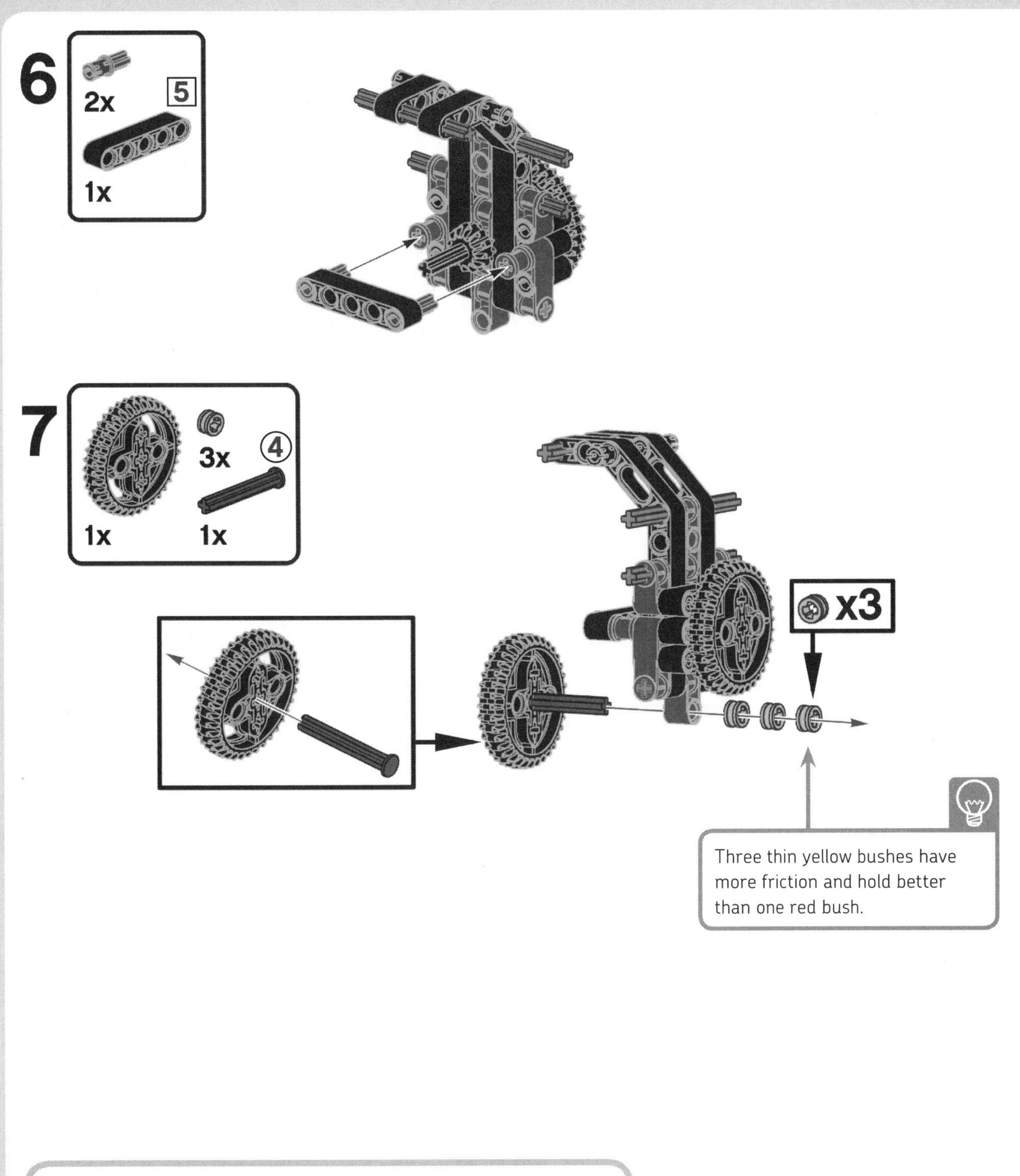
6
2x
5
1x
7
3x
4
1x
1x
x3
Three thin yellow bushes have more friction and hold better than one red bush.

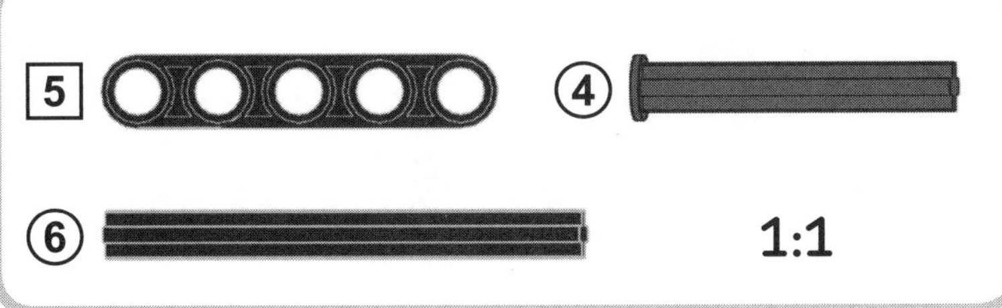
5
4
6
1:1

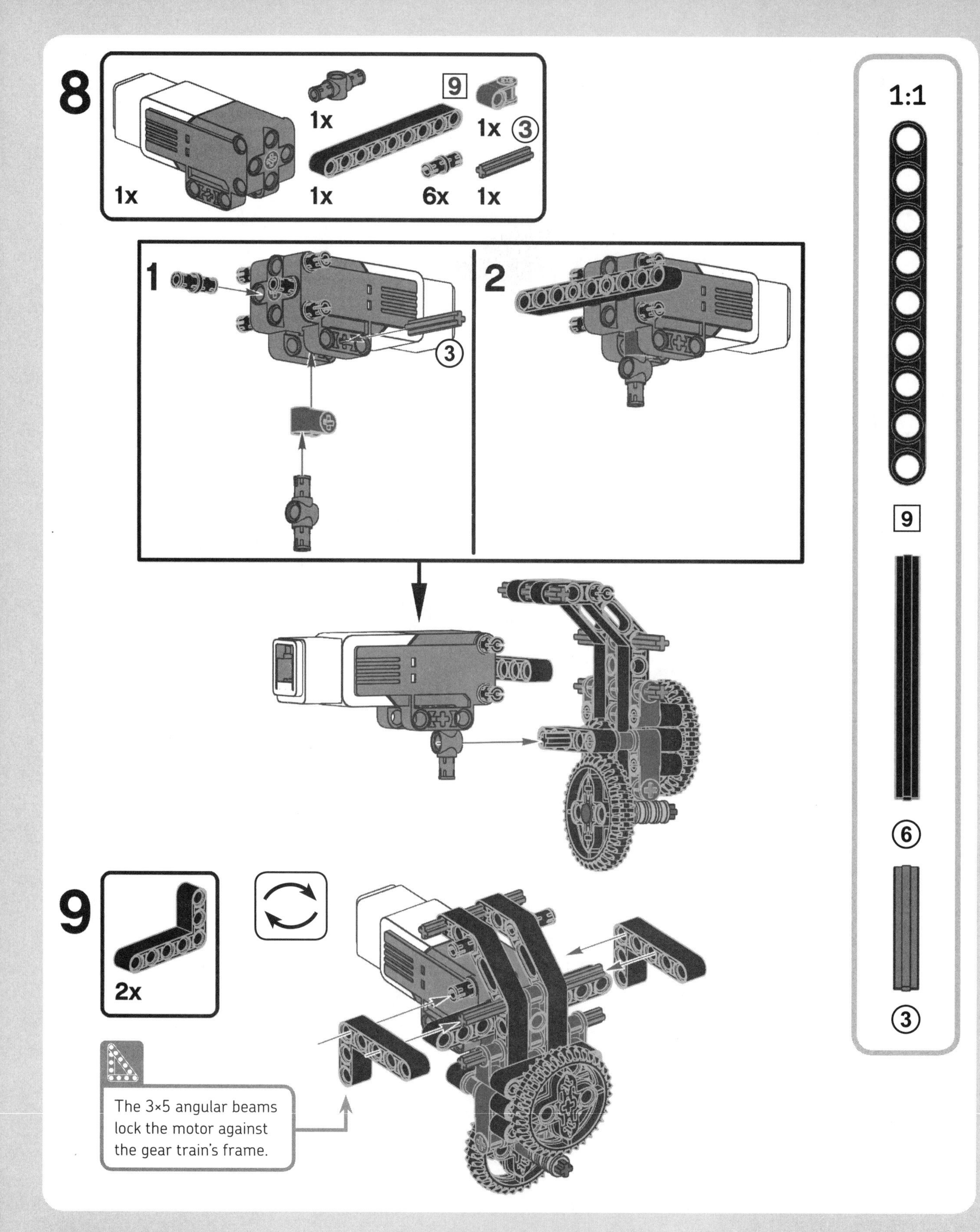
8
1x
1x
9
1x
3
1x
1x
6x
1x
1
3
2
9
2x
The 3×5 angular beams lock the motor against the gear train's frame.
1:1
9
6
3

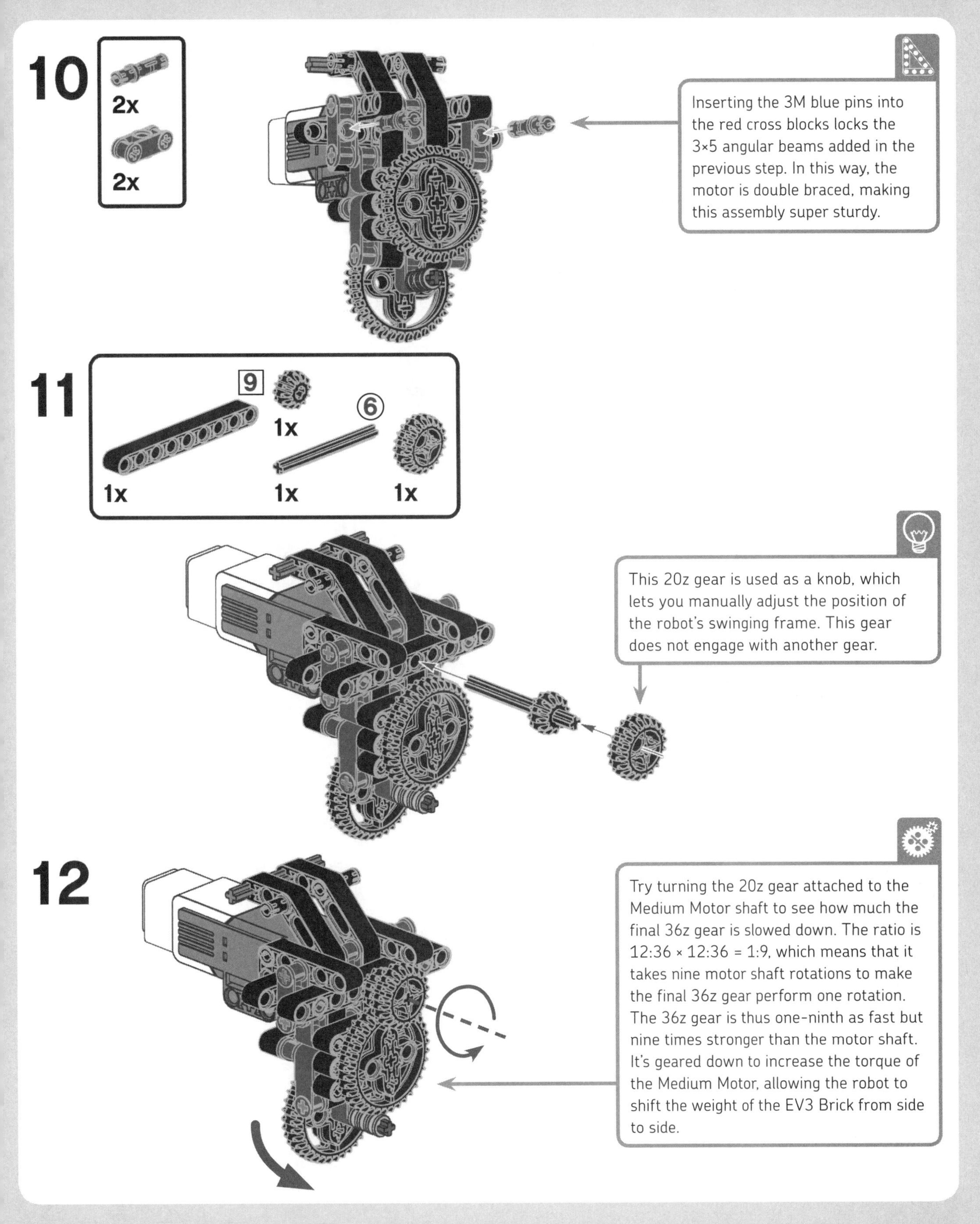
10
2x
2x
Inserting the 3M blue pins into the red cross blocks locks the 3×5 angular beams added in the previous step. In this way, the motor is double braced, making this assembly super sturdy.
11
9
1x
6
1x
1x
1x
This 20z gear is used as a knob, which lets you manually adjust the position of the robot's swinging frame. This gear does not engage with another gear.
12
Try turning the 20z gear attached to the Medium Motor shaft to see how much the final 36z gear is slowed down. The ratio is 12:36 × 12:36 = 1:9, which means that it takes nine motor shaft rotations to make the final 36z gear perform one rotation. The 36z gear is thus one-ninth as fast but nine times stronger than the motor shaft. It's geared down to increase the torque of the Medium Motor, allowing the robot to shift the weight of the EV3 Brick from side to side.

front bracket assembly

1

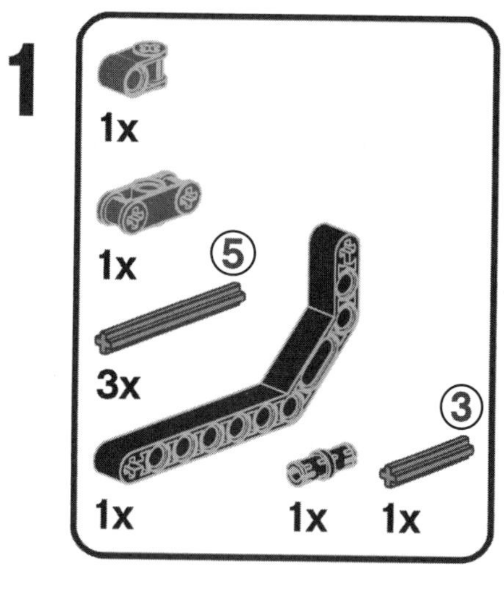

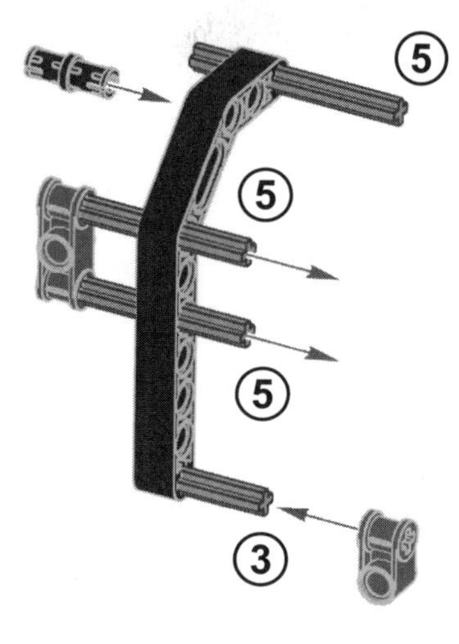

2

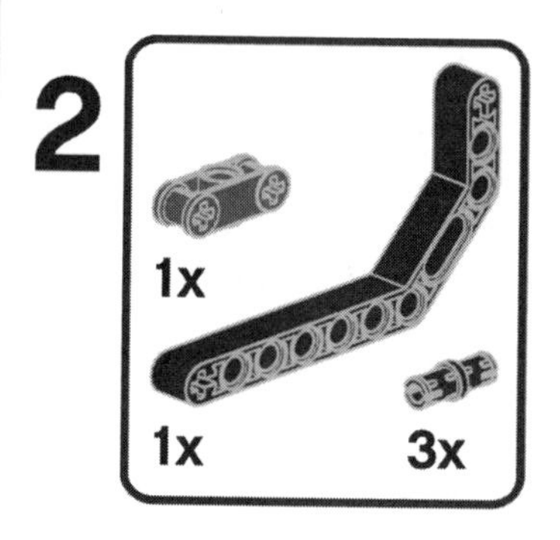

3

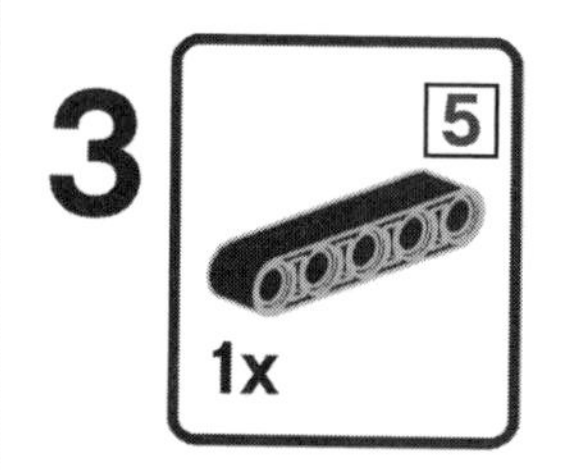

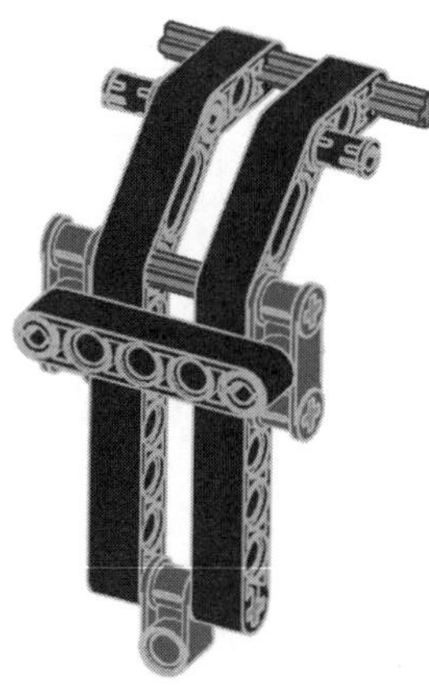

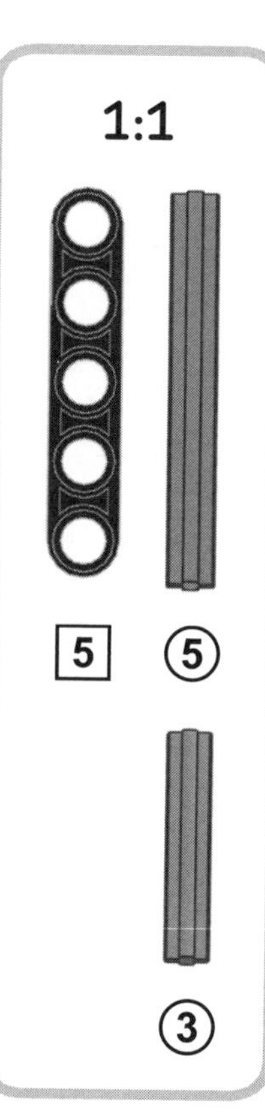

main assembly

9

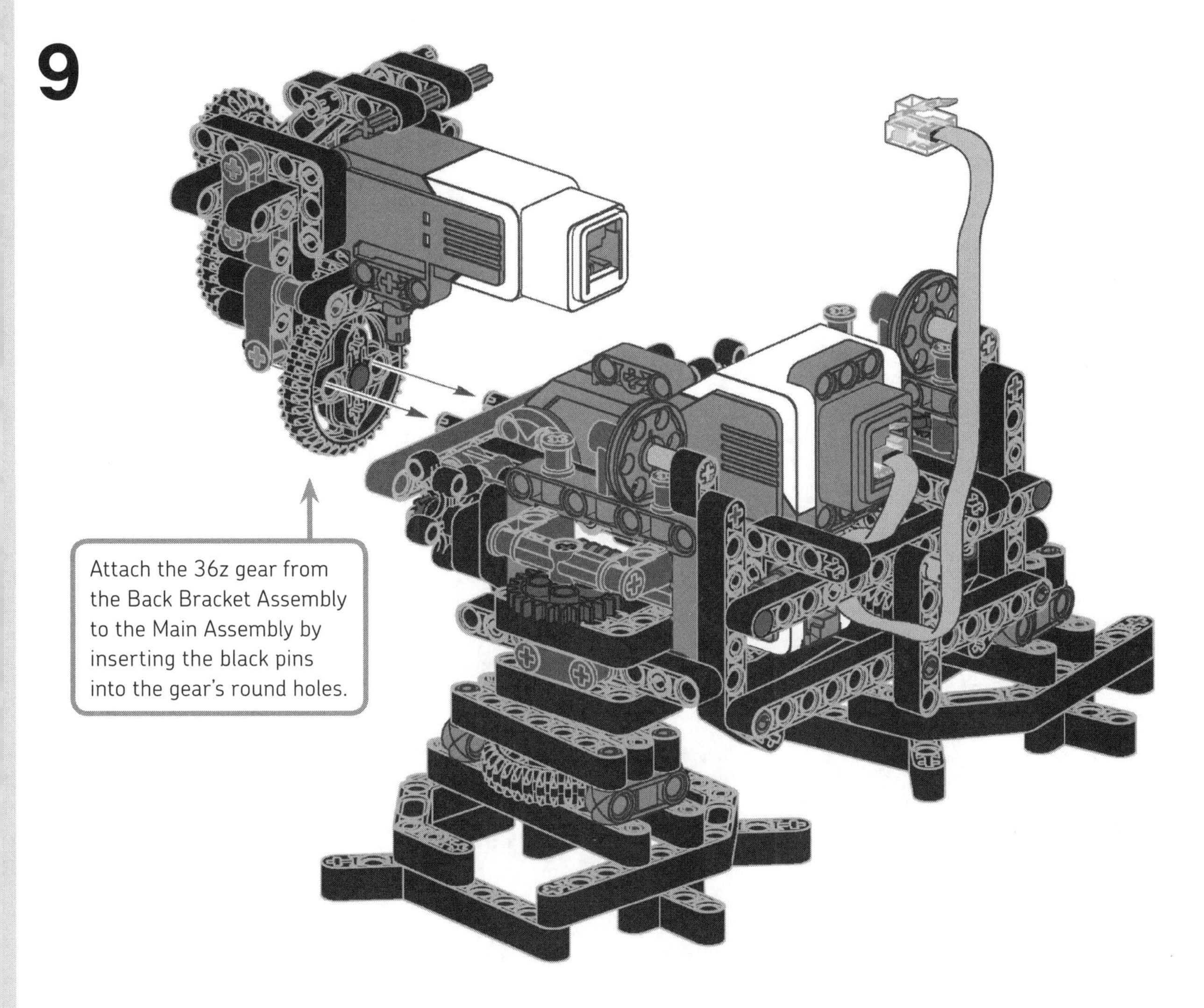

Attach the 36z gear from the Back Bracket Assembly to the Main Assembly by inserting the black pins into the gear's round holes.

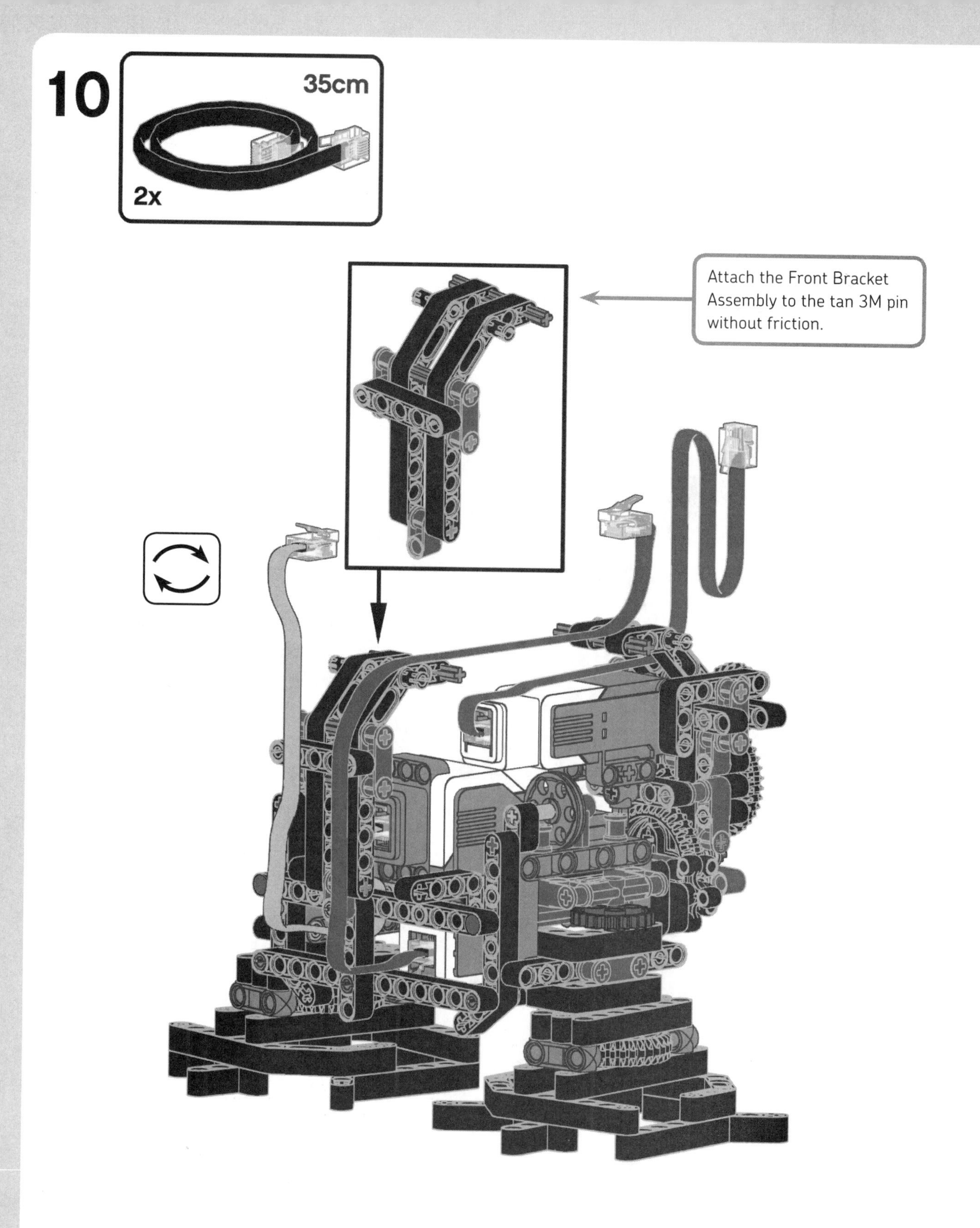
10
35cm
2x
Attach the Front Bracket Assembly to the tan 3M pin without friction.

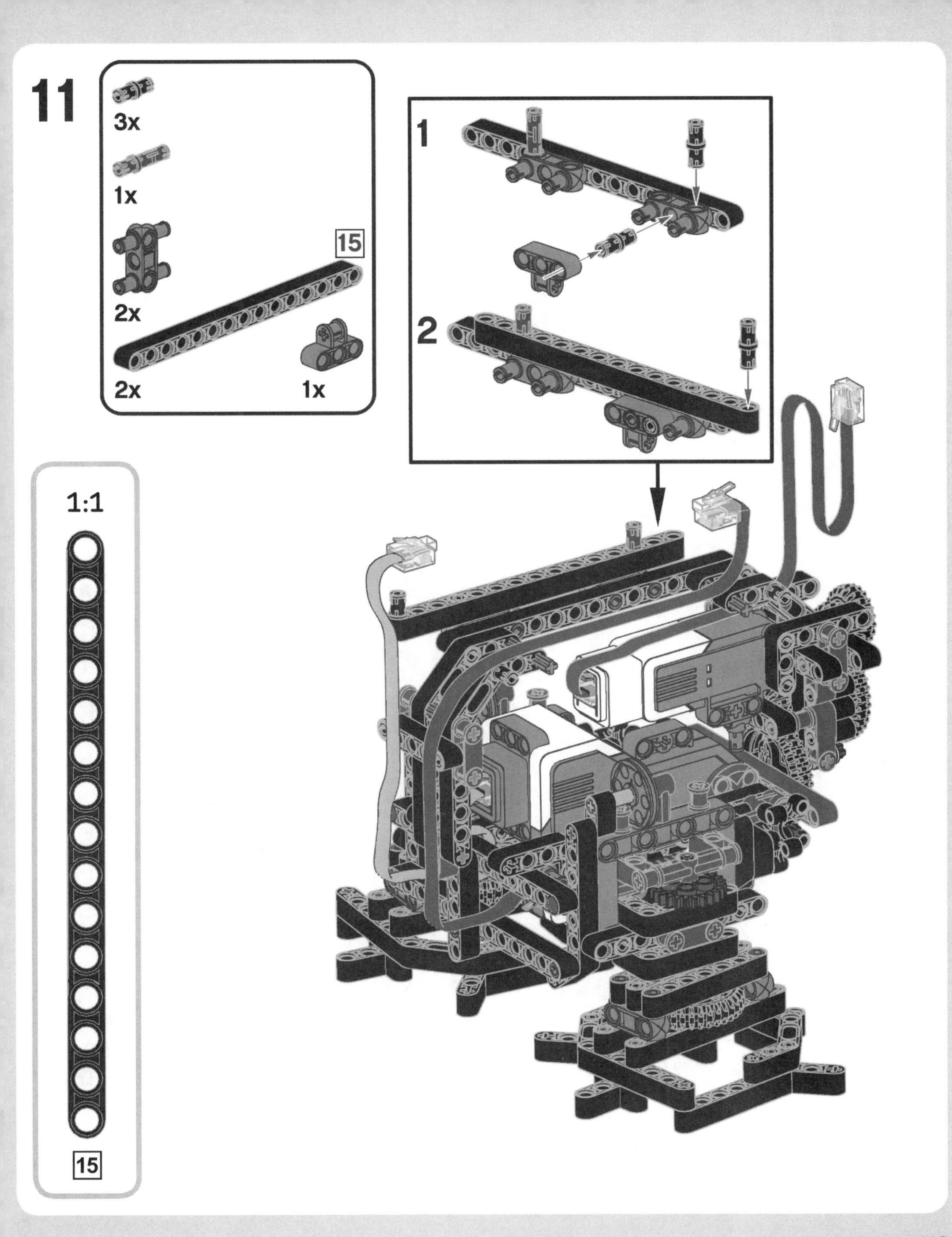
11
3x
1x
2x
15
2x
2x
1x
1
2
1:1
15

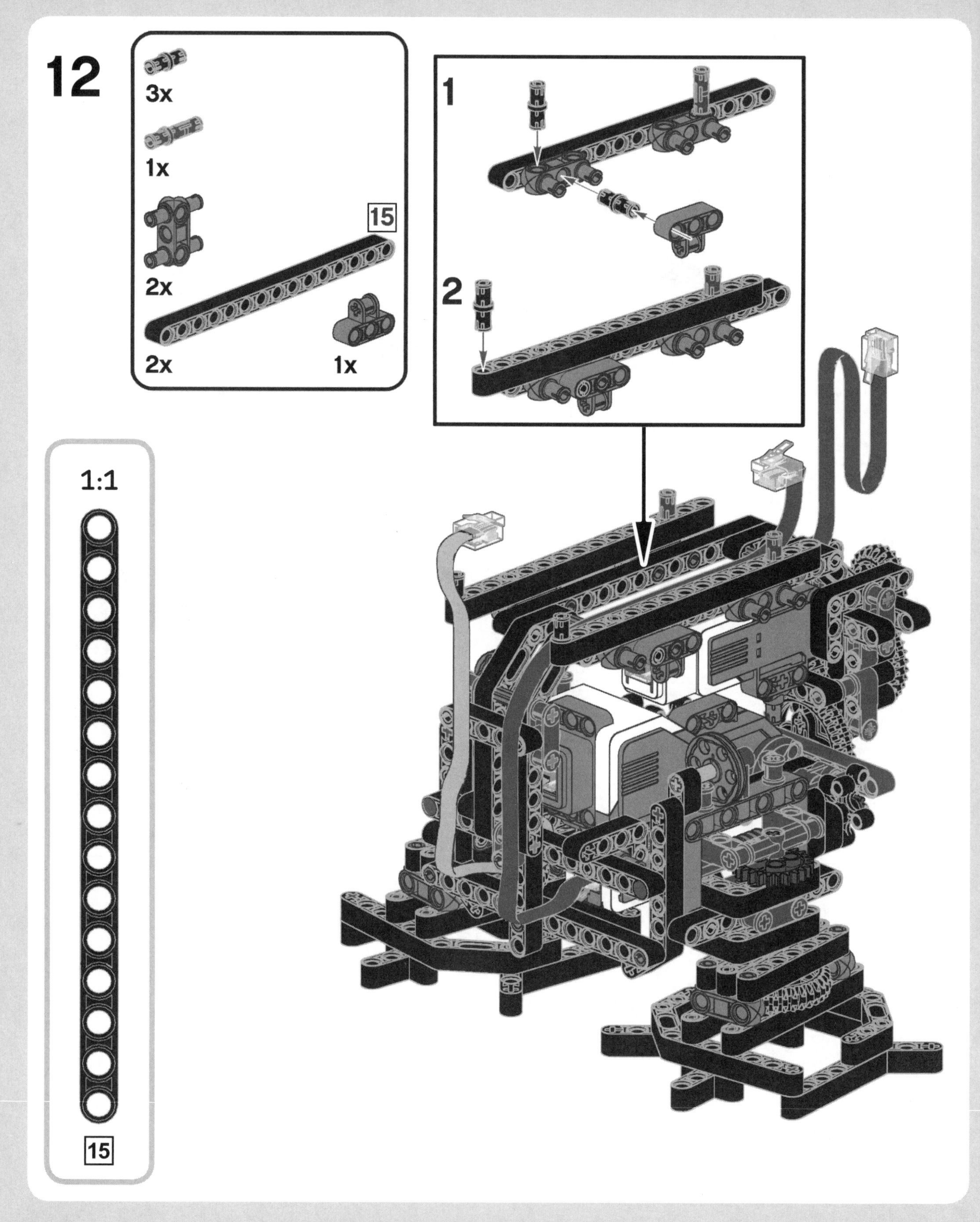
12
3x
1x
2x
15
2x
1x
1
2
1:1
15

13

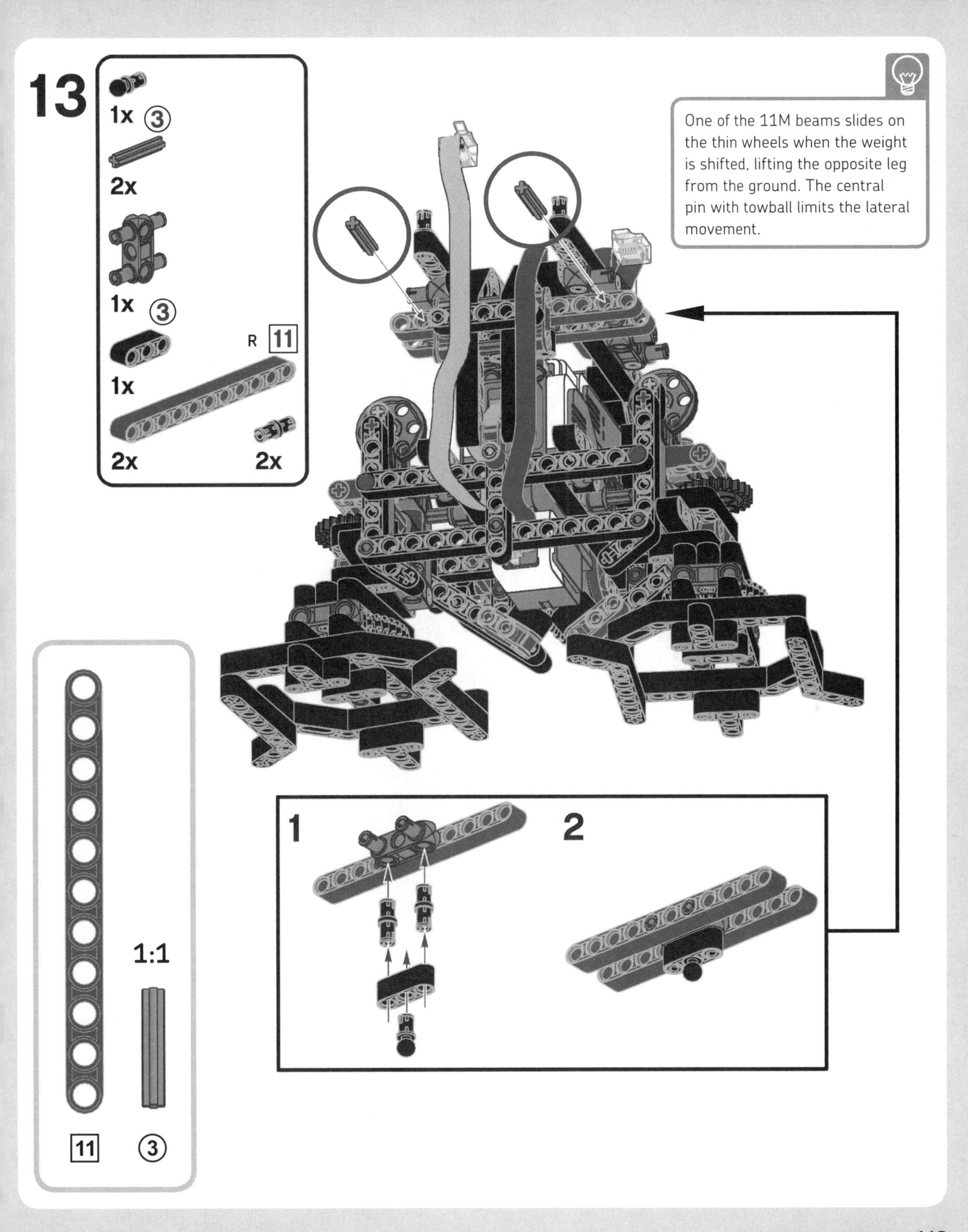
1x ③
2x
1x ③
R 11
1x
2x
2x
One of the 11M beams slides on the thin wheels when the weight is shifted, lifting the opposite leg from the ground. The central pin with towball limits the lateral movement.
1
2
1:1
11
③

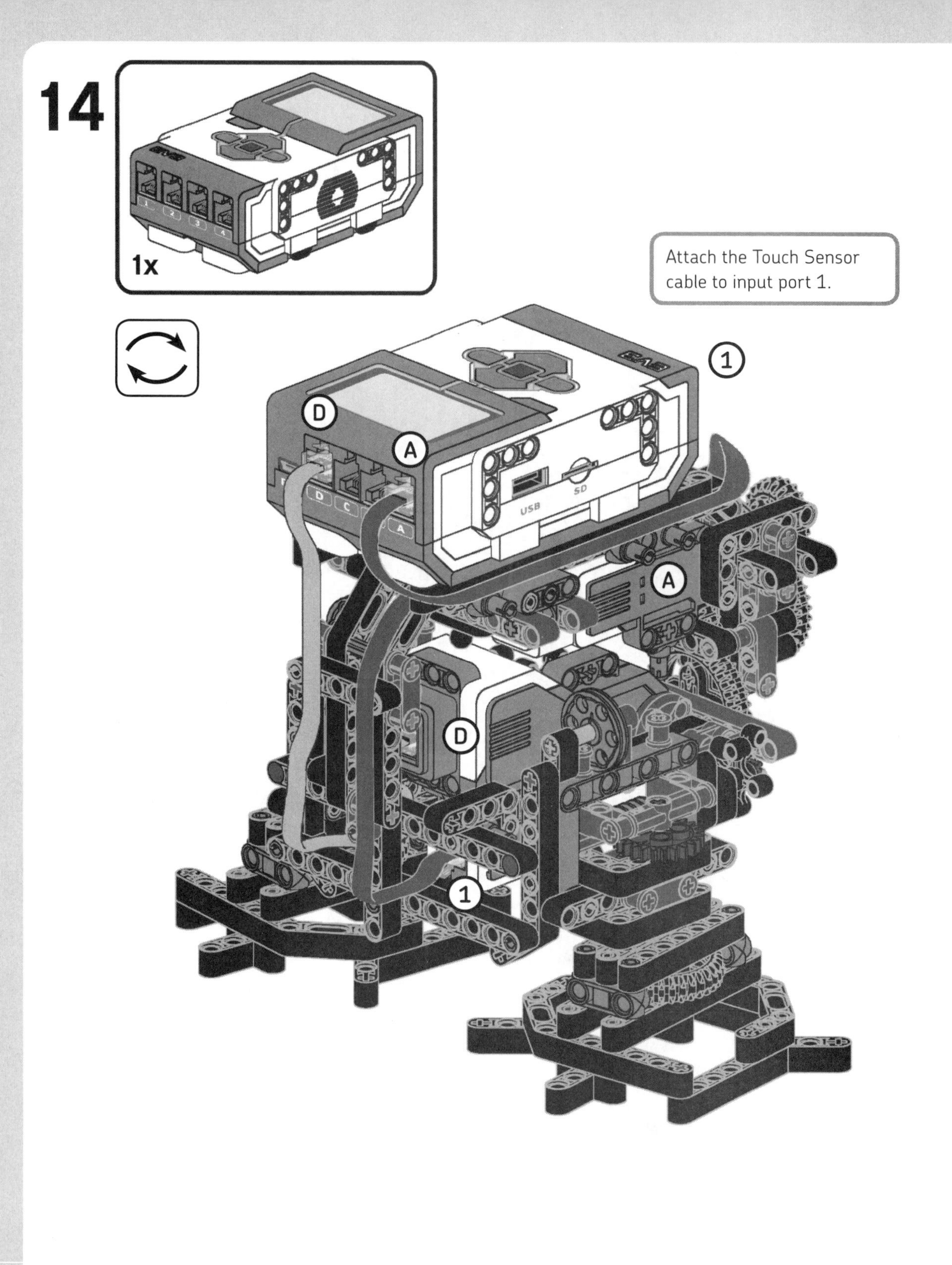
14
1x
Attach the Touch Sensor cable to input port 1.
1
D
A
A
D
1

neck assembly

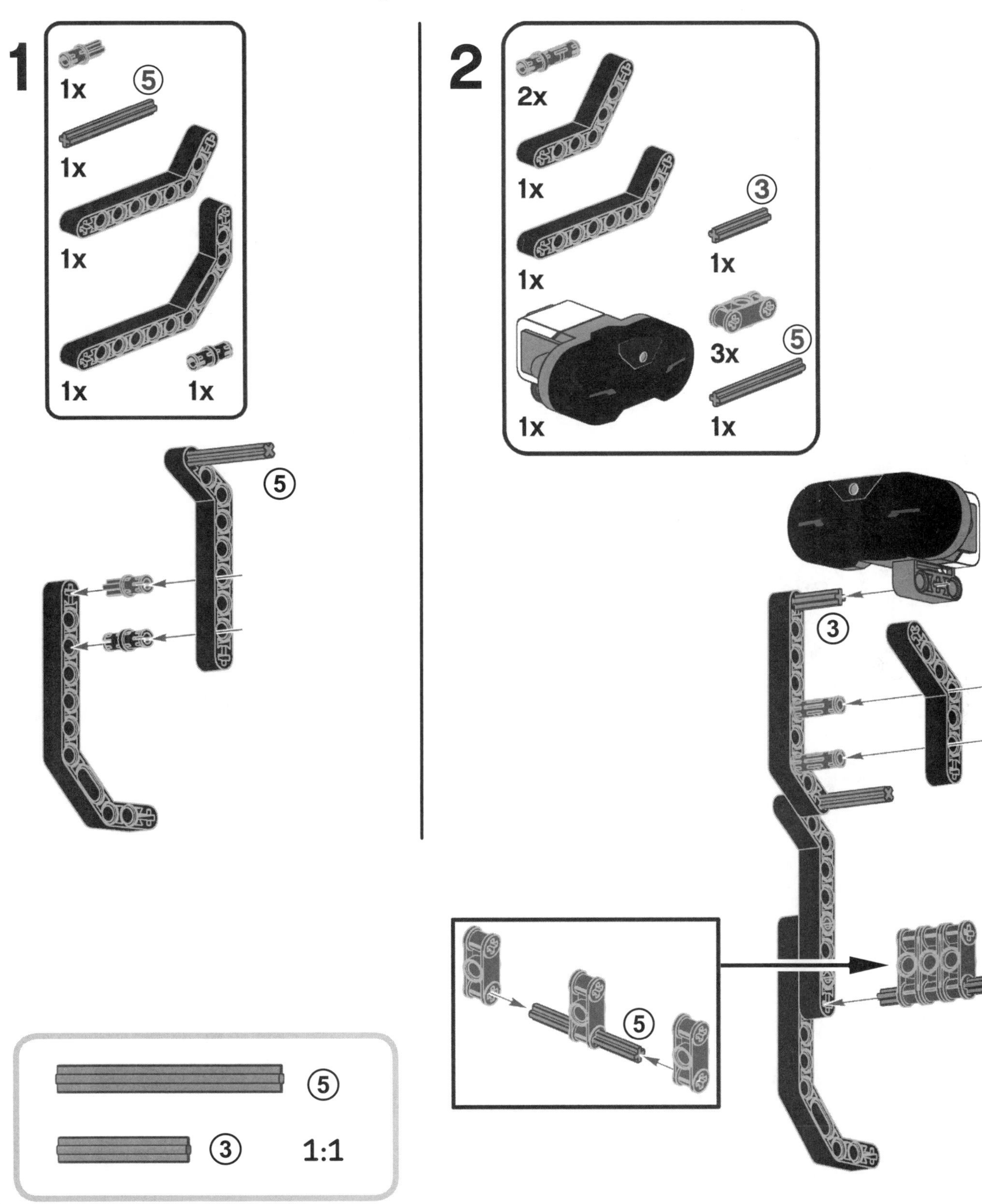

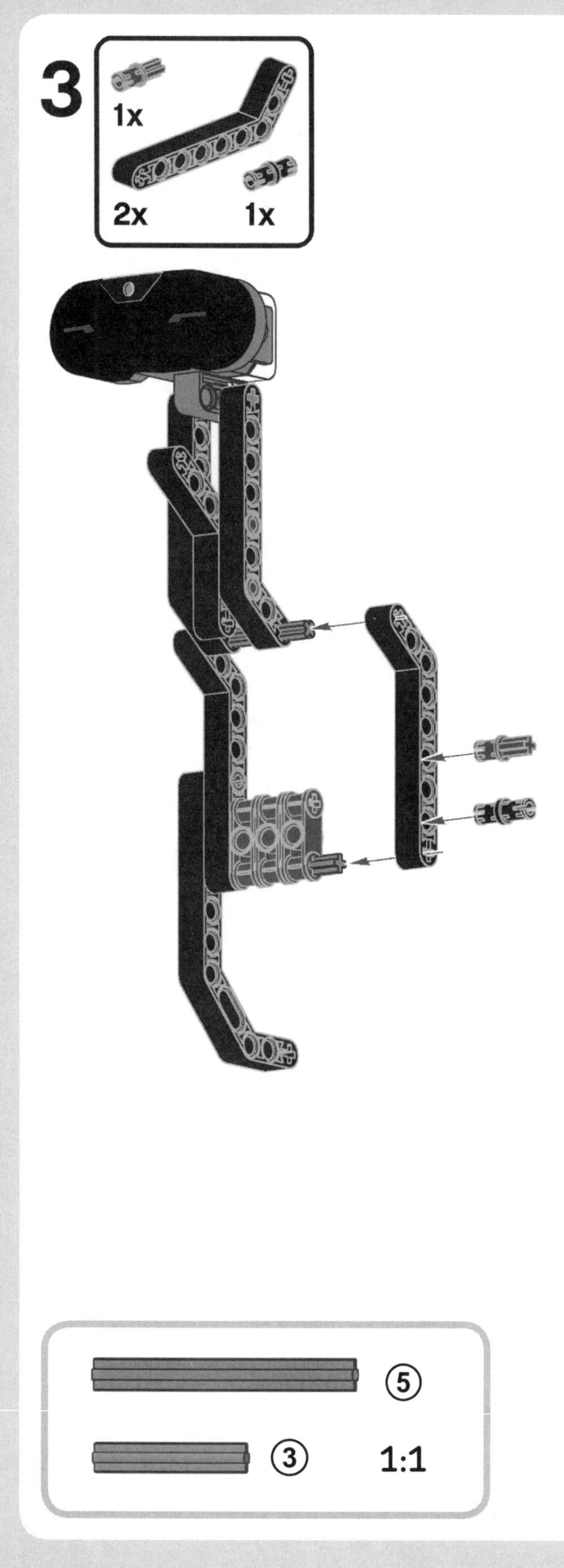
3
1x
2x
1x
5
3
1:1

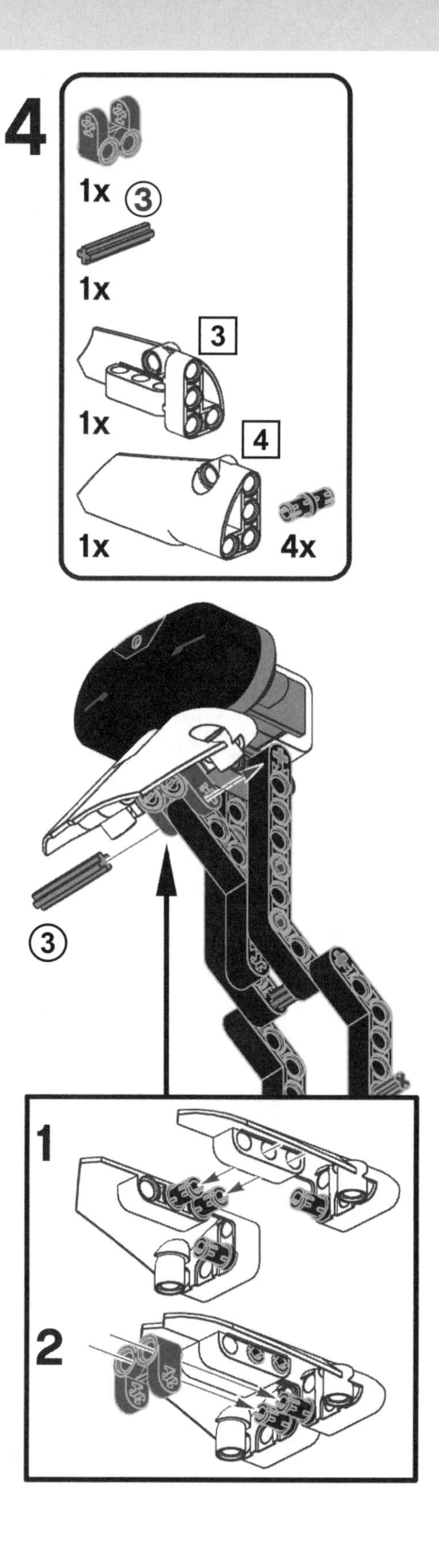
4
1x
3
1x
1x
3
1x
4
1x
4x
3
1
2

5

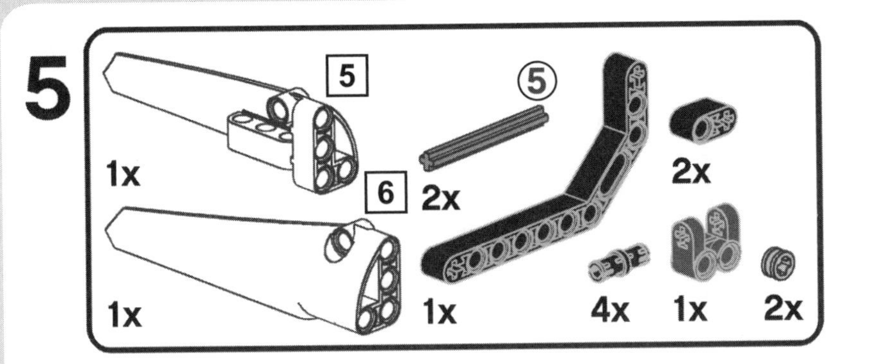

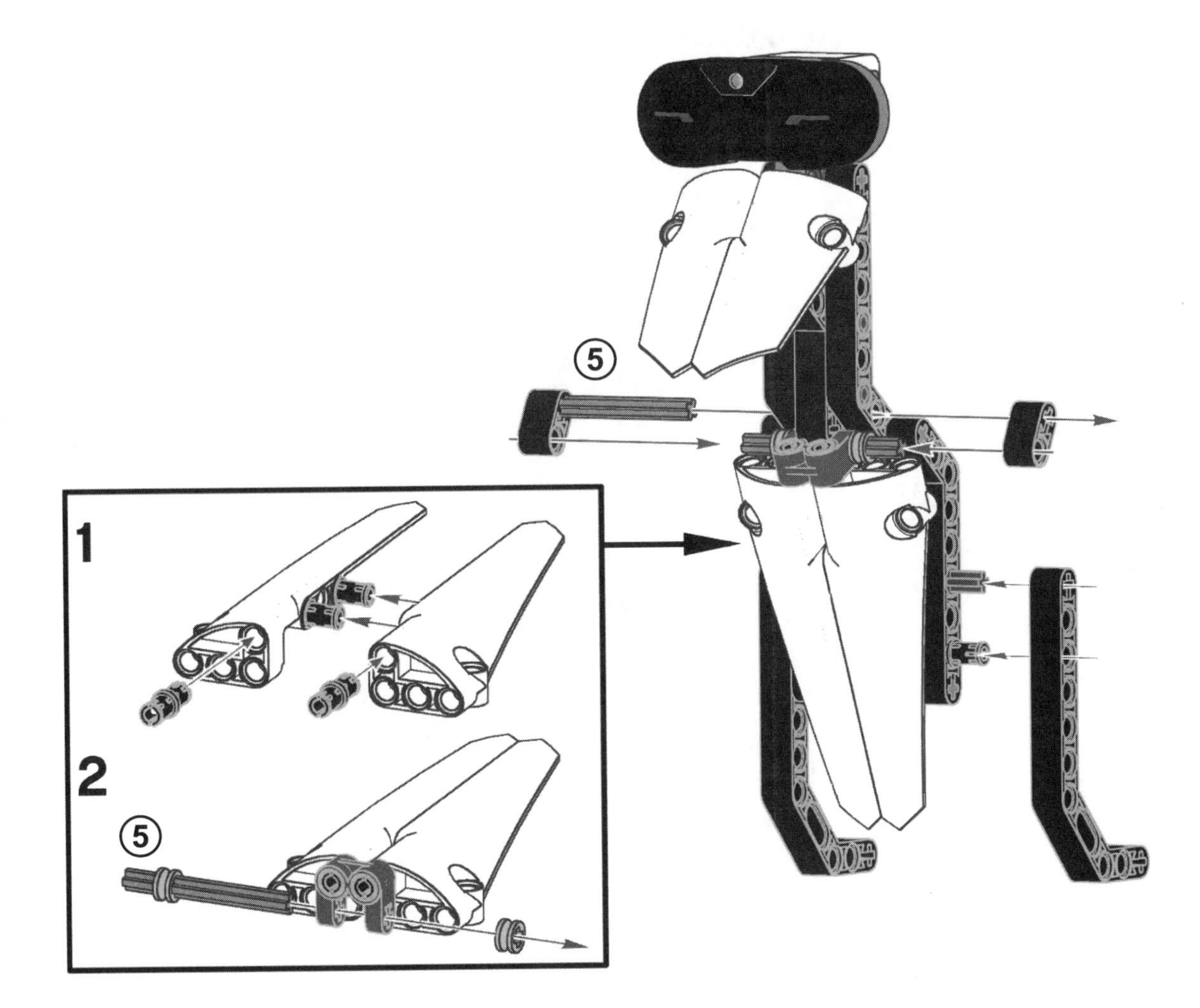

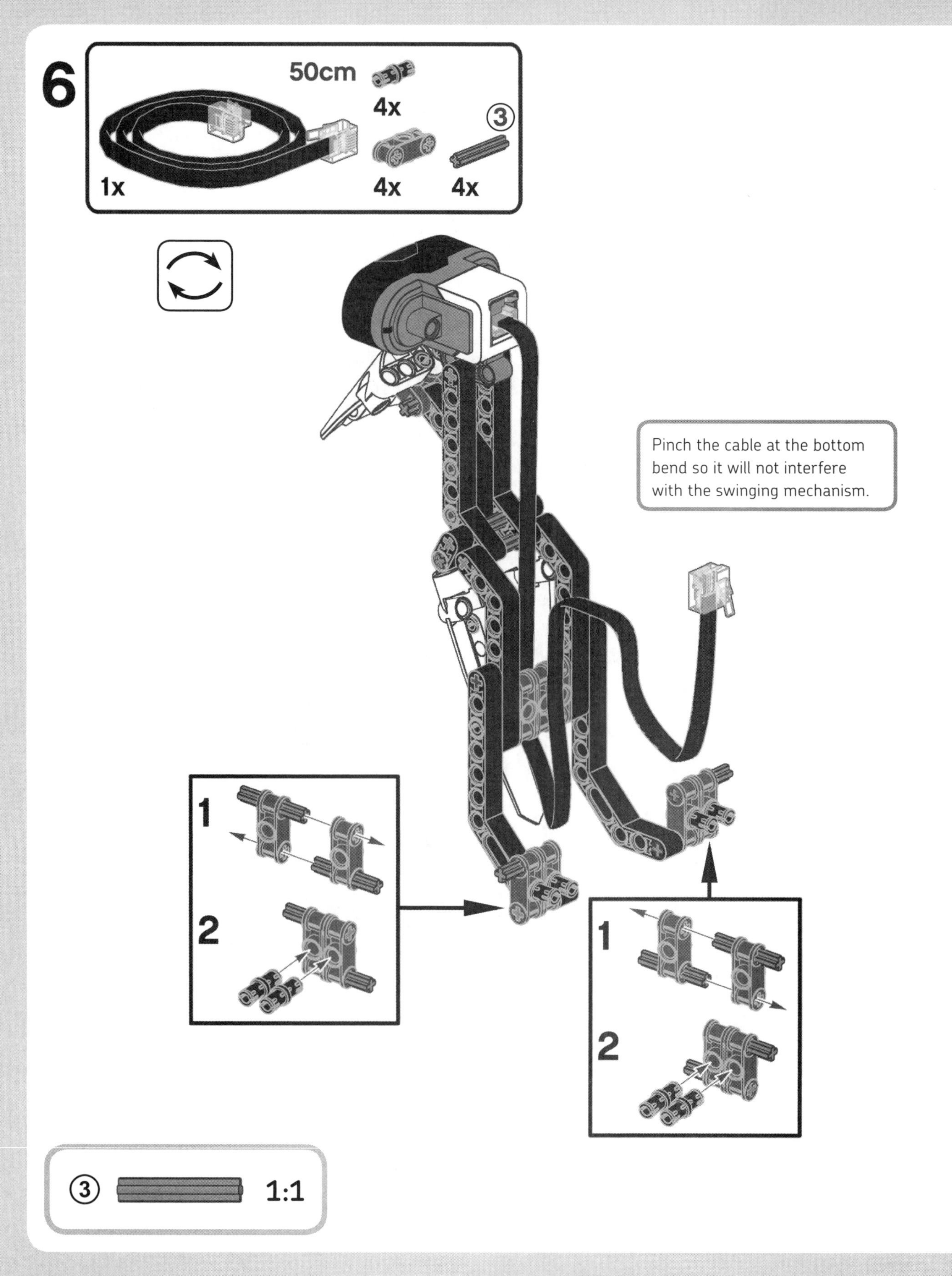
6
50cm
4x
③
1x
4x
4x
Pinch the cable at the bottom bend so it will not interfere with the swinging mechanism.
1
2
1
2
③
1:1

main assembly

15

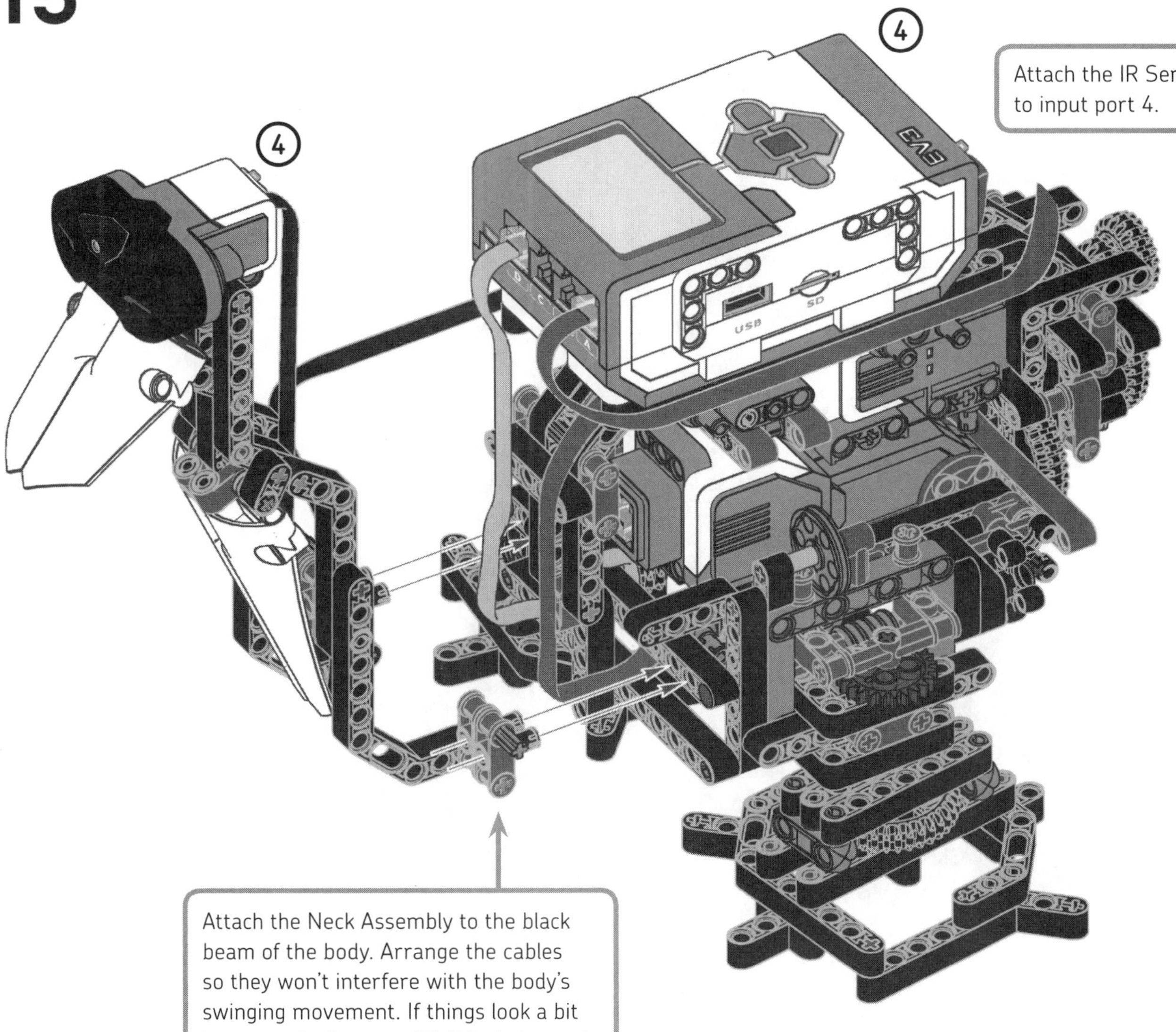

Attach the IR Sensor to input port 4.

Attach the Neck Assembly to the black beam of the body. Arrange the cables so they won't interfere with the body's swinging movement. If things look a bit insecure, don't worry: We'll lock the neck into place in the next step.

16

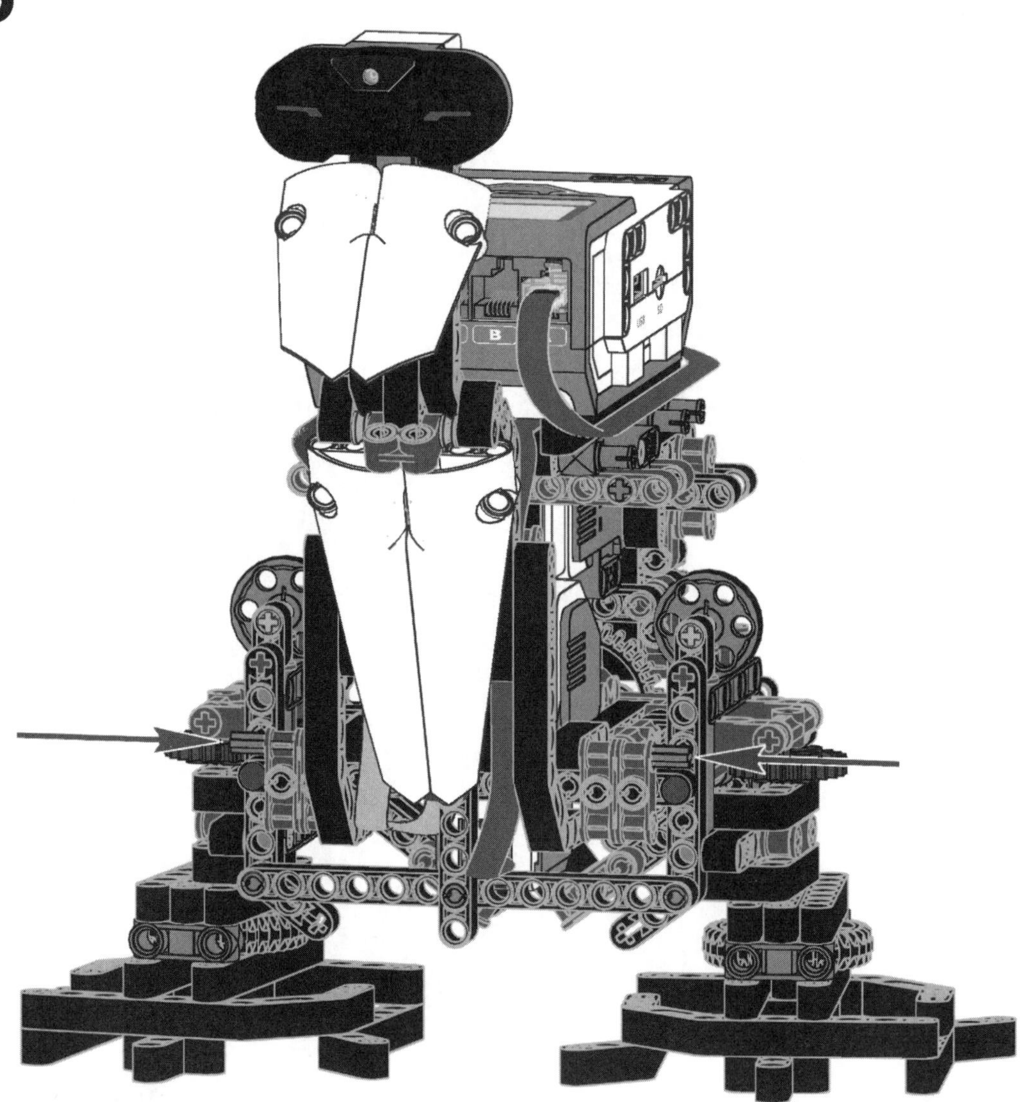

Push the 3M axles into the cross holes of the 2×4 black angular beams to lock the Neck Assembly to the Main Assembly.

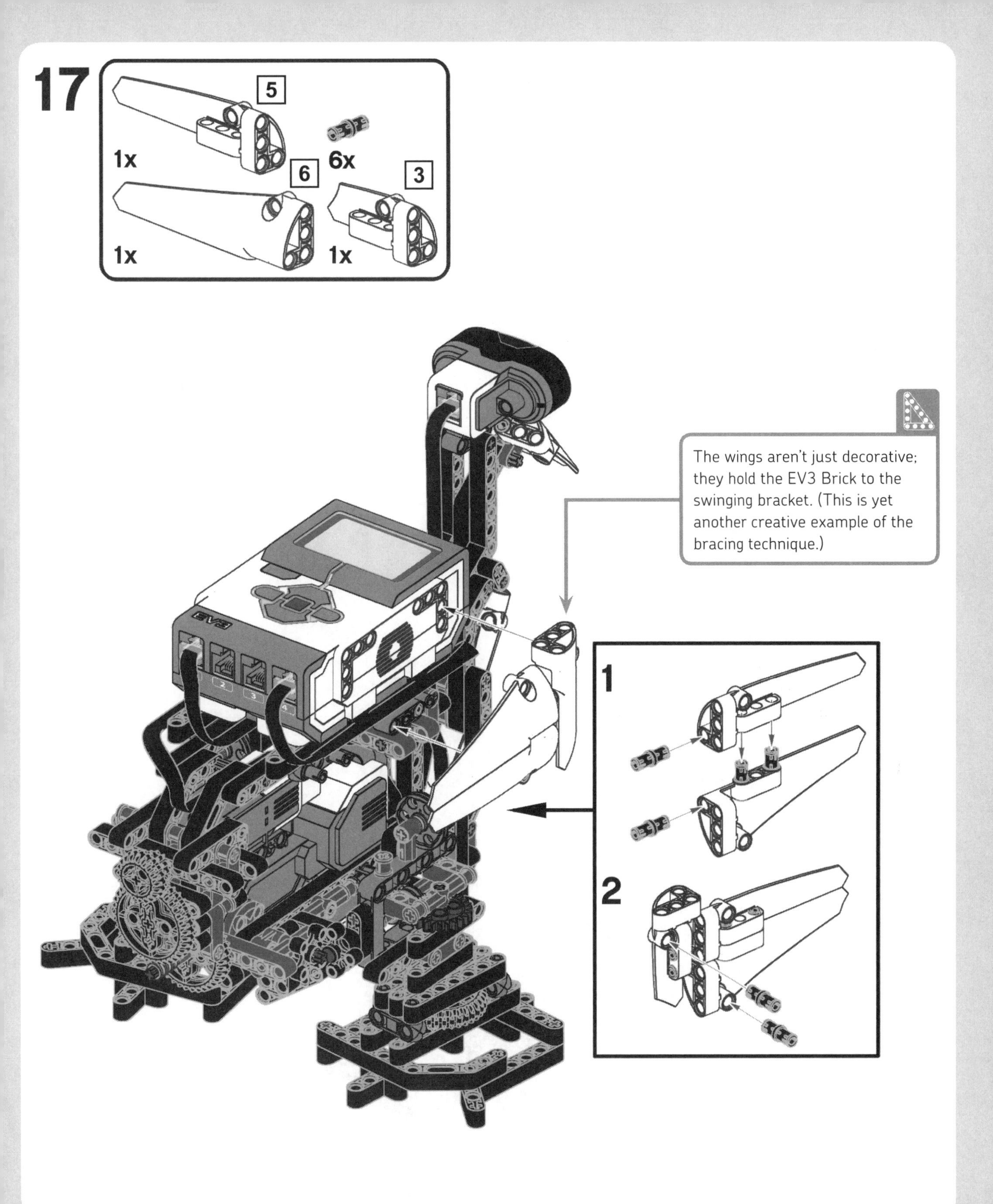
17
5
1x
6x
6
3
1x
1x
The wings aren't just decorative; they hold the EV3 Brick to the swinging bracket. (This is yet another creative example of the bracing technique.)
1
2

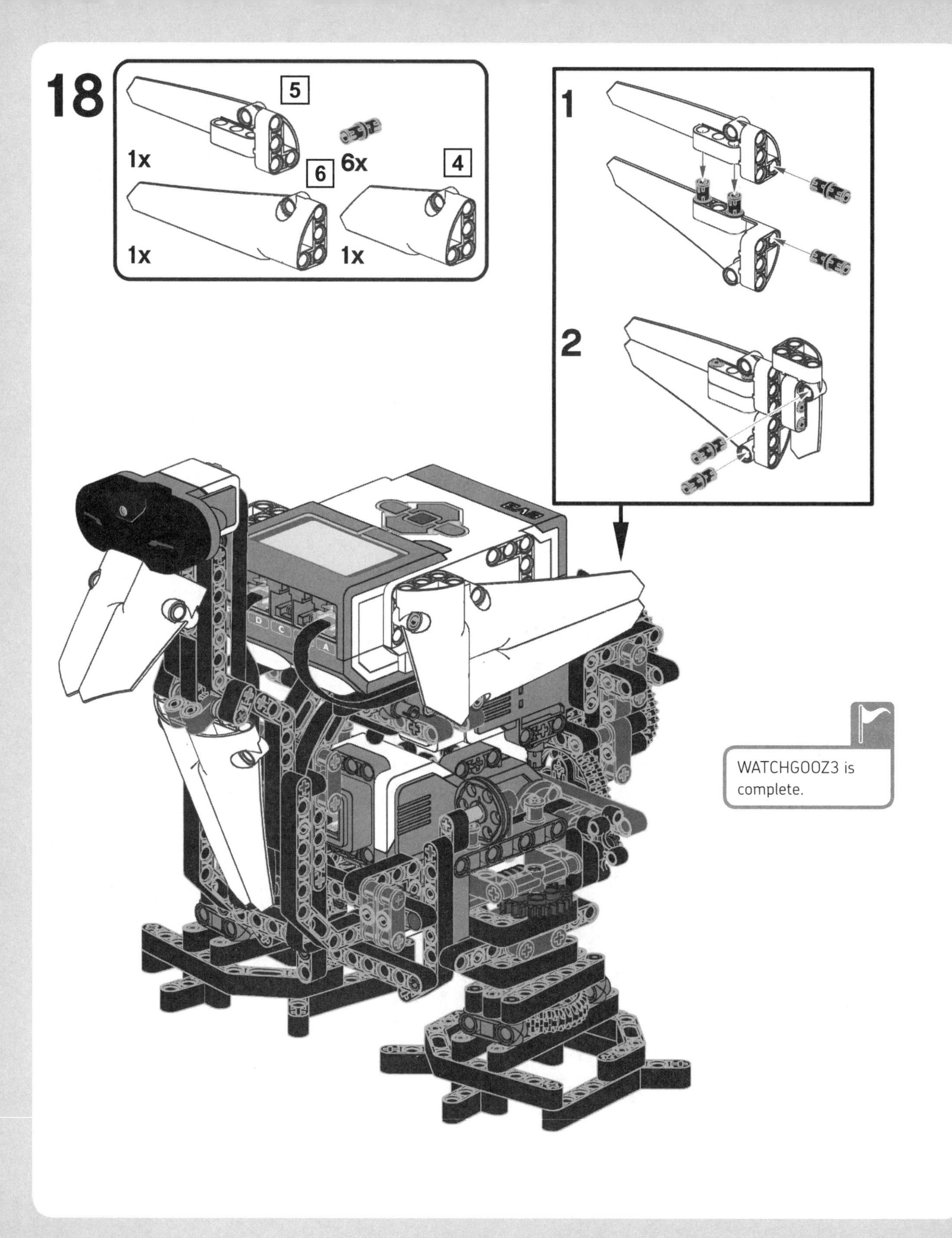
18
5
1x
6x
6
4
1x
1x
1
2
WATCHGOOZ3 is complete.

HIISSSSSS
QUAAACK!
GOTCHA!
BREEKK!
CLICK!
OFF
BLEEAARGH!!
09X1

10

programming WATCHGOOZ3

In this chapter, you'll learn to program the goose robot (WATCHGOOZ3) that you built in Chapter 9. But before using the EV3 Software, I'll show you how to program the robot with only the Brick Program App. That's right—this biped robot is designed so that it can be programmed to walk and avoid obstacles with just 16 programming blocks! Once you've finished creating the program in the Brick Program App, you'll import your finished program into the EV3 Software and learn how to use new blocks such as the Logic Operations block, the Timer block, and the Unregulated Motor block.

the brick program for WATCHGOOZ3

WATCHGOOZ3 uses two motors connected to ports A and D to shift its weight from side to side and to take steps by turning its ankles. The robot also has two sensors: a Touch Sensor and an IR Sensor.

When pressed, the robot's Touch Sensor tells the EV3 Brick that the robot's weight is fully shifted to one side (though the EV3 Brick can't determine which side). When the Touch Sensor is released, the robot knows its weight is almost vertical and both feet are touching the ground. The IR Sensor, which is used to create the shape of the goose's head, detects obstacles along the way.

When the robot is leaning (that is, balancing on one foot), it briefly turns the ankle that supports its weight to take a step forward. The key feature that allows the robot to avoid obstacles is a Wait IR Sensor block, which pauses the program while the ankle is turning until the obstacle is no longer in sight. At that point, the way ahead is clear, and WATCHGOOZ3 proceeds.

This software solution works because the robot's hardware was designed with the constraints of the Brick Program App in mind. As a rule, it's hard (if not impossible) to make a robot with badly designed hardware work well, even when it's running the best software. Good hardware design can make your programming much simpler, save you time, and make your robots work more reliably.

the program

As mentioned at the start of this chapter, you can program this robot using only the Brick Program App. So let's begin! Turn on the EV3 Brick and open the Brick Program App (the third tab and fourth app in the menu, as shown in Figure 3-6 on page 49). Then build the program shown in Figure 10-1.

how it works

This Brick Program uses 16 blocks, the maximum number of blocks allowed by the Brick Program App. The sequence, which repeats forever, is basically divided into two symmetrical parts (the two rows shown in Figure 10-1). Let's analyze the program block by block.

The first Medium Motor block turns on motor A at **–50 percent** power (the negative sign means reverse) to shift the EV3 Brick's weight to the right. Motor A runs until the Touch Sensor is **released** (first Wait Touch Sensor block) and then **pressed** again (second Wait Touch Sensor block). We need to wait for the sensor to be released first because if the weight were fully shifted to the left, the sensor would be pressed. If the program used only the Wait Touch Sensor block to wait for the pressed state, the motor would stop shifting the weight at once because the Touch Sensor would still be pressed. The program would continue, turning the ankles when the weight was still on the same side, resulting in the robot walking backward—not what we want!

NOTE A Wait Touch Sensor block does not pause the program if its ending condition is met immediately (in general, any Wait block works like this). For example, a block that should wait for a Touch Sensor to be released will not pause the program if the Touch Sensor is already released.

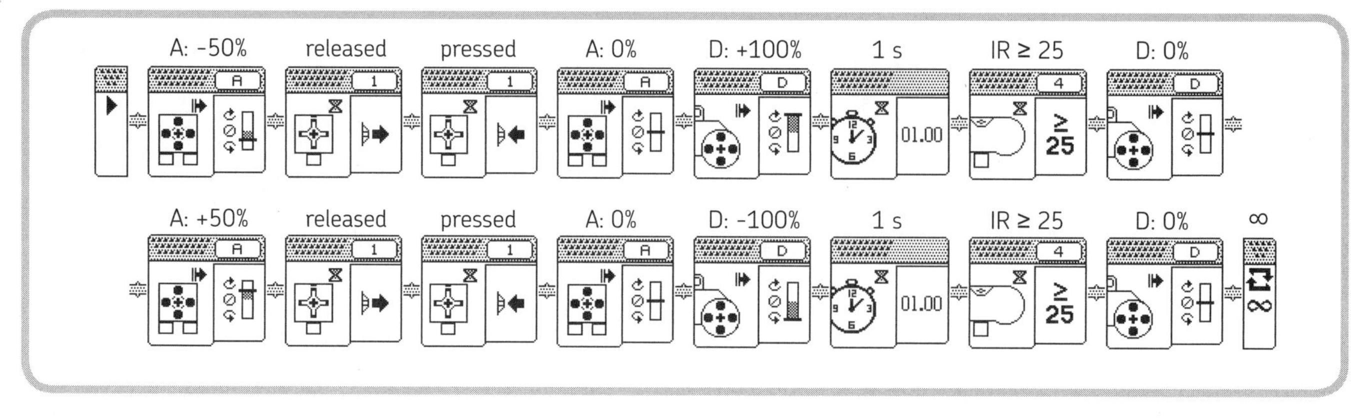

Figure 10-1: The Brick Program for WATCHGOOZ3

Once the Touch Sensor is released and pressed again, motor A is stopped (by setting the power to **0 percent**), and motor D is started by a Large Motor block set at **+100 percent** power (forward). This large motor turns the ankles to make the robot step forward. Now the Wait Time block pauses the sequence for **1 second** while motor D is still turning the ankles. If the robot detects an obstacle, the Wait IR Sensor block pauses the program until the Proximity reading goes above **25 percent**. If there is no obstacle, the Wait IR Sensor block lets the program continue and the ankle turns for just 1 second. In fact, following the Wait IR Sensor block, another Large Motor block stops the large motor D by setting the Power to **0 percent**.

This sequence is mirrored and repeated again, as shown in the second row of Figure 10-1. A Medium Motor block starts motor A at **+50 percent** power (forward this time) to shift the weight left. Then two Wait Touch Sensor blocks wait for the leg mechanism to release and then press the sensor again. Now the weight-shifting motor A is stopped, and the stepping motor D is started at **–100 percent** power (backward). The Wait Time and Wait IR Sensor blocks are used to make the robot take a step and eventually turn, as before. The last Large Motor block stops the stepping motor D, and the sequence repeats because the Loop block is set to Infinity (∞).

Once you've finished building the Brick Program shown in Figure 10-1, save it with the name *GOOSE* (if you don't remember how, see Chapter 3).

running and troubleshooting the robot

Before you run the program, rotate the black 20z gear attached to the axle of the Medium Motor shaft to position the swinging weight vertically, releasing the Touch Sensor and placing both of the robot's feet on the ground. Now run the program. The robot should begin to walk with a funny, goose-like, swinging gait.

If the weight-shifting mechanism isn't smooth or it gets stuck and you hear clicking from the back gear train (the sound of the gears disengaging), stop the program. Check the cables passing between the swinging frame and the neck and the Touch Sensor in the bottom of the robot.

The stiff cables in the front of the robot must not disturb its swinging movement (see step 15 on page 171). If they catch on some part while the weight is being shifted, bend them out of the way (pinch them so that they keep their shape). If the Medium Motor does not stop when the weight is completely shifted to one side, make sure that the Touch Sensor is pressed correctly by the lower levers.

importing and editing the program in the EV3 software

Now to import our program into the EV3 Software. First, create a new project. Then, import the *GOOSE* Brick Program into the EV3 Software (see "Importing a Brick Program" on page 76 for how to do this).

The resulting EV3 program should look like Figure 10-2. Rename it by double-clicking the program name tab and entering the name *BP_GOOSE*. Save the project as *myWATCHGOOZ3*.

NOTE The screenshots of the EV3 programs in this book have been edited for readability. For example, they will sometimes be split into many rows, as in the program shown in Figure 10-2.

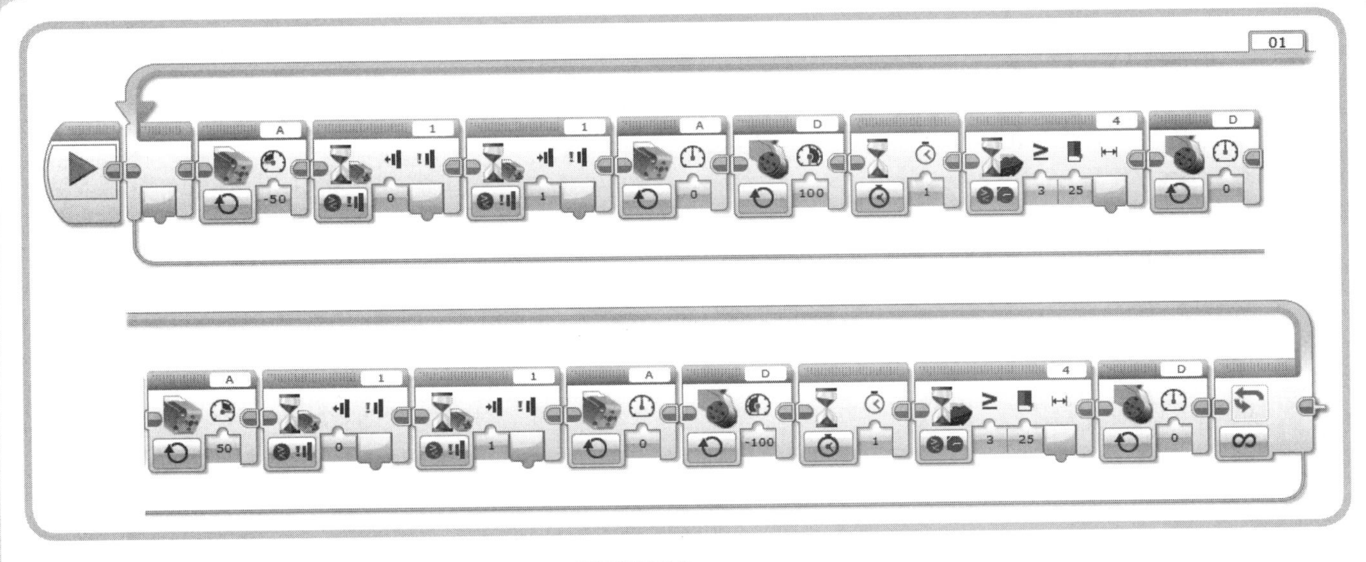

Figure 10-2: The EV3 Software equivalent of the Brick Program for WATCHGOOZ3

making a backup

Before you modify the program in Figure 10-2, make a backup copy. To do so, go to **Project Properties** (click the wrench icon at top left), go to the **Programs** tab, select the *BP_GOOSE* program, and click **Copy** and then **Paste** (the buttons are at the bottom of the list). The copied program name should be *BP_GOOSE2*. Double-click to open it in the Programming Area and then double-click the program name tab to rename it *BP_EDIT*. Now we'll continue editing it.

modifying the program

In the Brick Program App, we set the Medium Motor block parameter to 0 percent to stop the Medium Motor. In EV3 language, the equivalent is a Medium Motor block in Off mode with *Brake at End* set to **True**.

To make the robot's movements a bit smoother, change all Motor blocks currently in On mode with Power set to 0 percent to **Off** mode with *Brake at End* set to **False**. (Do this for the highlighted Medium Motor blocks and the Large Motor blocks.) Your program should now look like Figure 10-3. **Download and Run** the program to see how the robot moves differently.

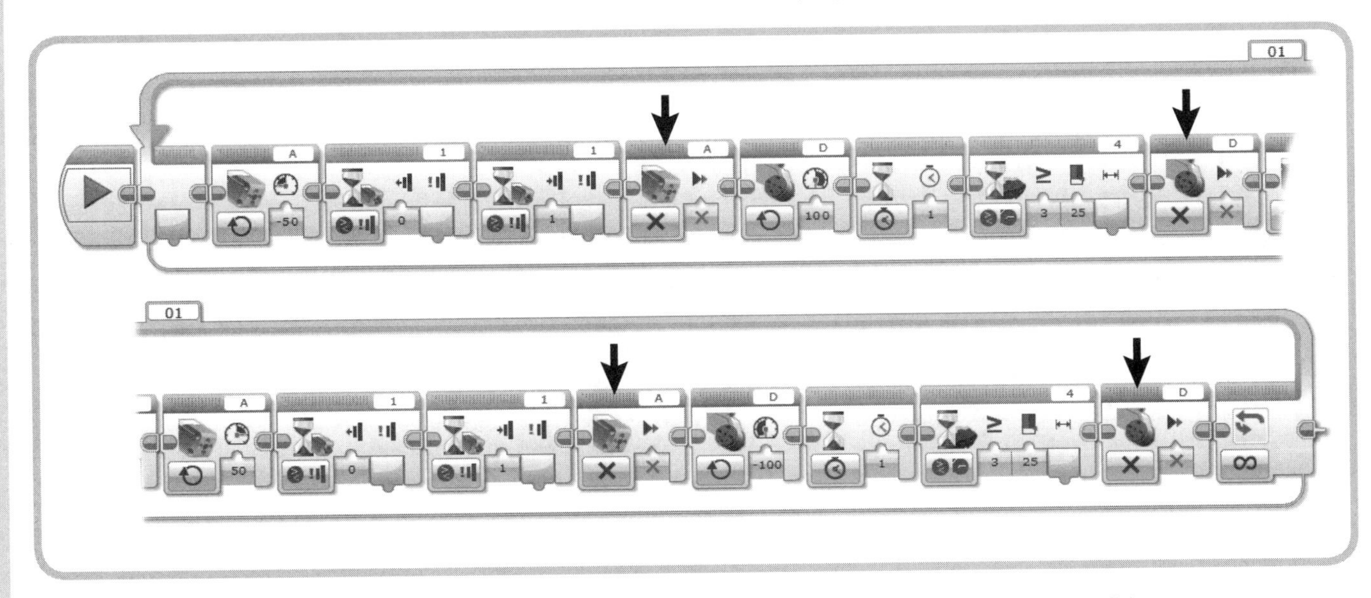

Figure 10-3: A slightly modified program for WATCHGOOZ3. Set the highlighted blocks to Off mode with the Brake at End input set to False.

creating My Blocks with the My Block Builder tool

The EV3 program we just made is simple but a bit bulky. Also, it's made up of two almost identical parts. To make this program more compact and readable, we'll group blocks into a single block using the My Block Builder tool. But first, back up the *BP_EDIT* program as discussed earlier and rename the new program *BP_EDIT_MB*. This new program should still look like Figure 10-3.

1. Drag a selection window to select the first eight blocks inside the Loop, as shown in Figure 10-4(a). The selected blocks should be highlighted with a cyan border.
2. With these blocks selected, click **Tools ▸ My Block Builder**. The My Block Builder dialog should appear, as shown in Figure 10-4(b).
3. In the My Block Builder dialog, enter a name (and a description, if you like) for your custom block. Also, choose an icon to remind you of the block's function. I suggest choosing the icon showing *two Large Motors* (fifth from left in the top row), since there is no icon showing the Medium and Large motors together and you can't customize the icons. You should see a preview of the block at the top of the dialog. Enter *RightStep* as the name of the block, and for the description, try something like *Shift the weight to the right and take a step*.
4. Click **Finish** to create the My Block.
5. The previously selected blocks have been automatically replaced by your new *RightStep* My Block, which should appear with a cyan header. Once created, the My Block is added to the last Programming Palette, and it is listed in the Project Content (My Blocks tab) in Project Properties

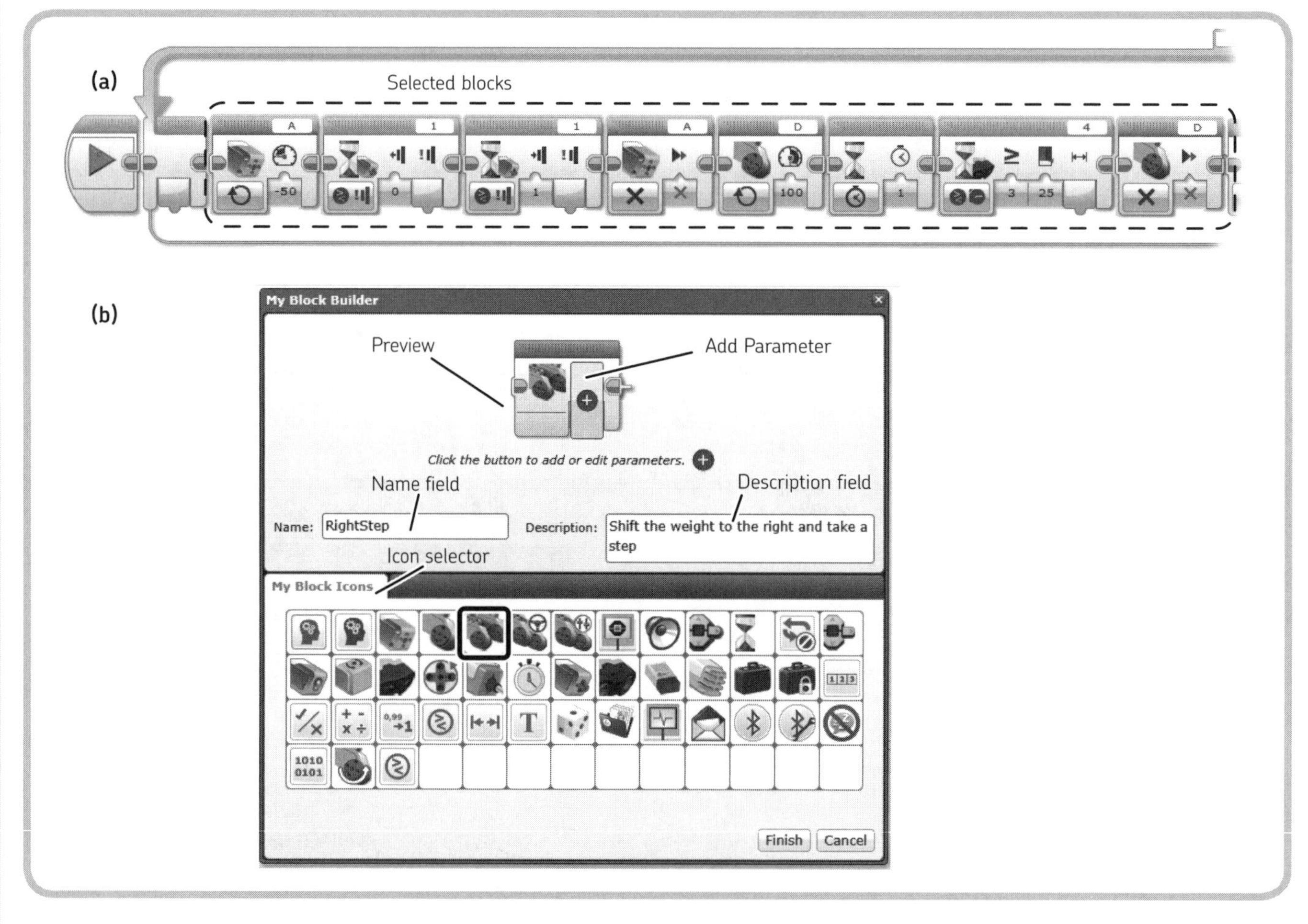

Figure 10-4: To create a My Block, select the blocks you want to include (a) and open the My Block Builder dialog (b).

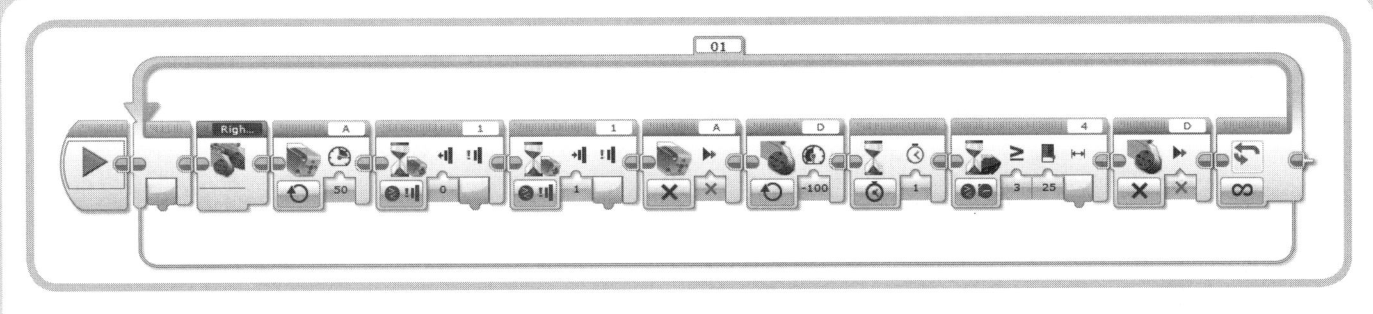

Figure 10-5: The program as it appears after creating your first My Block

(see Chapter 5 for information about Project Properties). After creating your My Block, the program should look like Figure 10-5.

6. Select the eight remaining blocks (but not your newly created My Block), and click **Tools ▸ My Block Builder**. Name the new My Block *LeftStep* and assign it the same icon you assigned to the *RightStep* My Block.

7. Click **Finish** to create the block. The program should look like Figure 10-6. Much cleaner, isn't it?

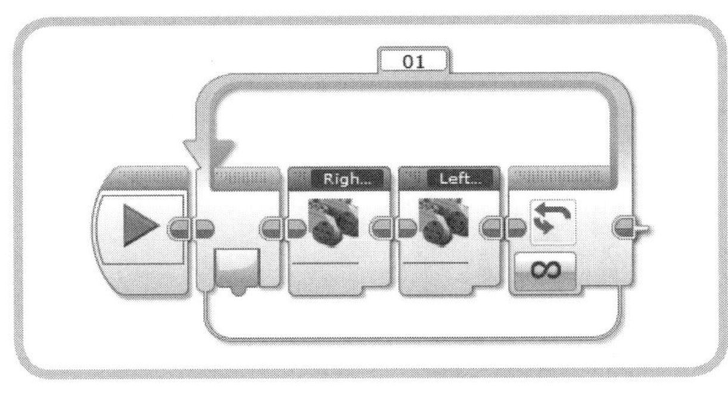

Figure 10-6: The program after creating both My Blocks

8. Run the program to make sure that the robot's operation is the same as before.

To see what's inside a My Block, double-click it. The *RightStep* My Block is shown in Figure 10-7.

WARNING Once you've created a My Block, you cannot edit its icon; its description; or the icon, name, and type of its inputs and outputs. You can edit only its content and its name (by opening it, double-clicking its name tab, and entering a new name).

creating My Blocks with inputs and outputs

Besides helping to clean up your programs and make them more readable, My Blocks are essential to grouping and reusing parts of your programs in more than one place in your projects, whether in the same or in other projects. (You can export and import My Blocks in Project Properties.)

For example, the program in Figure 10-3 is made up of two almost identical groups of eight blocks each; the only difference between the two groups is that different values are specified in the blocks' inputs. Instead of creating two different My Blocks, as in the previous section, we could make a single My Block with an input that specifies whether the robot should take a step to the right or left. You would put those two identical My Blocks into the main program and specify different inputs. Let's do it!

1. Go to **Project Properties**, select the **Programs** tab, and copy and paste the *BP_EDIT* program you made earlier. Name the copy *BP_EDIT_MB2*.

2. Using Figure 10-4(a) as reference, select the first eight blocks inside the Loop block and choose **Tools ▸ My Block Builder**. The dialog shown in Figure 10-4(b) should appear. Enter *Step* into the block's Name field and choose the icon showing *two Large Motors*. Now click the Add Parameter button—it looks like a plus sign in a circle. The Block Preview should look like the one in Figure 10-8.

3. Go to the **Parameter Setup** tab. Enter *Side* for the Name parameter. Choose **Input** as the Parameter Type and **Logic** from the Data Type drop-down menu. Set the Default Value to **True**.

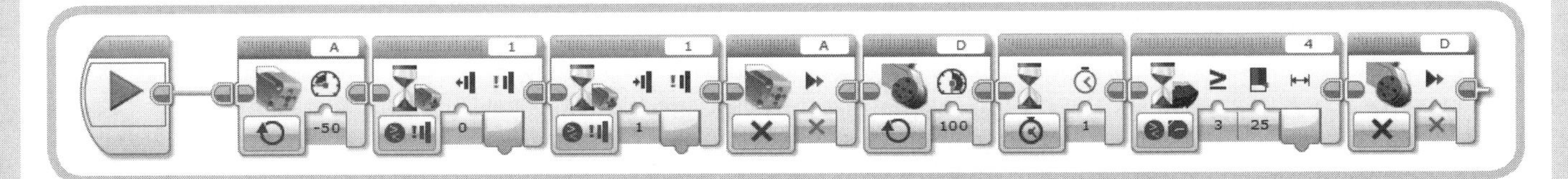

Figure 10-7: The content of the RightStep My Block

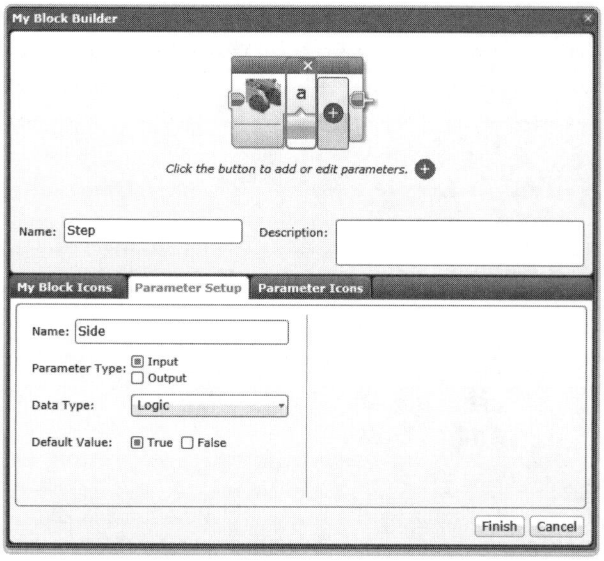

Figure 10-8: How to configure an input in the My Block Builder tool

4. Go to the **Parameter Icons** tab and select the fourth icon from the right in the top row (the true/false symbols icon), as shown in Figure 10-9.

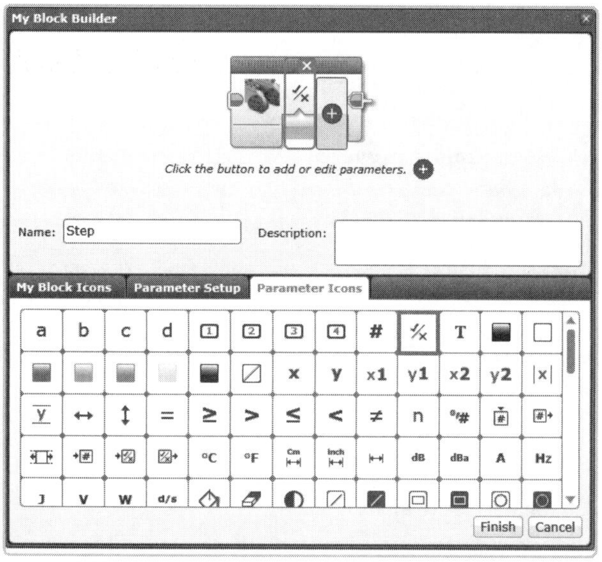

Figure 10-9: How to change the icon of a parameter in the My Block Builder tool

5. Click **Finish** to place the *Step* My Block into your program.
6. Delete the remaining eight blocks inside the Loop block, leaving only the *Step* My Block in the Loop.
7. Select and drag the *Step* My Block while holding the CTRL key to copy it; then drop it beside the other block. Now set the first block input to **True** and the second to **False** to use the same My Block to shift the weight from side to side. Your program should look like Figure 10-10.

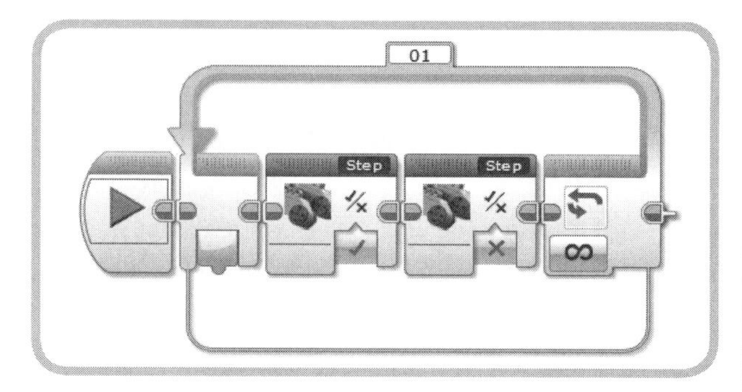

Figure 10-10: The finished program BP_EDIT_MB2, *with two My Blocks set with different parameters*

8. We're not done with the *Step* My Block quite yet, because the input is not connected to the blocks in the inner sequence of the My Block. Double-click either of its two instances to edit it. It should open, as shown in Figure 10-11. As you can see, the *Side* input has no Data Wire yet.
9. Add three Math blocks, as shown in Figure 10-12. Change the first to Advanced mode and enter **1-2*a** in its Formula field. This block will convert the logic value *True* to –1 (1 – 2 × 1 = –1) and *False* to 1 (1 – 2 × 0 = 1). We need it to switch the direction of the motors according to the value of the *Side* input parameter.
10. Change the second and third Math blocks to Multiply mode and set input **b** to **50** and **–100**, respectively.
11. Connect the *Side* logic input of the My Block to input **a** of the first Math block.
12. Drag Data Wires from the output of the first Math block to input **a** of the other Math blocks and connect their outputs to the Medium Motor and Large Motor blocks, as shown in Figure 10-12.
13. Now go to the main program *BP_EDIT_MB2* and run it to check the robot's operation.

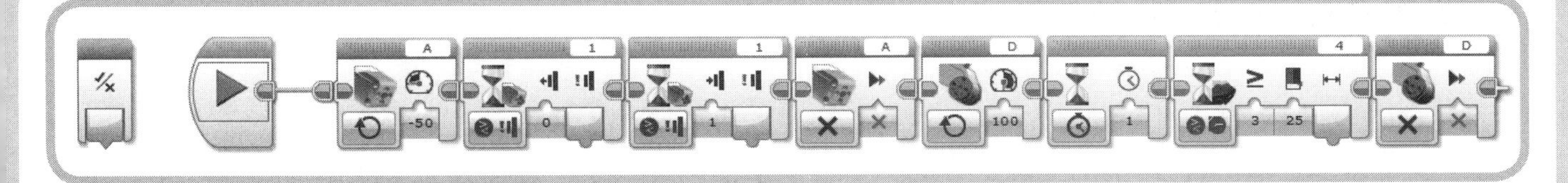

Figure 10-11: The Step *My Block before it is completed*

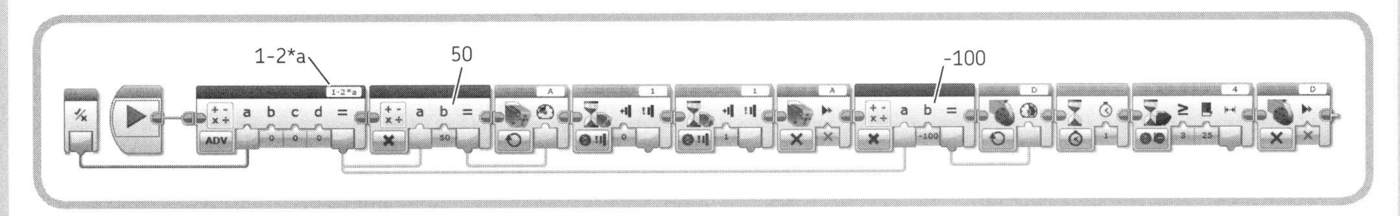

Figure 10-12: The finished Step *My Block*

automatically adding inputs and outputs to My Blocks

Now you'll learn how to use the My Block Builder tool to add inputs and outputs automatically when you create a My Block.

1. Open the *Step* My Block and select the first Math block, which is configured in Advanced mode (see Figure 10-13).

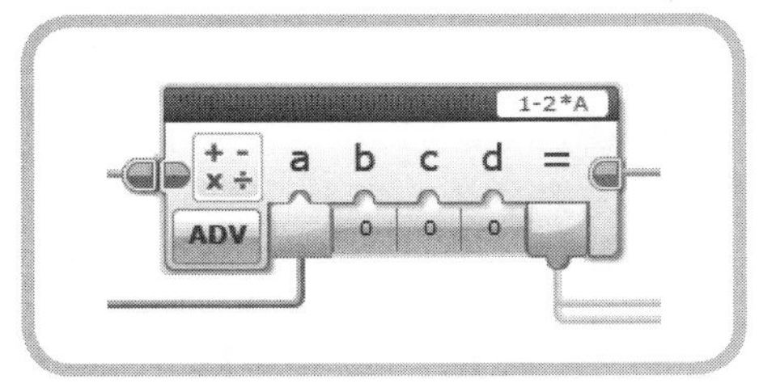

Figure 10-13: The selected block has incoming and outgoing Data Wires. When this block is selected to make a My Block, the My Block tool will add inputs and outputs automatically.

Figure 10-14: Configure the My Block with automatically assigned inputs and outputs.

2. Now open the My Block Builder tool to create a My Block. When your selection includes blocks with incoming or outgoing Data Wires, the My Block Builder dialog will automatically create specific data inputs and outputs of the correct type.
3. Change the My Block icon to look like the one shown in Figure 10-14 and enter the name *LogicToSign*. The My Block Builder tool has already configured the block's Logic input and Numeric output. Change the input name to *Input* and the output name to *Result*. Set the input Default Value to **True** and change the Parameters icons as shown in the My Block preview in Figure 10-14.
4. Click **Finish** to create the My Block. Your finished My Block should look like Figure 10-15.

Figure 10-15: The resulting LogicToSign *My Block, part of the* Step *My Block*

5. Now test the *BP_EDIT_MB2* program again to make sure that everything works as before.

additional configuration of a My Block

In addition to allowing you to add and remove parameters, the My Block Builder tool allows you to change the order of the parameters [Figure 10-16(a)]. Also, you can choose from among three styles for the Numeric inputs, as shown in Figure 10-16(b): Text input, Horizontal slider, and Vertical slider. You can also specify the maximum and minimum value of the input for the Slider styles. Figure 10-16 shows a fictional My Block with inputs and outputs of all possible data types.

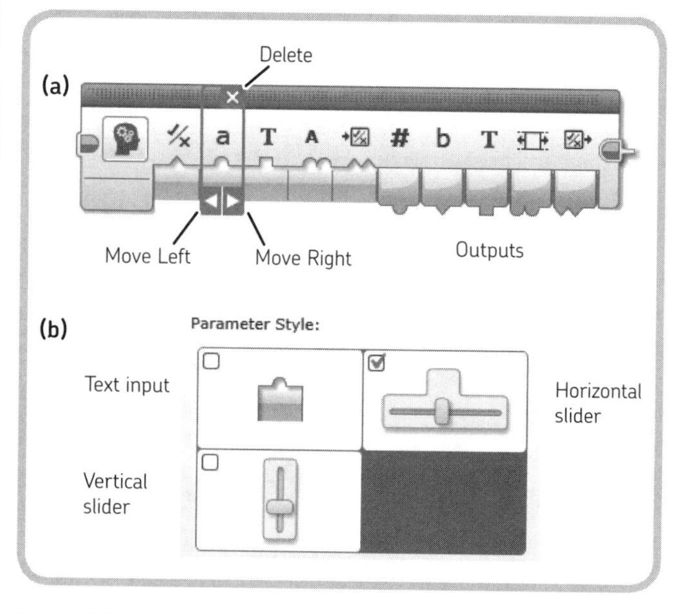

Figure 10-16: The My Block Builder tool's additional controls to delete and change the order of the parameters (a) and to change the style of a Numeric input (b)

NOTE A My Block replaces the blocks used to create it. To preserve the state of the program you are working on without replacing its blocks, you can use a temporary program to create My Blocks. Alternatively, you can click Undo after creating a My Block to bring the blocks you placed in the My Block back into the program; the My Block will remain in the project.

creating an advanced program

As explained in "Running and Troubleshooting the Robot" on page 178, to make the robot work correctly, you had to make sure that the weight was vertical, the Touch Sensor was released, and both feet were touching the ground. In this section, I'll show you how to make an initialization routine that will place the weight in the correct position automatically when the program first runs.

To create this program, you'll build the *ResetBody* My Block and add it at the beginning of the main program, before the Loop block. Then you'll make a modified version of the *Step* My Block that makes the robot turn in the same direction at all times, allowing it to find its way out of dead ends.

the ResetBody My Block

The *ResetBody* My Block (Figure 10-17) will be placed at the very beginning of the program.

Here's how the *ResetBody* My Block works.

* The Switch block in Touch Sensor Compare State mode (1) checks whether the Touch Sensor is pressed. If it is not pressed, the program continues with the main Loop block. If the Touch Sensor is pressed, we know that the weight is fully shifted to one side and the walking sequence can't start because we don't know which side the weight is shifted to. If we don't release the Touch Sensor before the main loop starts, the *Step* My Block (Figure 10-12) will turn on motor A, and the program might hang at the first Wait block, waiting for the Touch Sensor to be released. In that case, the motor would try to shift the weight, but since the weight would already be at its limit, the mechanism would be stressed; gears would disengage or even come apart. To avoid this potential problem, we need to move the weight to release the Touch Sensor.
* There is a new Advanced block inside the Switch block, the Unregulated Motor block (2). This block runs the motor at a certain power level, without regulating its speed. I use this block instead of a Medium Motor block so that if the motor sticks, the internal controller won't try to increase the power applied to the motor in vain. (When the motor is turned on but blocked, it overheats and drains the batteries very quickly.)
* Following the Unregulated Motor block, you see a Timer block in Reset mode (3). This block allows you to reset a timer so that you can measure the time elapsed from this reset.

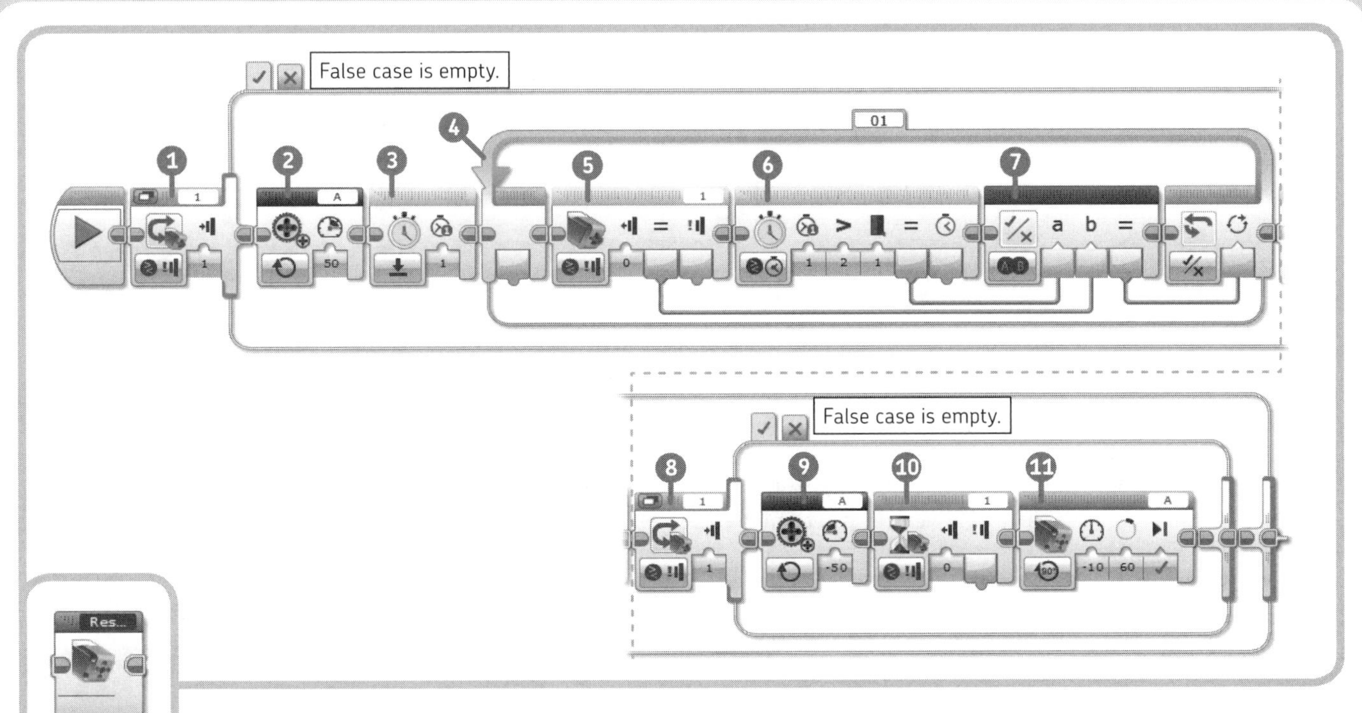

Figure 10-17: The ResetBody *My Block*

* The Loop block (4) that follows the Timer block is a waiting loop that runs until the Touch Sensor is released or until a specified time elapses, whichever occurs first.
  * The Loop continuously checks the Touch Sensor block output (5) (in Compare mode) and the Timer block output (6) (in Compare mode). As soon as the Touch Sensor block output becomes True (meaning that the Touch Sensor has been released) *or* the Timer block output becomes True (meaning that 1 second has elapsed), the Loop block ends. In this way, if the motor is running to bring the weight against the limit instead of lifting it to its vertical position, the Touch Sensor will remain pressed, but the motor will soon stop running, thereby avoiding high mechanical stress.
  * These two logic conditions are combined by a Logic Operations block in OR mode (7).
* At this point in the program, if the motor is turning in the direction needed to release the Touch Sensor, the weight will be almost vertical, and the Touch Sensor will be released. However, if the motor is running in the opposite direction, attempting to push the weight past its limit, the Touch Sensor will still be pressed. (The fact that the Touch Sensor is pressed only tells us whether the weight is fully shifted; it doesn't tell us which side it's shifted to.) The Switch block (8) contains blocks only in the True case; if the Touch Sensor is released, we're done.
* If the Touch Sensor is pressed, an Unregulated Motor block (9) turns Medium Motor A in the opposite direction to that of the other block (2).
* A Wait block in Touch Sensor Compare mode (10) waits for the Touch Sensor to be released. At this point, with the previous checks having been made, the motor is definitely turning in the direction needed to release the Touch Sensor.
* Once the Touch Sensor is released, Medium Motor block (11) turns the motor by a precise number of degrees to bring the weight up vertically, toward the center.

Now to build the *ResetBody* My Block. Add a new temporary program to your project and build the sequence shown in Figure 10-17. Add the blocks in the order specified by the numbers. (The Timer block is in the Sensors palette and the Logic Operations block is in the Data Operations palette.) Once you've finished, select the external Switch block (which includes all the other blocks) and use the My Block Builder tool to create a My Block called *ResetBody*. Have it display the Medium Motor icon, as shown in Figure 10-17. Once you've created the *ResetBody* My Block, delete it from the current program and use the empty canvas to create the next My Block, a modified version of the *Step* My Block.

creating the advanced My Block for walking

Now we'll create a modified version of the *Step* My Block to make the robot turn in the same direction at all times. This change will allow it to find its way out of dead ends.

Open the *Step* My Block by going to the *BP_EDIT_MB2* program and double-clicking the My Block. Select all blocks except the Start block and press CTRL-C (⌘-C) to copy them. Next, go back to the empty program you used before to build the *ResetBody* My Block (or create a new temporary program) and paste all the blocks into it by pressing CTRL-V (⌘-V).

With all the blocks still selected, open the My Block Builder tool and configure the My Block as follows:

* Enter *StepAdv* in the Name field.
* Set the icon to the two Large Motors icon.
* Click **Add Parameter** and then go to the **Parameter Setup** tab.
* Enter *Side* as the Parameter Name, and set the Parameter Type to **Input** and the Data Type to **Logic**. Set the Default Value to **True**.
* Go to the **Parameter Icon** tab and select the horizontal double-headed arrow icon (third row, second icon from the left).
* Click **Finish** to create the My Block.

Now, double-click the new *StepAdv* My Block and edit it as shown in Figure 10-18.

1. Drag a Data Wire from the *Side* input block to the *LogicToSign* My Block input.
2. Add a Wait block in Time mode and set the input to **0.2 seconds**.
3. Delete the Wait block in Time mode and Wait block in IR Sensor mode and add a Switch block in Infrared Sensor Compare Proximity mode.
4. Add a Wait block set in Time mode in the False case area of the Switch block and set its input to **1 second**.
5. Add a Timer block in Reset mode and choose Timer ID **1**.
6. Add a Sound block in Play File mode. Select a default sound or create a new sound (perhaps a squawk) using the Sound Editor tool.
7. Add a Large Motor block in On mode with its power set to **–100**.
8. Add a Loop block and change its ending input to **Logic**.
9. Add an IR Sensor block in Compare Proximity mode and set Compare Type to **Greater Than (2)** and Threshold to **30**.
10. Add a Timer block in Compare Time mode. Set its Timer ID to **1**, set Compare Type to **Greater Than (2)**, and set Threshold to **2** (seconds).
11. Add a Logic Operations block in AND mode.
12. Connect Data Wires from Sensor blocks 9 and 10 to Logic Operations block 11. Then drag the last Data Wire from the Logic Operations block output to the Loop control input.

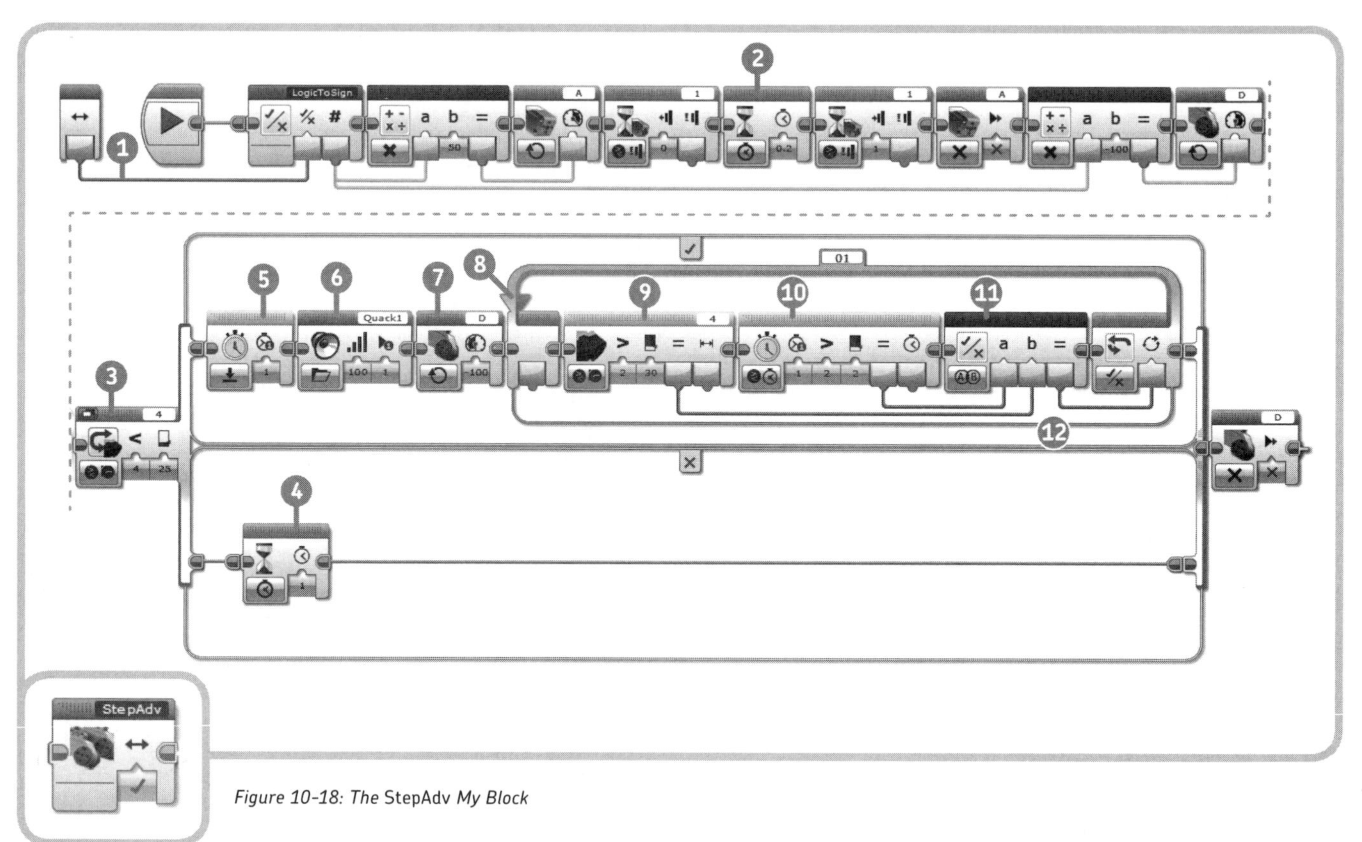

Figure 10-18: The StepAdv *My Block*

What are the differences between the *StepAdv* My Block and the *Step* My Block? Let's analyze them using Figure 10-18 as reference. We added a Wait block in Time mode (2) between the Wait blocks in Touch Sensor mode. We also replaced the Wait block in IR Sensor mode with a Switch block (3) to check for obstacles. If no obstacle is seen, the program waits 1 second [Wait block (4) in the False case] and then stops the Large Motor (the last block after the Switch block). If the IR Sensor detects an obstacle, the blocks in the True case are executed.

In the True case, the first Timer block resets Timer 1 (5), and the sound is played (6). Next, the Large Motor is turned On at full power to turn the robot's ankles (7), always in the same direction (regardless of the *Side* input value). Now, because it always turns left when it encounters obstacles, the robot can get out of a dead end and deal with corners.

Before stopping the Large Motor that turns the ankles, a Loop block (8) waits until the IR Sensor block (9) reads a proximity value greater than 30 percent *and* the Timer block (10) measures time greater than 2 seconds [the time elapsed since Timer 1 was reset by the Timer block (5)]. The two Sensor blocks' Logic outputs are combined by a Logic Operations block (11) in AND mode so that the Loop block will end only when the IR Sensor block's Proximity Value is greater than the specified threshold *and* 2 seconds have elapsed.

In the original *Step* My Block, the robot's ankles stop turning as soon as the IR Sensor no longer sees a nearby obstacle, which can happen after even a fraction of a second. I've added the timer here to make the robot turn for at least 2 seconds, even if an obstacle moves farther away before that time has elapsed.

the final program for WATCHGOOZ3

Once you've created the *ResetBody* and *StepAdv* My Blocks, delete all of the blocks in the temporary program that you used to create the My Blocks and build the program shown in Figure 10-19. You can find the My Blocks in the last cyan-colored tab of the Programming Palette. Notice that the first *StepAdv* My Block has its input set to **True**, while the second one has its input set to **False**.

Double-click the Program Name tab and enter the name *Wander* for the program. Then run the program to test it. WATCHGOOZ3 should be ready to be unleashed!

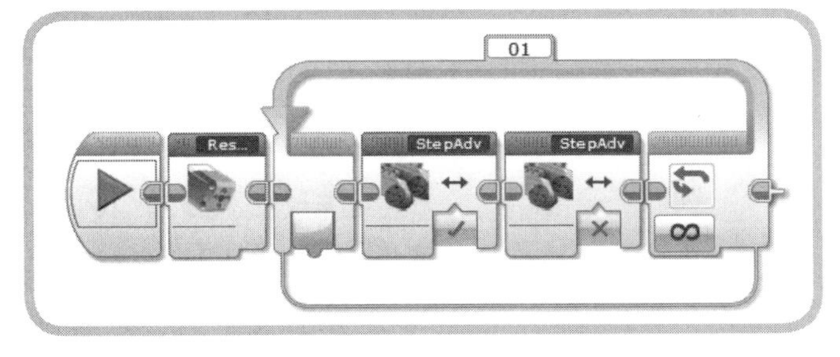

Figure 10-19: The final Wander *program for WATCHGOOZ3*

the logic operations block

The programs we've created in this chapter use the Logic Operations block (found in the Data Operations palette, red header) to perform logic operations (see Table 10-1) on the Logic Operation block's inputs to get a logic result as its output. The inputs and output can only have True or False values.

table 10-1: the logic operations

Mode	Inputs	Result
AND	A, B	True if both A and B are True; otherwise False
OR	A, B	True if either A or B or both are True; False only if both A and B are False
XOR	A, B	True if only A is True or only B is True; False if both are True or if both are False
NOT	A	True if A is False; False if A is True

the timer block

The Timer block in the Sensor palette gives you access to the EV3 Brick's internal clock. The EV3 Brick counts time in milliseconds, and it represents the time as a 32-bit unsigned integer variable. It can count up to $2^{32} - 1$ milliseconds, which is equal to 49 days, 17 hours, 2 minutes, 47 seconds, and 295 milliseconds of continuous operation. The Timer block uses this internal clock to provide you with eight independent timers that allow you to read the time in seconds (using Measure Time mode) elapsed from either the program start or the last reset. (You can reset a specific timer count using the Timer block in Reset mode, as we did in the *StepAdv* My Block.) In Compare Time mode, you can compare the current value of a timer against a threshold.

EXPERIMENT 10-1

Add the second Large Motor to the robot to make the head move. Be careful, as adding weight could upset your robot's balance. Create a program that moves the head when an object is detected.

DIGGING DEEPER: MOTOR SPEED REGULATION

The Unregulated Motor block simply turns on a motor without applying power regulation. The problem is that if friction or load applies a resisting torque to the motor shaft, the motor might not be able to overcome it. Because its power is constant, it won't automatically increase in response to resistance. The other Motor blocks use power regulation to keep the motor's speed constant even when resistance is applied to the shaft.

This regulation is done by a *Proportional Integrative Derivative (PID) controller*, which constantly updates the motor's power in an attempt to keep the difference (or *error*) between the desired speed and the actual speed near zero. The current speed is estimated using the built-in rotation sensor readings.

The PID controller's output is the sum of three things: a term that is proportional to the present error, a term that is proportional to the derivative of the error (this estimates how fast the error is changing to predict the future), and a term that is proportional to the integral of the error (this accumulates the errors, giving a view into the past). When the shaft is slowed by external resistance, the error between the desired speed and the actual speed increases. Then the controller's output increases as well, giving the motor sufficient power to overcome the resistance and thus reducing the speed error.

EXPERIMENT 10-2

Make WATCHGOOZ3 follow lines! Add a Color Sensor by attaching it to the 5M beam you added in step 8 on page 154. That 5M beam always remains vertical, no matter what direction the EV3 Brick leans toward, so the Color Sensor will always look down. Let it hang about 1 cm above the ground.

Making the program with the Brick Program App is easy. Start with the program in Figure 10-1. Remove the Wait Time blocks and replace the Wait IR Sensor blocks with two Wait Reflected Light Sensor blocks. I'll let you fill in their parameters to make WATCHGOOZ3 follow the edge of a black line.

conclusion

In this chapter, you've seen how, by combining good hardware with good software design, we can program WATCHGOOZ3 using only the Brick Program App. While creating more advanced programs in the EV3 Software, you learned how to create and configure My Blocks, the purpose of the Unregulated Motor block, and how to use the Logic Operations and Timer blocks.

YOU'RE QUICK! BRAVO!
YOU DIDN'T MENTION THIS!
HAHAHA!
YOU'RE NOT GOING TO KEEP WEARING THAT, ARE YOU?
HERE YOU ARE, AN OFFICIAL EV3L SCIENTIST SWEATER!
THAT WAS PRETTY GROSS.
YOU'RE STILL IN ONE PIECE, AREN'T YOU?
DO YOU WANT TO GIVE UP?
NO WAY!
THE NEW THREADS SUIT YOU!
10X1

BEING AN APPRENTICE DOES NOT MEAN CLEANING FLOORS, SOLVING RIDDLES IN THE DARK, AND TAKING CARE OF CLOCKWORK ANIMALS!
FAIR ENOUGH, YOUNG MASTER! HOW DO YOU THINK I GOT THIS SCAR?
WHAT'S THE GRIME ON MY COAT?
I HAD TO VERIFY THAT YOU COULD FACE THE REST!
RIP!
SO NO MORE WHINING!
COME ON, IT'S TIME TO BUILD SOMETHING NEW! SINCE YOU LIKE REMOTE CONTROLS SO MUCH, YOU WILL BUILD A SUP3R CAR MODEL, WITH ITS OWN R3MOTE! IT'S A VERY SPECIAL CAR, YOU KNOW...
IT IS "SUP3R," NOT "SUPER," ISN'T IT?
GOOD BOY, YOUR DICTION IS GETTING BETTER!
10X2

building the SUP3R CAR

In this chapter, you'll build the SUP3R CAR, an evil-looking armored vehicle! You'll also build the R3MOTE, a handy two-lever remote control. Both models are shown in Figure 11-1. Like a real rear-wheel-drive car, the SUP3R CAR has front wheels that steer and rear wheels that drive. In this sense, it's different from ROV3R, which uses two motors running at different speeds to change direction or turn on the spot. ROV3R's turning radius can be zero, while the SUP3R CAR has a larger turning radius.

The SUP3R CAR uses a Medium Motor for steering and two Large Motors for driving. You'll learn all about programming the SUP3R CAR in Chapter 12, but now, let's start building!

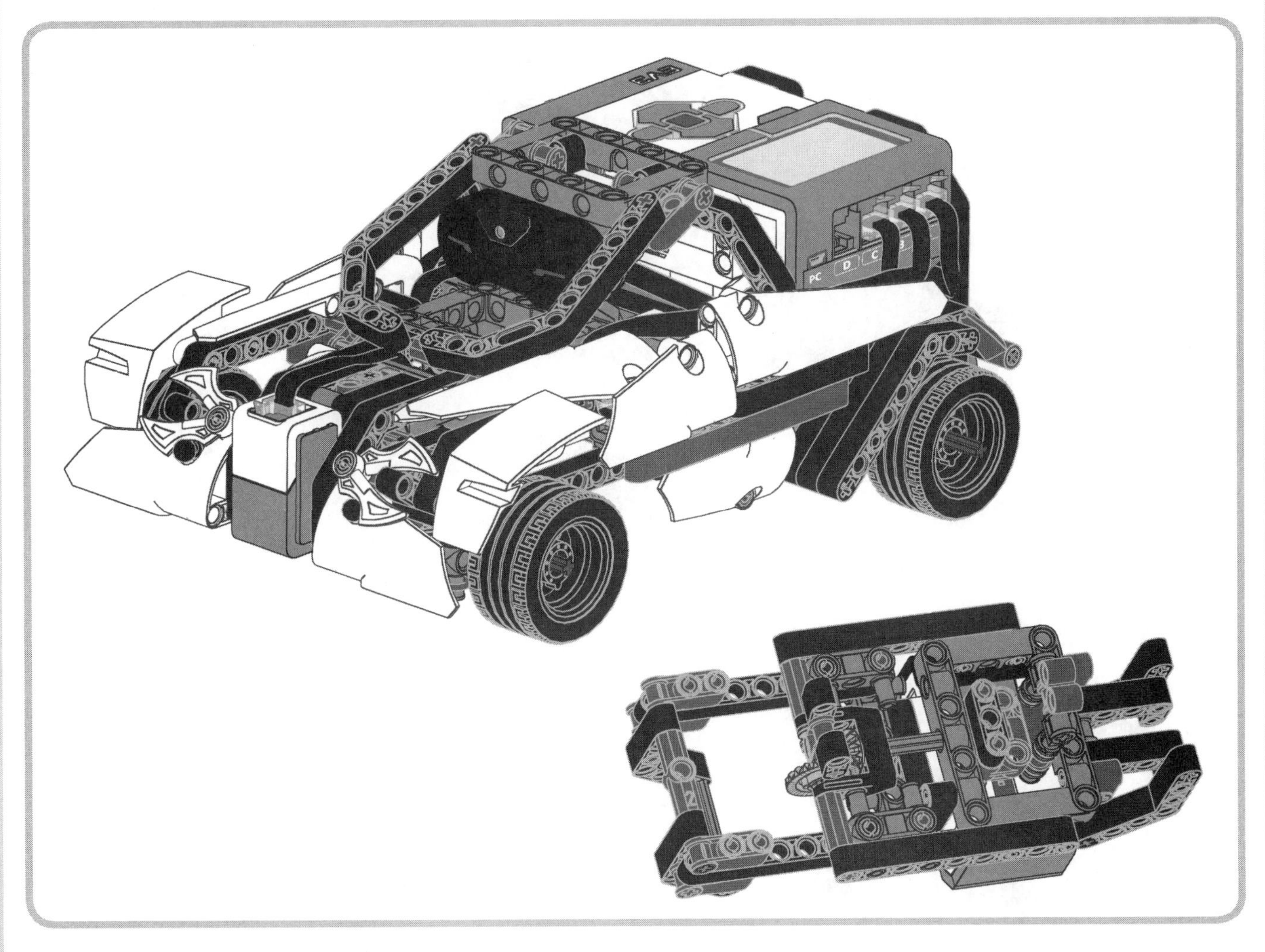

Figure 11-1: The SUP3R CAR and the R3MOTE

main assembly

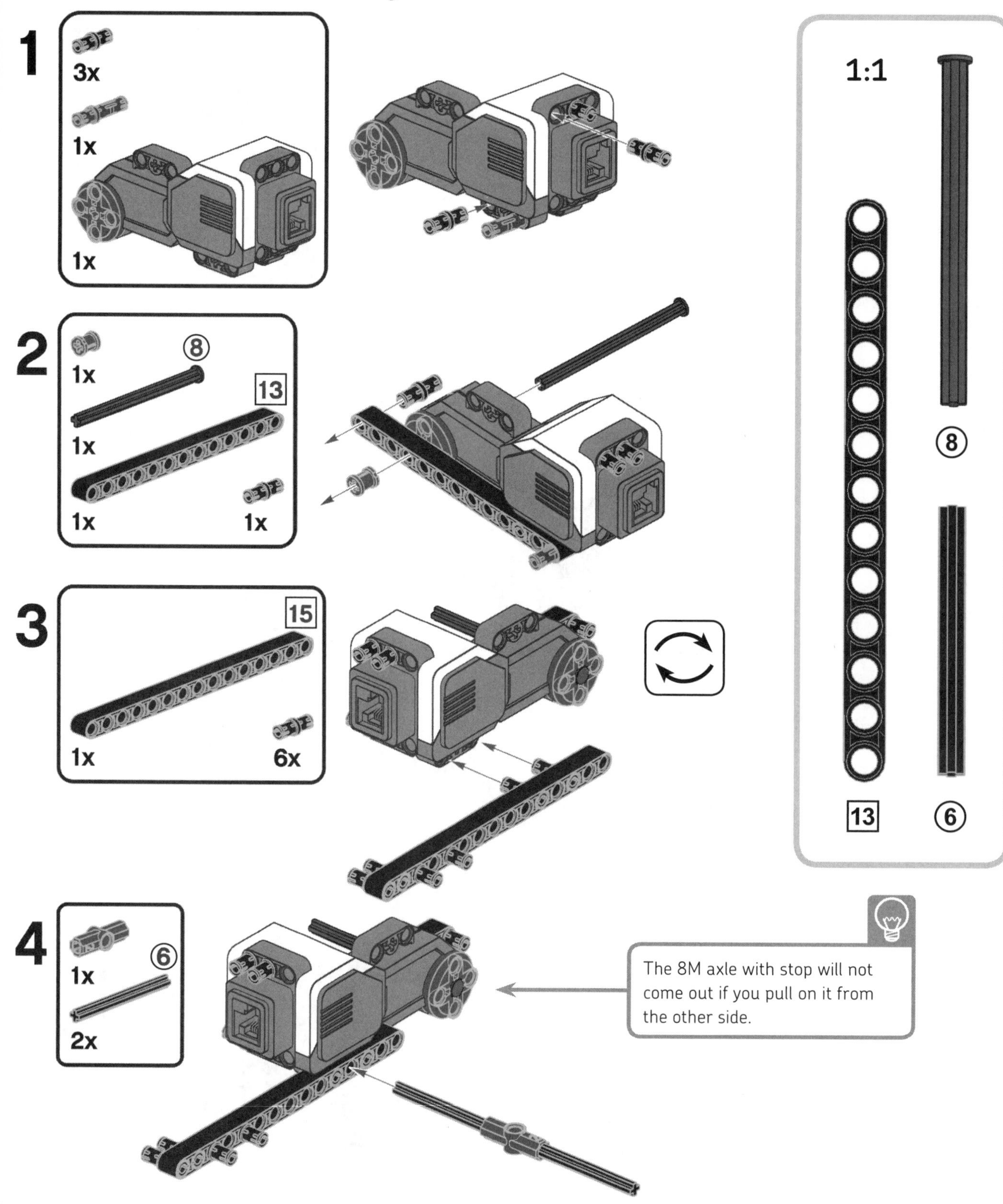

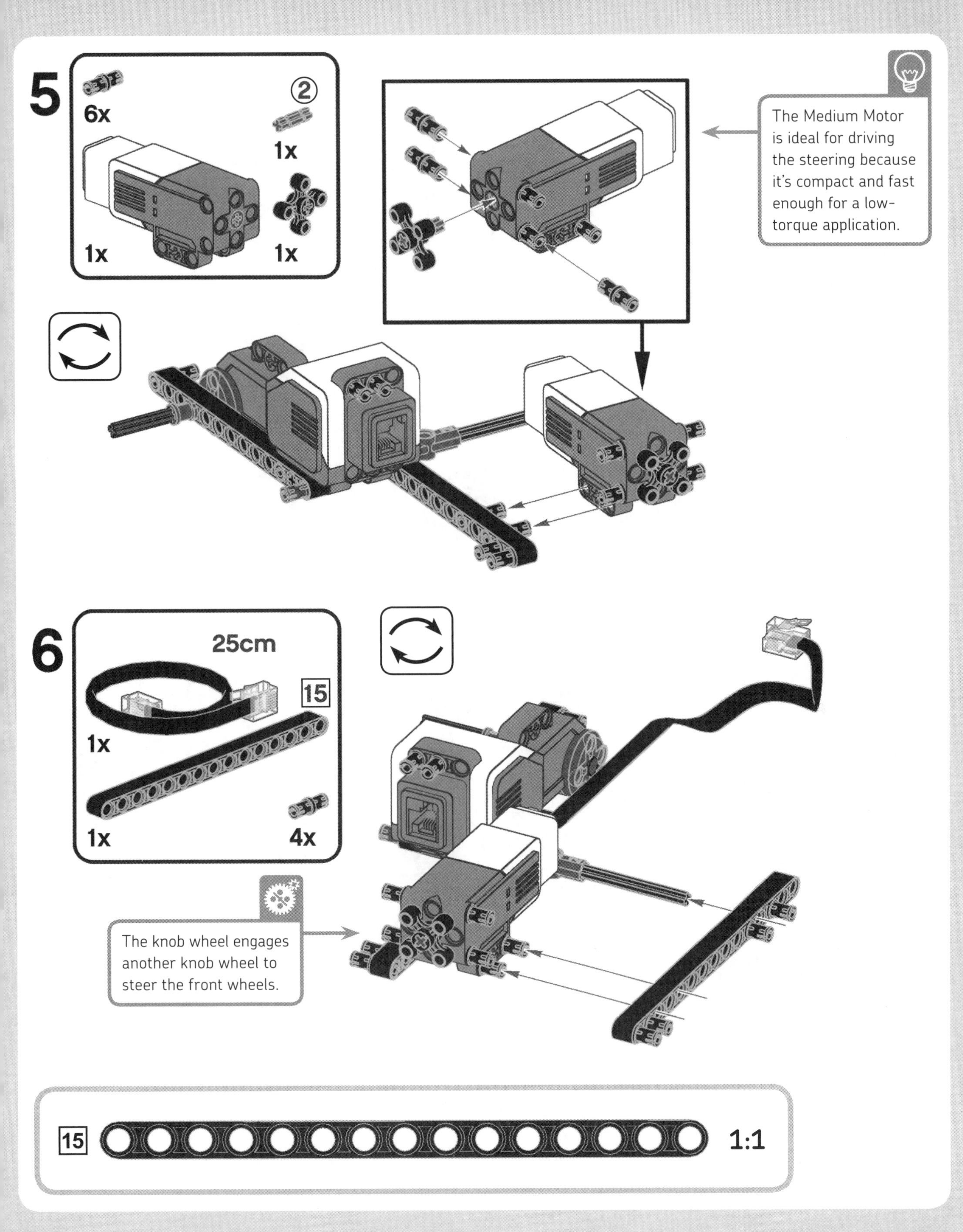
5
6x
2
1x
1x
1x
The Medium Motor is ideal for driving the steering because it's compact and fast enough for a low-torque application.
6
25cm
15
1x
1x
4x
The knob wheel engages another knob wheel to steer the front wheels.
15
1:1

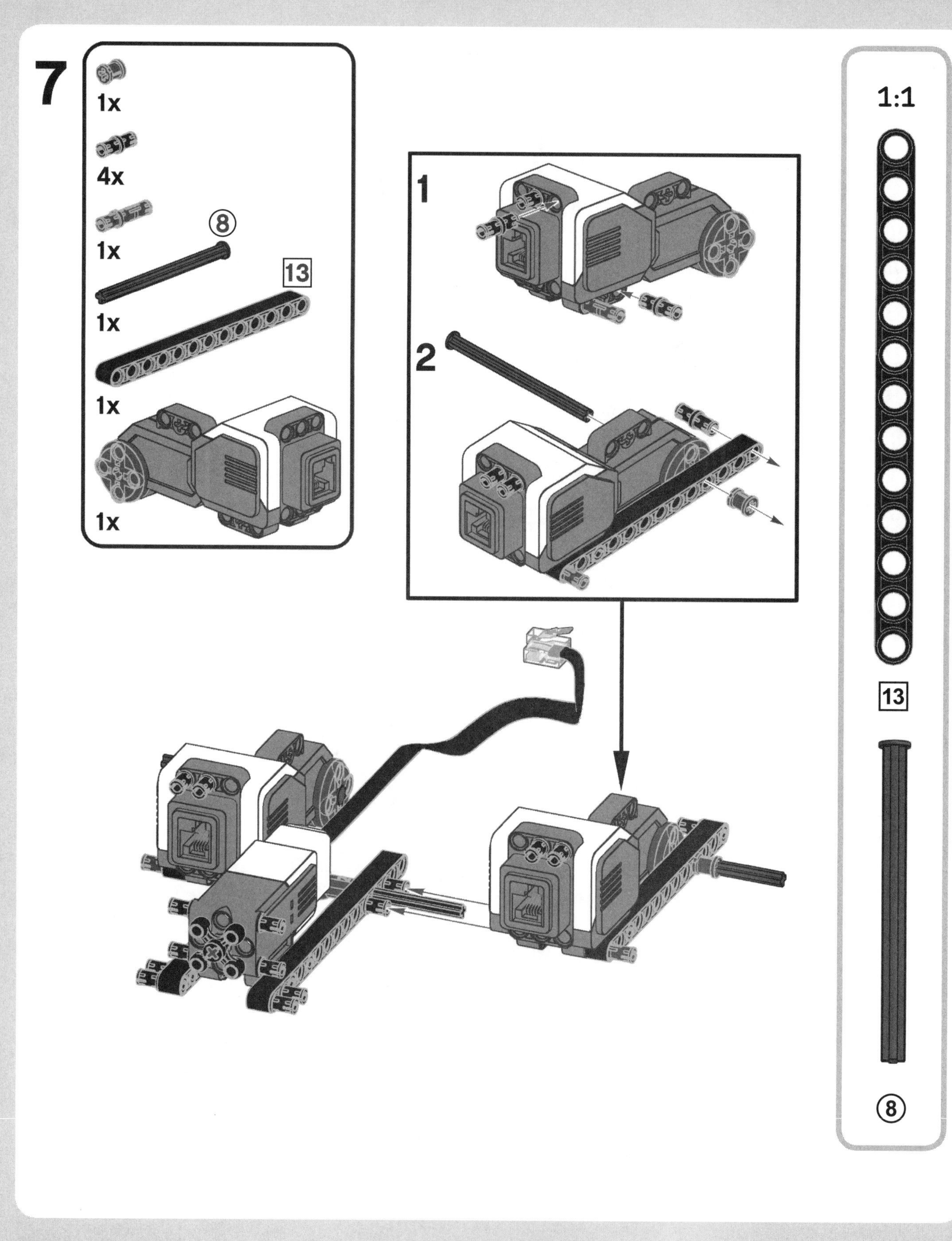
7
1x
4x
1x
8
1x
13
1x
1x
1x
1
2
1:1
13
8

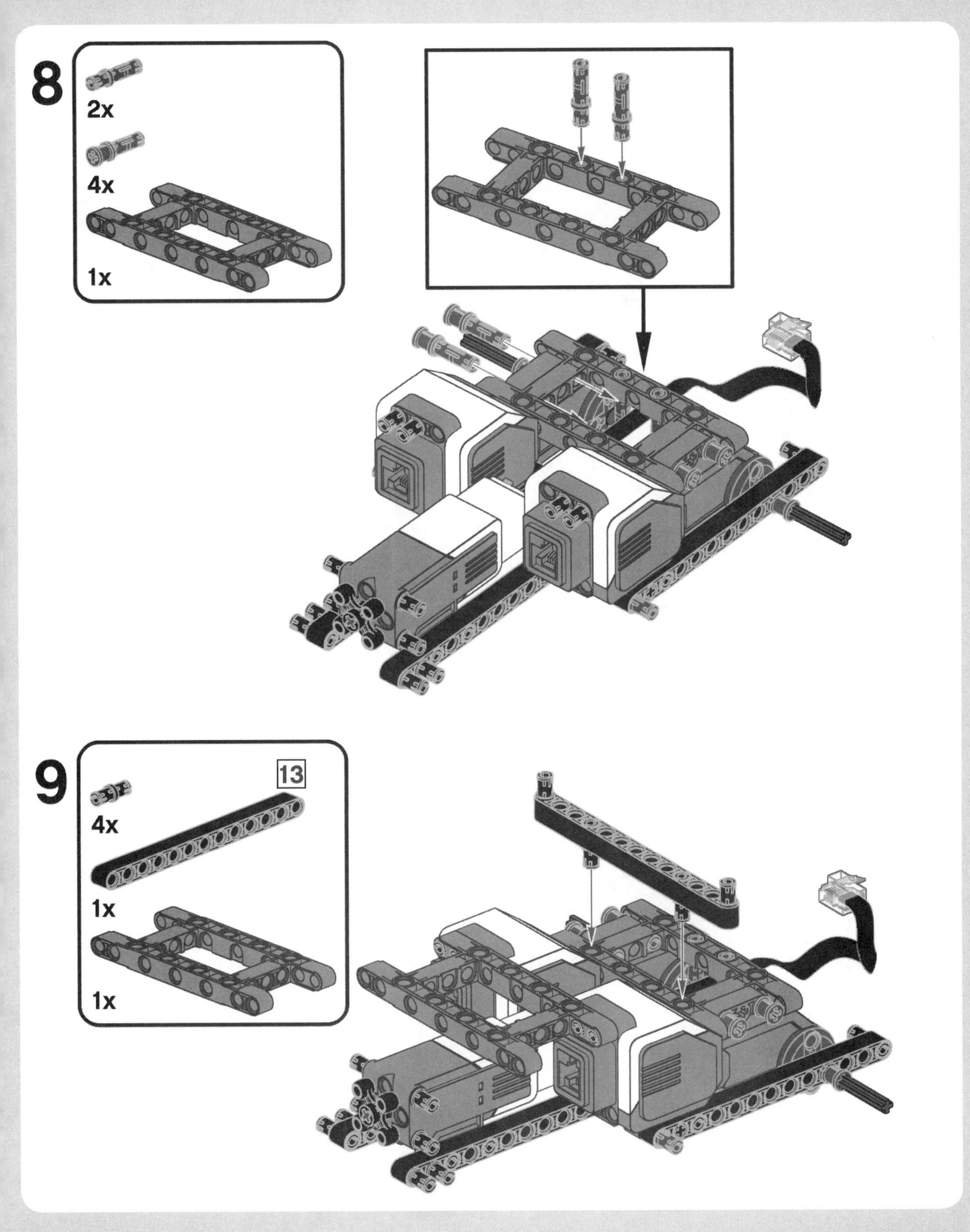
8
2x
4x
1x
9
13
4x
1x
1x

hood assembly

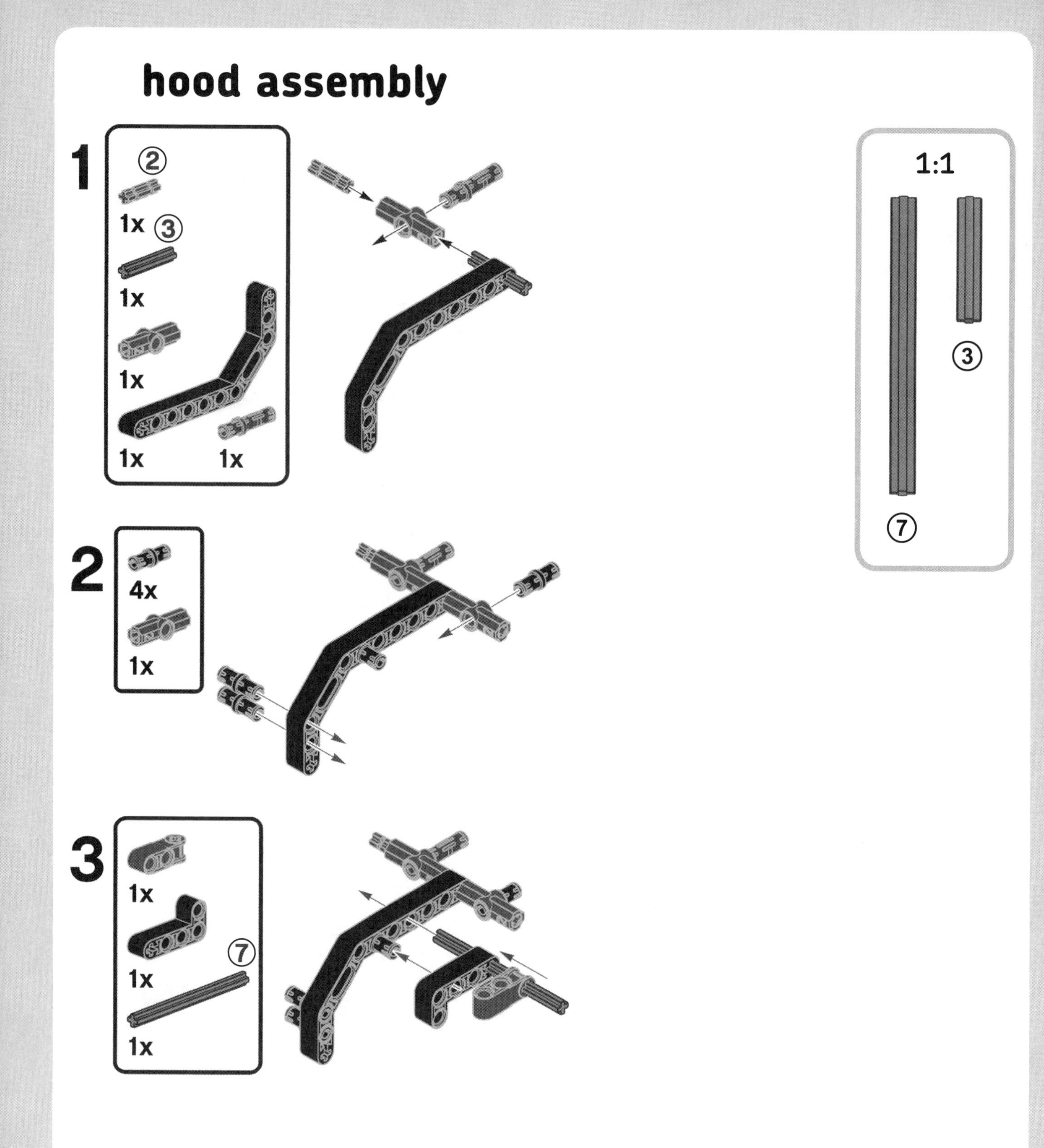

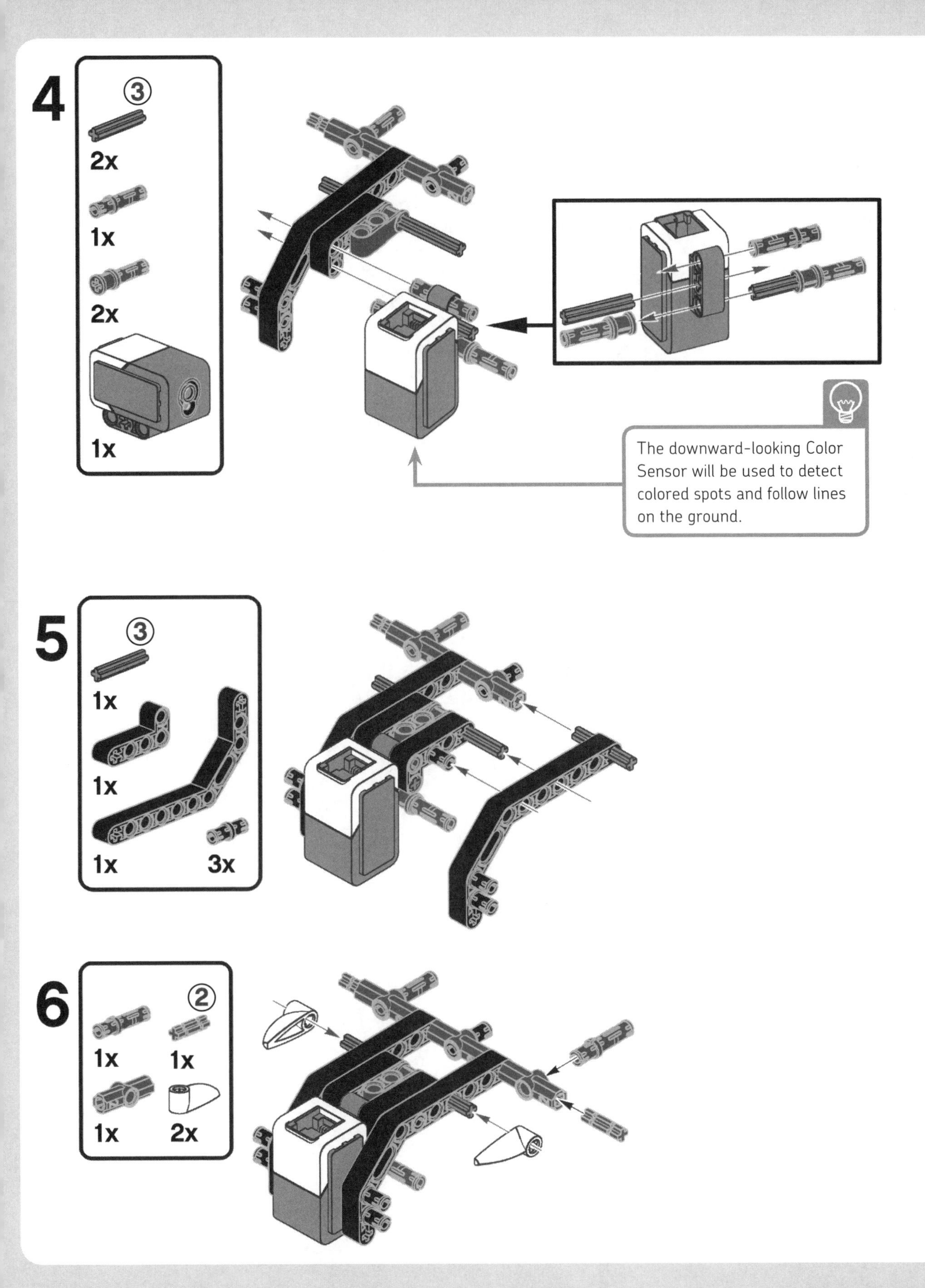
4
2x
1x
2x
1x
The downward-looking Color Sensor will be used to detect colored spots and follow lines on the ground.
5
1x
1x
1x
3x
6
1x
1x
1x
2x

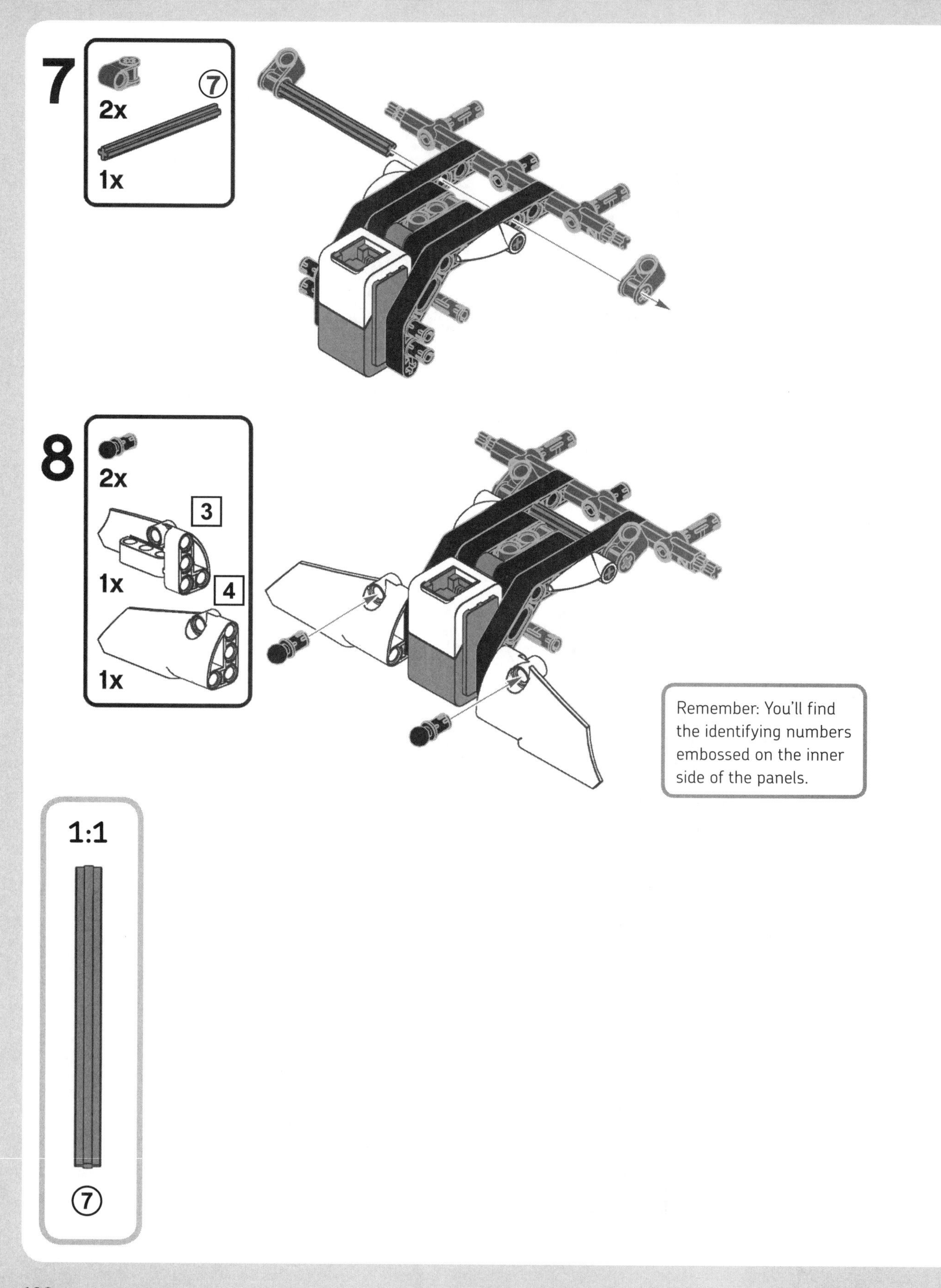
7
2x
7
1x
8
2x
3
1x
4
1x
Remember: You'll find the identifying numbers embossed on the inner side of the panels.
1:1
7

main assembly

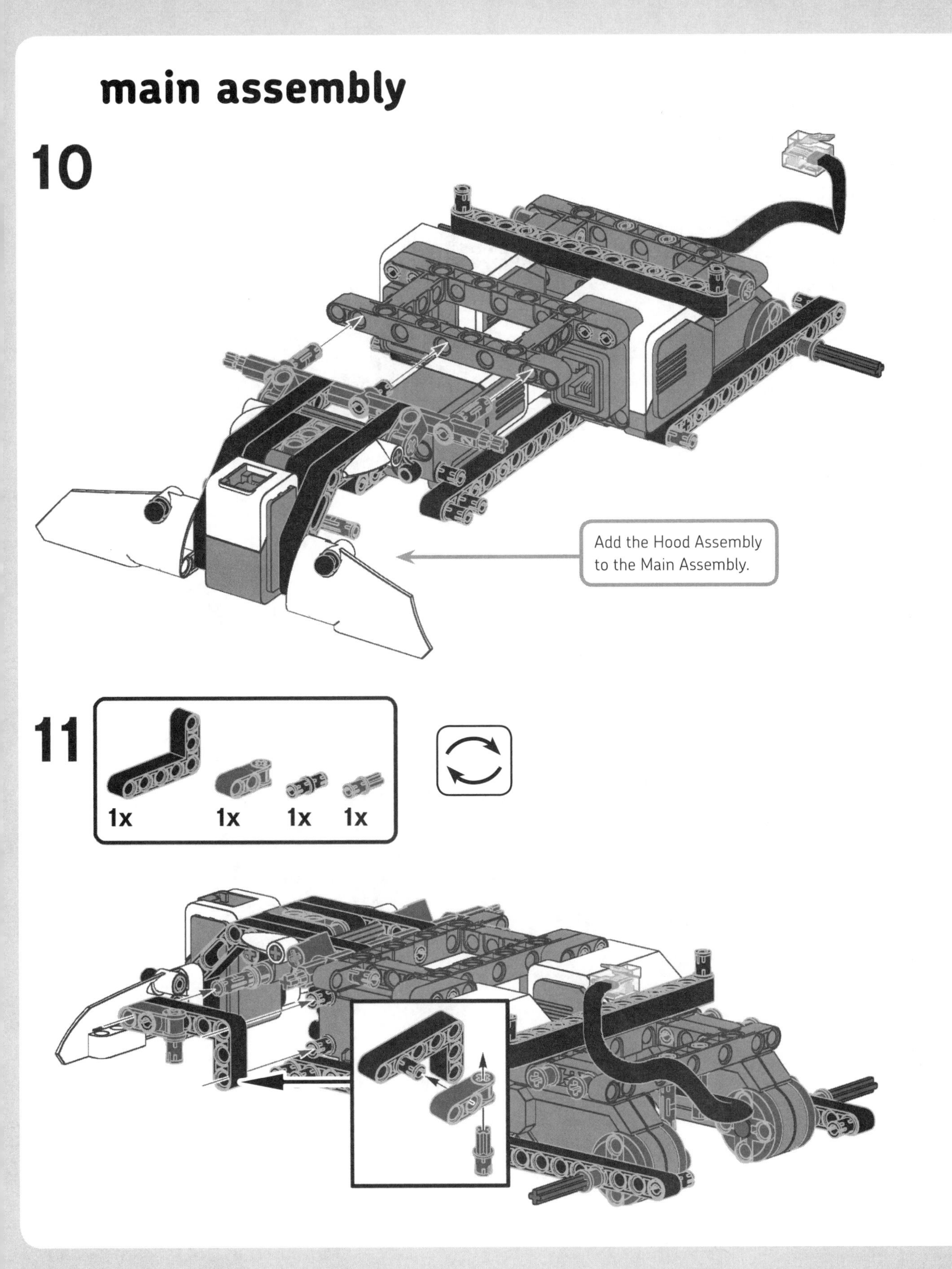

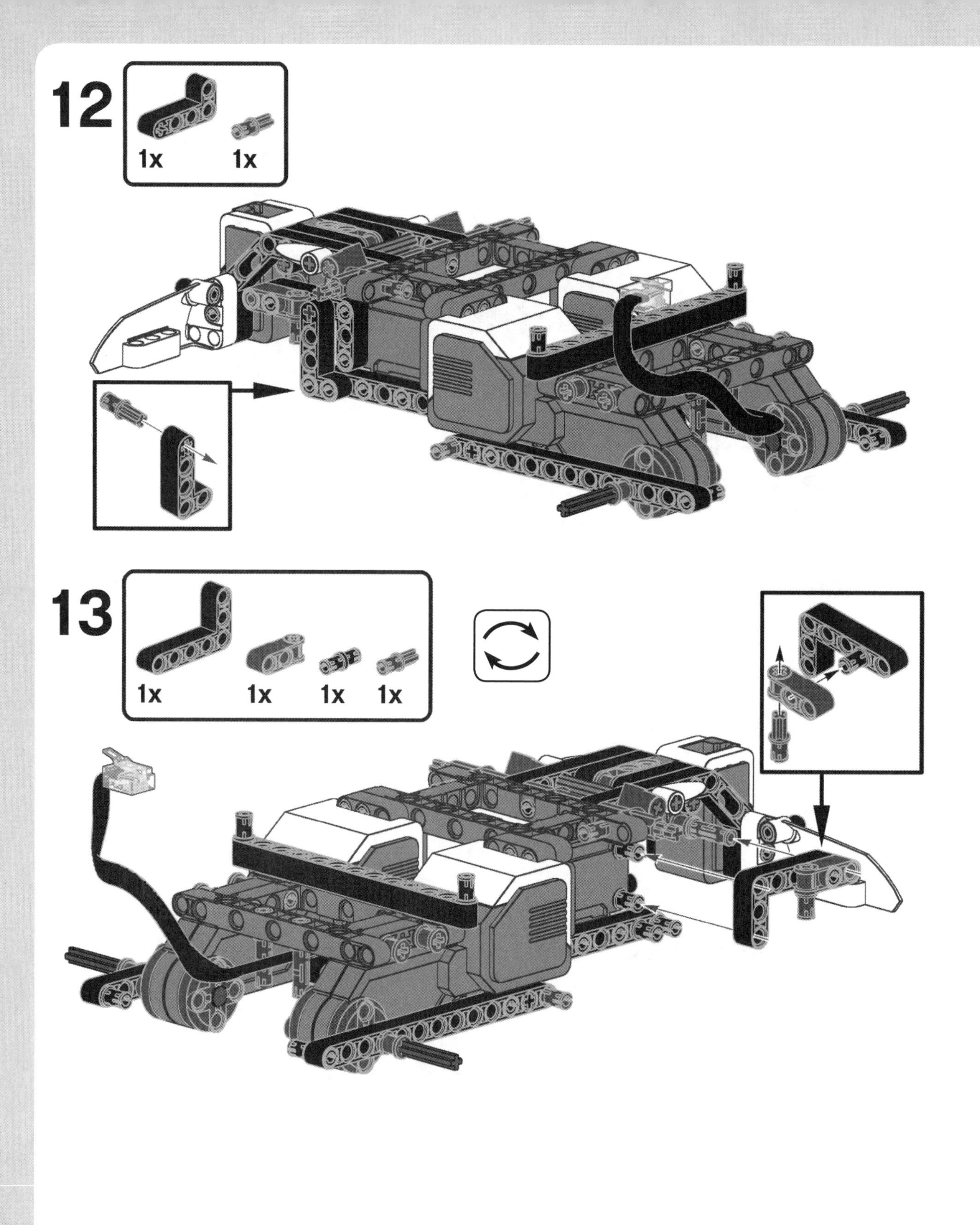
12
1x
1x
13
1x
1x
1x
1x

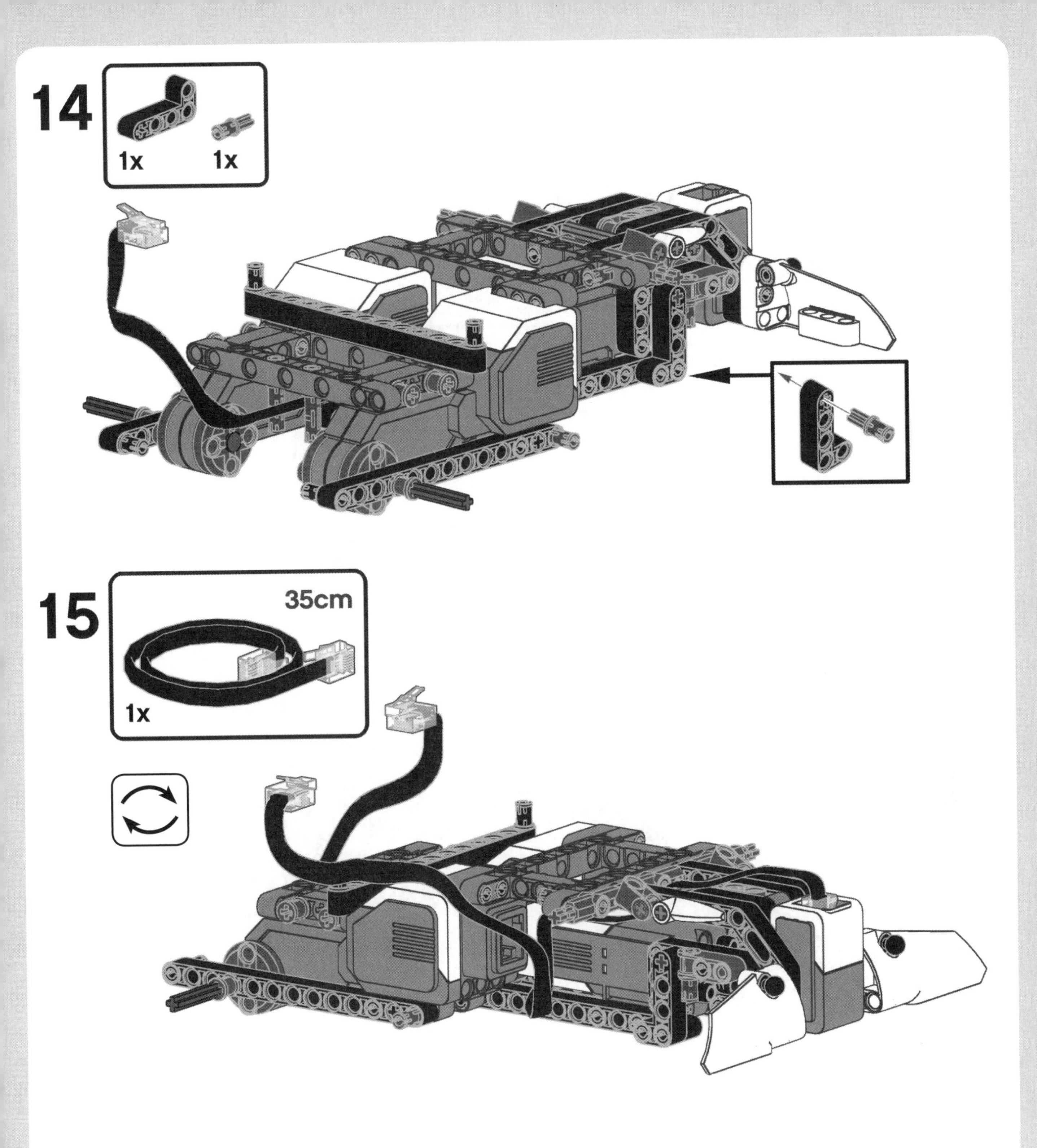
14
1x
1x
15
35cm
1x

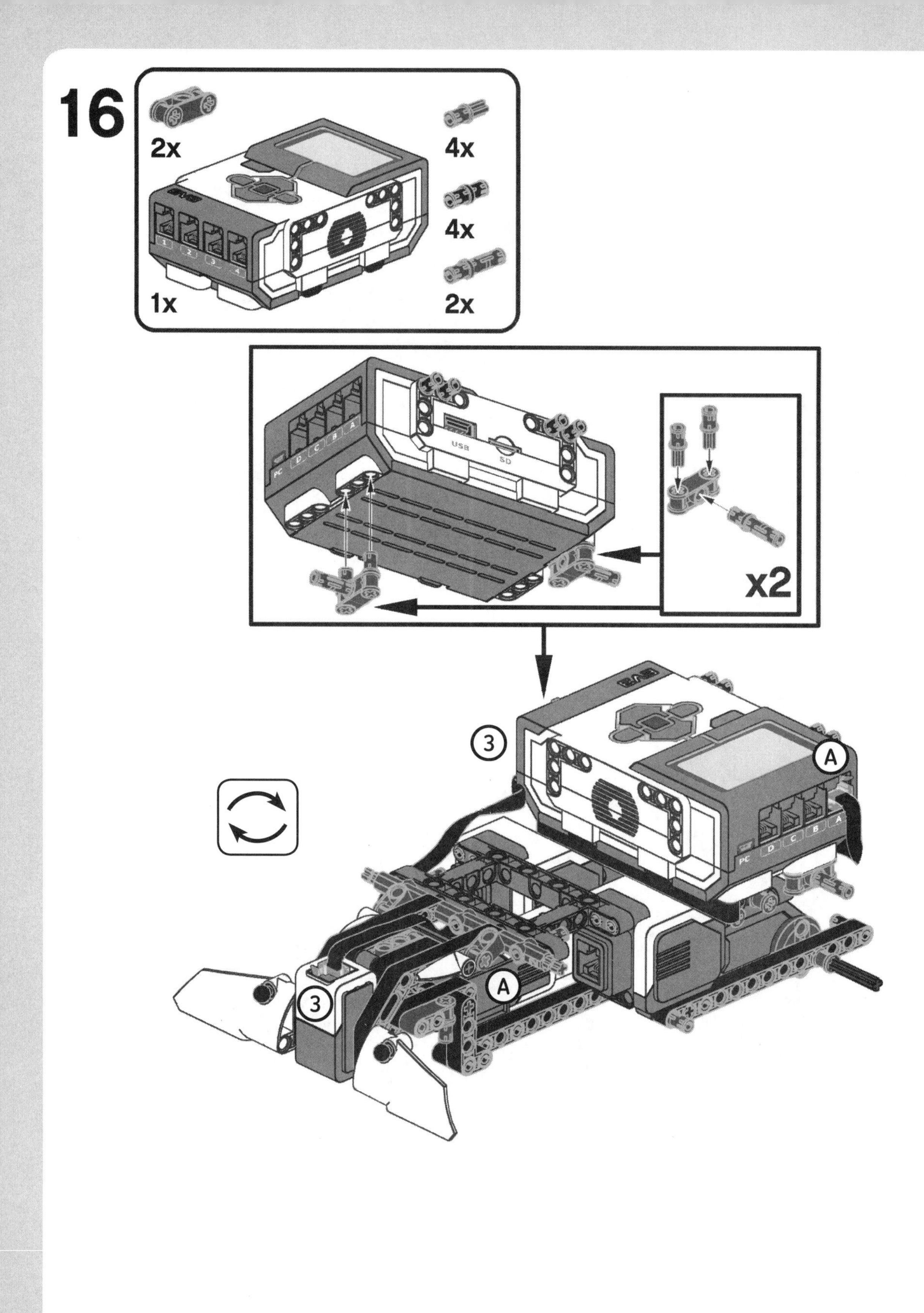
16
2x
4x
4x
1x
2x
USB
SD
PC
D
C
B
A
x2
3
A
3
A
PC
D
C
B
A

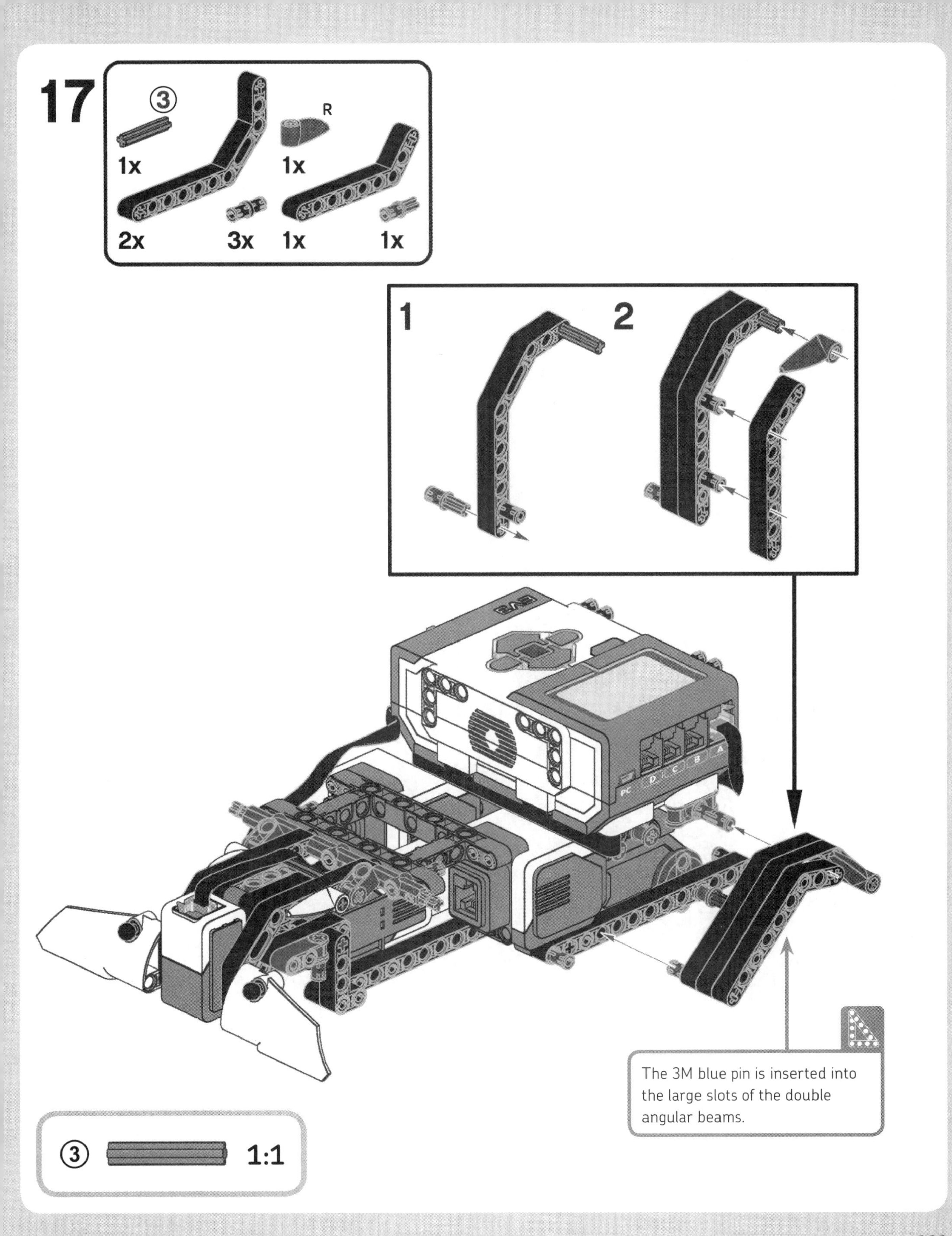
17
3
R
1x
1x
2x
3x
1x
1x
1
2
PC
D
C
B
A
The 3M blue pin is inserted into the large slots of the double angular beams.
3
1:1

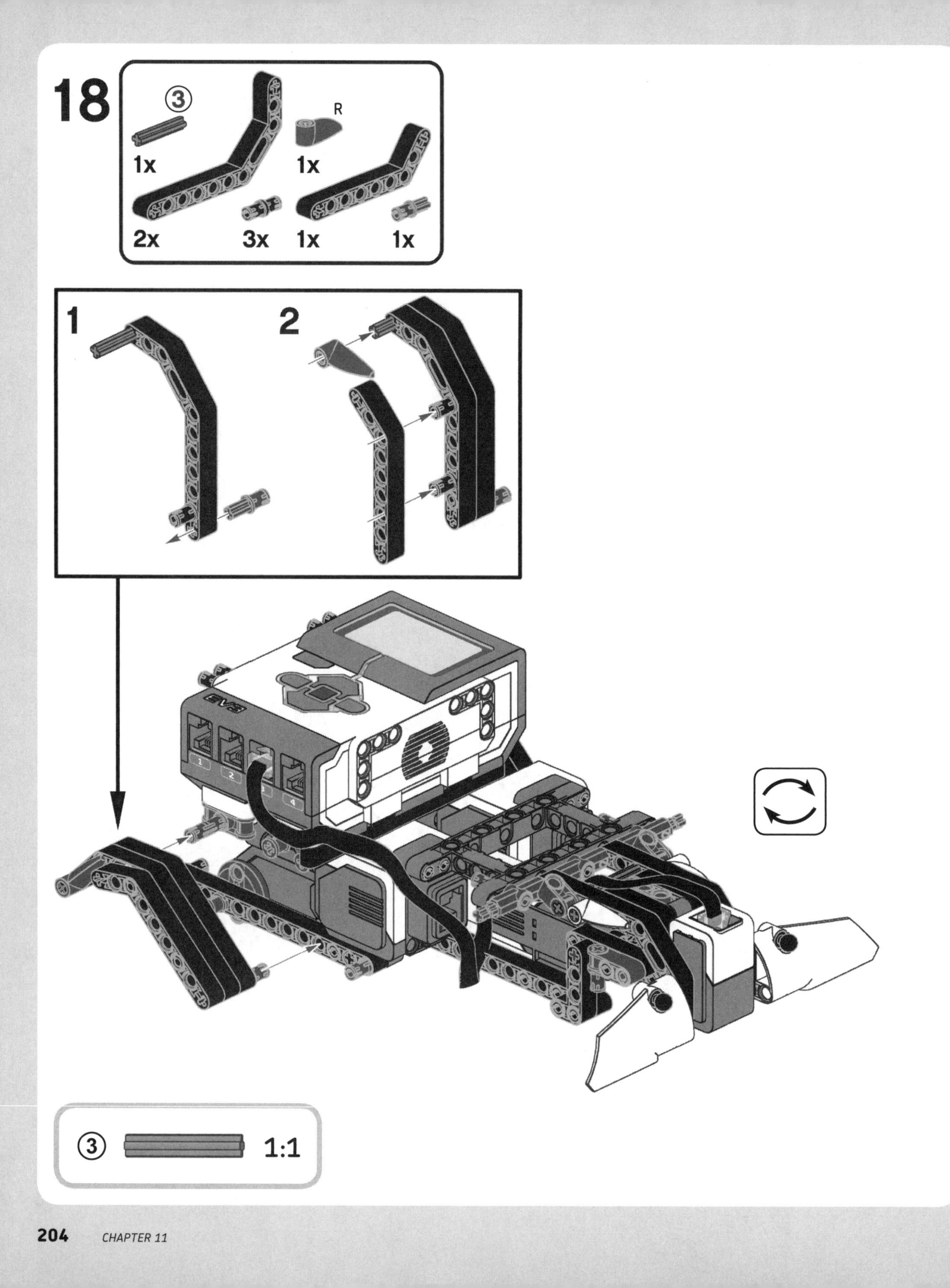
18
3
R
1x
1x
2x
3x
1x
1x
1
2
3
1:1

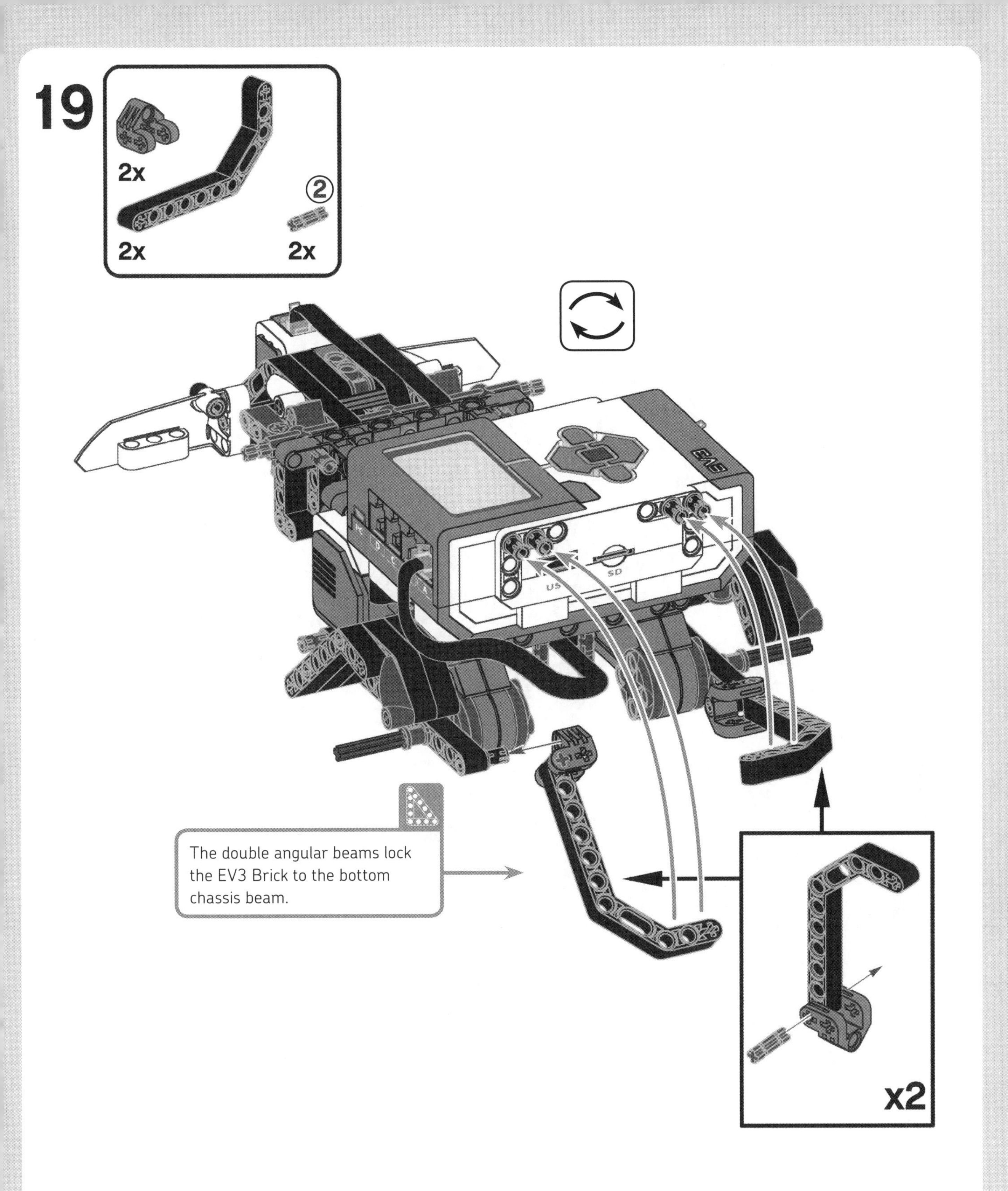
19
2x
2x
2
2x
The double angular beams lock the EV3 Brick to the bottom chassis beam.
x2

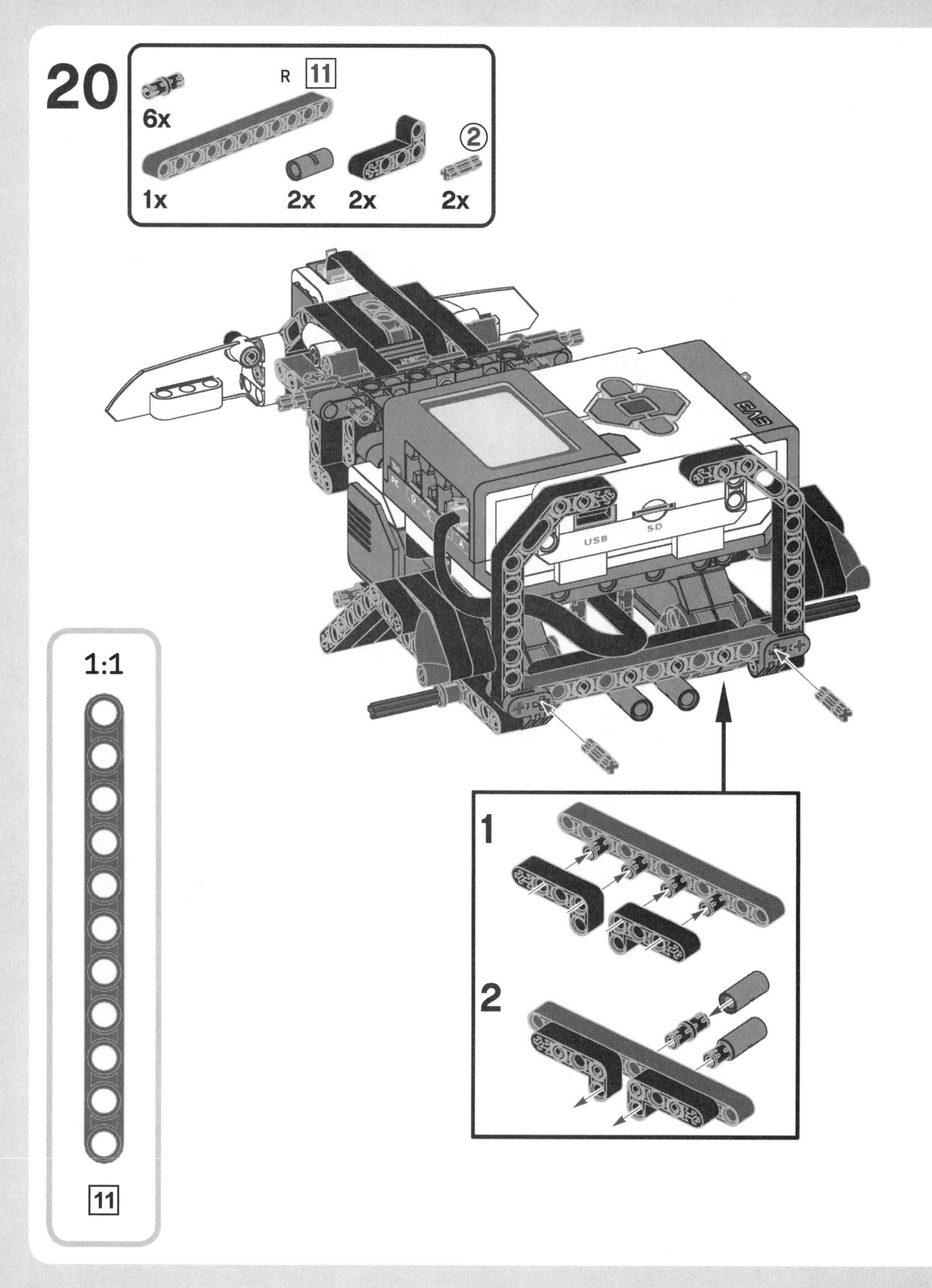
20
R
11
6x
1x
2x
2x
2x
2
USB
SD
1
2
1:1
11

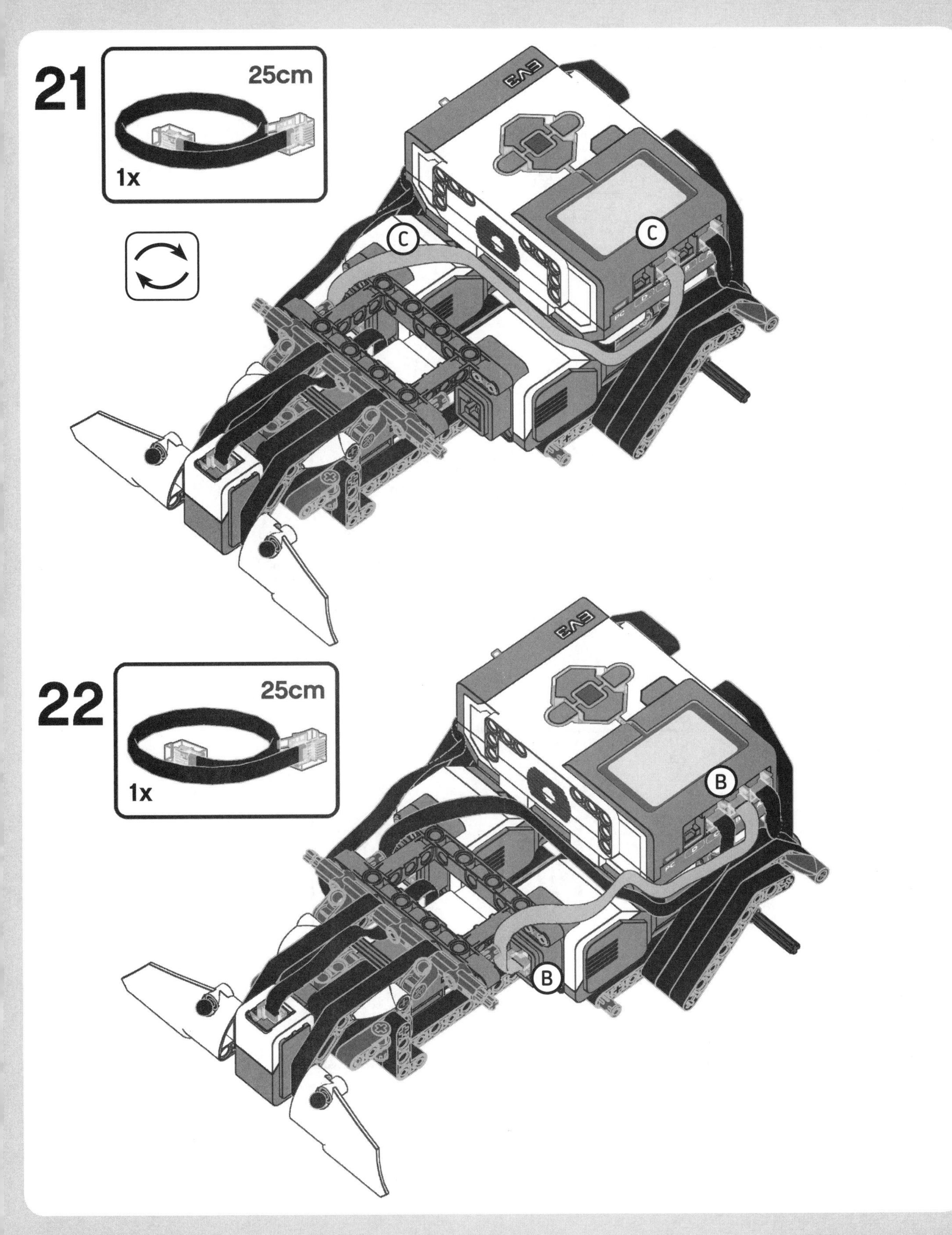
21
25cm
1x
C
C
22
25cm
1x
B
B

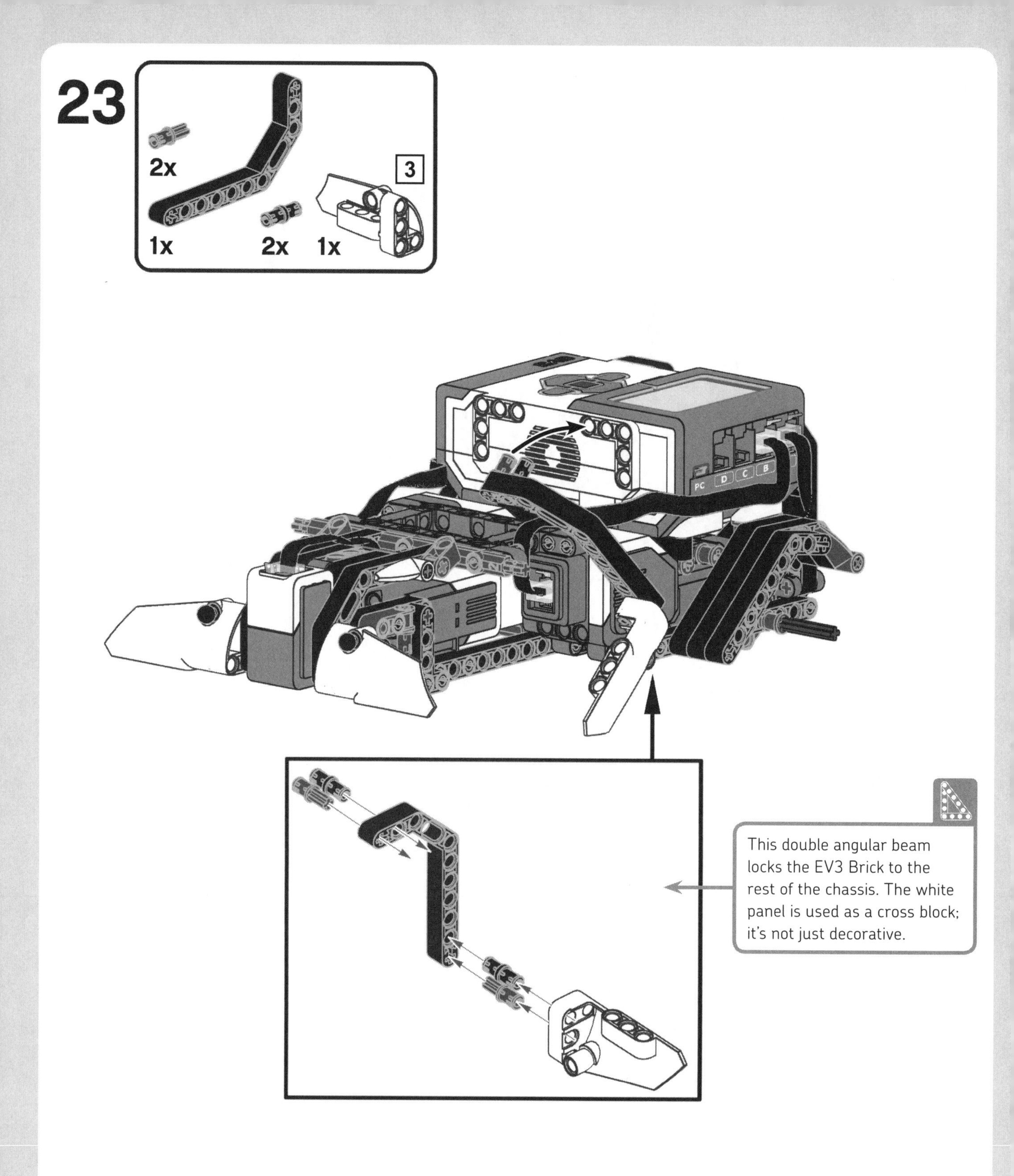
23
2x
3
1x
2x
1x
PC
D
C
B
This double angular beam locks the EV3 Brick to the rest of the chassis. The white panel is used as a cross block; it's not just decorative.

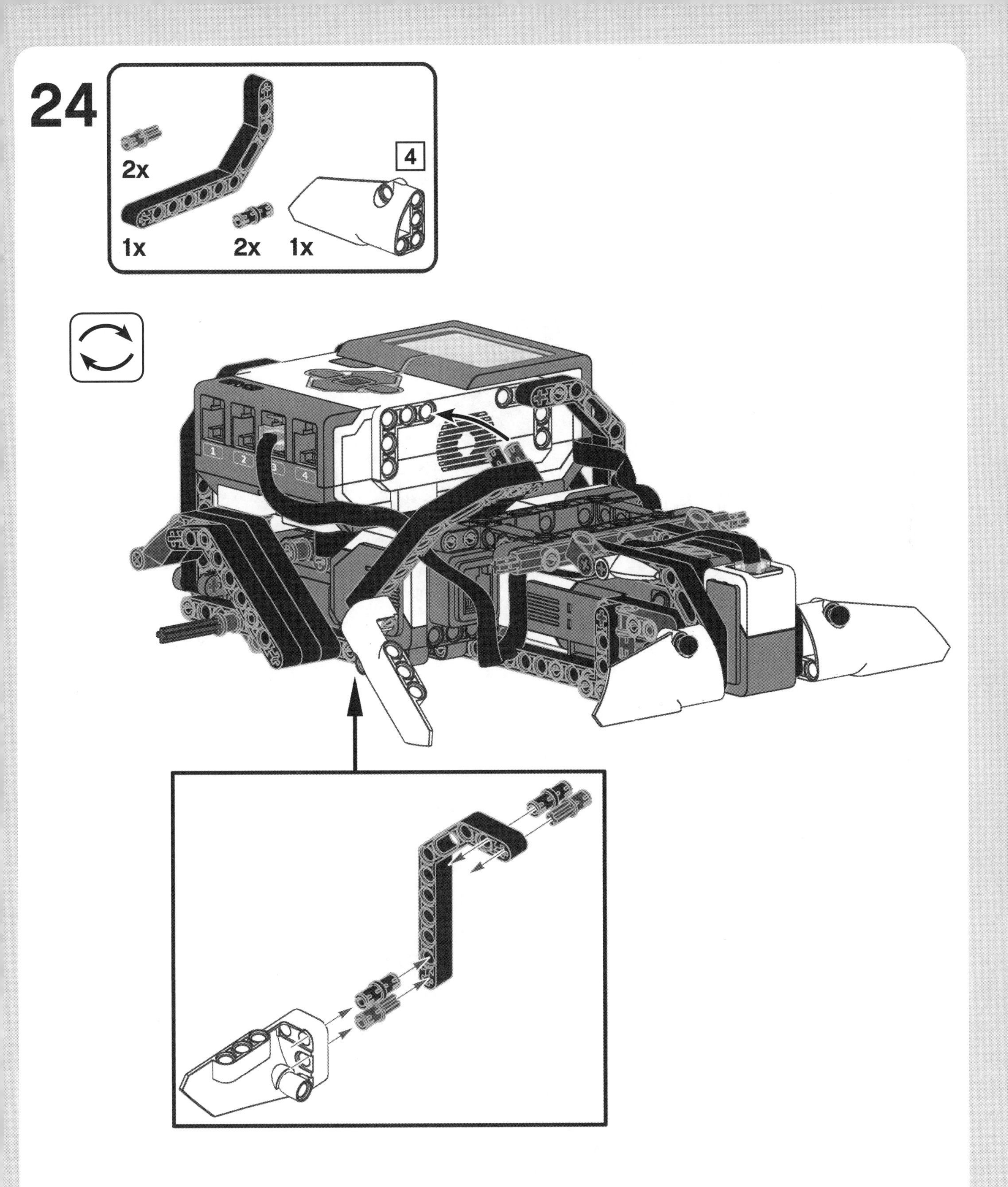
24
2x
1x
2x
1x
4
1
2
3
4

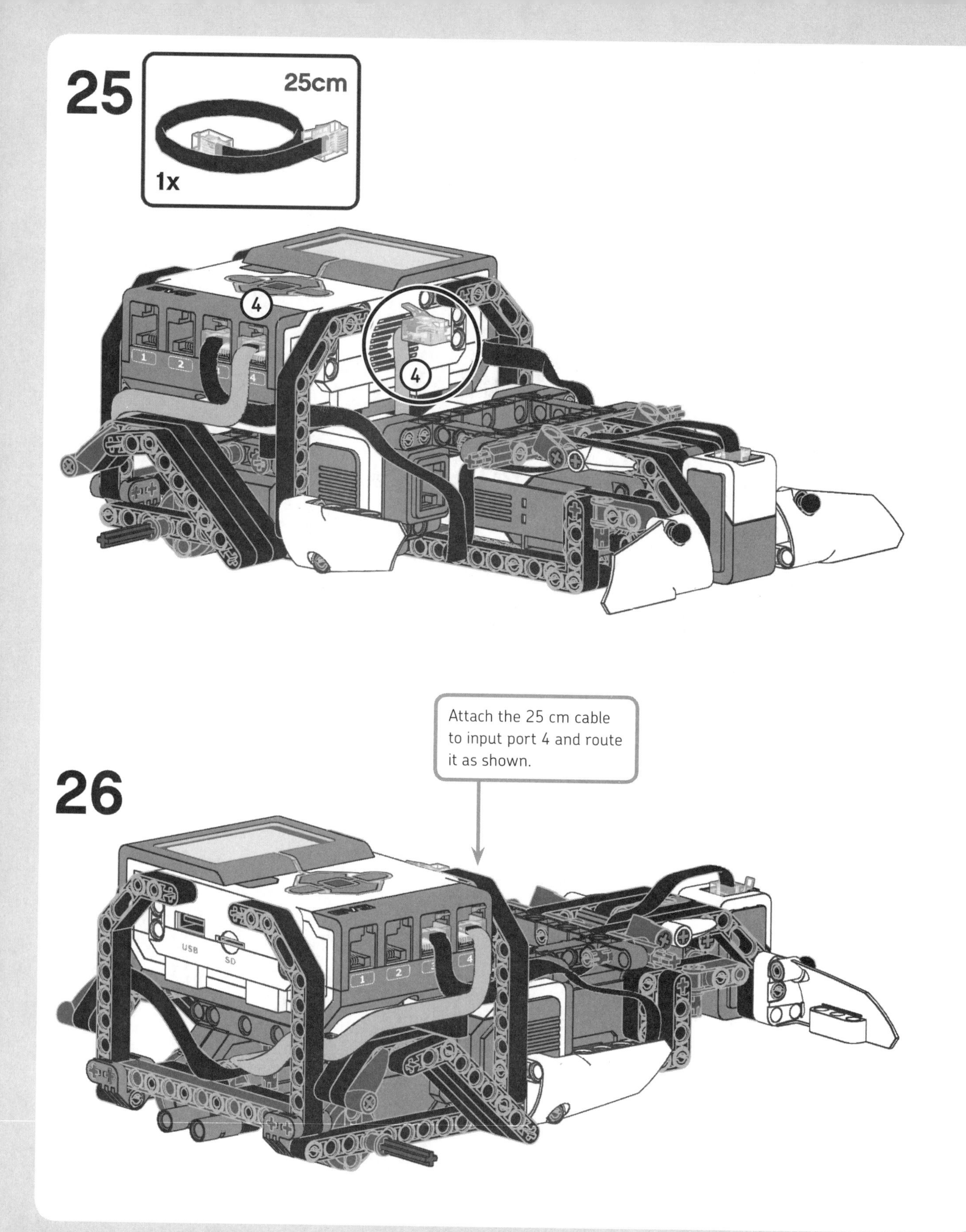
25
25cm
1x
4
4
Attach the 25 cm cable to input port 4 and route it as shown.
26
USB
SD
1
2
4

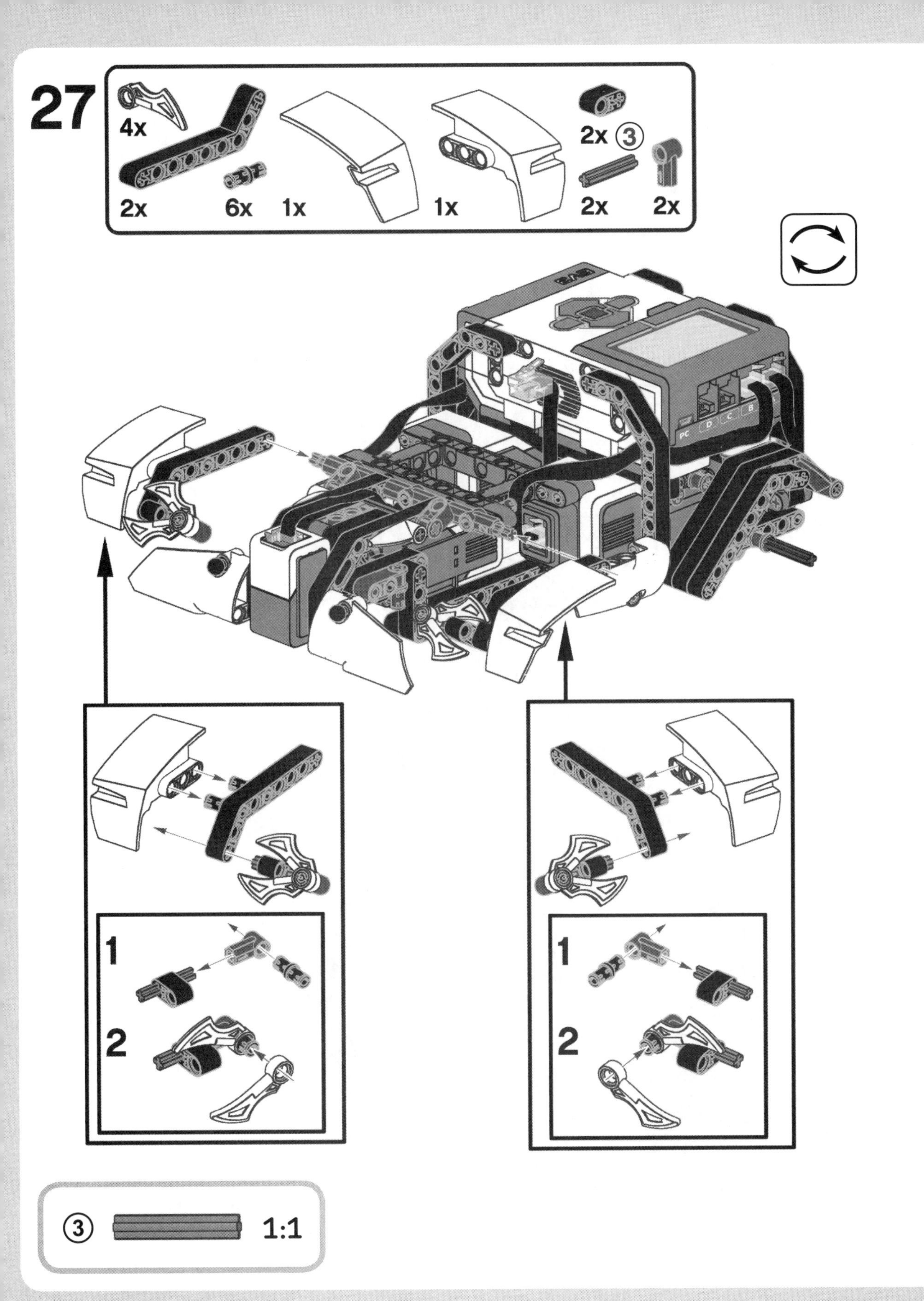
27
4x
2x
6x
1x
1x
2x ③
2x
2x
1
2
1
2
③
1:1

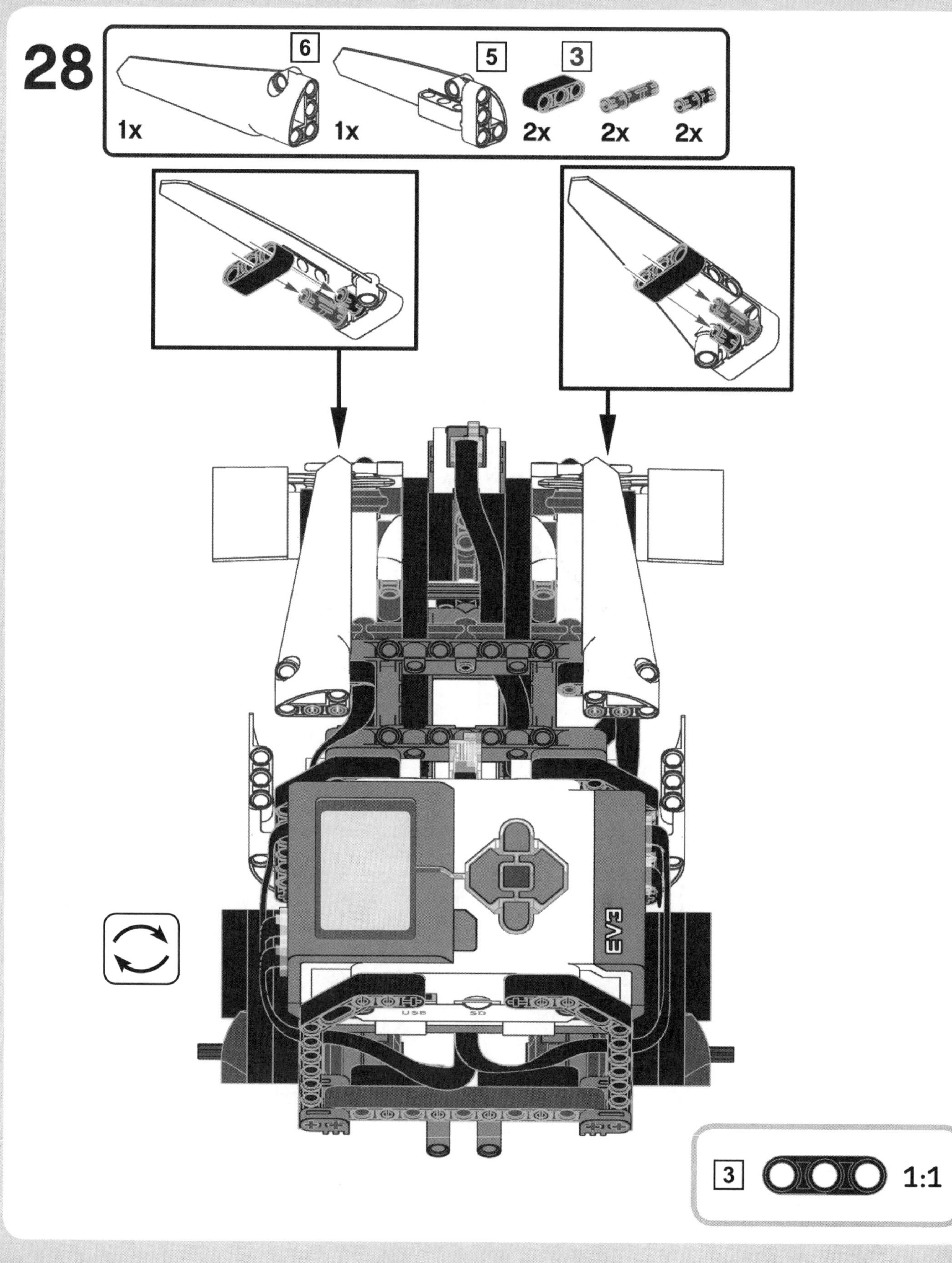
28
6
5
3
1x
1x
2x
2x
2x
EV3
USB
SD
3
1:1

car roof assembly

1

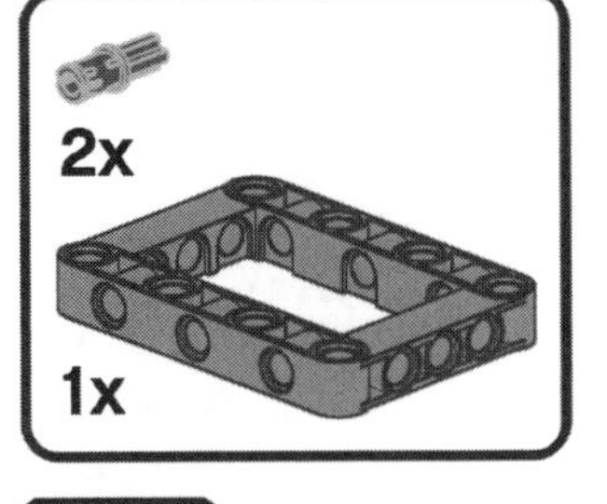

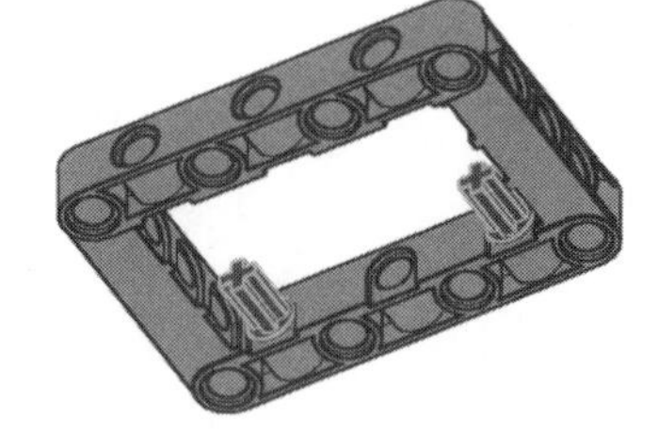

2

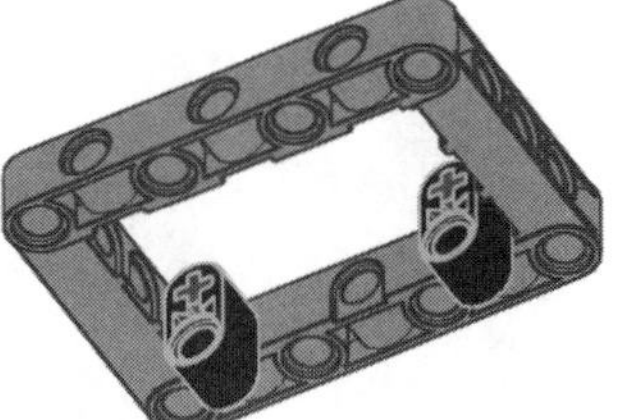

3

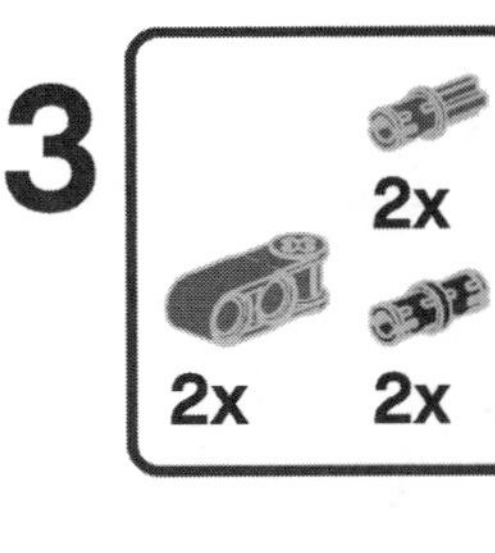

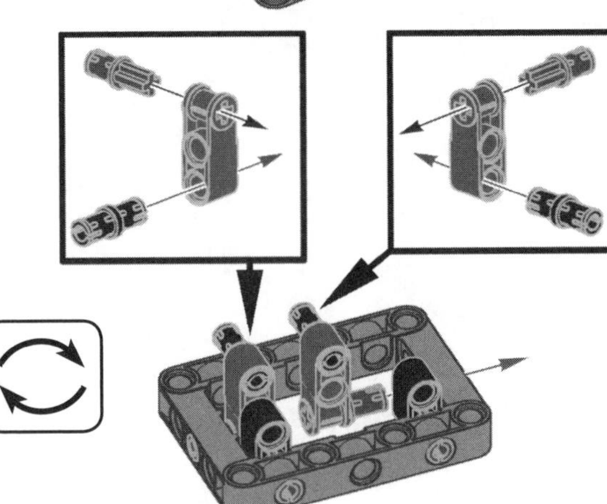

4

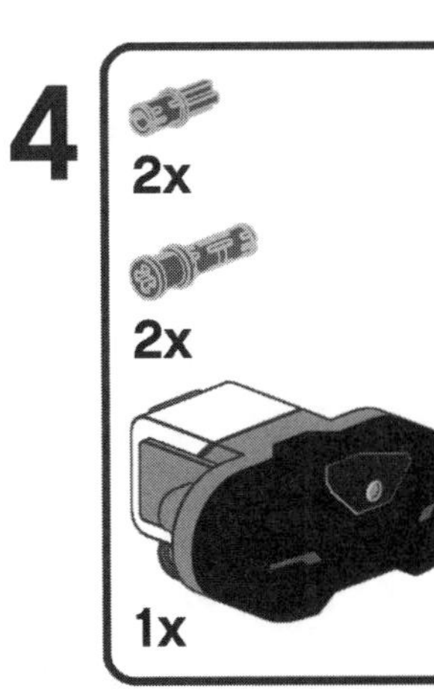

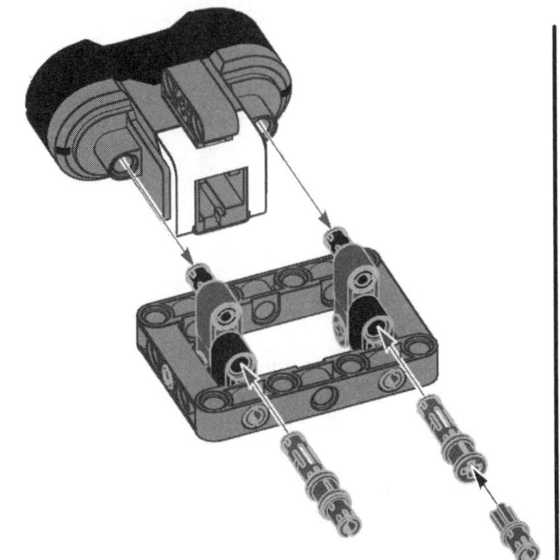

5

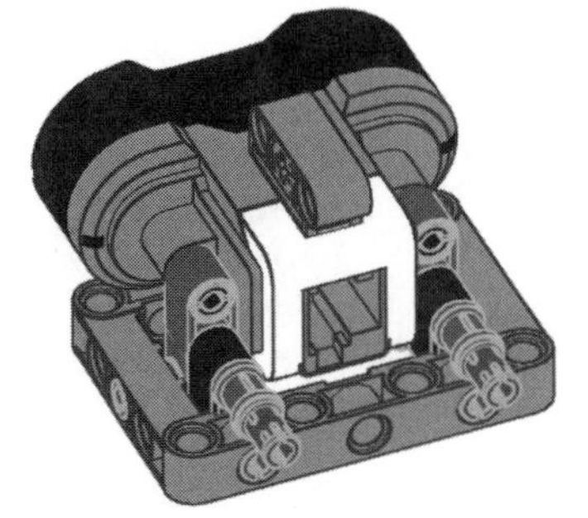

main assembly

29

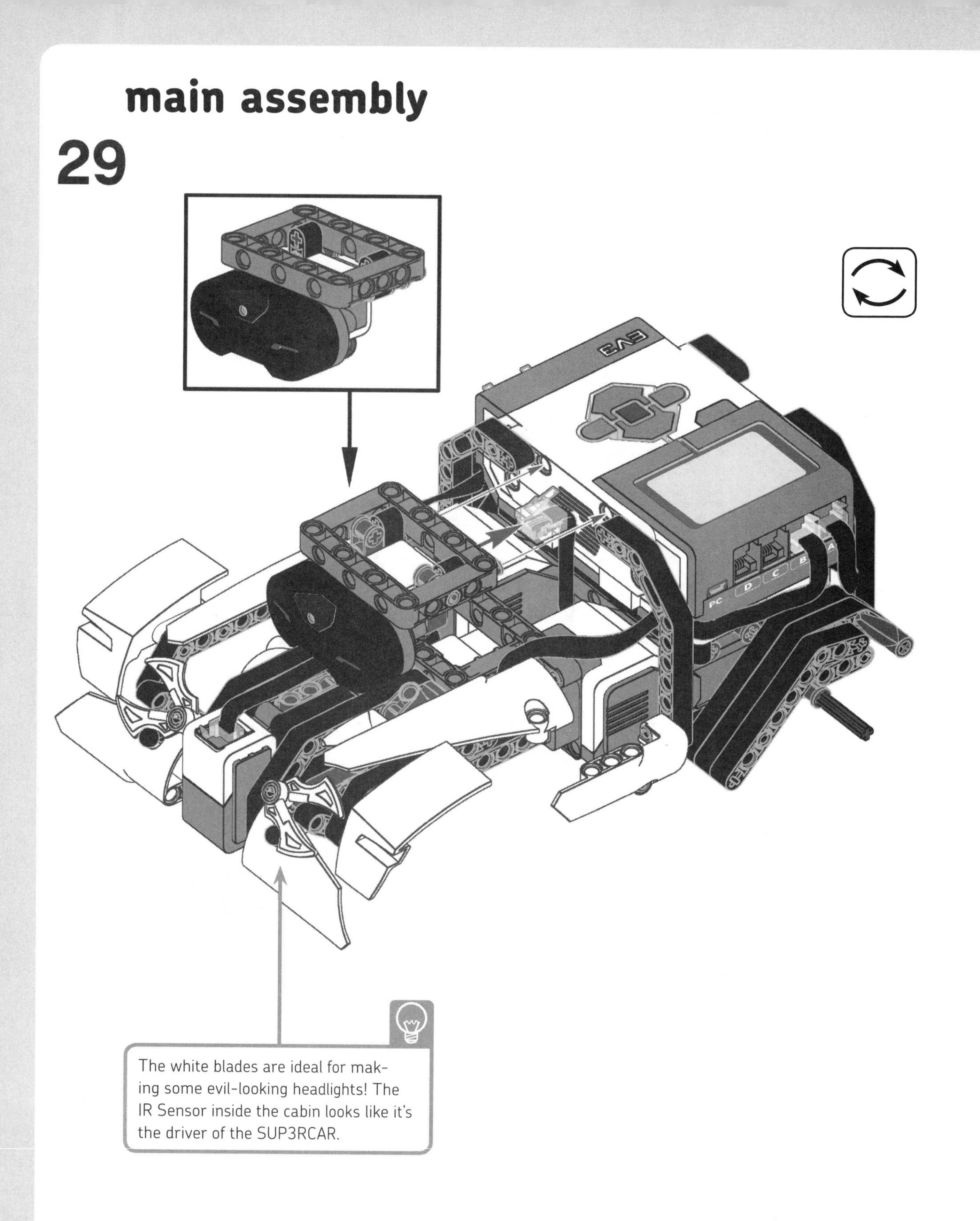

The white blades are ideal for making some evil-looking headlights! The IR Sensor inside the cabin looks like it's the driver of the SUP3RCAR.

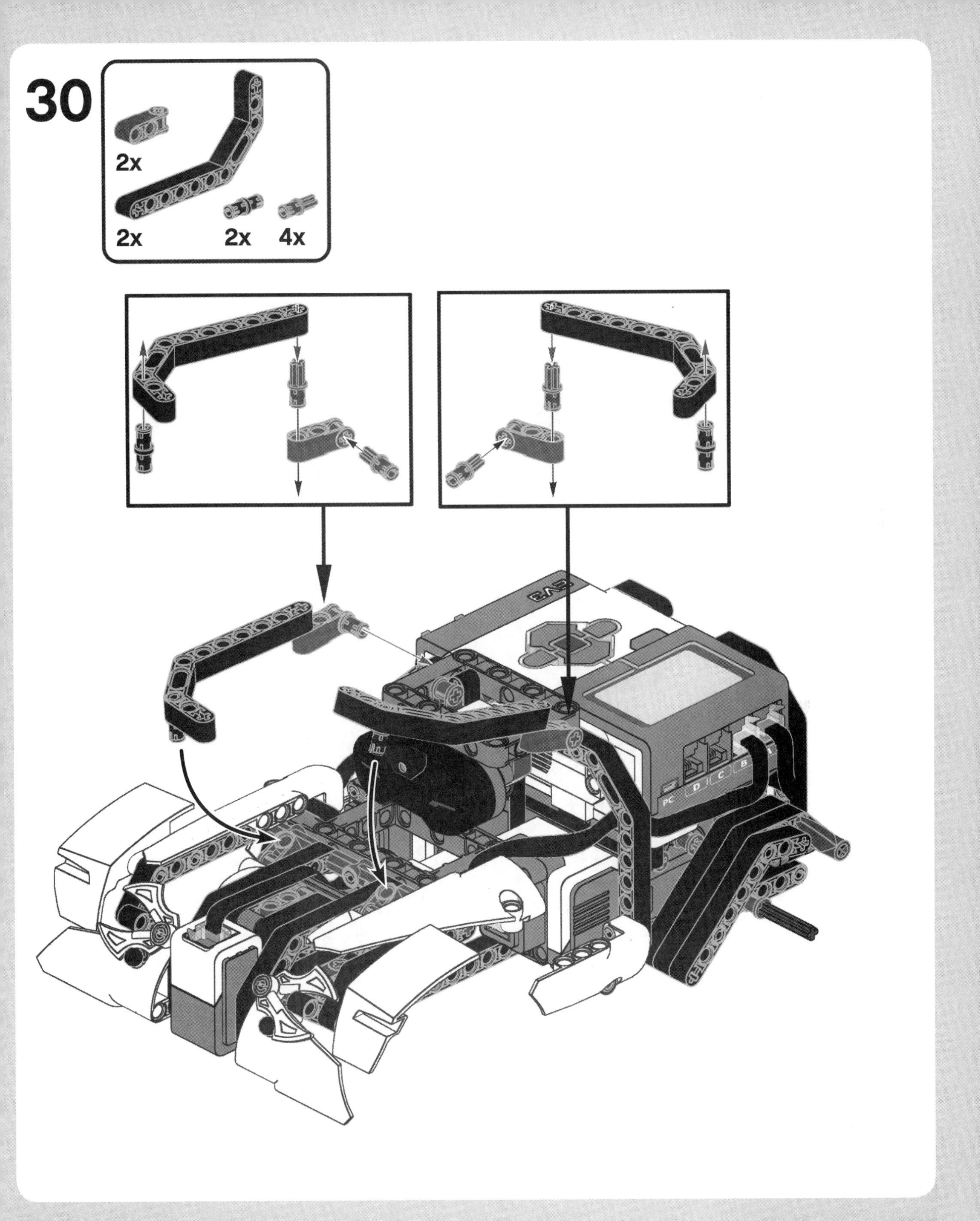
30
2x
2x
2x
4x
EV3
PC
D
C
B

steering assembly

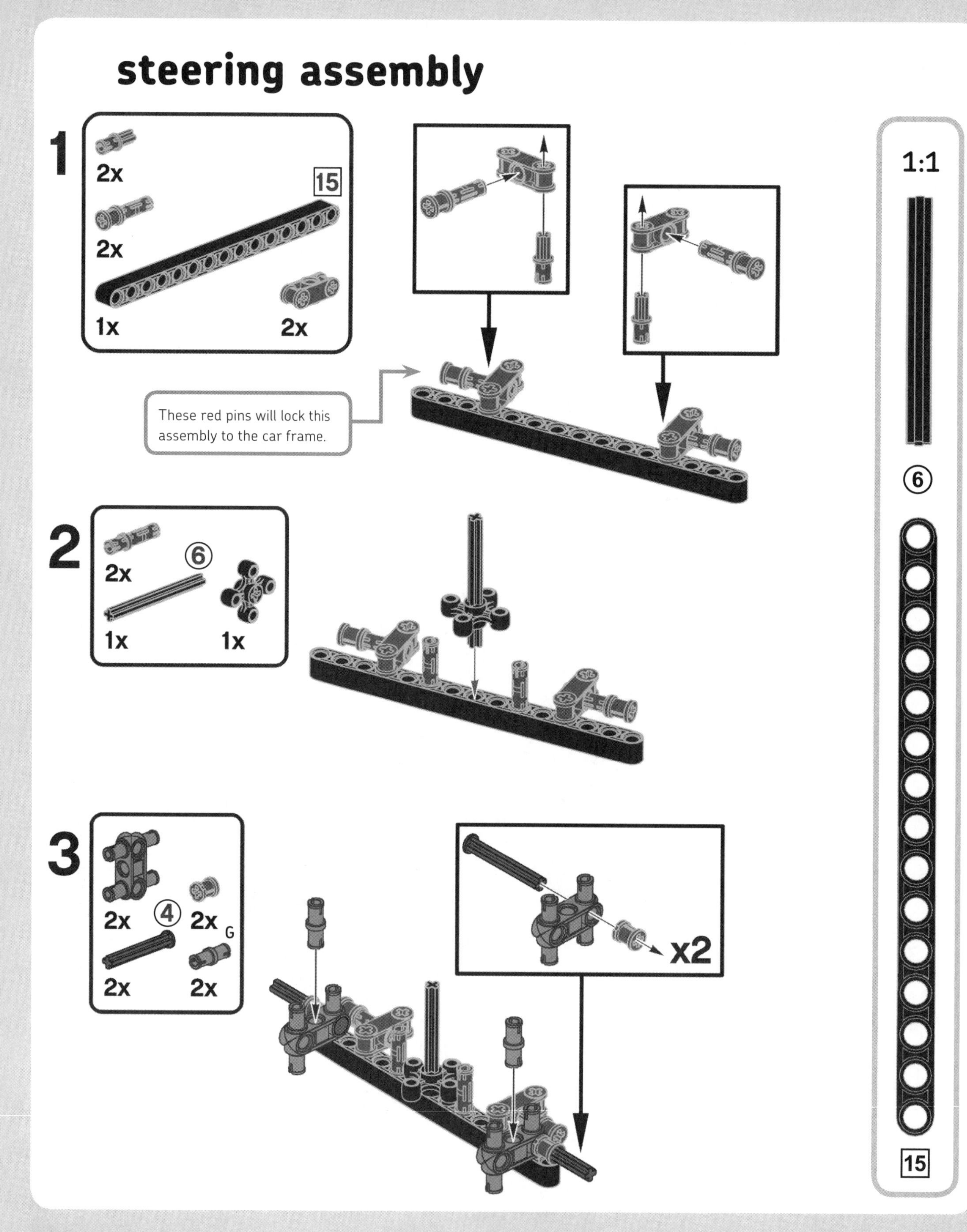

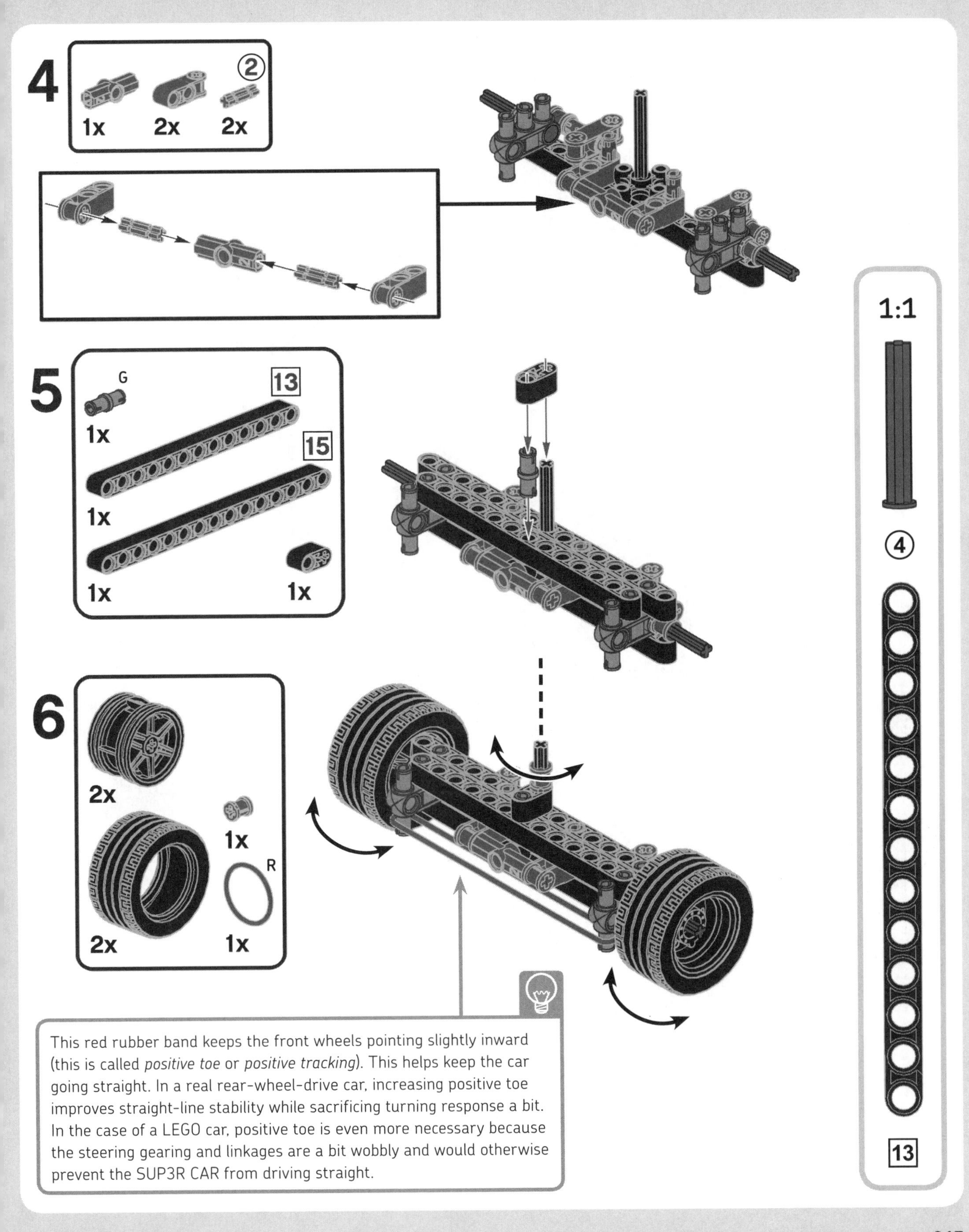

This red rubber band keeps the front wheels pointing slightly inward (this is called *positive toe* or *positive tracking*). This helps keep the car going straight. In a real rear-wheel-drive car, increasing positive toe improves straight-line stability while sacrificing turning response a bit. In the case of a LEGO car, positive toe is even more necessary because the steering gearing and linkages are a bit wobbly and would otherwise prevent the SUP3R CAR from driving straight.

main assembly

31

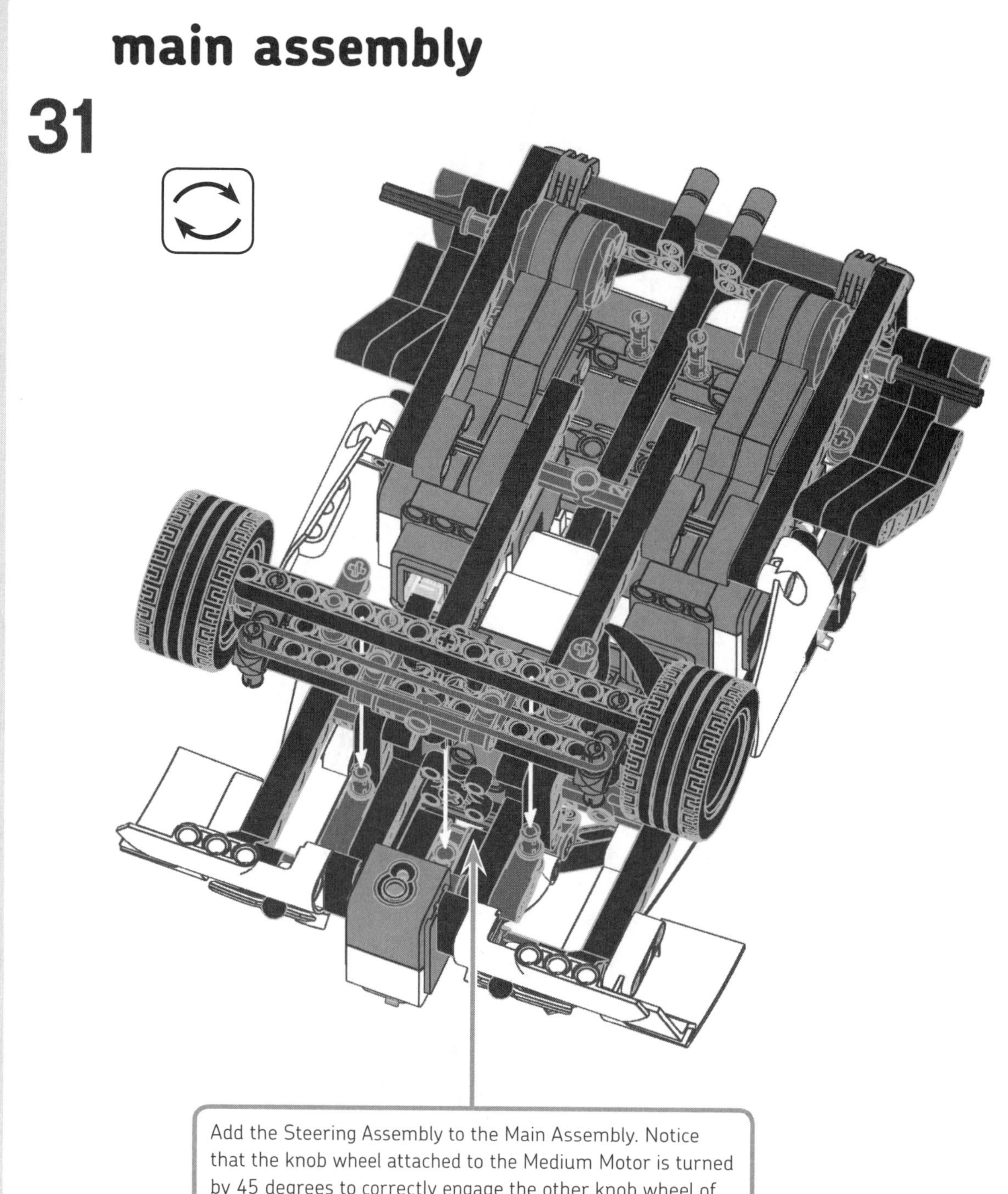

Add the Steering Assembly to the Main Assembly. Notice that the knob wheel attached to the Medium Motor is turned by 45 degrees to correctly engage the other knob wheel of the Steering Assembly. After you've inserted the Steering Assembly, push the two red 3M pins with stop into the car frame to lock the assemblies together.

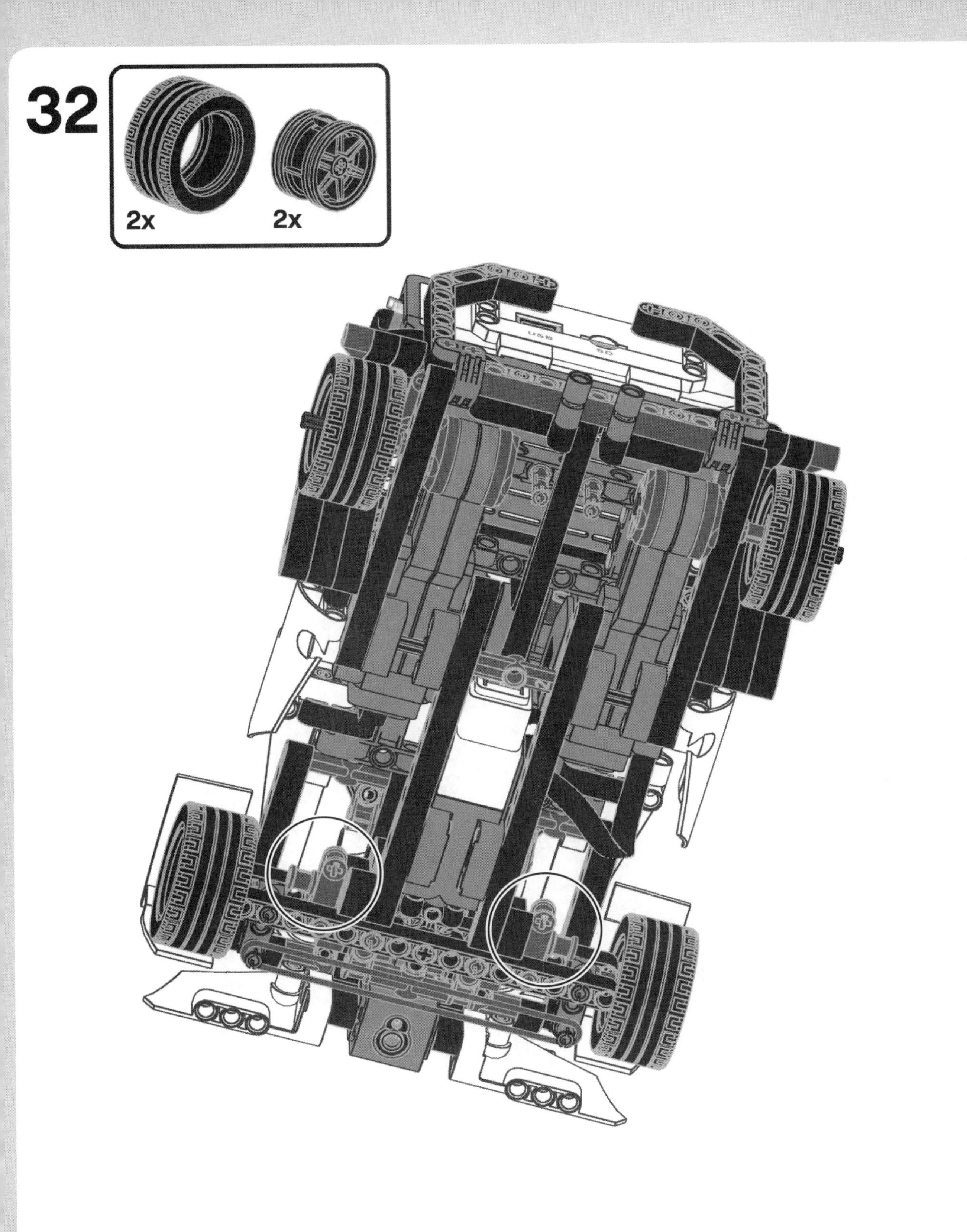
32
2x
2x

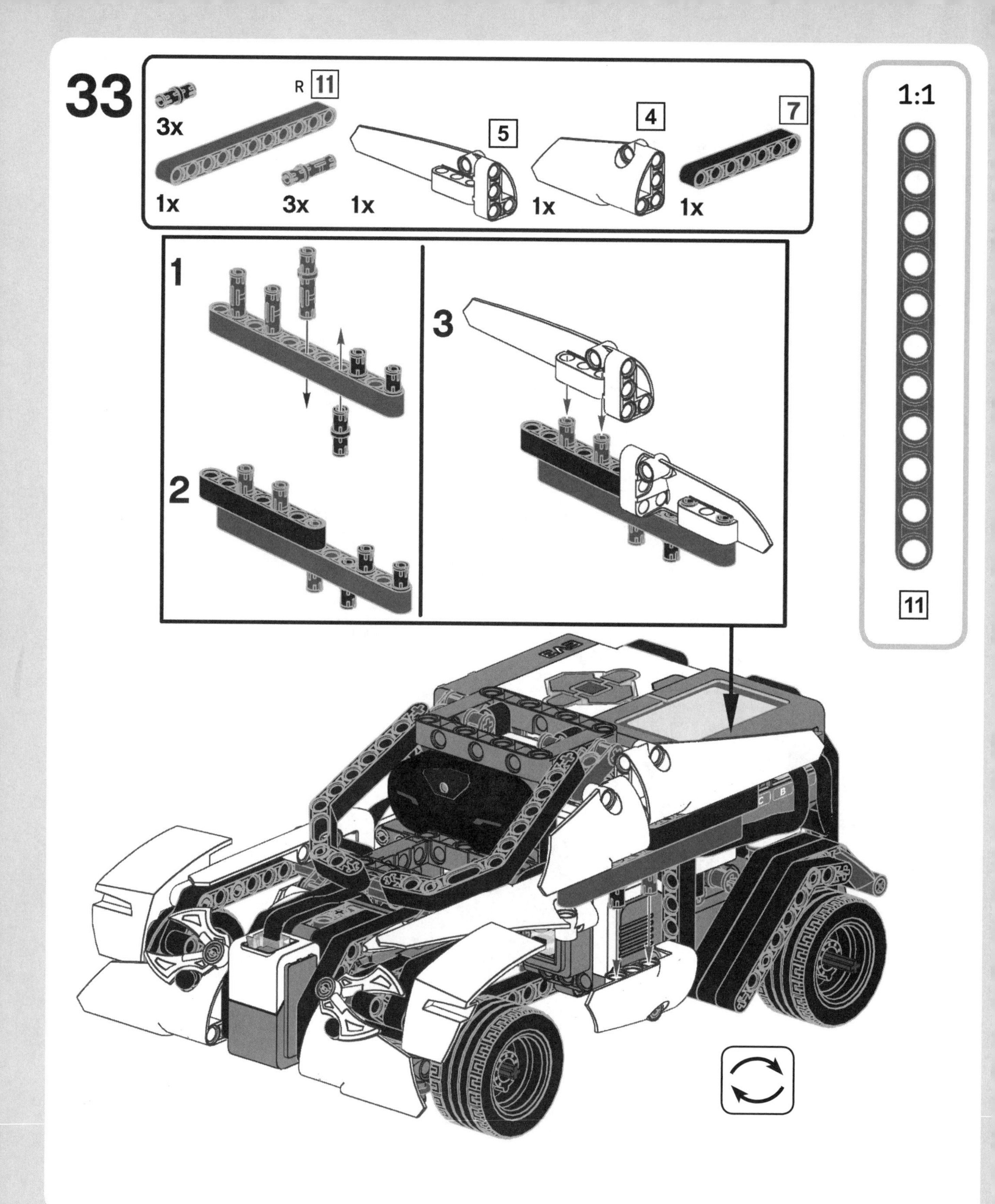

33
3x
R 11
1x
3x
1x
5
1x
4
1x
7
1
2
3
1:1
11

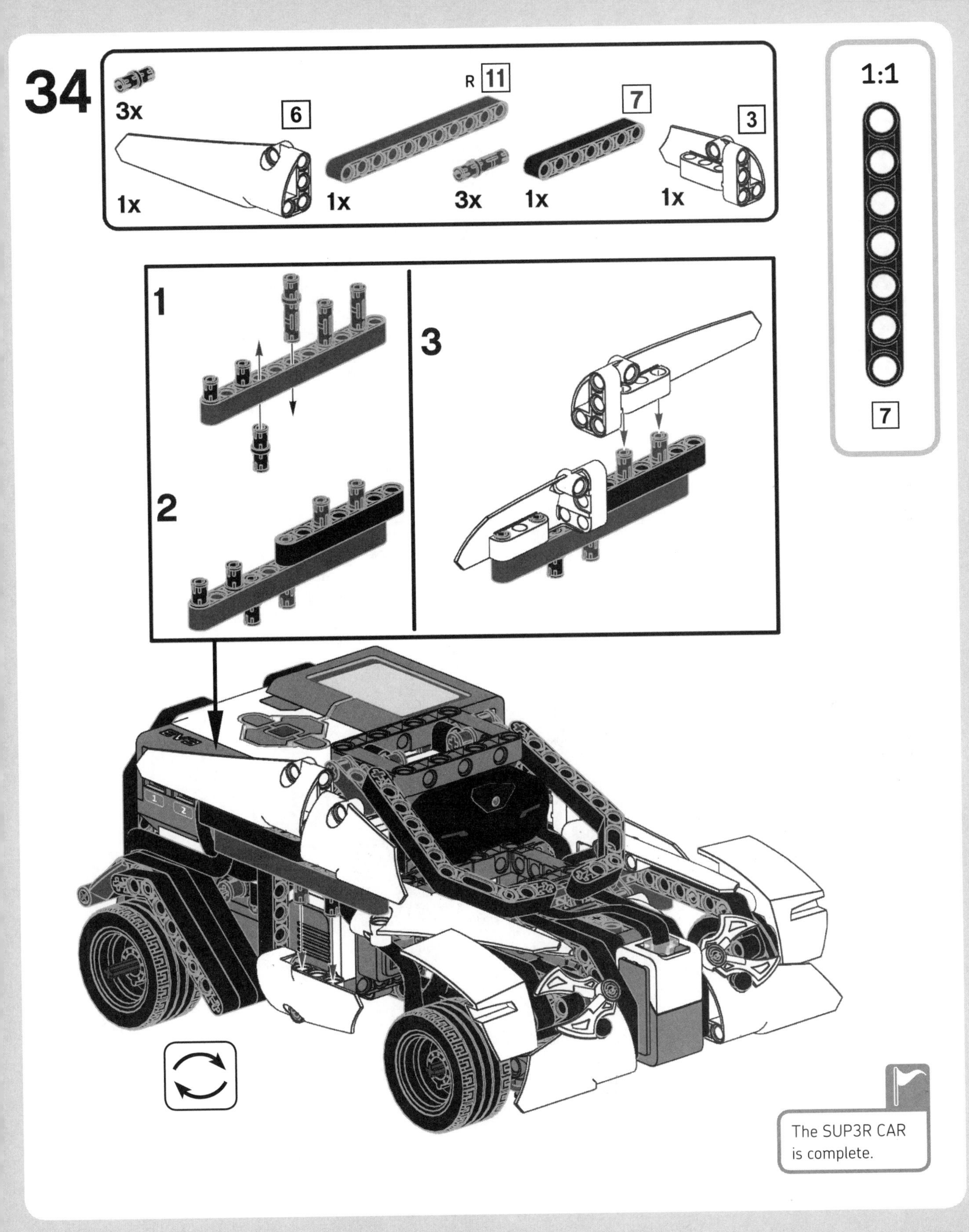
34
3x
6
R 11
7
3
1x
1x
3x
1x
1x
1:1
7
1
2
3
The SUP3R CAR
is complete.

building the R3MOTE

1

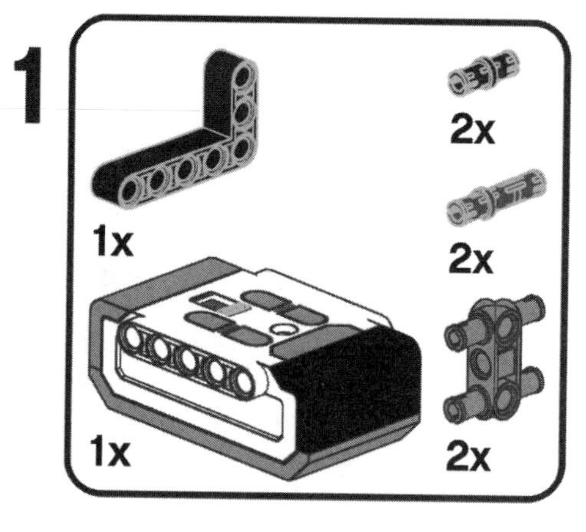

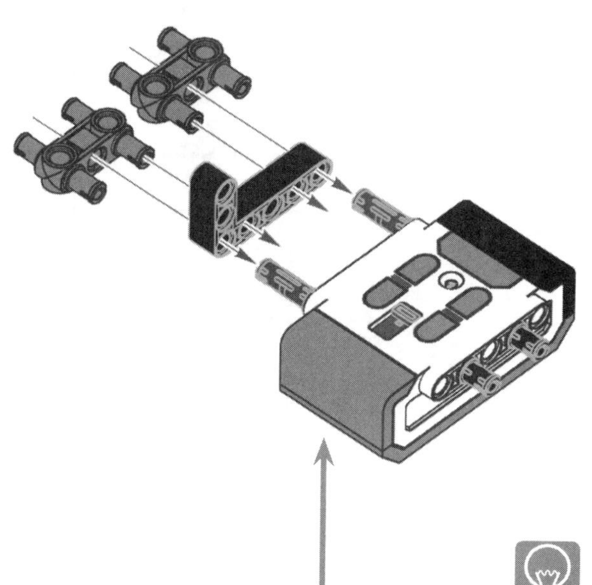

The Remote IR Beacon has tiny buttons and would be uncomfortable and counterintuitive to use with a steering car. This ergonomic R3MOTE solves the problem!

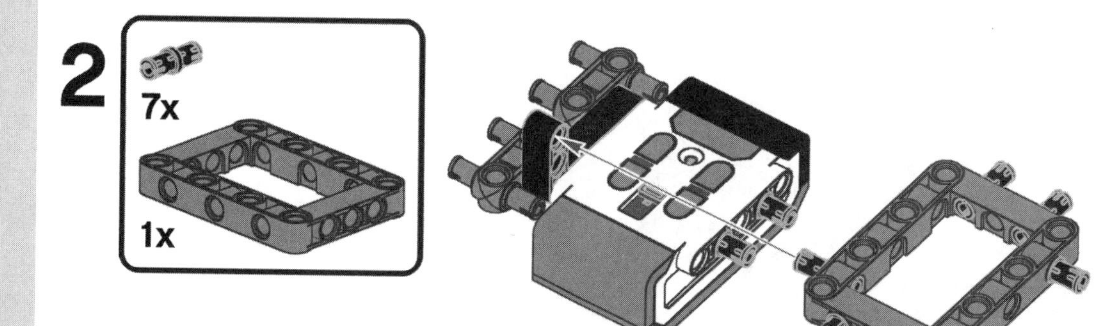

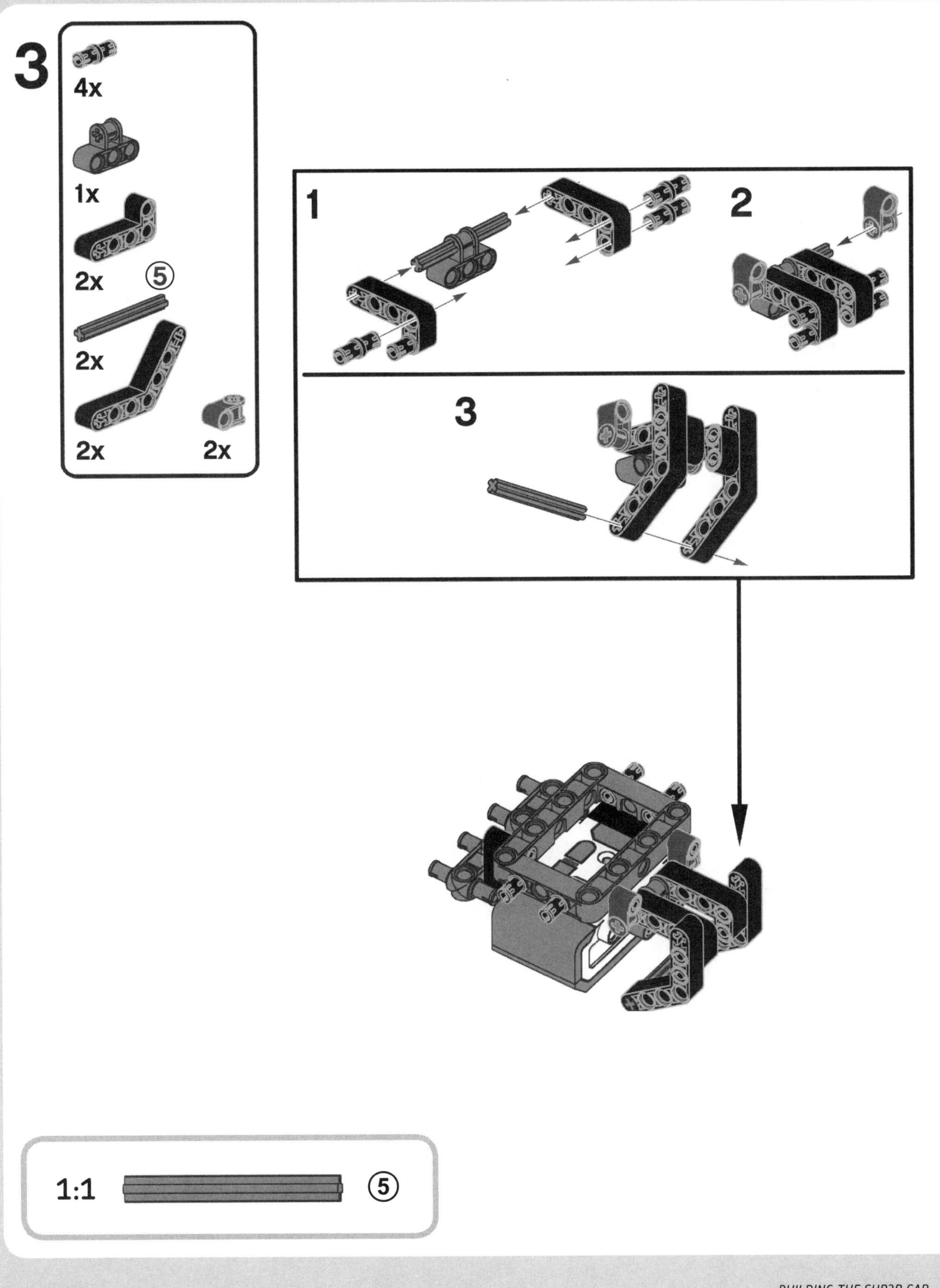
3
4x
1x
2x
5
2x
2x
2x
1
2
3
1:1
5

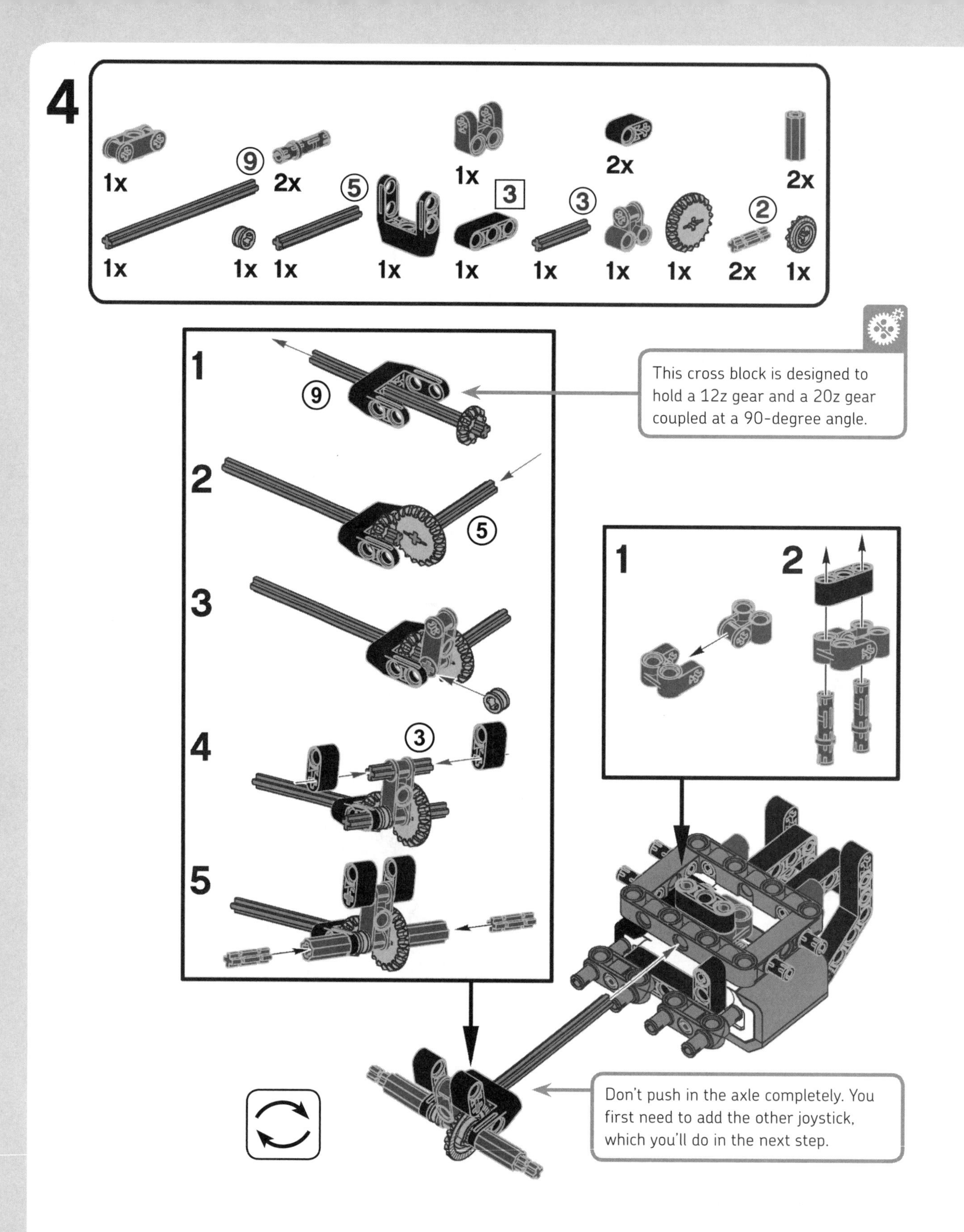
4
1x
9
2x
5
1x
3
3
2x
2x
2
1x
1x
1x
1x
1x
1x
1x
1x
2x
1x
1
9
2
5
3
4
3
5
1
2
This cross block is designed to hold a 12z gear and a 20z gear coupled at a 90-degree angle.
Don't push in the axle completely. You first need to add the other joystick, which you'll do in the next step.

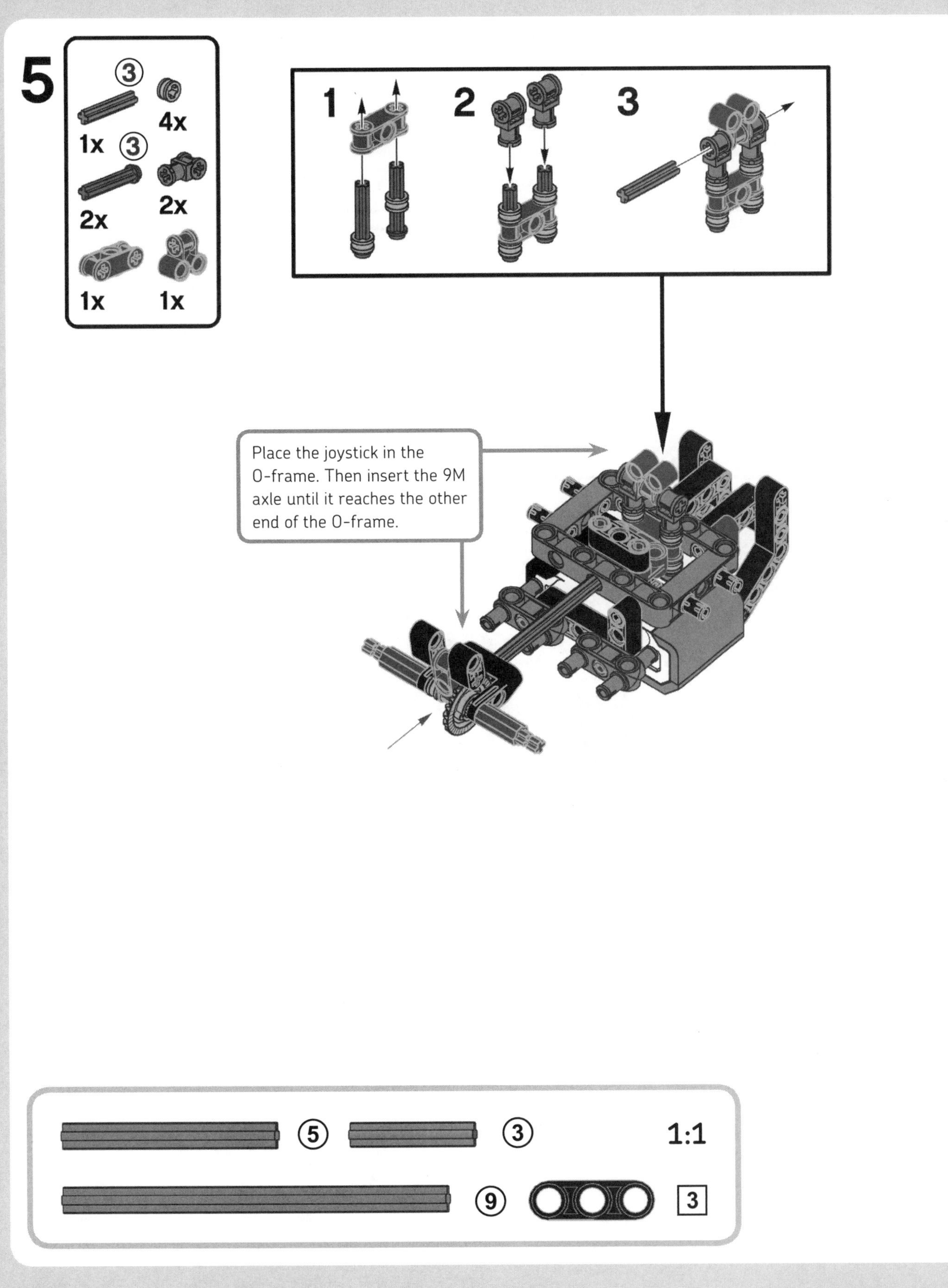
5
3
1x
3
2x
1x
4x
2x
1x
1
2
3
Place the joystick in the O-frame. Then insert the 9M axle until it reaches the other end of the O-frame.
5
3
1:1
9
3

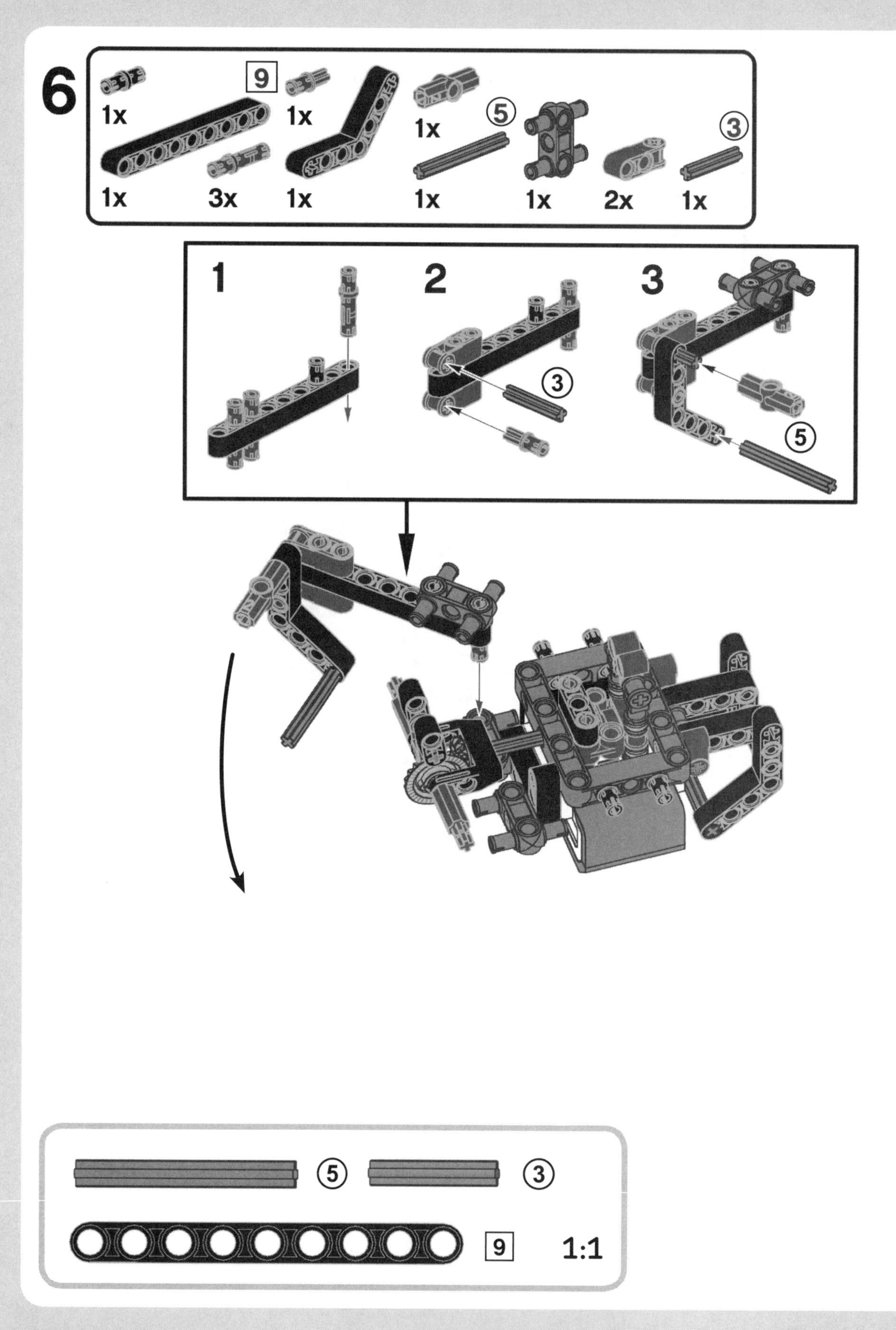
6
1x
9
1x
1x
5
1x
3
1x
3x
1x
1x
1x
2x
1x
1
2
3
3
5
5
3
9
1:1

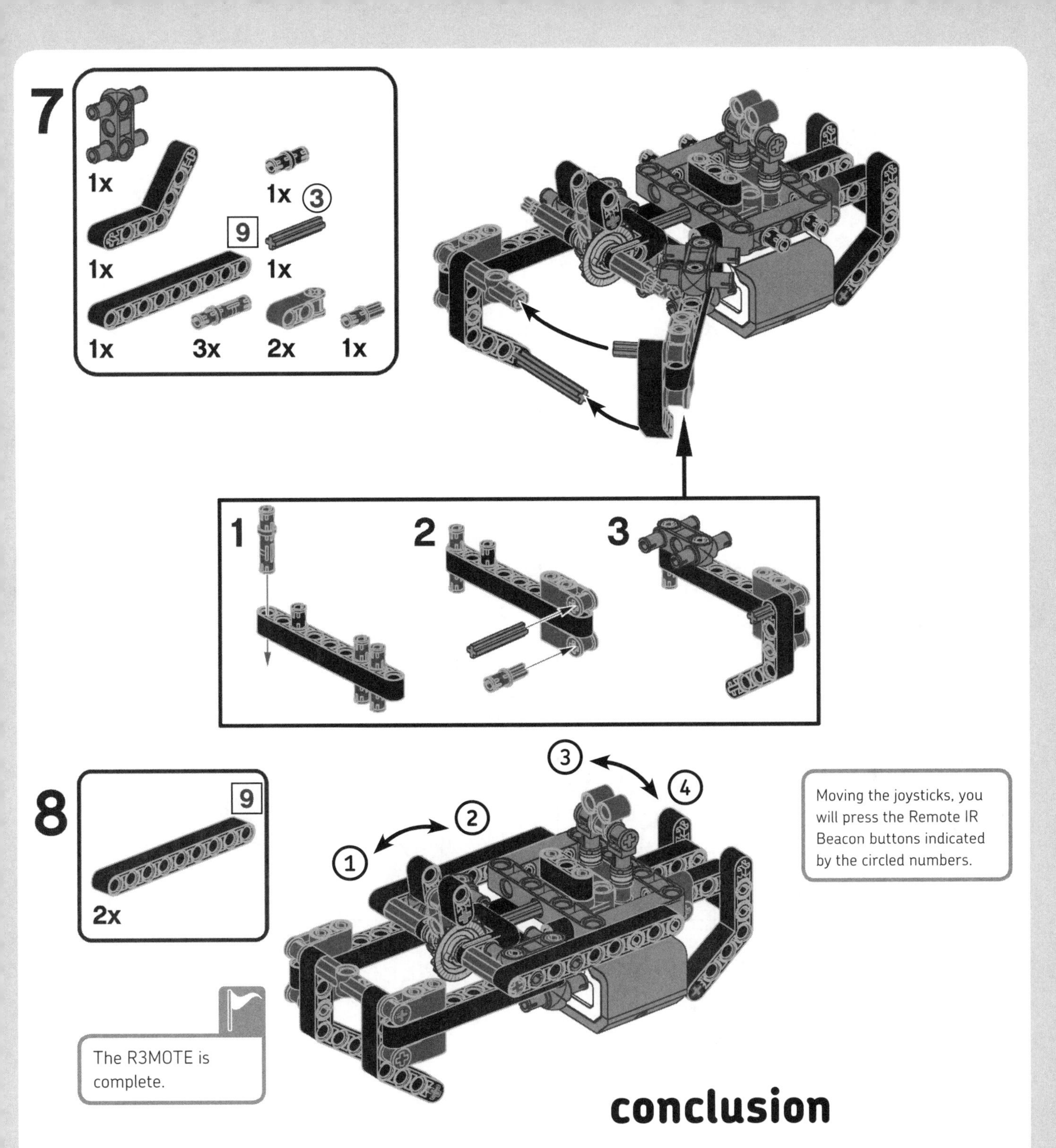

conclusion

In this chapter, you learned to build the SUP3R CAR and its R3MOTE. The SUP3R CAR is not just a remote-controlled car. It's an autonomous vehicle that can drive itself by following lines on the ground, or it can even follow you if you walk around wearing the Remote IR Beacon! You'll learn how to program it in the next chapter.

ACK! EXCUSE ME!
GROWL!
YOUR STOMACH IS RIGHT— IT'S LUNCHTIME! WE HAVE TWO OPTIONS: WE COULD GO EAT THE INFAMOUS SANDWICHES OF THAT CAFÉ DOWN THE BLOCK, OR...
OR?
NAH, LEAVE IT! THE ALTERNATIVE IS A LOT OF DIRTY WORK. YOU WON'T LIKE IT. AND I DON'T KNOW IF YOU'RE EVEN UP TO THE TASK!
SURE, I CAN DO IT!
GOOD BOY...
HERE'S WHAT HAPPENED...
ONCE, HERE INSIDE EV3L THERE WAS A CANTEEN. OUR INDIAN COOK MADE INCREDIBLE FOOD! BUT THEN...
...THE RESEARCH LAB CREATED SOME VEGAMECHANIC ROBOTS THAT WERE CAPABLE OF PHOTOSYNTHESIS AND COULD COMMUNICATE WIRELESSLY, MAKING A COMPLEX NETWORKING LIFE-FORM.
11X1

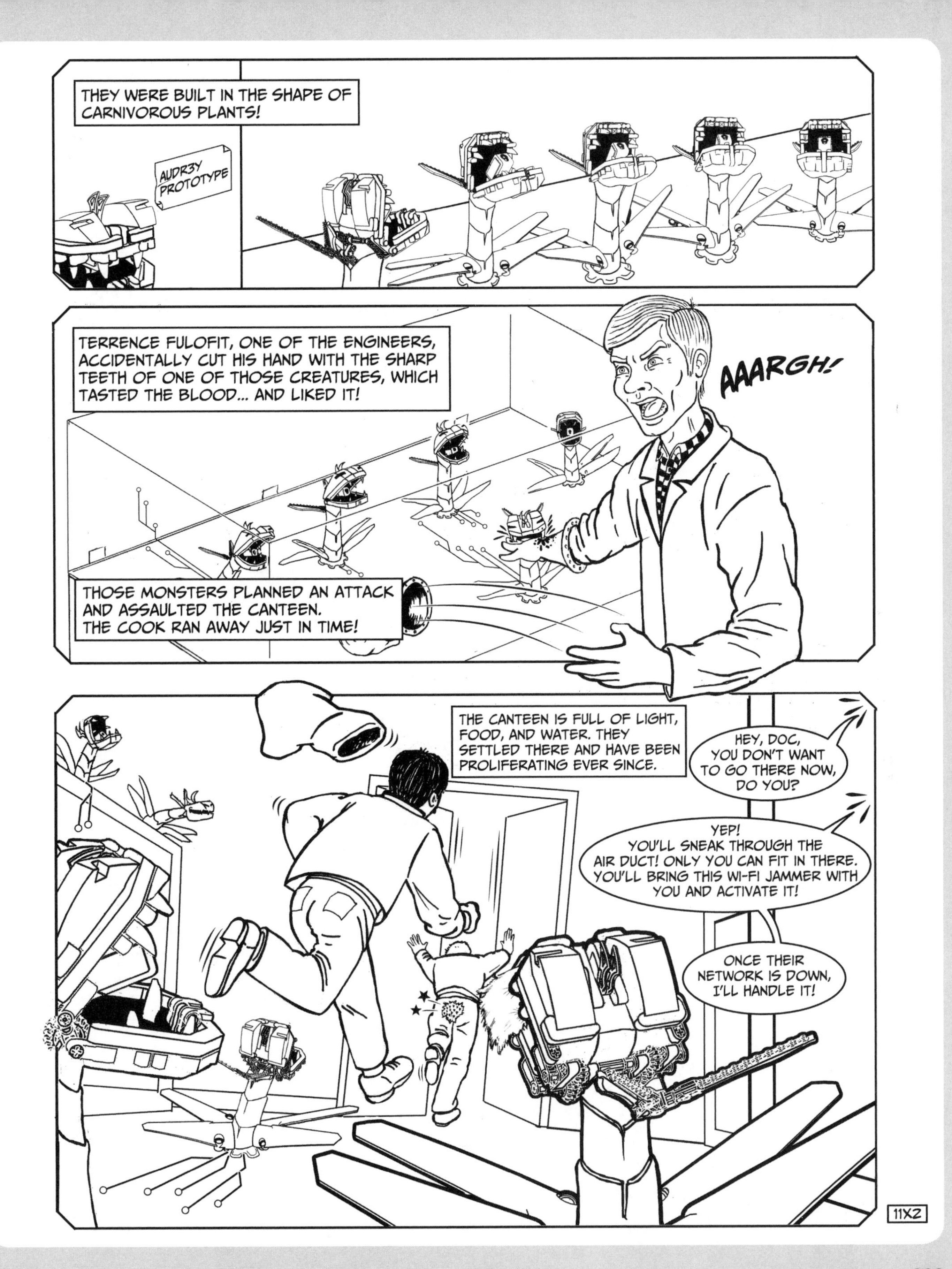
THEY WERE BUILT IN THE SHAPE OF CARNIVOROUS PLANTS!
AUDR3Y PROTOTYPE
TERRENCE FULOFIT, ONE OF THE ENGINEERS, ACCIDENTALLY CUT HIS HAND WITH THE SHARP TEETH OF ONE OF THOSE CREATURES, WHICH TASTED THE BLOOD... AND LIKED IT!
AAARGH!
THOSE MONSTERS PLANNED AN ATTACK AND ASSAULTED THE CANTEEN. THE COOK RAN AWAY JUST IN TIME!
THE CANTEEN IS FULL OF LIGHT, FOOD, AND WATER. THEY SETTLED THERE AND HAVE BEEN PROLIFERATING EVER SINCE.
HEY, DOC, YOU DON'T WANT TO GO THERE NOW, DO YOU?
YEP! YOU'LL SNEAK THROUGH THE AIR DUCT! ONLY YOU CAN FIT IN THERE. YOU'LL BRING THIS WI-FI JAMMER WITH YOU AND ACTIVATE IT!
ONCE THEIR NETWORK IS DOWN, I'LL HANDLE IT!
11X2

12

programming the SUP3R CAR

In this chapter, you'll program the SUP3R CAR that we built in Chapter 11. First you'll create a simple program to make the car go around by itself, avoiding obstacles. Next, you'll create a program to remotely control the car with the R3MOTE that we also built in Chapter 11. We can use the same remote as a beacon for the car to follow, and you'll also make a program to do that. As you build your programs, you'll learn new programming concepts like how to configure the Switch block to handle multiple cases and how to use variables and arrays. You'll also learn how to execute multiple sequences of blocks simultaneously, how to interrupt a loop, and how to stop a program entirely.

But before discussing the programming aspects, I'd like to briefly introduce you to the mechanics of a steering car.

electronic vs. mechanical differentials

Unlike ROV3R, which steers by changing the relative speed of its motors, the SUP3R CAR can be steered by moving its front wheels—like a real car. As shown in Figure 12-1, when a car steers, its driven wheels turn at different speeds. For instance, when a car is traveling on a circular path, its outer wheel has to travel a greater distance than its inner wheel. Because the wheels are attached to the car traveling at speed *V*, but one is farther from the center of the circle than the other, they must travel different distances in the same amount of time. As a result, the outer wheel speed, V_R, ends up being greater than the inner wheel speed, V_L. The *differential* is a mechanical device with gears that allows the engine speed to be distributed to the wheels independently when one wheel is turning more slowly than the other. Without the use of a differential, one wheel could lose traction with the ground and slip when the car is turning.

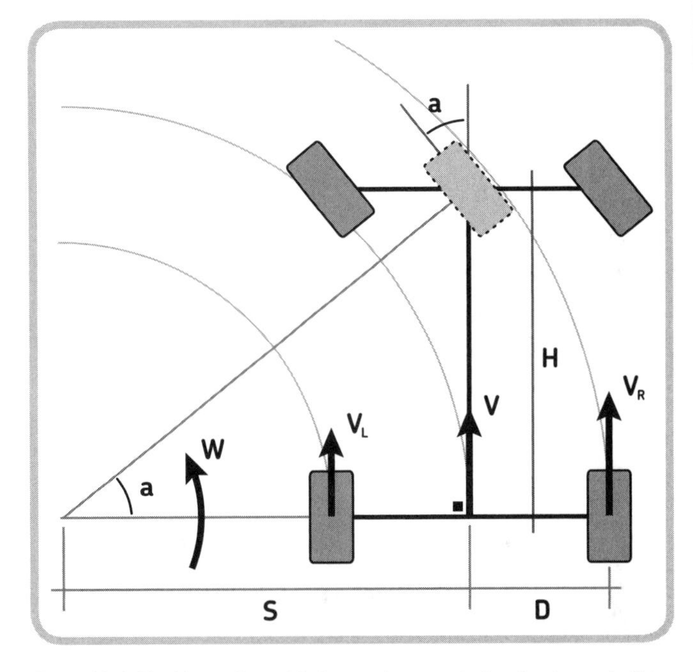

Figure 12-1: The kinematic model of a steering car. a *is the steering angle;* S *is the turning radius;* D *is the distance between the driving wheel and the center of the vehicle;* H *is the distance between the front and back wheels;* W *is the vehicle's angular speed;* V *is the speed of the car; and* V_L *and* V_R *are the speed of the left and right driving wheels, respectively. (The dotted wheel in the middle is the single turning wheel of the equivalent tricycle model.)*

Figure 12-2 shows a LEGO differential assembly extracted from a car model (part ID: 4525184, design ID: 62821, part name: Differential 3M z28).

However, the 31313 set comes without such a LEGO differential, so I designed the SUP3R CAR to have an independent motor for each driving wheel. When the SUP3R CAR steers, the speeds of its driving wheels should be precisely controlled, as if it had a mechanical differential. In fact, this is an *electronic differential*, one that we'll implement with software (in the *Drive* My Block).

Figure 12-2: A LEGO Technic differential assembly. The differential is not included in the EV3 set.

using variables

Up to now, you've carried data around your programs using Data Wires, but there is another way to do this: variables. A *variable* is a location in the computer memory that can hold data of a certain type. In the EV3 system, variables can contain five data types, as described in Chapter 6: Numeric, Logic, Text, Numeric Array, and Logic Array.

To manage variables, you'll use the Variable block. (It has a red header and is found in the Data Operations palette.) Figure 12-3 shows the various controls of the Variable block. Using the Mode Selector, you can choose to write (a) or to read (b) to a variable of each of the five types.

To add a variable, first select the variable type using the Mode Selector. Then click the **Variable Name** field and select **Add Variable**, as shown in Figure 12-3(c). Now enter the name of the variable in the New Variable dialog and click **OK** (d). The name can be a single letter, a word, or any sequence of letters and numbers (and spaces). I suggest you use short names that help you remember the meaning of a particular variable, because long names may not fit in the small Variable Name field at the top of the Variable block. For example, to create a numeric variable, first set the block to Read (or Write) Numeric and then click the Variable Name field and select Add Variable. The name *spd* would be a good name for the car speed variable.

The Variable block's appearance changes according to its mode. For example, when it's set in Write Numeric mode, the block has a numeric input where you can enter or carry a value with a Data Wire (e); in Read Numeric mode (f), it has an output from which you can read the stored value. Keep in mind that whenever you write to a variable, you overwrite any existing value. For example, if you place three Variable blocks in Write Numeric mode and have them write into the same variable the values 1, 2, 3 in sequence, the content of the variable will be 3.

DIGGING DEEPER: COMPUTING WHEEL SPEEDS FOR AN ELECTRONIC DIFFERENTIAL

Using the kinematic model of a steering car shown in Figure 12-1 as reference, let's compute the speeds of a car's driving wheels when the car is turning. (The model of the car with two turning wheels is equivalent to a tricycle model that has a single front turning wheel.)

Our goal is to find two equations that relate the car's speed and the steering angle of its front wheels to the speed of its left and right rear driving wheels. Since the speed of the EV3 motors under a constant load is proportional to their power, those two speeds will become the Power inputs for the motors that drive those wheels. (See "Digging Deeper: How 'Power' Relates to Speed" on page 297.)

Let H be the distance between the rear and the front wheels and D be half the distance between the rear wheels. Let V be the car's speed and a be the steering angle. S is defined as the car's turning radius. Using the geometric relationships, and knowing that the angle marked with a black square is a right angle (90°), we have $S = H / \tan(a)$. The steering angle a must be converted into radians: This is done by multiplying the angle in degrees by $(\pi / 180)$, which in this case comes to approximately 0.017. Also, for small values of angle a [rad], $\tan(a)$ can be approximated with the value of a itself, so we can write the approximate equation $S \approx H / (0.017 \times a)$.

When the car is turning, $W = V / S$, where W is the angular speed, V is the car's speed, and S is the turning radius. The speed of the outer wheel (the right wheel in Figure 12-1) is $V_R = W \times (S + D)$. Using some substitutions and simple manipulations, we end up with $V_R = V / S \times (S + D) = V \times (1 + D / S)$, which finally yields $V_R = V \times (1 + D \times 0.017 \times a / H)$. Similarly, for the inner wheel (the left wheel in Figure 12-1), we get $V_L = V \times (1 - D \times 0.017 \times a / H)$. We'll implement these equations using Math blocks in Advanced mode in the *Drive* My Block (page 236).

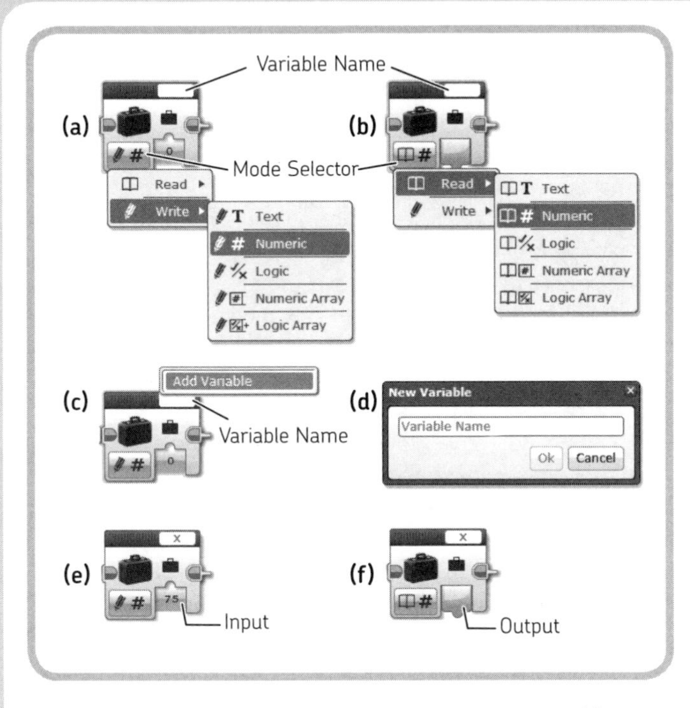

Figure 12-3: The Variable block. Select different data types in Write (a) and Read (b) modes; to add a variable (c), enter the new variable name in the New Variable dialog (d); to write into a numeric variable, select Write Numeric mode (e); and to read a numeric variable, select Read Numeric mode (f).

NOTE All of a project's variables are accessible from each of its programs. These are called *global variables*.

You can manage variables (copy, paste, delete, and add them) from the Variables tab in the Project Content area in Project Properties. You'll see various examples of how to use variables in this chapter and the following chapters.

using arrays

In the previous chapters, you used three data types: numbers, logic values, and text. In Chapter 6, I briefly mentioned another data type: arrays. In this chapter, I'll describe them in detail and show you how to use them.

In computer science, an *array* is a data type that describes a collection of elements that can be selected with an *index*, a number that indicates a specific element in the array. An empty array has zero elements. It exists but occupies no memory space. The first element of any array has index 0. The last element of an array of length N is at index $N - 1$ (for example, the seventh element in an array is at index 6).

You might think of arrays as a kind of elevator: You can access a floor (the *element*) by selecting the number of the floor (its *index*). You can leave some stuff on each floor by writing data, or you can pick it up by reading data. EV3 offers both numeric and logic arrays that can contain numeric and logic values, respectively. You can write and read data into arrays by using the Variable block in combination with the Array Operations block.

using the variable block with numeric and logic arrays

The operations available on a numeric array are shown in Figure 12-4. (These apply similarly to logic arrays.) As you can see, the Variable block in Write Numeric Array mode allows you to clear all elements in an array (a) or add or remove elements using the drop-down menu (b). (Remember to give the array a name!)

using the array operations block

The Array Operations block, also shown in Figure 12-4, allows you to write (c) and read (d) elements of an array at a specified index, append elements to the end of the array (e), and retrieve the array's length (f).

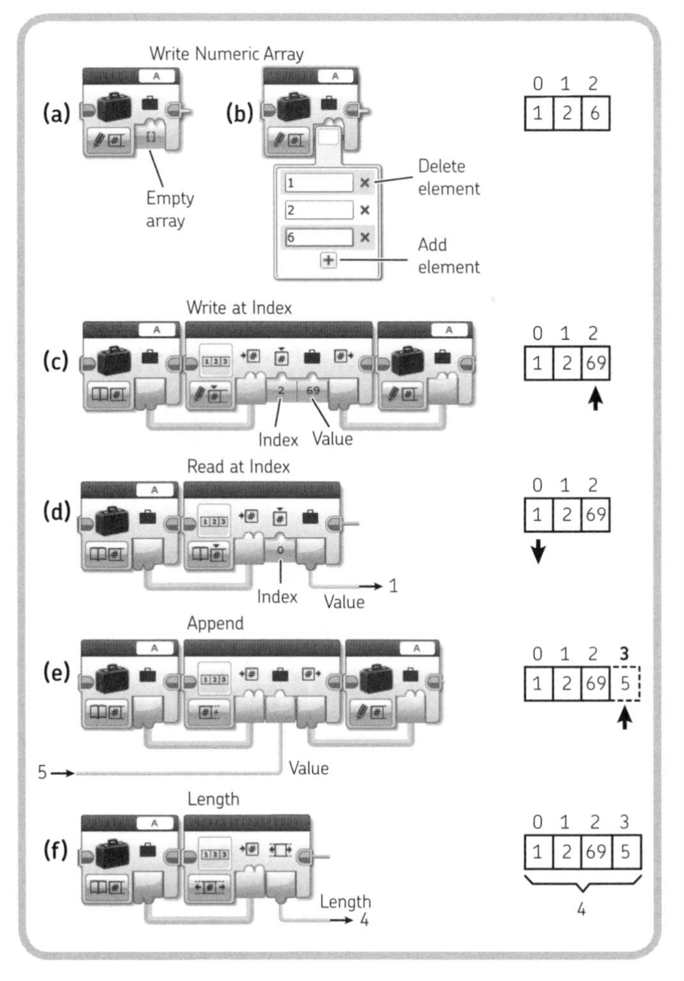

Figure 12-4: Managing arrays using the Variable and Array Operations blocks

In each of its modes, the Array Operations block needs an array type variable to work on. In the Write at Index and Append modes, the block outputs an array variable. Usually, you will overwrite the array used as the input for the operation, but you can also write the result to another array variable.

For example, in Figure 12-4(c), the Array Operations block in Write at Index mode takes a numeric array A as input and outputs an array that is written into A again. The value 69 is written at index 2 (the third element because the first element has index 0). This array should already have at least three elements, or the program will abort.

WARNING If you try to read values at indices that are outside an array's boundaries, your program will abort, and the EV3 Brick will show an error message. If you write a new value at an index that is beyond the array size, the array positions before the new indexed value will be filled with zeros or garbage values. For example, if your array contains [1,2] and you write 3 at index 4, the resulting array will be [1,2,*x*,*y*,3], where *x* and *y* are garbage values.

As with any block, the values for the Index and Value inputs of the Array Operations block can be entered as fixed values or passed as dynamic values that change at runtime using Data Wires.

NOTE You can use the Array Operations block in Append mode to create arrays whose number of elements is limited only by the amount of RAM in the EV3 Brick. For example, I made a test program on the EV3 Brick that appended data to an array in a loop while showing a counter. After many tens of thousands of elements were appended, I stopped the program. I estimate that with the 64MB of RAM of the EV3 Brick, you can create a numeric array with about 2 million elements.

using the switch block with multiple cases

In previous chapters, you saw how the Switch block in Compare mode can be used to choose to execute one of two possible logic cases, depending on whether its Logic test is True or False. When used in Numeric mode, Text mode, and its various Measure modes, the Switch block will let you choose which of the possible cases to execute, according to a test. For example, a choice can be made based on the color measured by the Color Sensor (Color Sensor Measure Color mode) or the command received by the IR Sensor (Infrared Sensor Measure Remote mode), as you'll see when you make the program to remotely control the SUP3R CAR .

In Numeric mode, you can execute a numbered case according to the Numeric input. In Text mode, you can execute a labeled case that corresponds to the text in the Text input. In Numeric, Text, and Measure modes, the Switch block shows multiple cases, as well as some additional controls (see Figure 12-5) that allow you to add or remove a case and select the value for cases. The Default button lets you set which case to execute by default if the sensor's test value doesn't match any case in the Switch block. You can resize each case independently using the Resize handles, and you can change the way the Switch block is displayed using the Flat/Tabbed Selector.

running parallel sequences (multitasking)

The programs that we've created so far have had one sequence of blocks running at a time. But what if you need your robot to perform different actions in parallel? For example, perhaps you want your robot to roam around while playing a sound and updating text on the EV3 Brick display. Programs that handle multiple tasks at once are *multitasking*, and they need to have multiple tasks (sequences) running at the same time.The EV3 programming language lets you create multiple sequences of blocks running in parallel by either placing multiple Start blocks in a program or connecting parallel sequences of blocks with a Sequence Wire. You'll see this in action when we create the *Drive* My Block and the *BeaconSiren* program.

In multitasking programs, be careful when using shared resources for different tasks, as resource conflicts may arise! For example, using two Motor blocks to drive the same motor from two different parallel sequences will lead to unexpected behaviors. Also, writing to the same variable from parallel sequences will make the content of that variable unpredictable, as you won't know which sequence was the last to write to it.

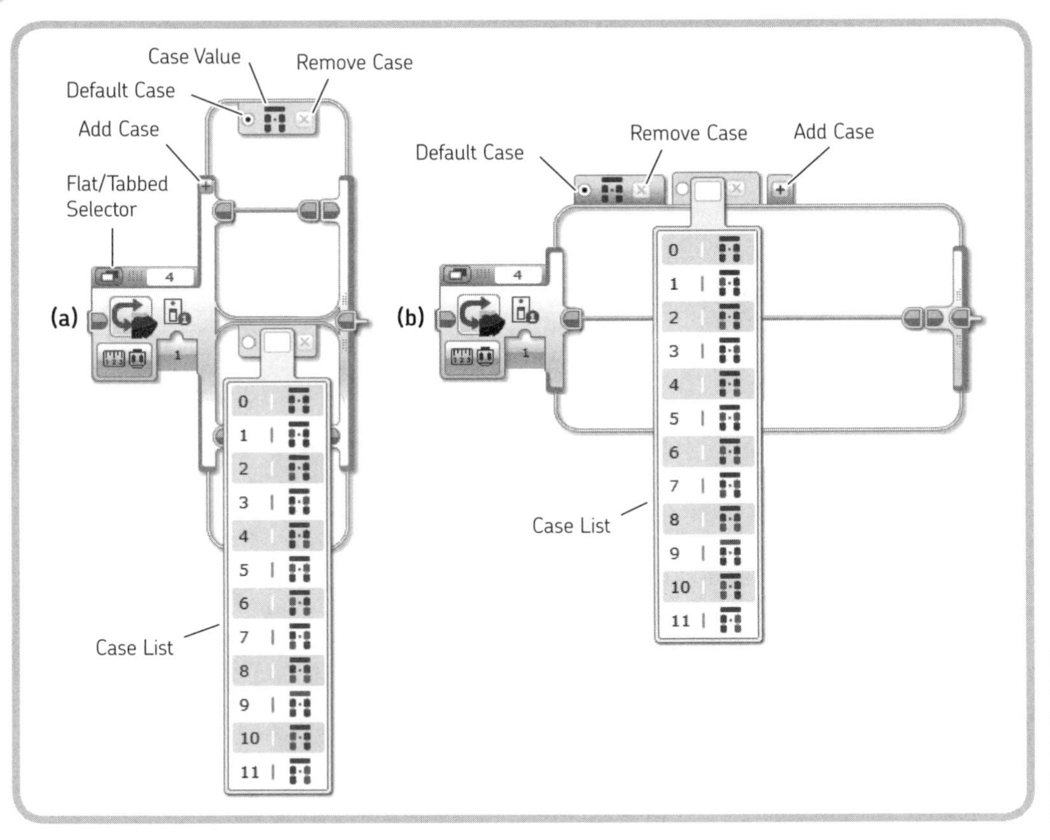

Figure 12-5: The Switch block in one of the Measure modes shows multiple, numbered cases: Flat View (a) or Tabbed View (b). Here the Switch block is set in Infrared Sensor Measure Remote mode.

building the My Blocks

Now to make the SUP3R CAR's programs! Before making the final programs, we need to create some My Blocks. For each My Block, I'll list the sequence of blocks and the final My Block showing the default input values and the icons used. To begin, create a new project and save it as *mySUP3RCAR*. Then create each My Block using the following figures as references.

the ResetSteer My Block

The *ResetSteer* My Block is used at the beginning of each program to center the steering wheels. Build the sequence shown in Figure 12-6 and create a My Block that looks like the one shown in the figure.

The Sound block plays a *Motor start* sound, and the Medium Motor is powered at low speed for a short time to steer the front wheels as far as possible. Once the wheels reach their limit, they're steered back to center by a precise number of degrees. This central position will be the reference point for successive motor movements.

The Motor Rotation block in Reset mode resets the rotation count of Medium Motor A to zero. The *Steer* My Block, which we'll create next, needs this initial reset to turn the motor at absolute angles (referencing the center position at which the rotation count was reset).

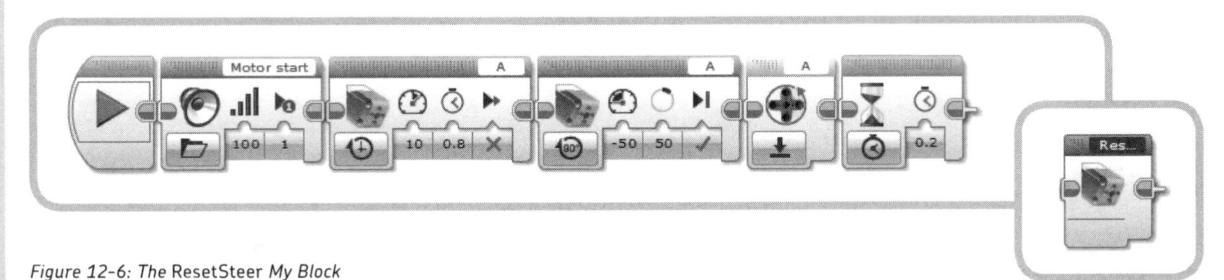

Figure 12-6: The ResetSteer *My Block*

the Steer My Block

The *Steer* My Block wraps around a Medium Motor block to make the motor turn at an absolute angle. Normally, if you set a Motor block to On for Degrees mode, the Degrees input specifies the number of degrees the motor should rotate with respect to the current shaft position. But in this case, the *Angle* input of this My Block makes the motor shaft move to an *absolute position* determined by the last time its rotation count was reset (regardless of its actual position). For example, a value of 0 brings the motor shaft to the position it had when it was reset, and a value of 30 brings it 30 degrees forward with respect to the 0 position. If the rotation count is currently 30, an input value of –30 brings the shaft to a –30 degree angle; it doesn't simply rotate the shaft by 30 degrees backward, as a Motor block would.

I use the Motor Rotation block (in Measure Degrees mode) to read the current motor rotation count and subtract this value from the desired Angle input. The result is used for the Motor Block Degrees input. The Medium Motor block is executed only if the difference between the current and the desired angle is greater than 1 degree. To consider either a positive or negative difference, I've used a Math block in Absolute Value mode with a Compare block.

Create this My Block using Figure 12-7 as reference. The block has four inputs: *Port*, *Power*, *Angle*, and *Brake at End*. (Notice that the Data Wires can pass through the Switch block.)

NOTE As explained in Chapter 6, Data Wires can pass through a Switch block only if it is shown in Tabbed View. You cannot pass Data Wires through a Switch block shown in Flat View.

the Drive My Block

The *Drive* My Block simultaneously steers the front wheels and runs the driving motors. As shown in Figure 12-8, it has two numeric input parameters: *Power* and *Steer*.

When creating this My Block, set the Default value for the **Power** input to **50**, set Parameter Style to **Vertical Slider**, and limit the Power input to between **–75** and **75**. Set the Default value for the **Steer** input to **0**, set Parameter Style to **Horizontal Slider**, and limit the Steer input to between **–30** and **30**.

Two Math blocks in Advanced mode compute the power to be applied to the driving motors. The Math blocks use the formulas for the electronic differential, as explained in "Digging Deeper: Computing Wheel Speeds for an Electronic Differential" on page 232. The equations in the Math blocks have a minus sign because the driving motors are facing backward. The motor ports in the Move Tank block are C+B, not B+C, to avoid Data Wire entanglement.

NOTE To invert the direction of rotation of a motor, you can use the *Invert Motor* block (Advanced palette, blue header). Any programming block following the Invert Motor block that would normally make the motor turn clockwise will make the motor turn counterclockwise, and vice versa. This block is useful when your robot's driving motors are mounted backward (as in the SUP3R CAR) or when the direction of the wheels' rotation has been inverted by gears (as discussed in Chapter 8). In this case, because we're already using the Math block, it is more efficient to use the negative sign to invert the motor's rotation direction.

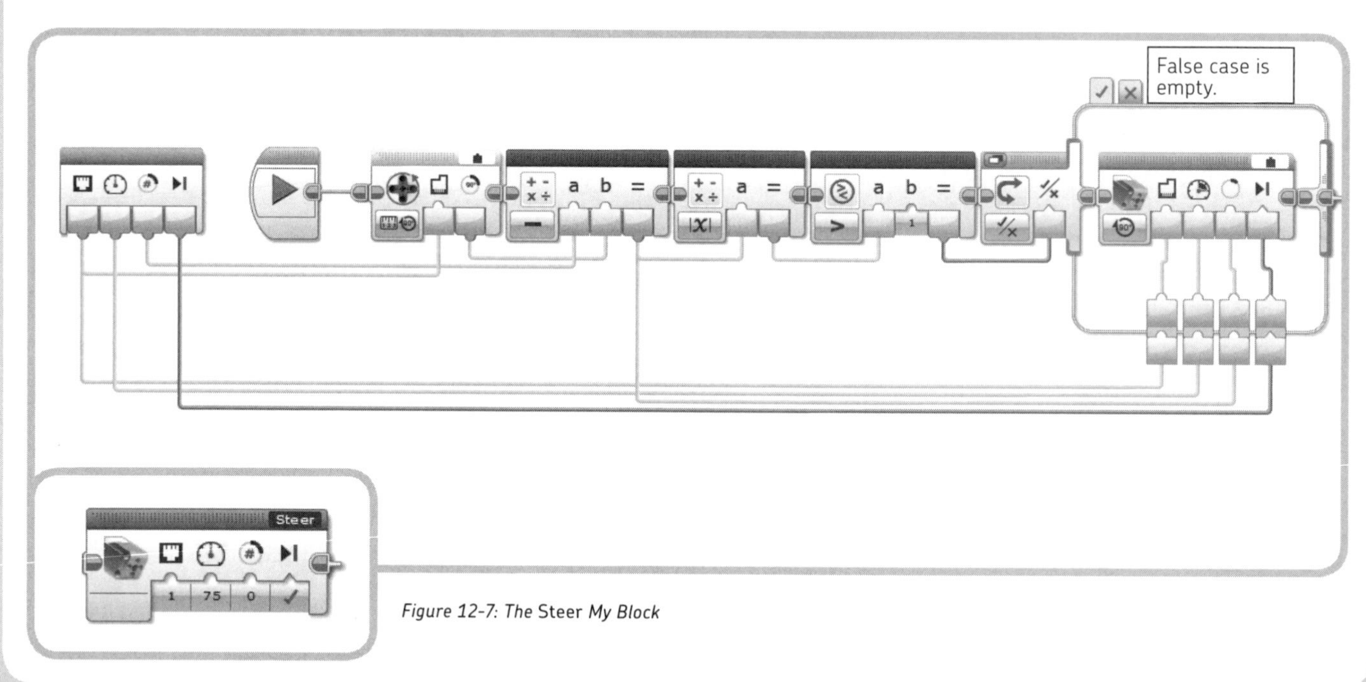

Figure 12-7: The Steer My Block

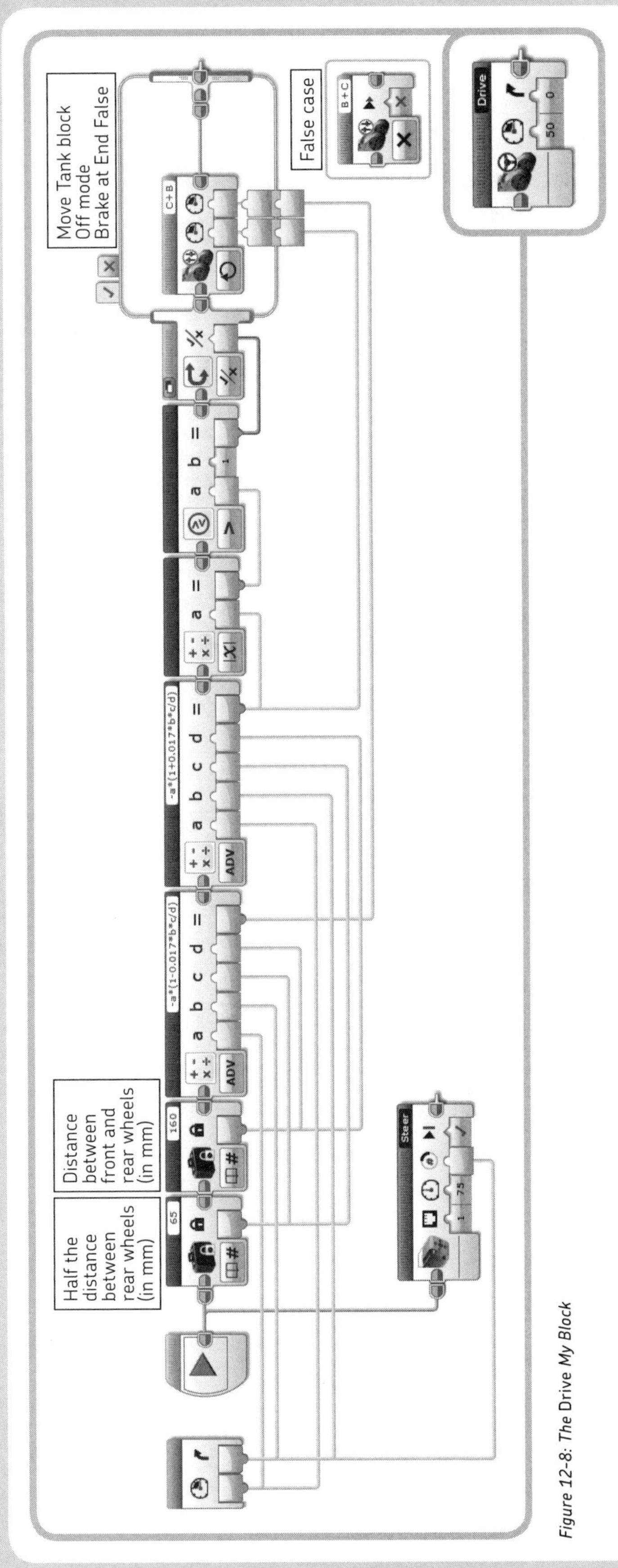

Figure 12-8: The Drive My Block

The car's dimensions (the distance *H* between the rear and the front wheels and half the distance *D* between the rear wheels) are set in two Constant blocks.

The Move Tank block is executed in the True case of the Switch block if the absolute value of the calculated power for the left motor is greater than 1. The False case of the Switch block has a Move Tank block in Off mode, with the *Brake at End* input set to False. The Switch block is set in Tabbed View to pass the Data Wires that are carrying the left and right motor power values through it.

This My Block executes the *Steer* My Block in parallel with the main sequence. If the block were placed in the main sequence, the driving motors would have to wait for the steering motor to finish before starting or changing speed. Using Figure 12-9 as reference, connect the *Steer* My Block with a Sequence Wire in parallel with the other sequence.

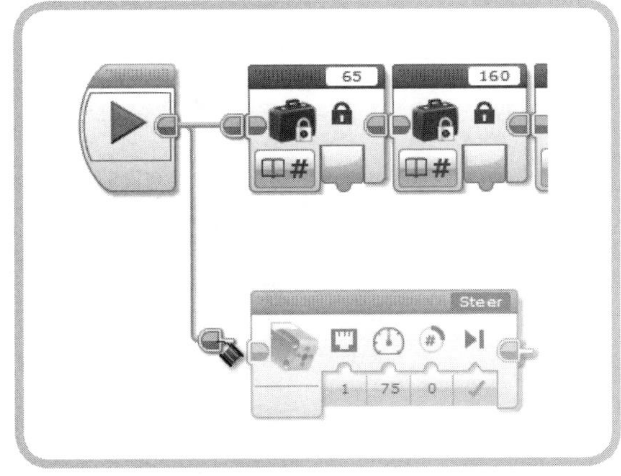

Figure 12-9: Click and drag a Sequence Wire from the Sequence Plug Exit of the block before the blocks you want to execute in parallel.

the ReadRemote2 My Block

The *ReadRemote* My Block, shown in Figure 12-10, uses a Switch block in Infrared Sensor Measure Remote mode to set the variables *spd* and *str* according to the buttons pressed on the R3MOTE. The *spd* variable uses the values 0 (motors are stopped) and 1 or –1 (go forward or backward). The *str* variable uses the values 0 (steering wheels are centered) and 1 or –1 (steer left or right). Both of these variables are initially set to 0.

Inside the cases of the Switch block, these variables are updated according to the various combinations of the remote buttons. Buttons 1 and 2 control the steering, and 3 and 4 control the forward/backward direction. For example, when you move the R3MOTE joysticks to go forward while turning left, the combination sent by the Remote IR Beacon is 5 (buttons 1 and 3 pressed together, as per Figure 6-1 on page 85). In the case corresponding to combination 5, the *spd* variable is set to 1 (forward), and the *str* variable is set to 1 (left).

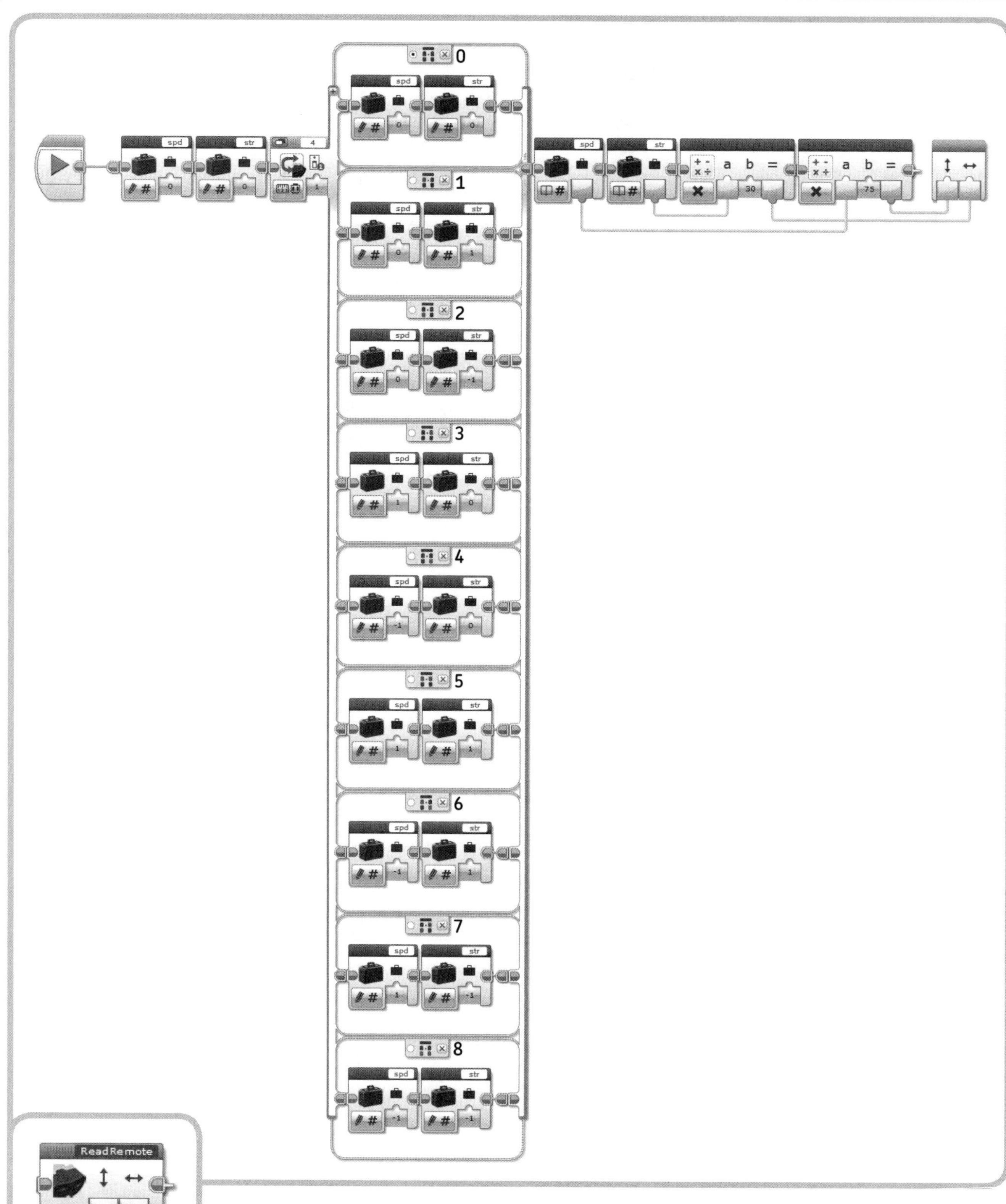

Figure 12-10: The ReadRemote *My Block uses a Flat Switch to handle all the possible IR Remote commands.*

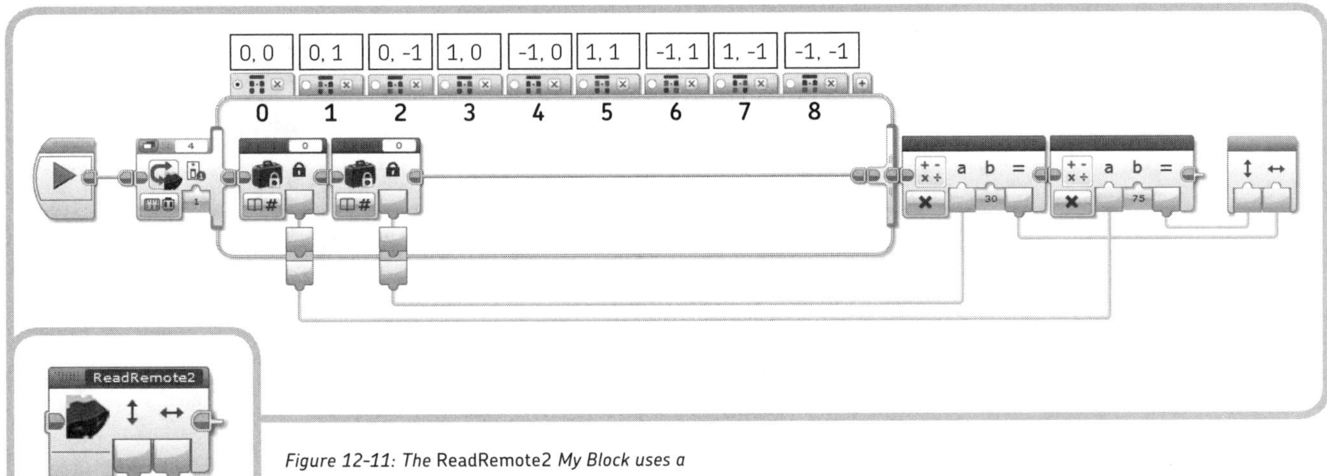

Figure 12-11: The ReadRemote2 *My Block uses a Tabbed Switch to handle all IR Remote commands.*

After the Switch block, the *spd* and *str* values are multiplied by 75 (to get the maximum 75 percent power for the driving motors) and 30 (to make the steering motor turn by 30 degrees forward or backward), respectively. The results are carried to the My Block outputs *Speed* and *Steer*.

As you can see in Figure 12-10, the sequence of the *ReadRemote* My Block is simple but quite bulky. The same functionality can be implemented without variables, using Constant blocks inside a Tabbed Switch block (Figure 12-11). The Switch block is in Infrared Sensor Measure Remote mode. In each case, instead of setting the variables for speed and steer, Constant blocks like the ones visible in the first case provide those values to the Math blocks outside the Switch block via Data Wires. On top of each Case tab, a comment indicates which values you should set for the hidden Constant blocks.

The Tabbed Switch block shows the Data Wire tunnels for each case. Connect the output of each Constant block in the other cases to these tunnels.

You can create just the *ReadRemote2* My Block, because that's the one you'll use in the *RC_switch* program.

programming the car to drive around

Having created the needed My Blocks, let's make a program to let the SUP3R CAR drive around without bumping into obstacles. Build the *DriveAround* program shown in Figure 12-12.

This program is pretty simple. First, the *ResetSteer* My Block centers the steering wheels. Then, in a Loop (Unlimited mode), the *Drive* My Block commands the car to go straight (Steer input is set to 0) at 50 percent power, until the IR Sensor reads a proximity value less than 35 percent (first Wait block). Then, another *Drive* My Block commands the car to back up while turning (Power –50, Steer –30), until the measured proximity from the obstacle exceeds 45 percent. The last Wait block makes the program wait a bit before switching direction from backward to forward.

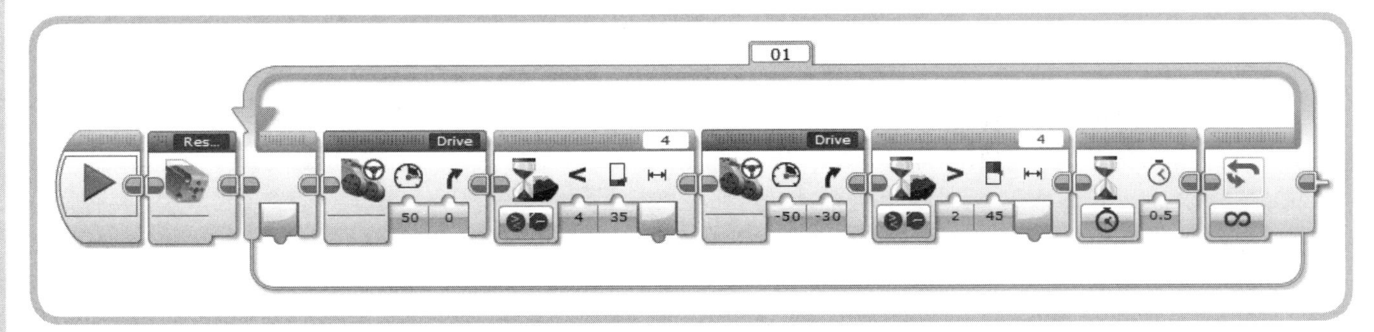

Figure 12-12: The DriveAround *program*

programming the car for remote control

Now we'll make the *RC_switch* program that will receive commands from the R3MOTE, using Figure 12-13 as reference. Once the *ResetSteer* My Block centers the steering wheels, a Display block in Image mode displays the Remote IR Beacon image on the EV3 Brick screen. In the Loop (Unlimited mode), the *ReadRemote2* My Block directly provides the *Steer* and *Power* values to the *Drive* My Block via Data Wires.

The Switch block in Numeric mode uses the input value from the *Power* output of the *ReadRemote2* My Block to change the color of the Brick Status Light. Because case number 1 (which changes the color of the light to red) is set as the Default case, this case will be executed when the input is equal to 1 but also whenever the input is different from 0. The Default case could have any value, because all we want to do is distinguish between zero and nonzero values.

Setting the Wait block to **0.05 seconds** prevents the steering motor from jiggling. This jiggling occurs if the *Steer* My Block inside the *Drive* My Block is called too often; the *Steer* My Block would try to update the motor position before it reached the commanded angle.

using arrays to clean up the ReadRemote My Block

While the *ReadRemote* and *ReadRemote2* My Blocks as described previously will do the job, we can do the same thing more elegantly with arrays. Specifically, we can use the commands that the Remote IR Beacon sends (according to the buttons pressed) as an index to read from two arrays containing the speed and steering values (–1, 0, or 1 as for the previous My Blocks). The *RC_arrays* program is shown in Figure 12-14. When it starts, two Variable blocks in Write Numeric Arrays mode create and fill the arrays *str_array* and *spd_array*. Using the drop-down menu shown in Figure 12-4(b), fill the arrays as indicated in Table 12-1.

The *ReadRemoteA* My Block is shown in Figure 12-15. As you can see, the IR Sensor block's Measure Remote mode provides the *Button ID* as *Index* input via Data Wires to two Array Operations blocks so as to read the *spd_array* and *str_array* values.

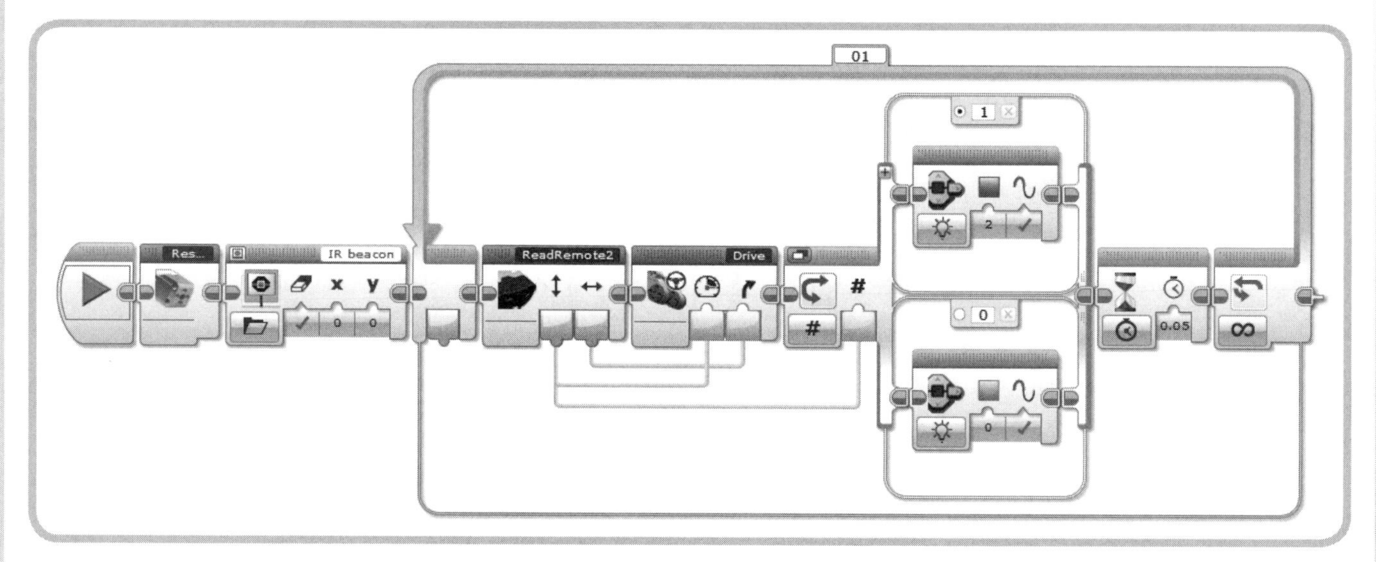

Figure 12-13: The RC_switch *program that allows the car to be remotely controlled*

In Chapter 6, I listed all the possible commands sent by the remote (numbers from 1 to 11). Since the remote's command is used as index for the array, we must be sure that the arrays have at least the number of elements that the index can reach. Although the Remote Button combinations 9 (Beacon Mode on), 10 (buttons 1 and 2 pressed together), and 11 (buttons 3 and 4 pressed together) can't be produced by moving the R3MOTE joystick, we should fill the arrays at those indices as well. If the arrays had just nine elements (indexed from 0 to 8) and the IR Sensor received commands 9, 10, or 11, the Array Operations blocks would try to access elements that were out-of-bounds, making the program abort. To avoid that, we fill the arrays with zeroes at indices 9, 10, and 11.

Build the *RC_arrays* program shown in Figure 12-14 and the *ReadRemoteA* My Block shown in Figure 12-15, using Table 12-1 to fill the arrays.

table 12-1: the content of the *str_array* and *spd_array* arrays

Index (Remote Button ID)	str_array	spd_array
0	0	0
1	1	0
2	-1	0
3	0	1
4	0	-1
5	1	1
6	1	-1
7	-1	1
8	-1	-1
9	0	0
10	0	0
11	0	0

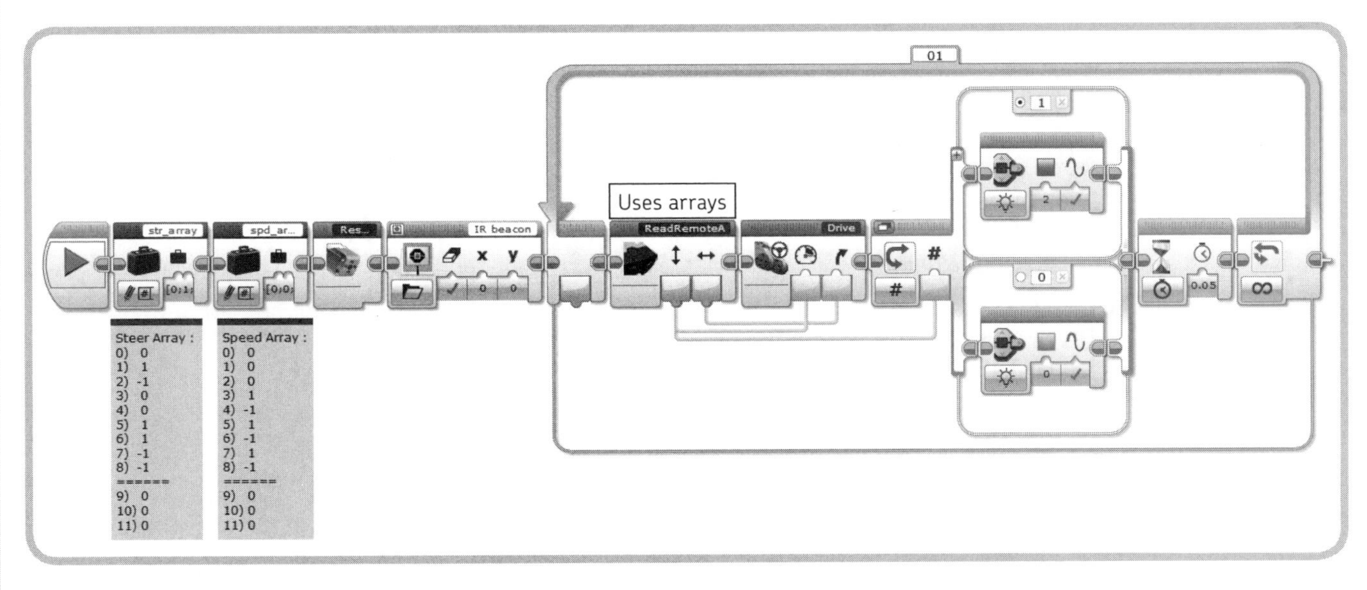

Figure 12-14: The RC_arrays *program is a different way to implement the* RC_switch *program, which remotely controls the car.*

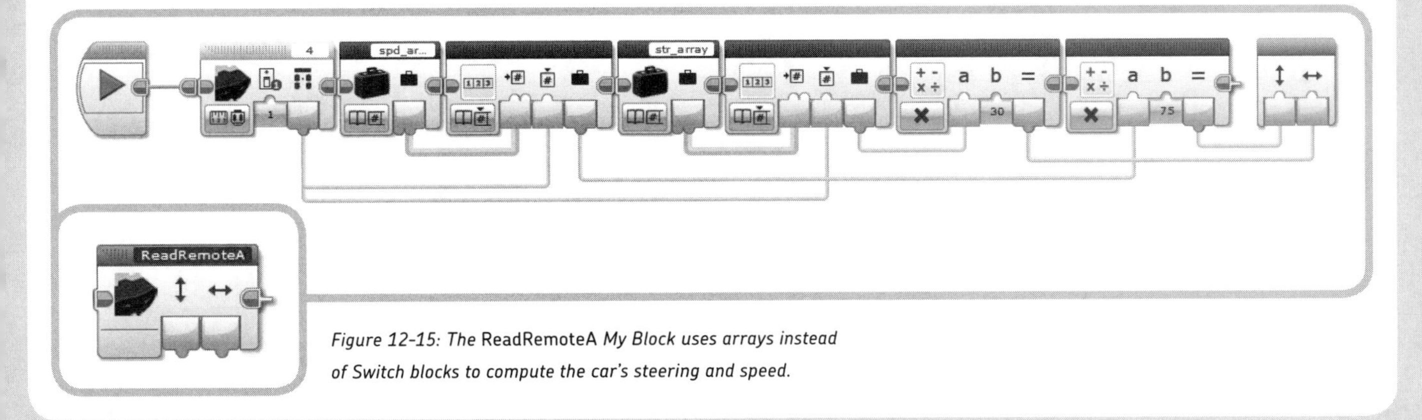

Figure 12-15: The ReadRemoteA *My Block uses arrays instead of Switch blocks to compute the car's steering and speed.*

programming the car to follow the beacon

The SUP3R CAR can be programmed to follow the Remote IR Beacon (set in Beacon Mode). The program is similar to the beacon-following program that we created for ROV3R in Chapter 6 (Figure 6-9 on page 91), but it is necessarily more complex because it needs to deal with the more involved steering of the SUP3R CAR. Before building the main *FollowBeacon* program, you'll need to create the My Blocks as described in the following sections.

the Sign My Block

The *Sign* My Block (Figure 12-16) implements the *sign* math function. It accepts a number as input and returns 1 if the number is positive (greater than zero); 0 if the number is exactly equal to zero; and -1 if the number is negative (less than zero). Give the input and output any name that you like (for example, *Input* and *Result*).

As you can see, the *Sign* My Block uses two Switch blocks, one nested inside the other. The outer one, together with a Compare block, checks whether the input number is greater than or equal to zero. If the value is False, then a Constant block outputs -1 through the Data Wire tunnel. If the value is True, a further check is done: If the number is exactly equal to zero, a Constant block outputs 0; otherwise, another Constant block (in the False case of the nested Switch) outputs 1.

the Saturation My Block

The *Saturation* My Block limits the input value so that it lies between a maximum (*Max*) and minimum (*Min*). Build it using Figure 12-17 as reference. The original input value is stored in a temporary variable called *_v*. If the value stored in *_v* is greater than *Max*, its value is overwritten with *Max*; values less than *Min* are overwritten with *Min*.

the ReadBeacon My Block

Now we'll build the *ReadBeacon* My Block, as shown in Figure 12-18. This My Block generates the drive and speed commands (with the usual values of -1, 0, or 1) to make the SUP3R CAR follow the beacon. The *Heading* and *Proximity* outputs of the IR Sensor block (in Measure Beacon mode) are saved into the *heading* and *prox* numeric variables, respectively.

The *Detected* Output of the IR Sensor block is True when the beacon is detected; otherwise it is false. Sometimes, however, this Output returns True even when the beacon is hidden from the IR Sensor's sight, but the *Proximity* Output returns 100. For this reason, we want the car to drive toward the beacon only if the *Detected* Output is True *and* the *Proximity* value is less than 100; we make this happen with the Compare and Logic Operations blocks. The result of the Logic Operations block is carried to the input of a Switch block. When the beacon

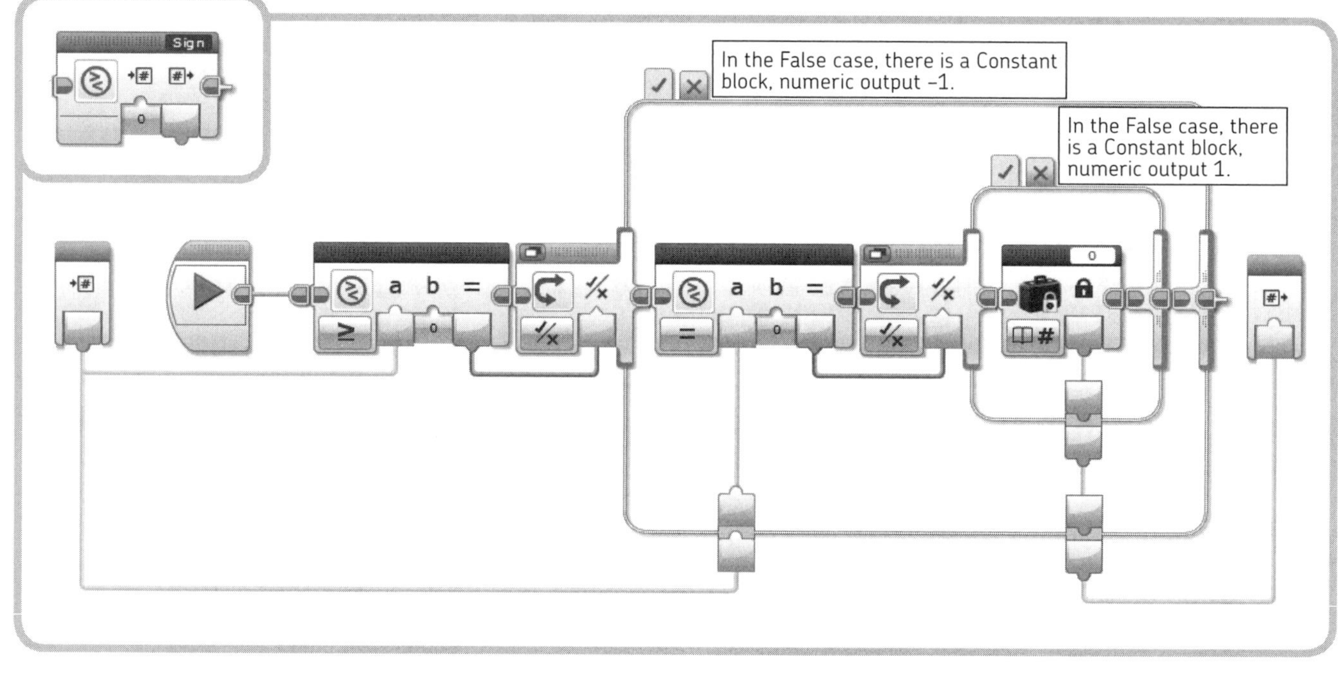

Figure 12-16: The Sign *My Block implements the sign math function.*

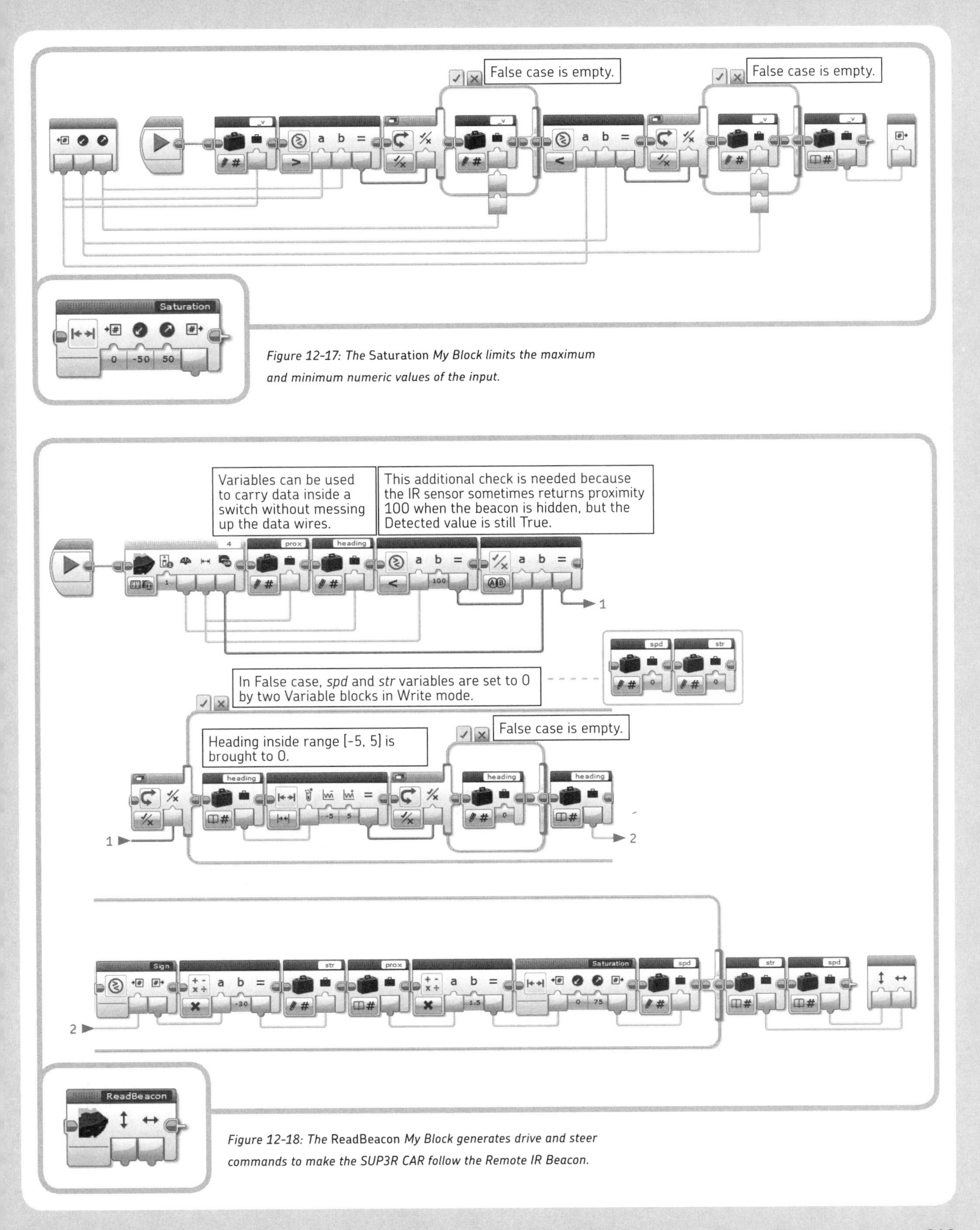

Figure 12-17: The Saturation My Block limits the maximum and minimum numeric values of the input.

Figure 12-18: The ReadBeacon My Block generates drive and steer commands to make the SUP3R CAR follow the Remote IR Beacon.

is not in sight (False case of the Switch block), the variables *spd* and *str* are both set to 0. When the result is True, the value of the *heading* variable is used to compute the steering value, which is then saved to the *str* variable.

The Range block checks whether the heading to the beacon, stored in the *heading* variable, is between –5 and 5. If so, the value is zeroed out. The goal here is to create a dead zone around zero so that the car goes straight instead of zigzagging if the heading to the beacon is small. We just need to know if the beacon is seen at the right or the left, so we use the *Sign* My Block to compute the sign of the heading value. The value obtained overwrites the *heading* variable and is then multiplied by –30 to compute the final *str* value. (We multiply by a negative number to make sure that the car will steer toward the beacon.)

The proximity value stored in the *prox* variable is multiplied by 1.5, constrained between 0 and 75, and used to update the *spd* variable. The speed is limited to 75 so that the electronic differential formula $V \times (1 + D \times 0.017 \times a / H)$ (see "Digging Deeper: Computing Wheel Speeds for an Electronic Differential" on page 232) won't produce a value larger than the maximum allowed speed for an EV3 motor, even for the maximum steering angle. For a steering angle equal to 35, the faster wheel speed would be $75 \times (1 + 65 \times 0.017 \times 35 / 160) \approx 93$, which will be safely less than 100 (that is, less than the full motor speed).

the range block

The Range block (Figure 12-19) tests whether an input numeric value is within a range specified by the Lower Bound and Upper Bound. You can choose to test whether a number is inside a range by choosing Inside mode or outside a range by choosing Outside mode. In both modes, the Range block includes the boundaries in its test. For example, testing whether 25 is inside the range 25–50 would return True, testing whether 50 is inside the range 25–50 would return True, and testing whether 50 is outside the range 25–50 would return False.

This block does the same job as two Compare blocks and a Logic Operations block. For example, in Inside mode, it tests whether the input value is greater than or equal to the lower bound *and* less than or equal to the upper bound.

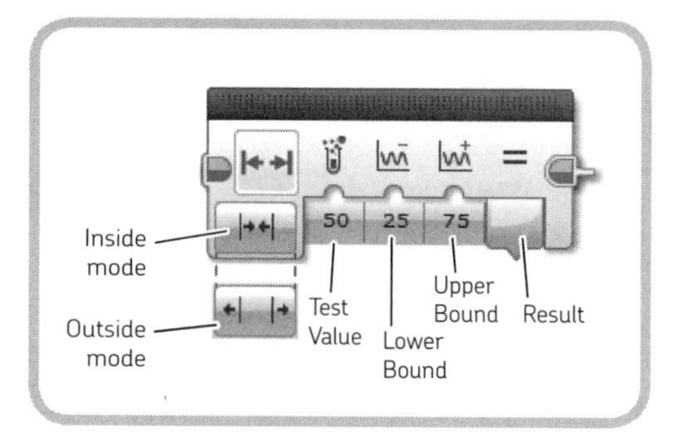

Figure 12-19: The Range block

the FollowBeacon program

Finally, we can build the *FollowBeacon* program, using Figure 12-20 as reference. This program first displays the *Target* image file and centers the turning wheels. Next, the *ReadBeacon* My Block provides the *Power* and *Steer* values to the *Drive* My Block. These blocks and a Wait block are inside a Loop block set in Color Sensor Compare Color mode (with red [5] specified as the input). When the Color Sensor detects red, the program stops the motors, plays a *Motor stop* sound, and ends.

EXPERIMENT 12-1

How about using this program to play a game? By waving the beacon, try to drive the SUP3R CAR across the EV3 Test Pad without crossing any red lines. If you cross a red line, the car stops and you lose. (To change the color the car must avoid, change the loop's ending condition.)

You can expand on this game by drawing more complicated and winding line patterns on large sheets of paper and trying to drive across these new terrains.

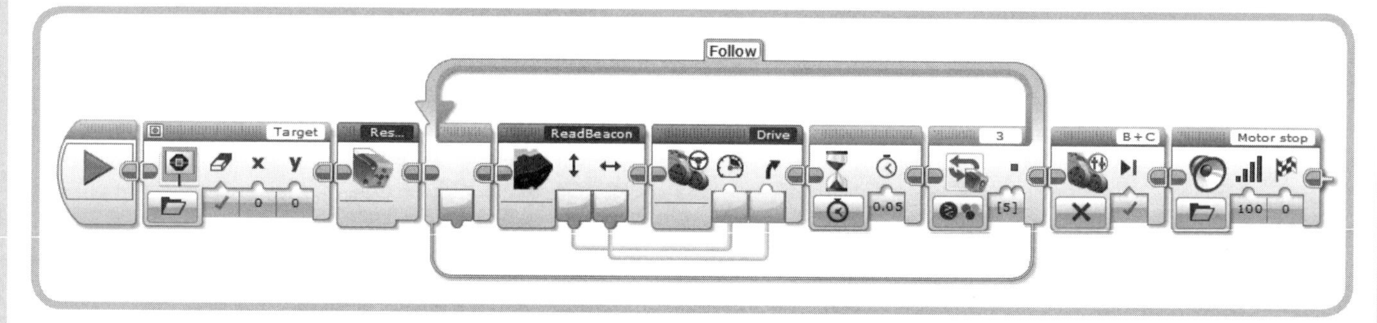

Figure 12-20: The FollowBeacon *program allows the SUP3R CAR to drive toward the Remote IR Beacon.*

NOTE **Since the Remote IR Beacon is built inside the R3MOTE, you can put it into Beacon Mode by pressing its Beacon Mode button with a Technic beam, axle, or panel.**

adding a siren effect to the SUP3R CAR

In "Running Parallel Sequences (Multitasking)" on page 234, I mentioned that you can make multiple sequences of blocks run in parallel when you use multiple Start blocks in the same program. In this section, we'll use a parallel sequence to add a light-and-sound siren effect to the *FollowBeacon* program.

1. Go to Project Properties and copy and paste the *FollowBeacon* program. This creates a program called *FollowBeacon2*; rename it to *BeaconSiren* (double-click to open it; then double-click the Program tab and enter the new name).
2. Add the two parallel sequences with two new Start blocks, as shown in Figure 12-21.
3. Change the name of the Loop block in the second sequence to *Siren* and the Loop block in the third sequence to *Lamp*.
4. Using Figure 12-21 as a reference, add a Loop Interrupt block and a Stop Program block to the main sequence.
5. Select the *Siren* name from the Loop Interrupt block Name field. When the *Follow* loop ends, the Loop Interrupt block will also end the *Siren* loop in order to play the Motor stop sound file without causing a *resource conflict* (which would occur if the *Siren* loop and the main loop tried to access the EV3 loudspeaker resource at the same time).

WARNING **You may incur a *resource conflict* each time you use the same resource (the same motor, the display, or the loudspeaker) from multiple sequences running in parallel. Try to avoid such situations because when resource conflicts occur, the EV3 Brick will behave unpredictably.**

If we didn't place a Stop Program block at the end of the first sequence, the *Lamp* loop would continue to run forever and the program would not stop as expected.

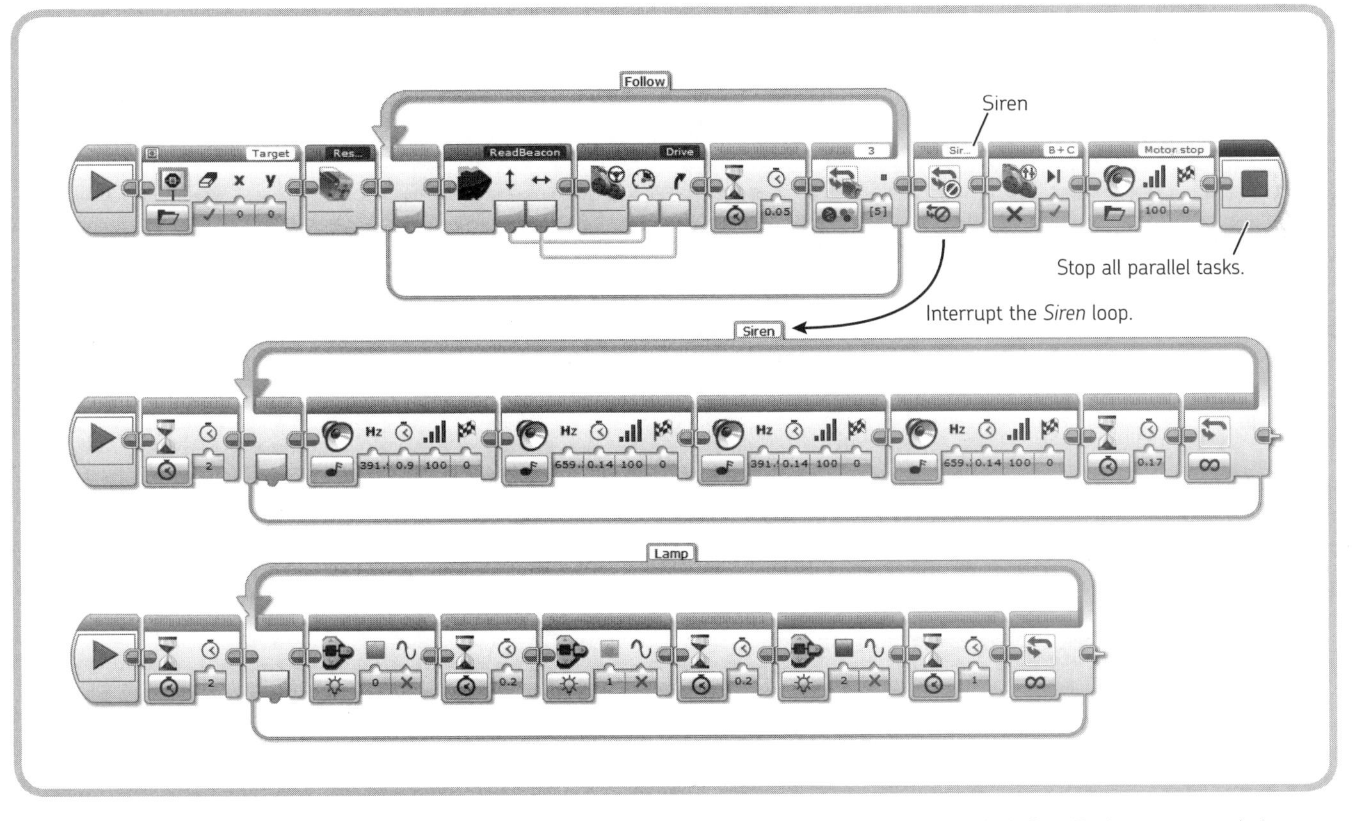

Figure 12-21: The BeaconSiren *program adds a light-and-sound siren effect to the* FollowBeacon *program by using multiple Start blocks to execute tasks in parallel.*

the loop interrupt block

The Loop Interrupt block (Figure 12-22) can end the execution of a Loop block at any time (overriding its ending condition), regardless of the block being executed. Every loop is labeled. To specify which one to interrupt, select its name from the heading of the Loop Interrupt block's Name field, which lists the names of all the Loop blocks of the project. If two or more Loop blocks share the same name, the Loop Interrupt block will end all of them. (You can use the Loop Interrupt block from within the loop you want to end or from a sequence running in parallel.)

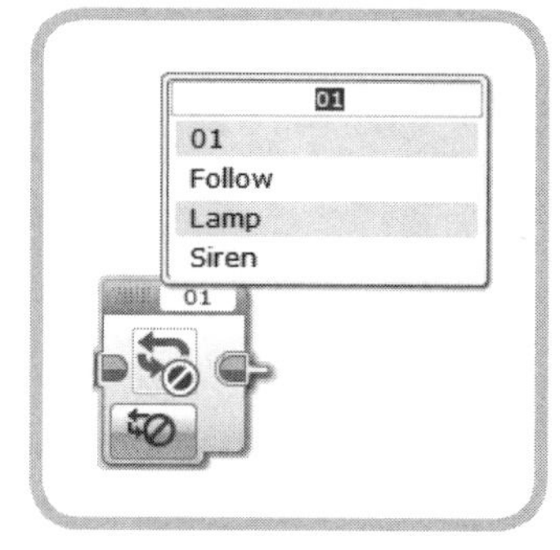

Figure 12-22: The Loop Interrupt block makes the specified loop end at once.

NOTE The Loop Interrupt block will also allow you to end a sequence that needs to be executed just once. Just put the sequence that you want to interrupt inside a Loop block in Count mode with Count set to 1. Also, give the loop a name to uniquely identify it in the sequence. The sequence executes only once and can be stopped at any time.

the stop program block

The Stop Program block (Figure 12-23) is an optional block that can be placed only at the end of a programming sequence (it has no Sequence Exit Plug). It terminates all sequences at once, ending the program. You'll find it in the Advanced palette (blue heading). Programs with just one running sequence don't need this block because the program will end as soon as the sequence ends. But if you have multiple sequences running in parallel, this block is useful for ending a program in any sequence.

Figure 12-23: The Stop Program block terminates a program at once.

EXPERIMENT 12-2

The SUP3R CAR has the Color Sensor mounted in front, facing downward. This positioning is ideal for following lines on the ground. Can you make a line-following program for the SUP3R CAR? Since the car has a limited steering angle, its turning radius is quite large, allowing the car to follow only lines with wide curves. (In contrast, ROV3R can follow lines with tight curves because it can steer in place.)

conclusion

Wow, this chapter has introduced a lot of new programming concepts! You learned how a car differential works, how to work with variables and arrays, how to use Switch blocks with multiple cases, how to make programs with multiple sequences running in parallel, how to use the Range block, how to interrupt a loop, and how to stop a whole program.

CLICK!
EEEEEEEEEEEEEEEEEEEEEEEEEK!
12X1

CRACK!
IT'S TIME FOR GARDENING!
CHOKE, FILTHY BEASTS!
FLOP! FLOP!
YUM!
BLEURB!
EEK!
NOM NOM NOM
GREAT! WHAT DID YOU FEED THEM?
I GAVE THEM THE SANDWICHES FROM THAT DREADFUL CAFÉ! THANK GOODNESS THEY'RE SO HARD TO SWALLOW!
0x400DF00D
12X2

13

building the SENTIN3L

In this chapter, you'll build the SENTIN3L, a badass security robot. It walks on three legs and has two wicked blaster cannons (see Figure 13-1). As you build, you'll learn how to create mechanisms that transform rotation into alternating motion. These mechanisms are particularly useful when designing walking robots. You'll also learn how to assemble structures diagonally within the constraints of LEGO geometry and how to assemble curved structures like the SENTIN3L's elegant back shield.

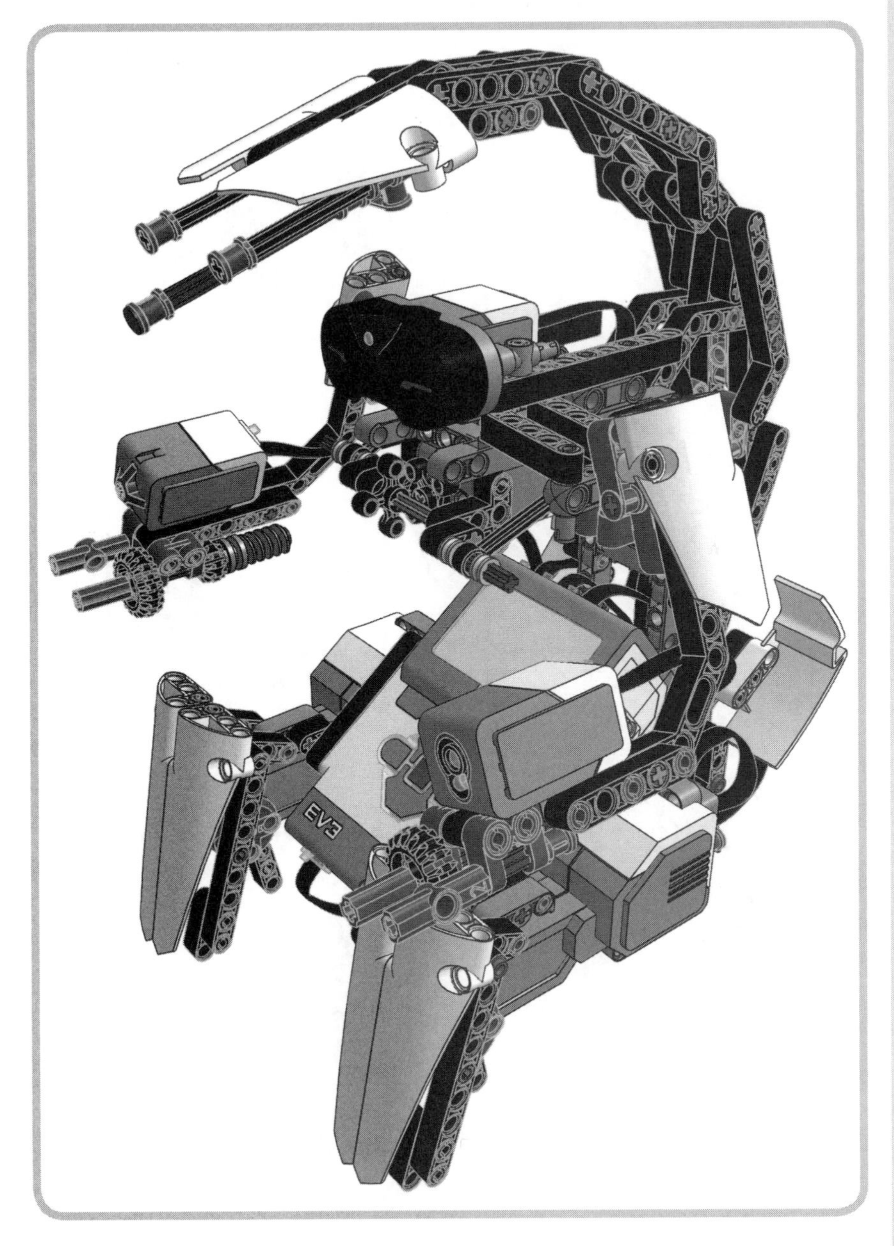

Figure 13-1: The SENTIN3L

main assembly

1

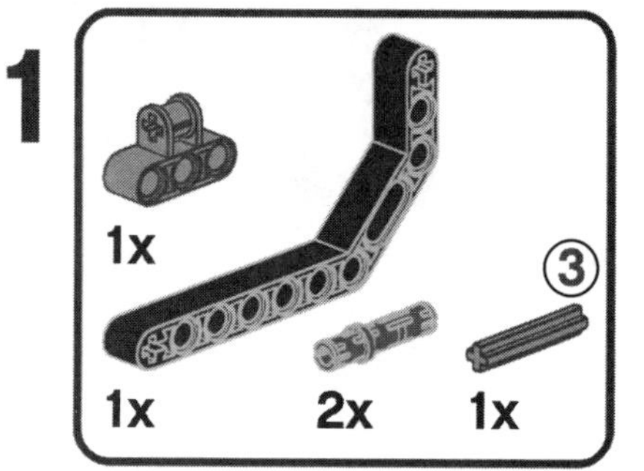

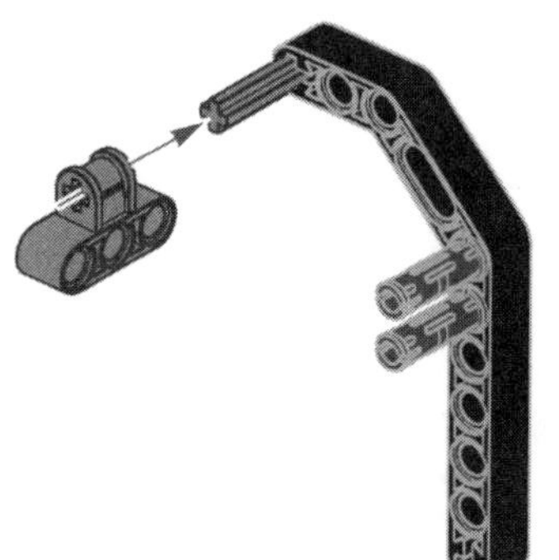

2

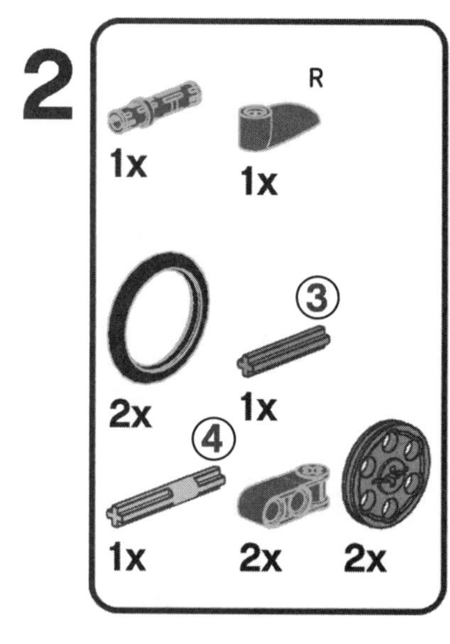

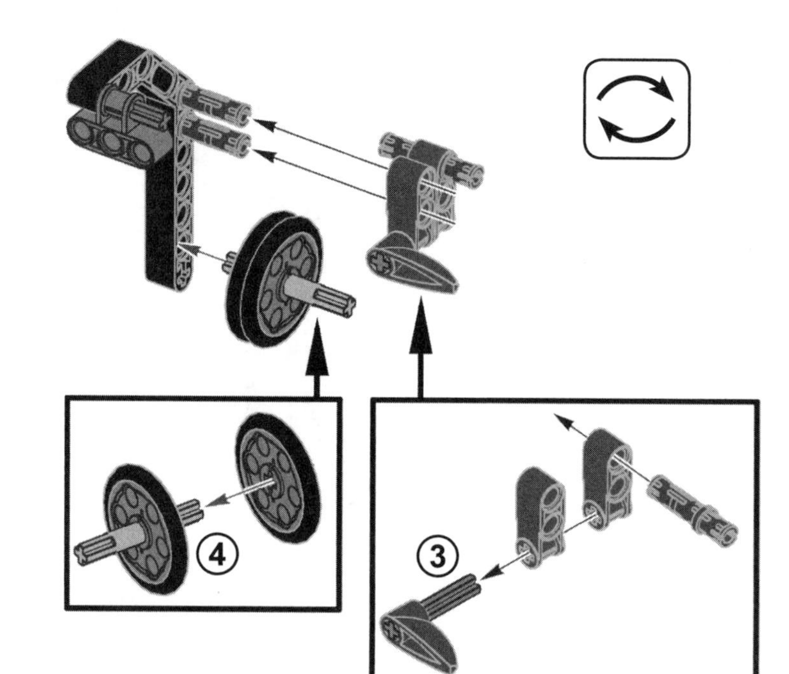

3

1x

1x

1:1

③ ④

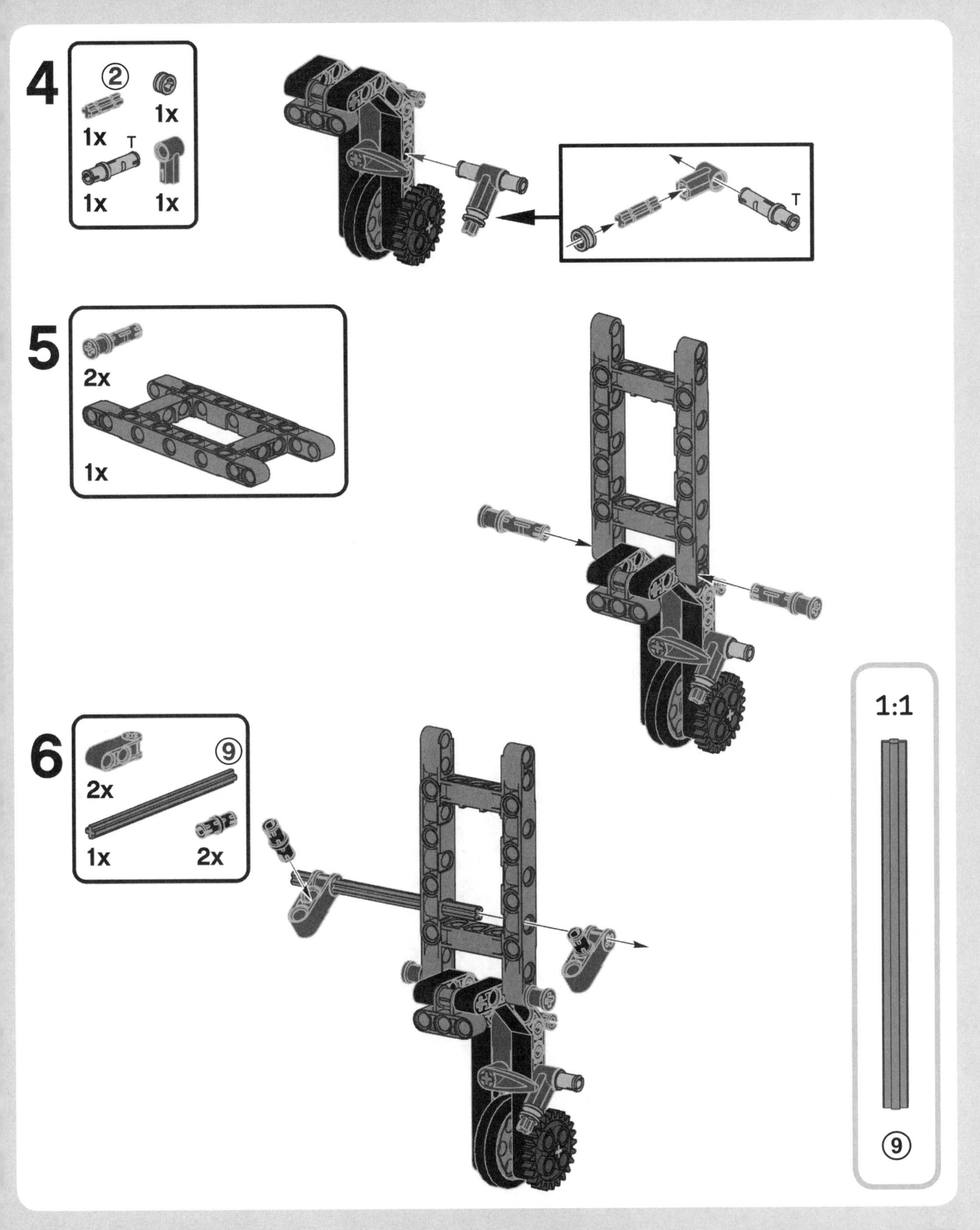
4
②
1x
1x
T
1x
1x
T
5
2x
1x
6
2x
⑨
1x
2x
1:1
⑨

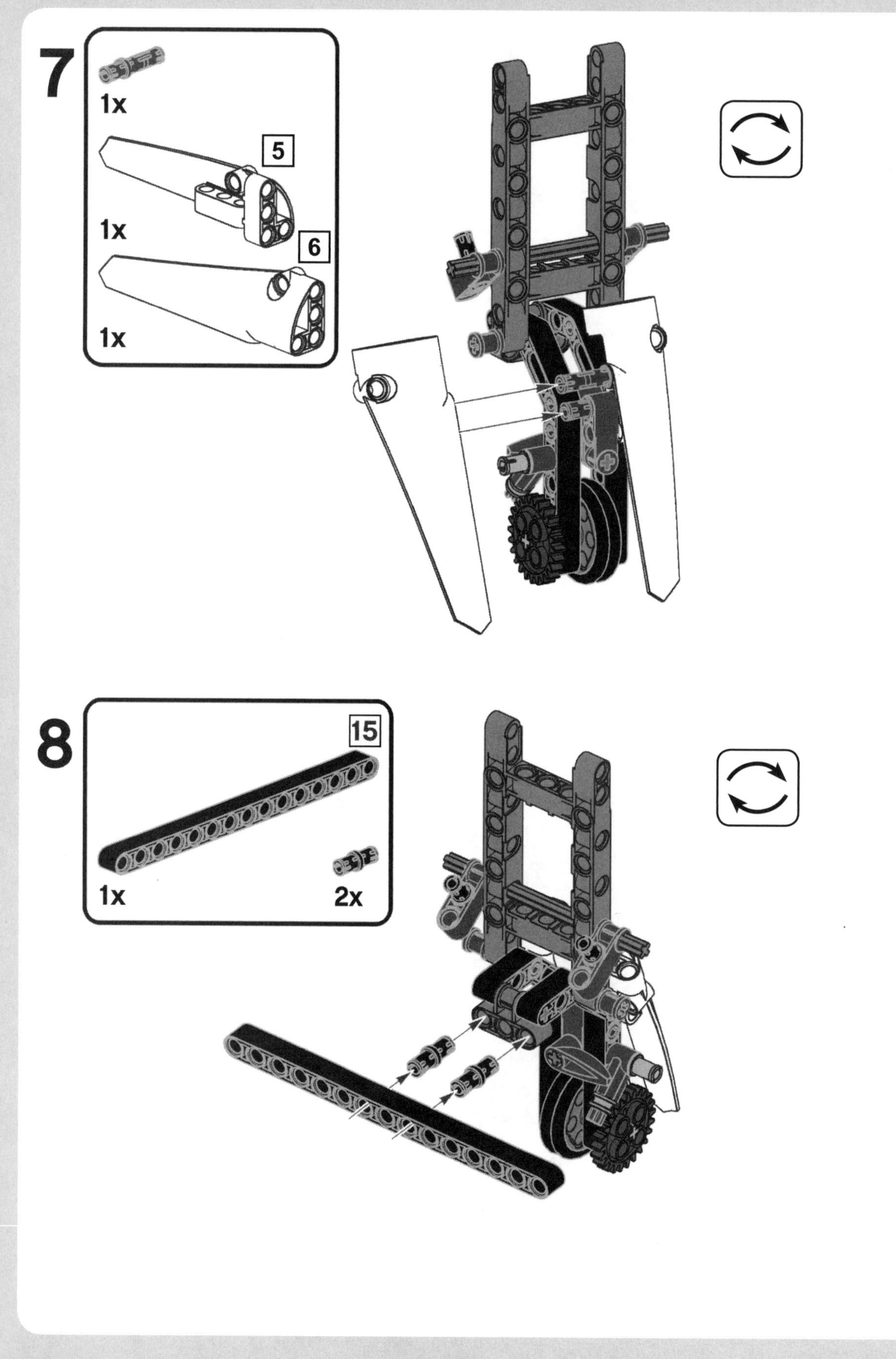
7
1x
5
1x
6
1x
8
15
1x
2x
1:1
15

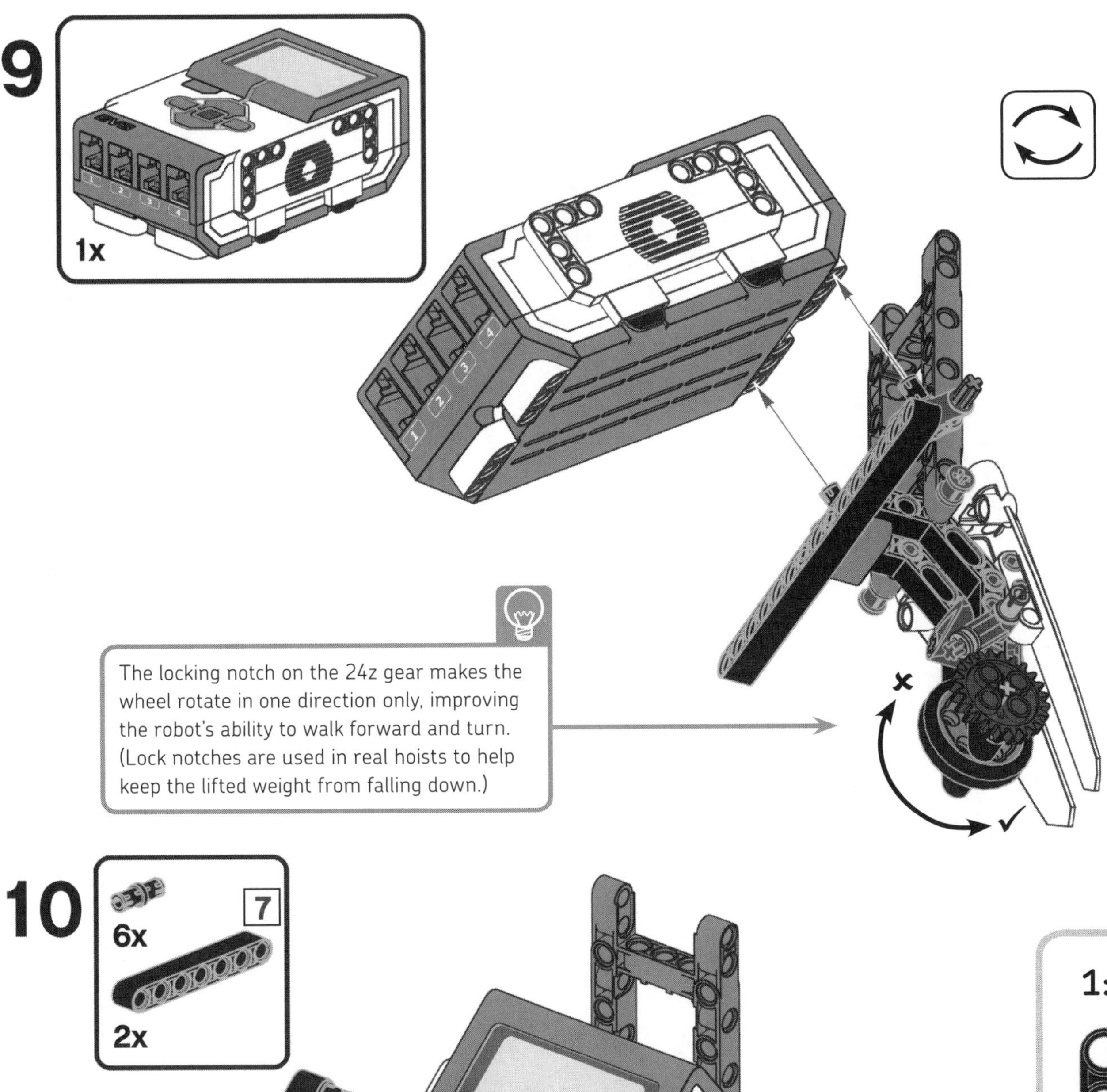

The locking notch on the 24z gear makes the wheel rotate in one direction only, improving the robot's ability to walk forward and turn. (Lock notches are used in real hoists to help keep the lifted weight from falling down.)

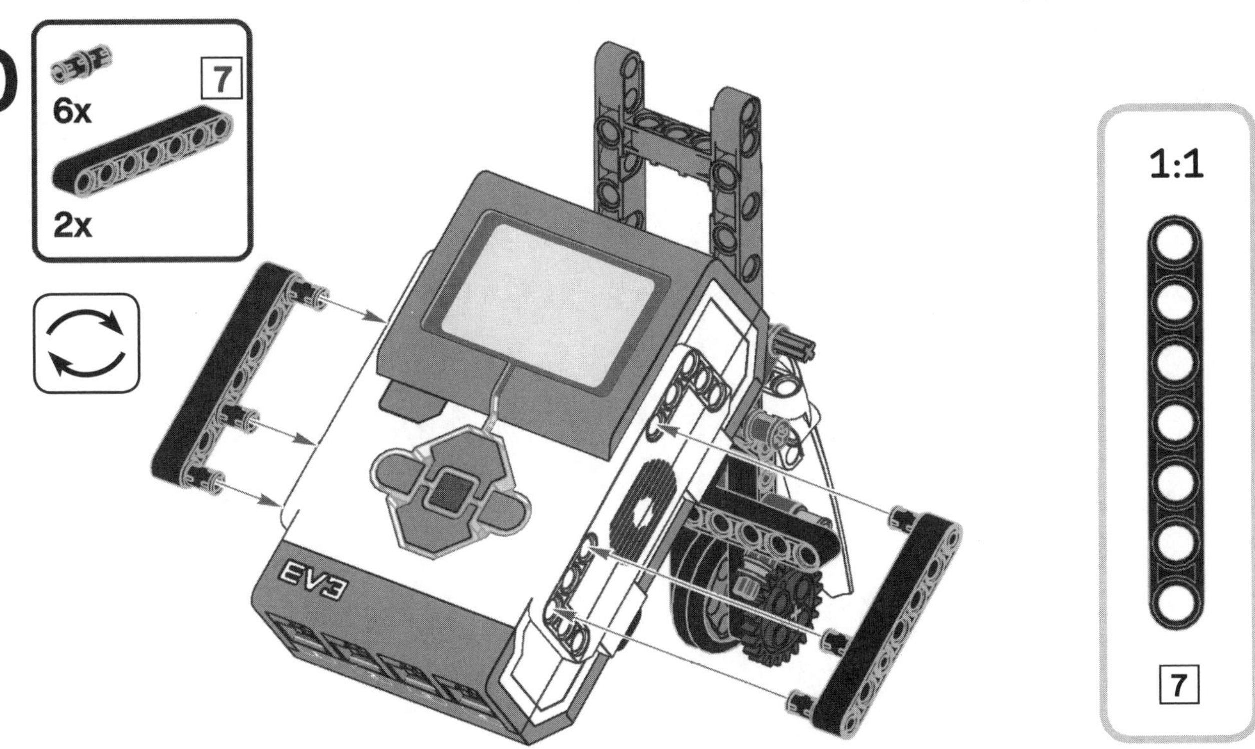

right leg assembly

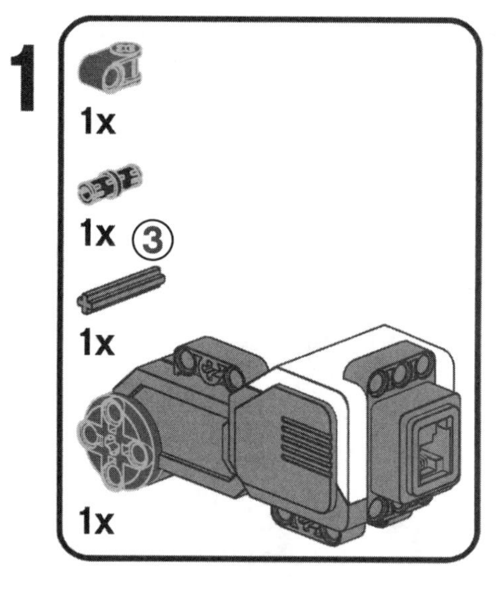

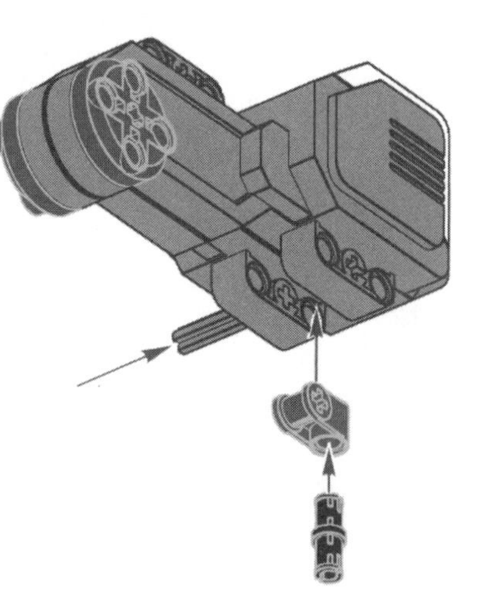

1:1

3

5

9

3

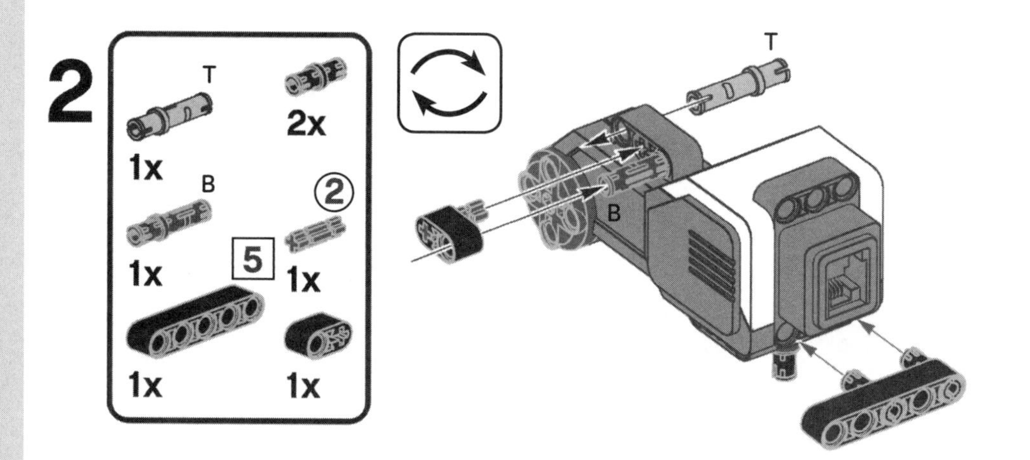

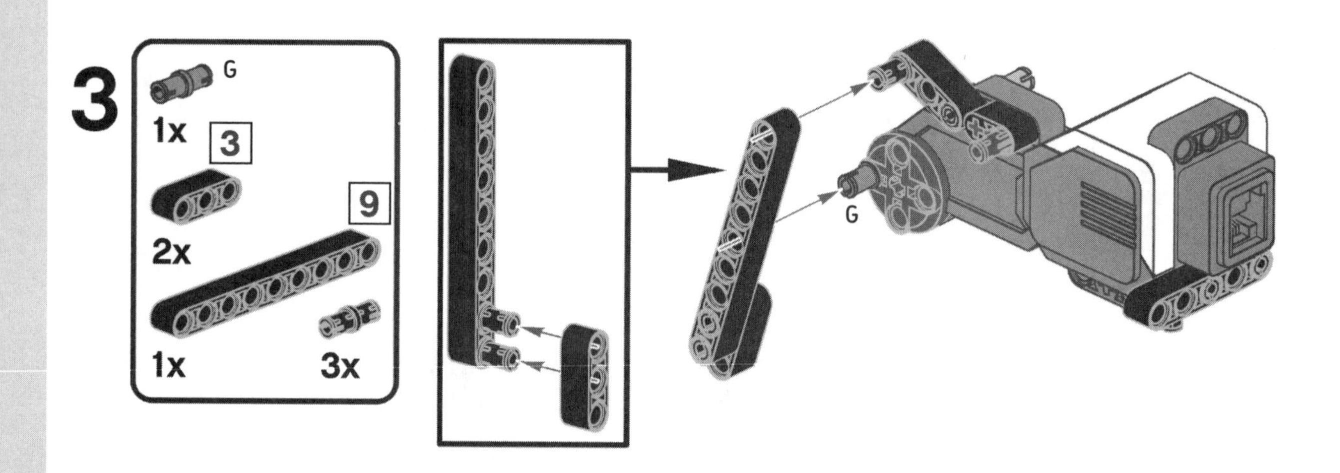

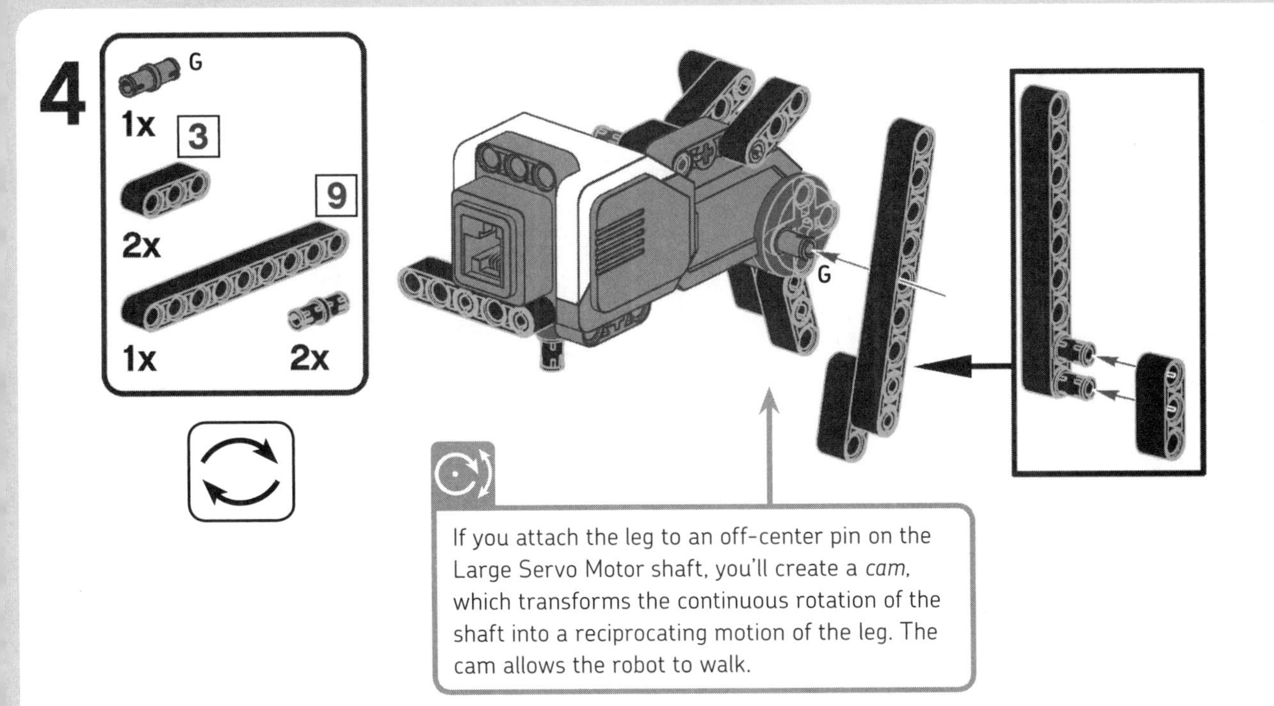

If you attach the leg to an off-center pin on the Large Servo Motor shaft, you'll create a *cam*, which transforms the continuous rotation of the shaft into a reciprocating motion of the leg. The cam allows the robot to walk.

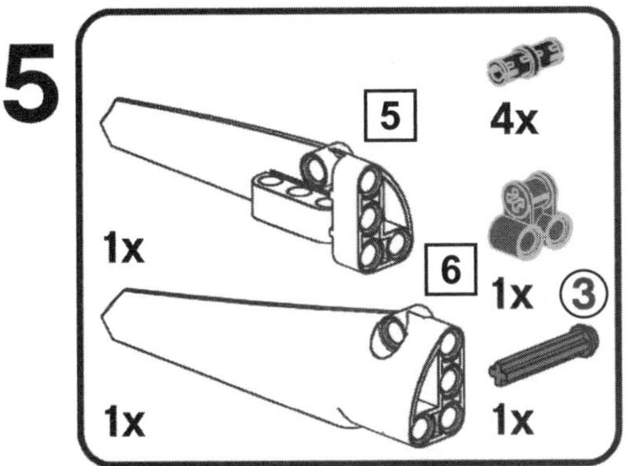

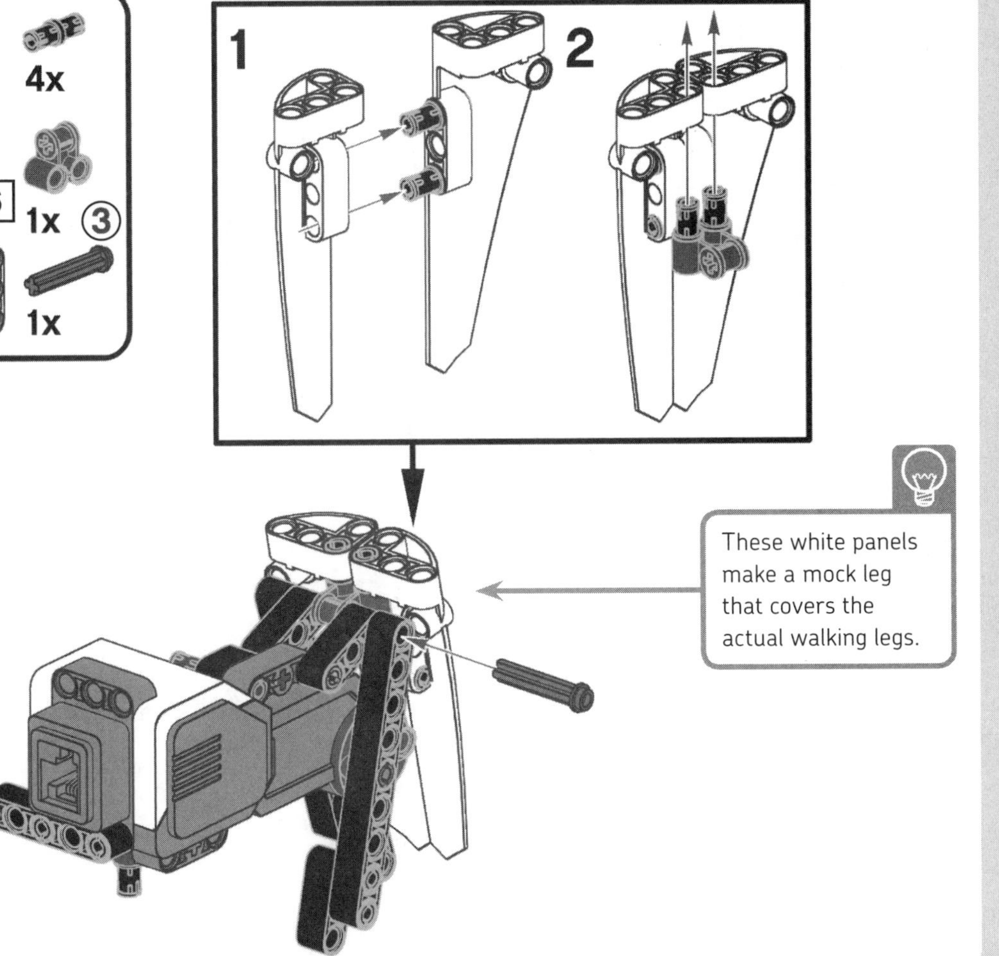

These white panels make a mock leg that covers the actual walking legs.

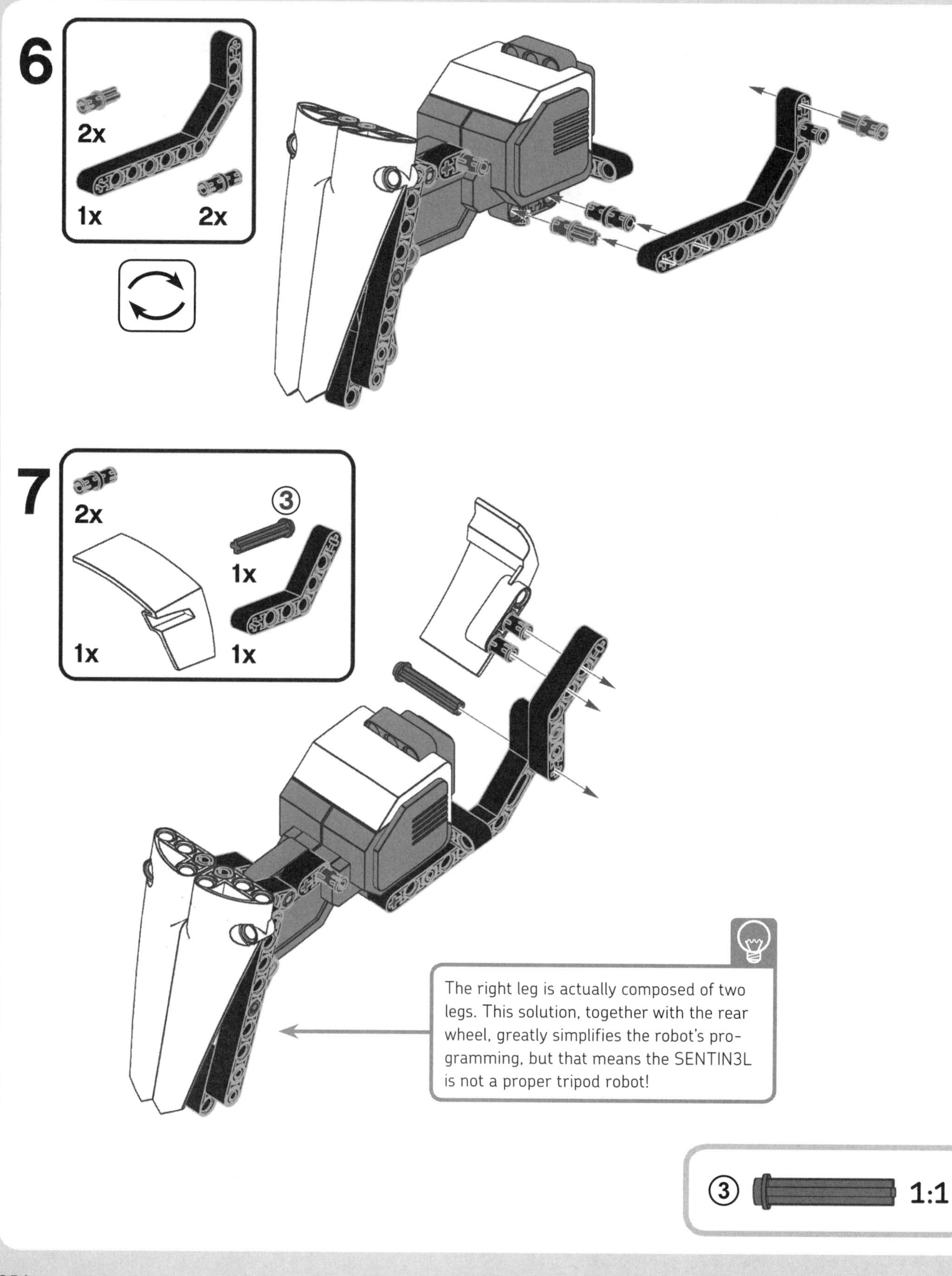
6
2x
1x
2x
7
2x
3
1x
1x
1x
The right leg is actually composed of two legs. This solution, together with the rear wheel, greatly simplifies the robot's programming, but that means the SENTIN3L is not a proper tripod robot!
3
1:1

main assembly

11

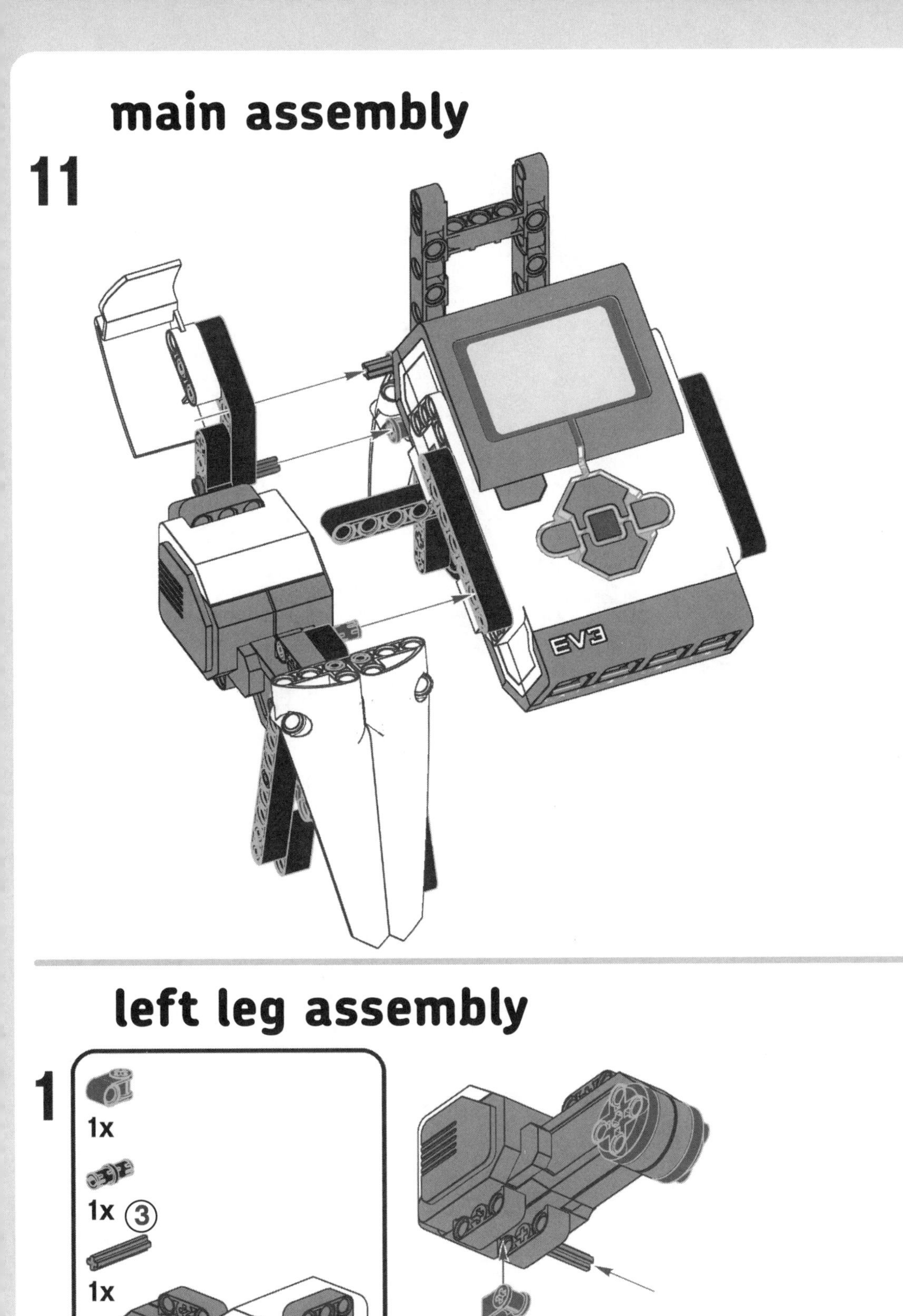

left leg assembly

1

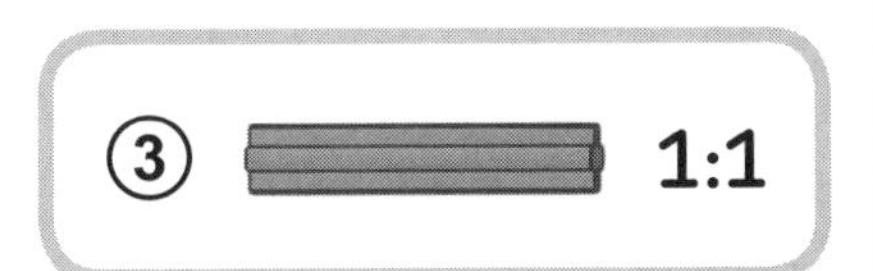

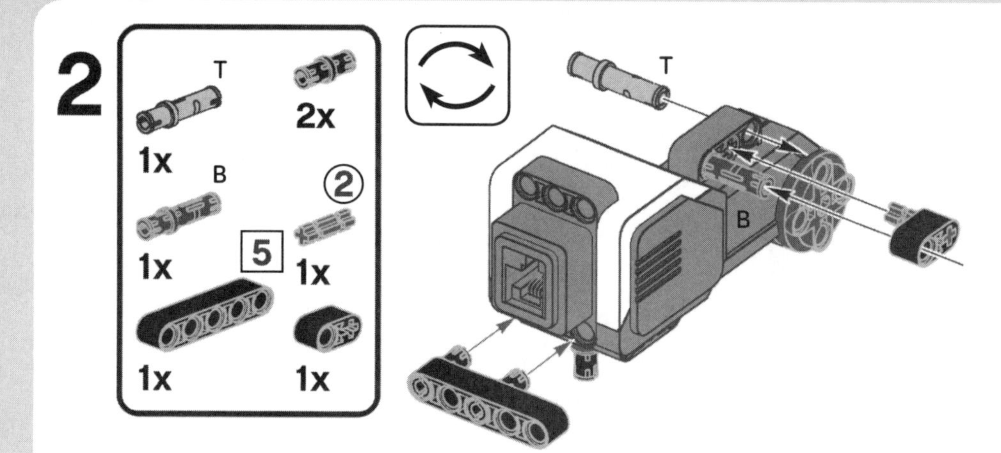
2
T
1x
2x
B
1x
2
5
1x
1x
1x
T
B

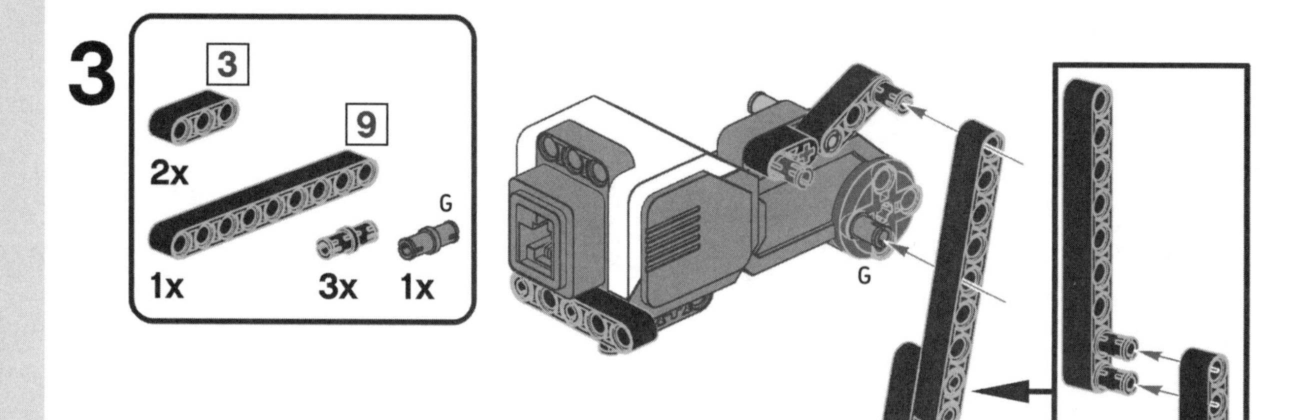
3
3
2x
9
1x
3x
G
1x
G

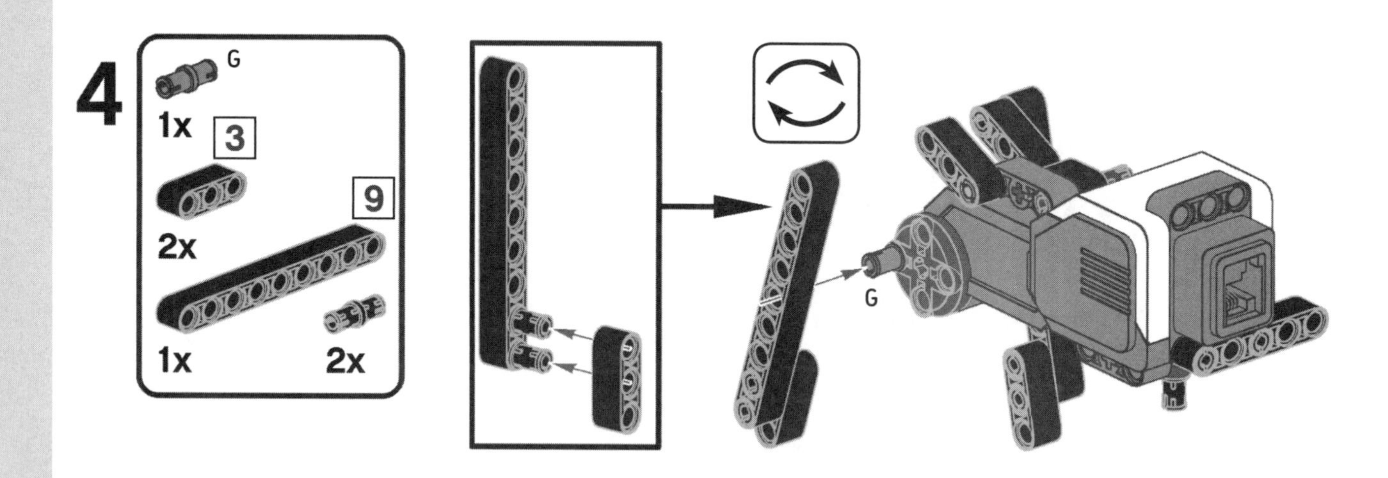
4
G
1x
3
2x
9
1x
2x
G

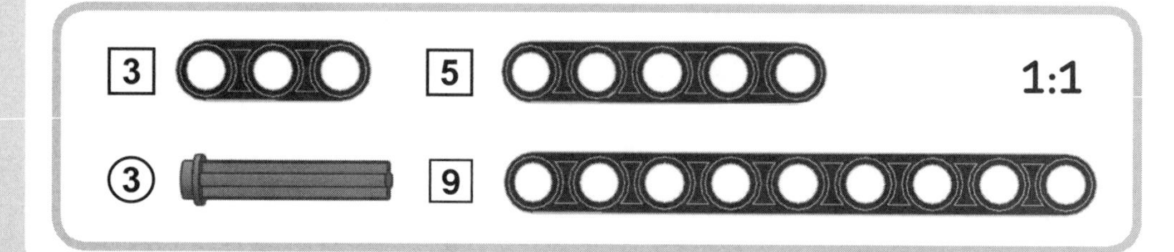
3
5
1:1
3
9

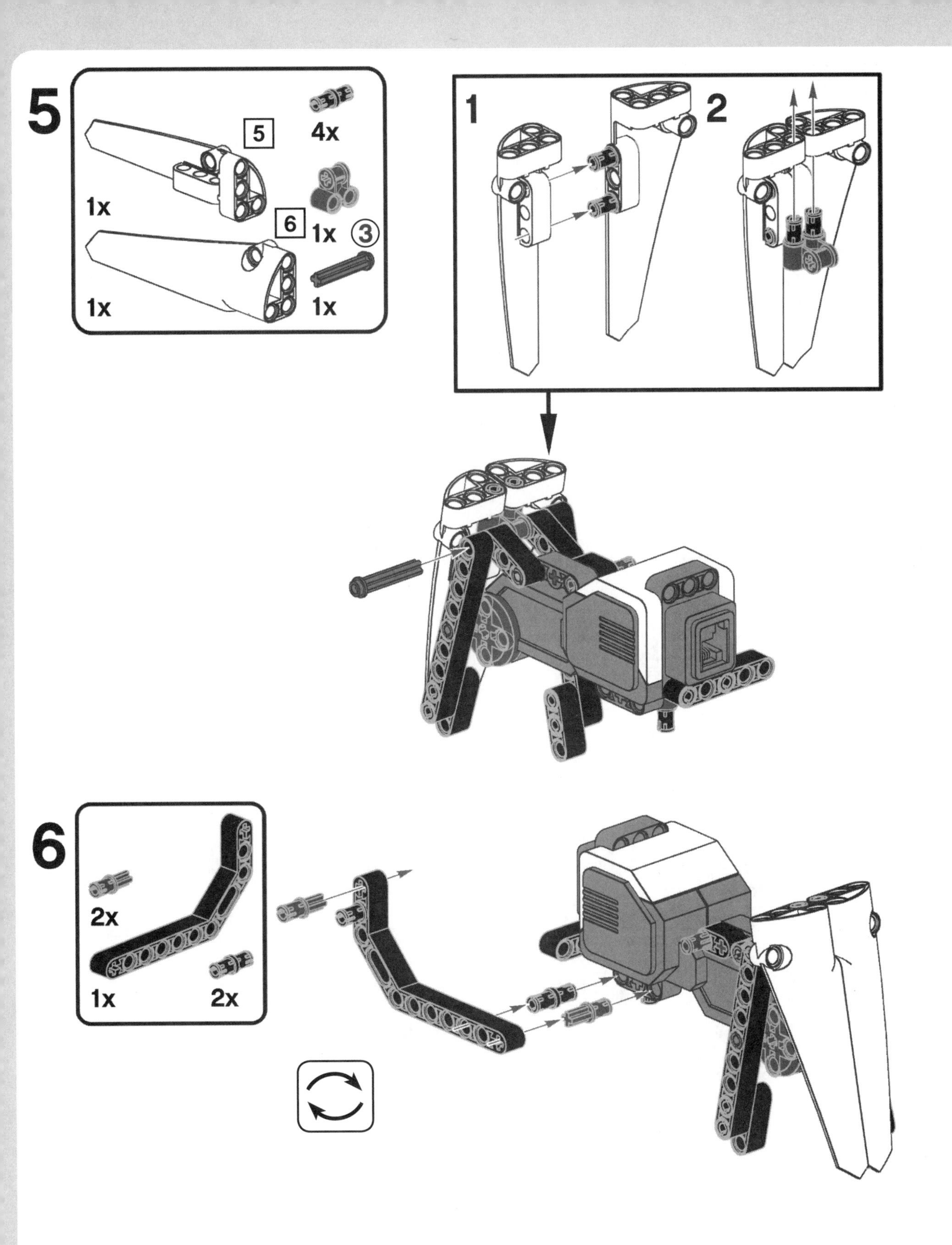
5
5
4x
1x
6
1x
3
1x
1x
1
2
6
2x
1x
2x

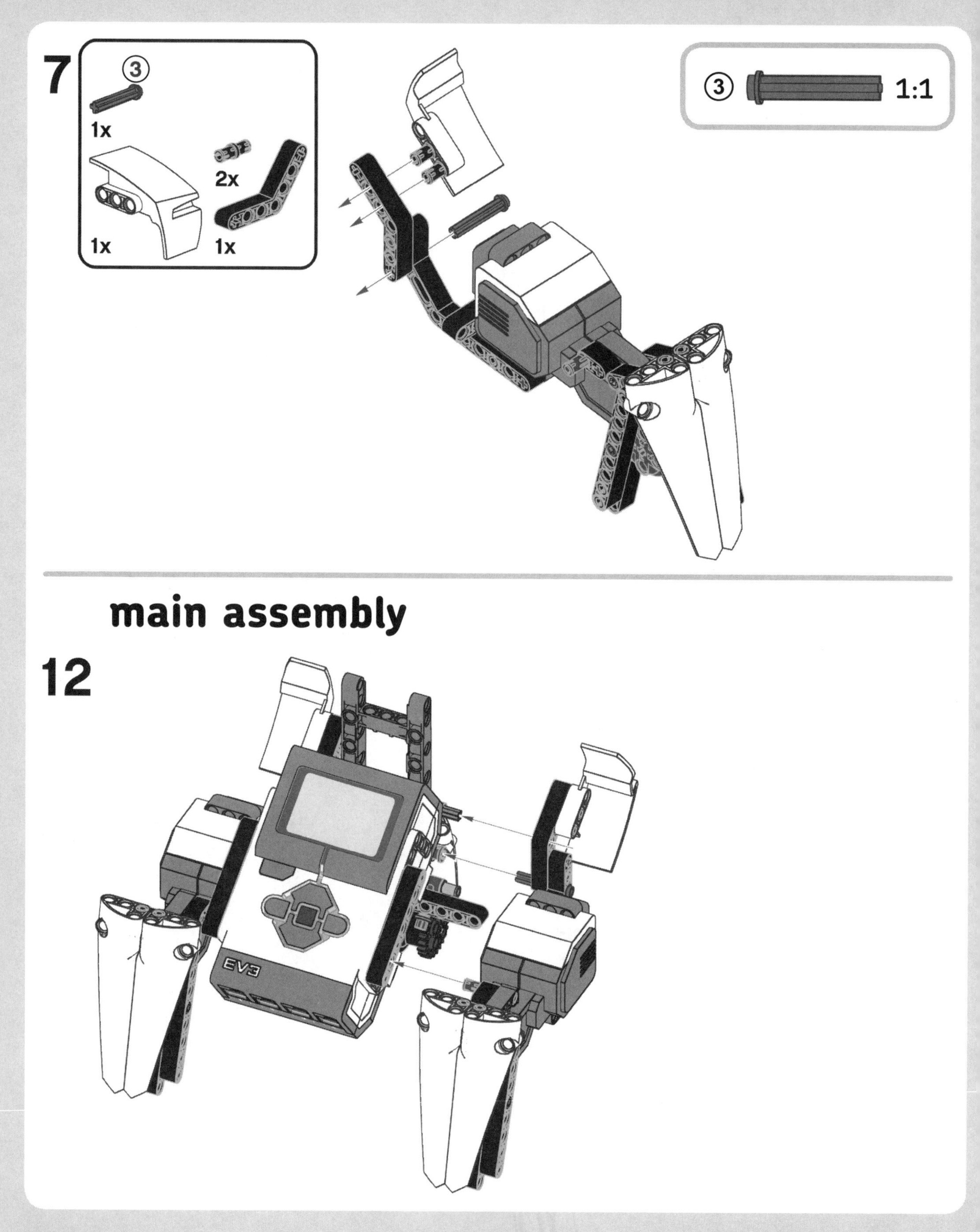
7
3
1x
2x
1x
1x
3
1:1
main assembly
12
EV3

The Pythagorean Theorem states that the area of the square of the hypotenuse (the side opposite the right angle, indicated with a black square) is equal to the sum of the squares of the two legs (the two sides that meet at a right angle). The bigger triangle (whose sides are 6M, 8M, and 10M, as shown in the figure) shows this relation. Even though the hypotenuse of the smaller triangle is not a whole number (it's slightly shorter than 8M), the assembly still works, and the parts are flexible enough that we don't have to force them too much to make things fit.

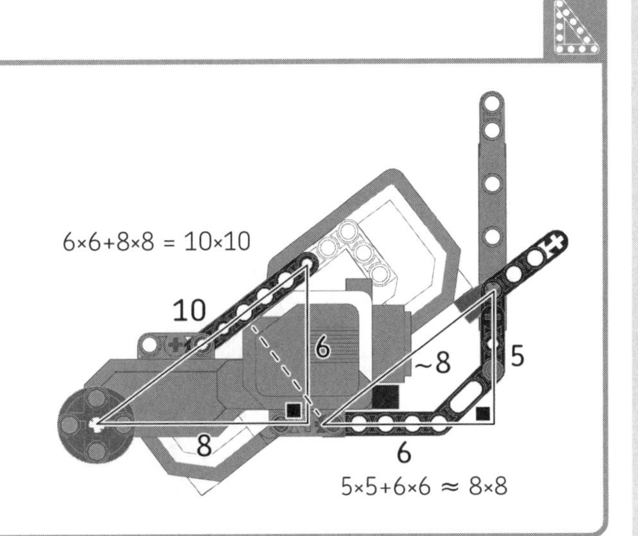

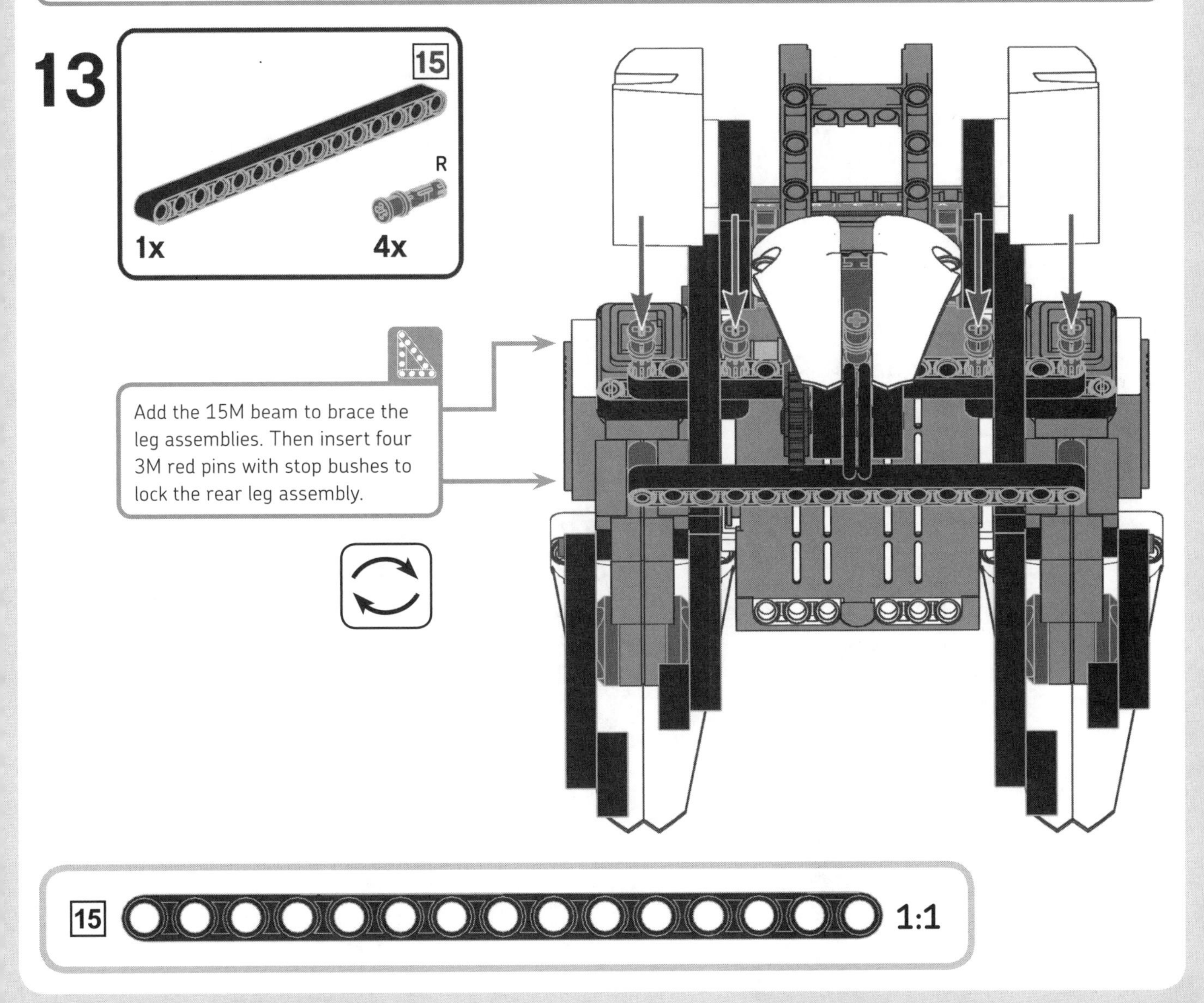

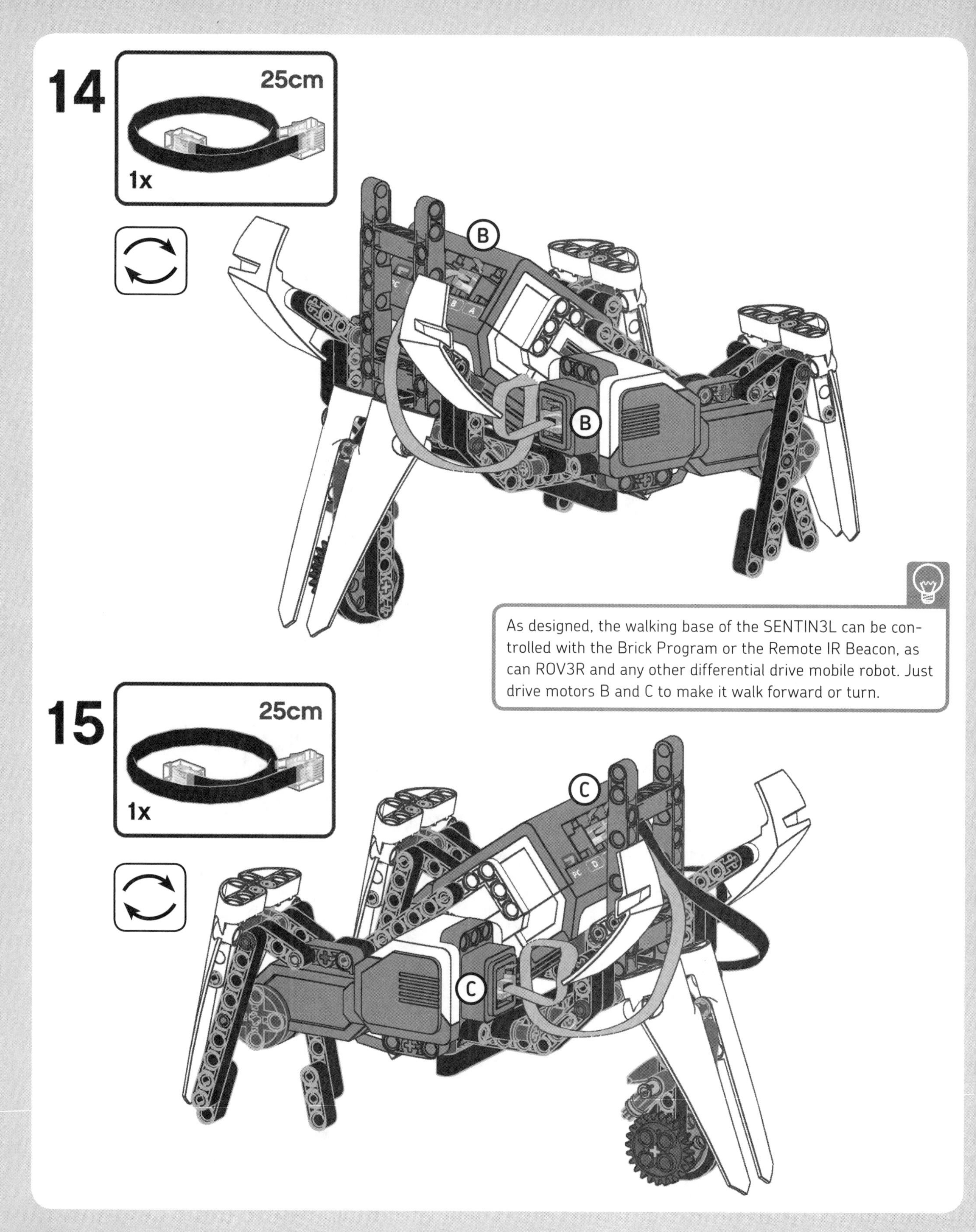

As designed, the walking base of the SENTIN3L can be controlled with the Brick Program or the Remote IR Beacon, as can ROV3R and any other differential drive mobile robot. Just drive motors B and C to make it walk forward or turn.

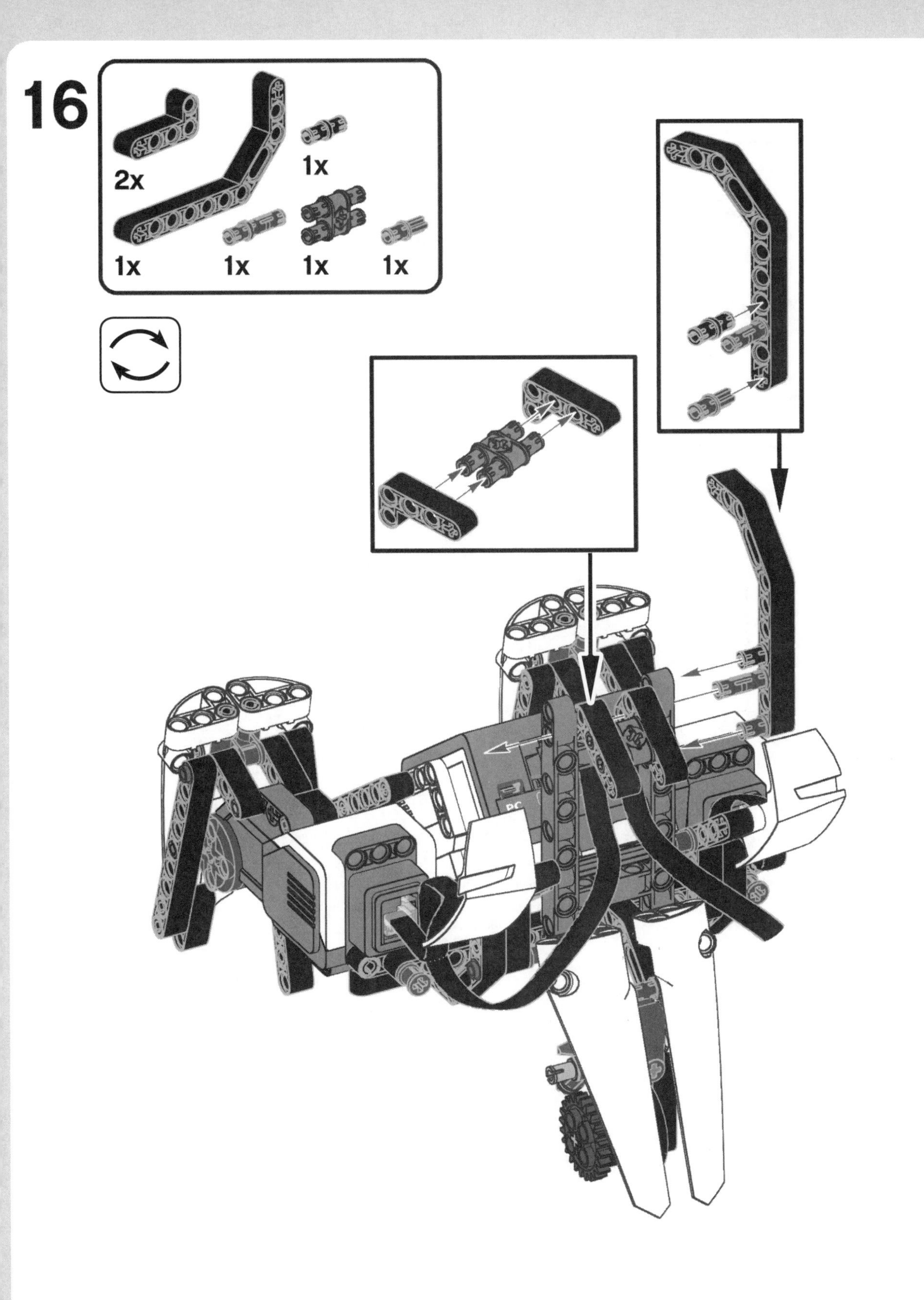
16
2x
1x
1x
1x
1x
1x

chest assembly

1

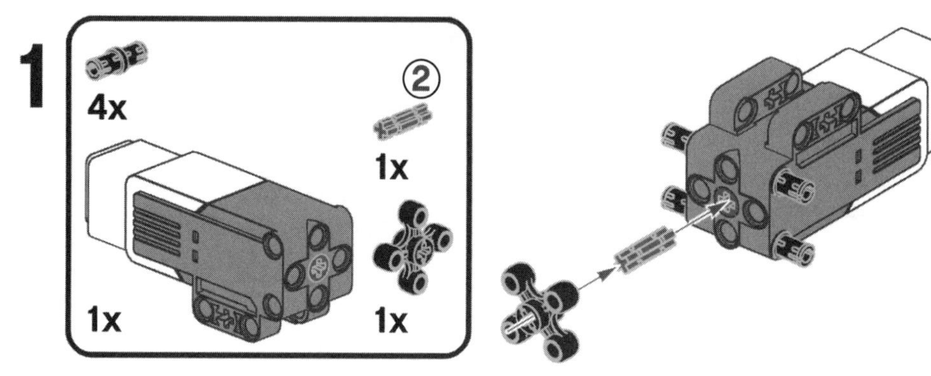

2

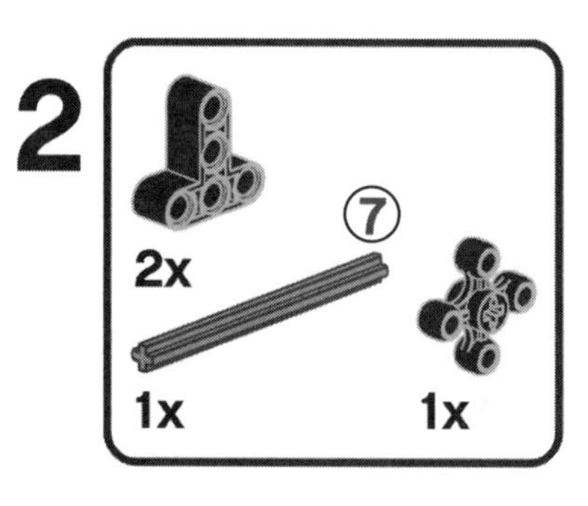

3

2x ④

2x 2x

You can easily reuse this assembly in your creations. If you choose to use knob wheels, the driving axle will come out from the side rather than the front of the motor.

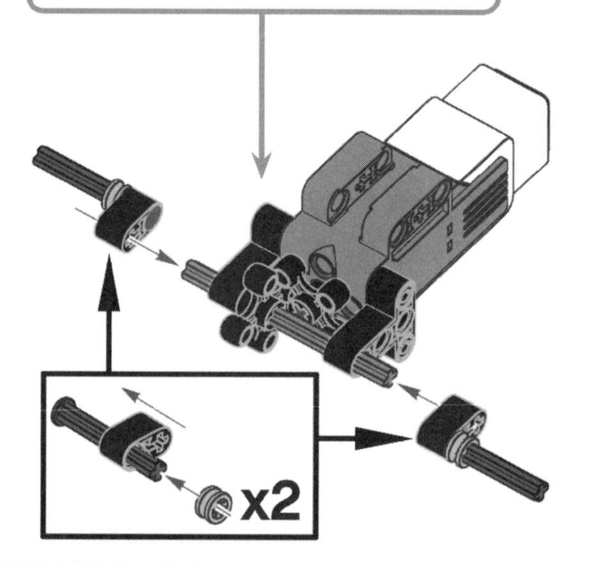

Don't insert the 2M beams all the way onto the 7M axle; leave some space between them and the T beams. Otherwise, the flat heads of the 4M axles with stop might catch on the holes of the T beams when the pieces rotate.

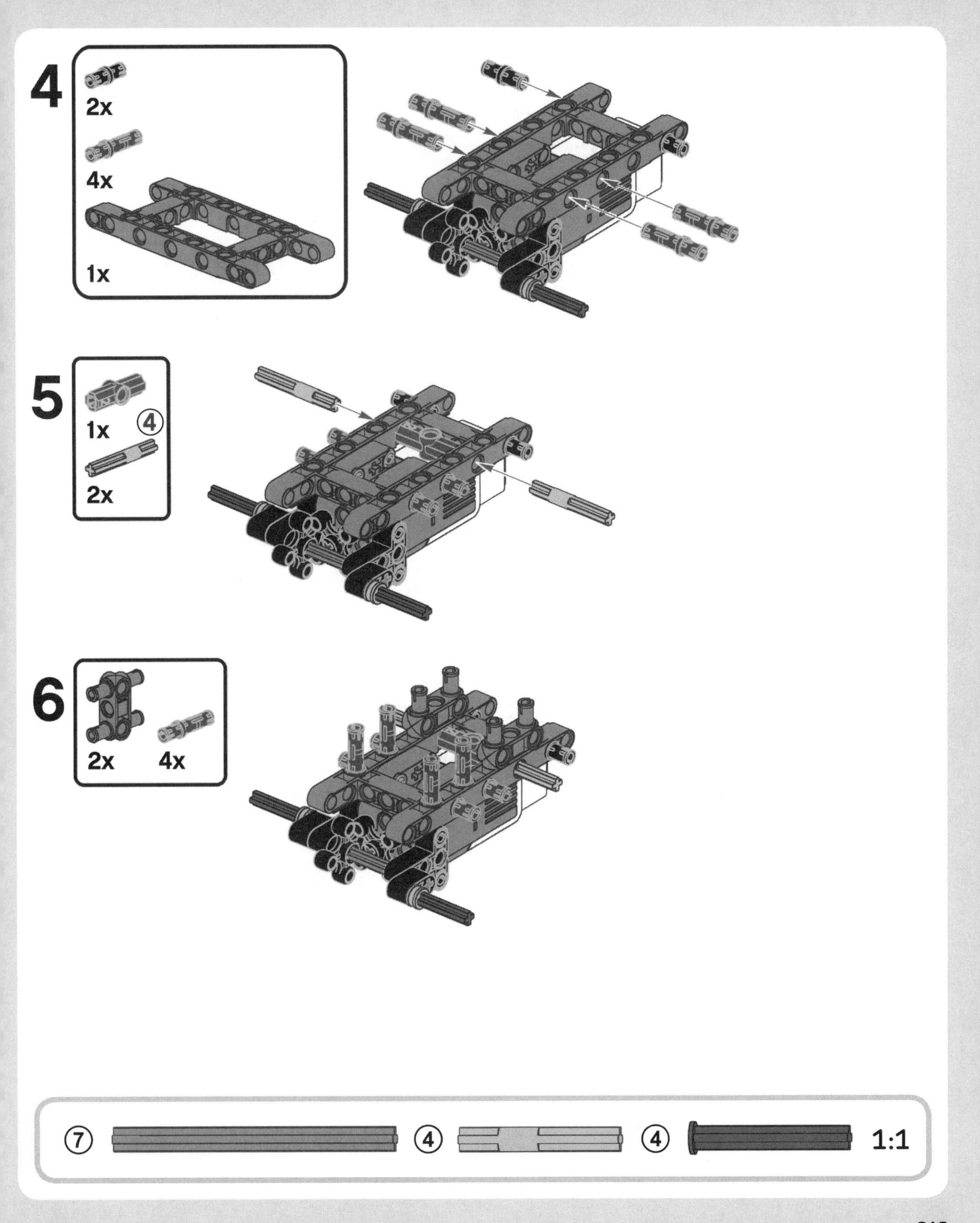
4
2x
4x
1x
5
1x
4
2x
6
2x
4x
7
4
4
1:1

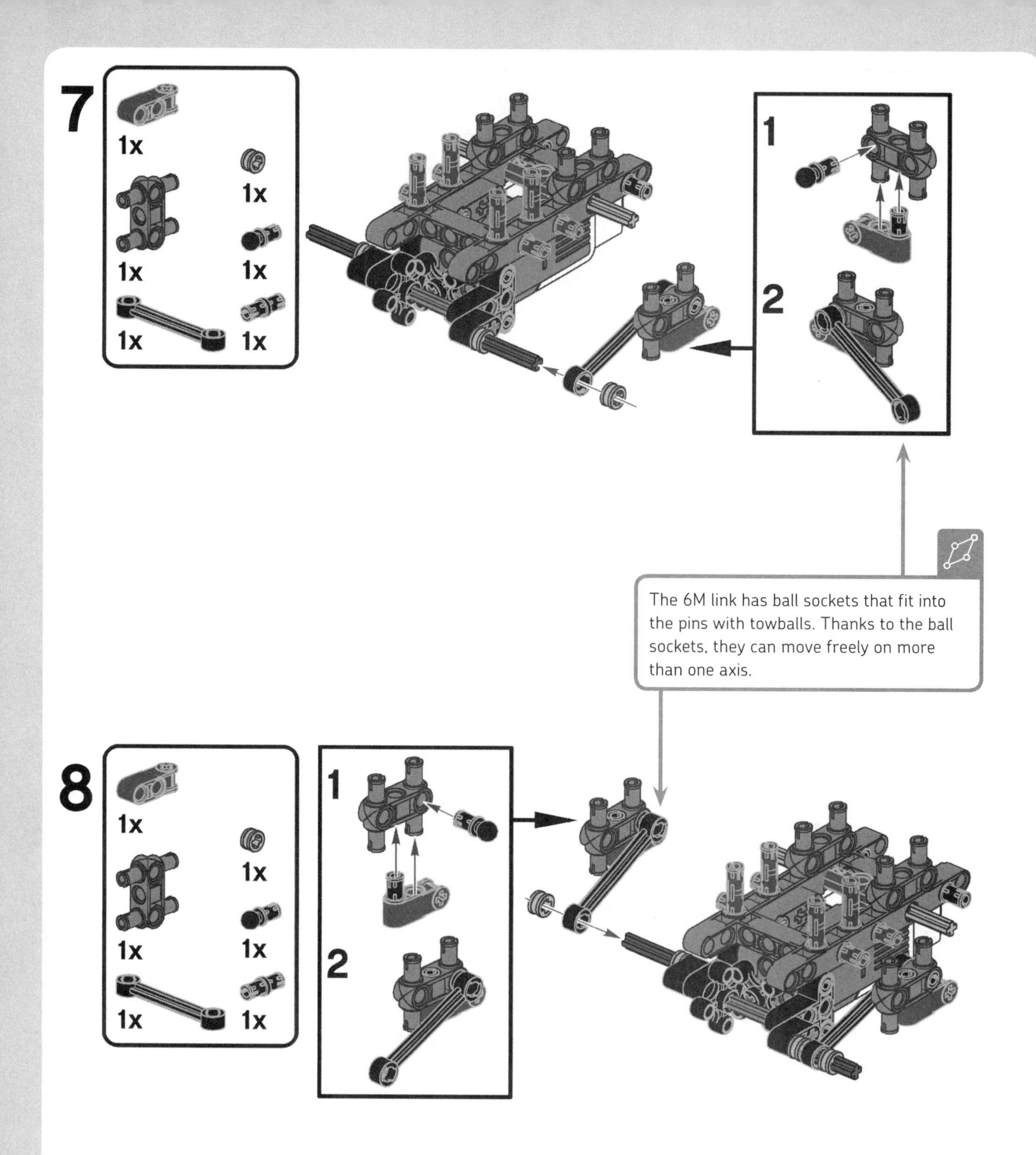
7
1x
1x
1x
1x
1x
1x
1x
1
2
The 6M link has ball sockets that fit into the pins with towballs. Thanks to the ball sockets, they can move freely on more than one axis.
8
1x
1x
1x
1x
1x
1x
1x
1
2

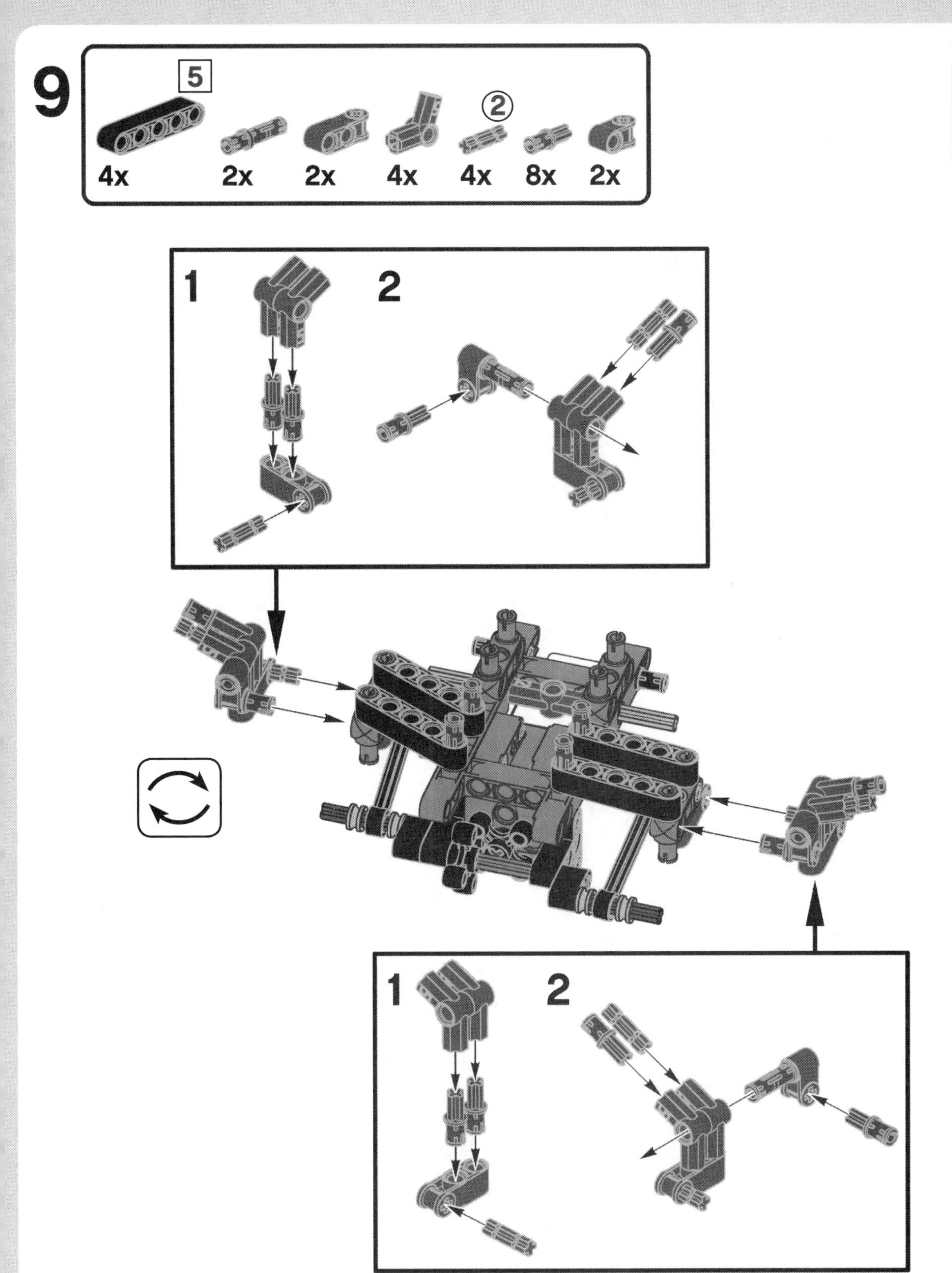
9
5
2
4x
2x
2x
4x
4x
8x
2x
1:1
5
1
2
1
2

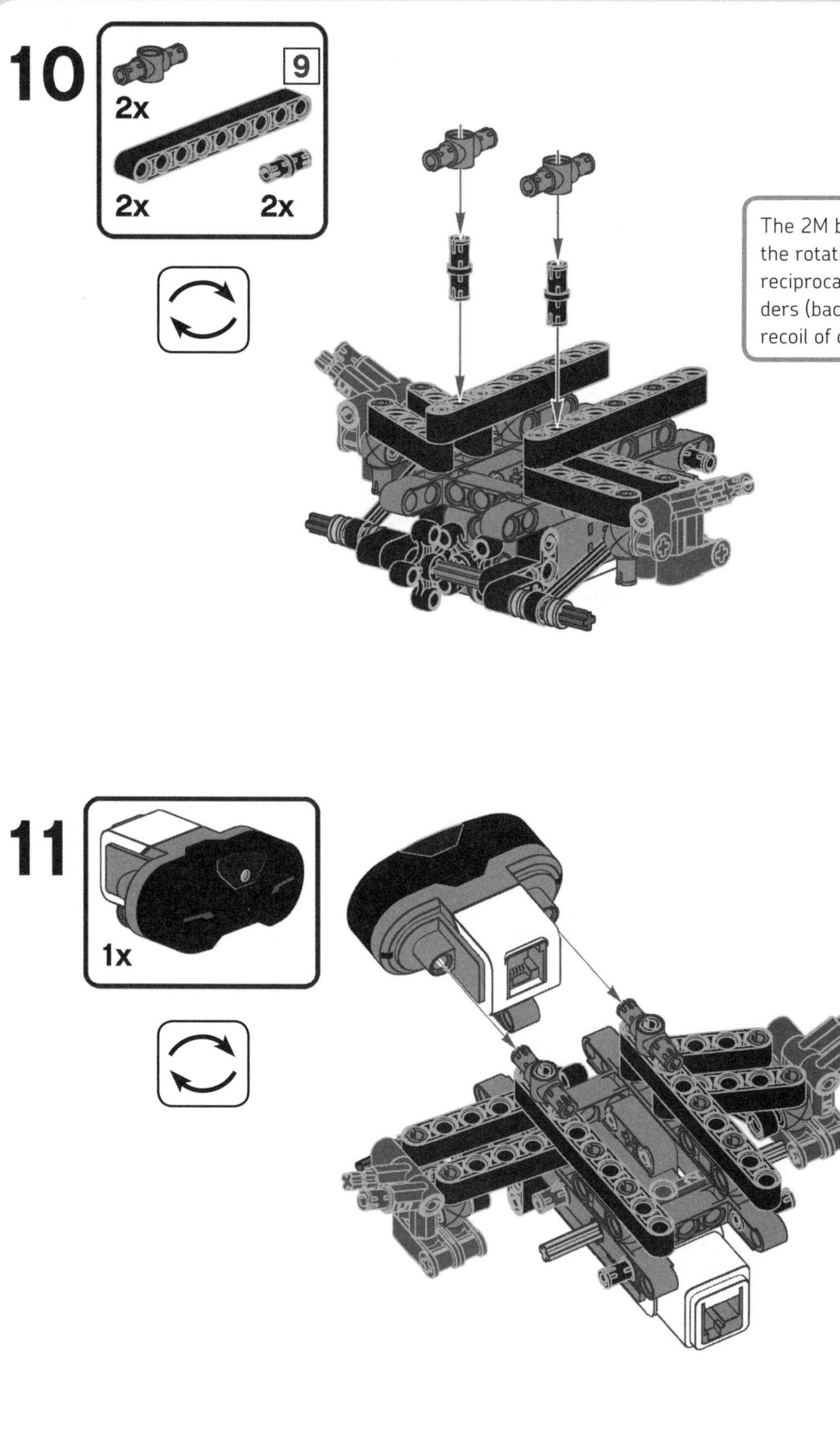

The 2M beams are cams that transform the rotation of the motor shaft into a reciprocating motion of the robot's shoulders (back and forth), which simulates the recoil of cannons.

1:1

9

main assembly

17

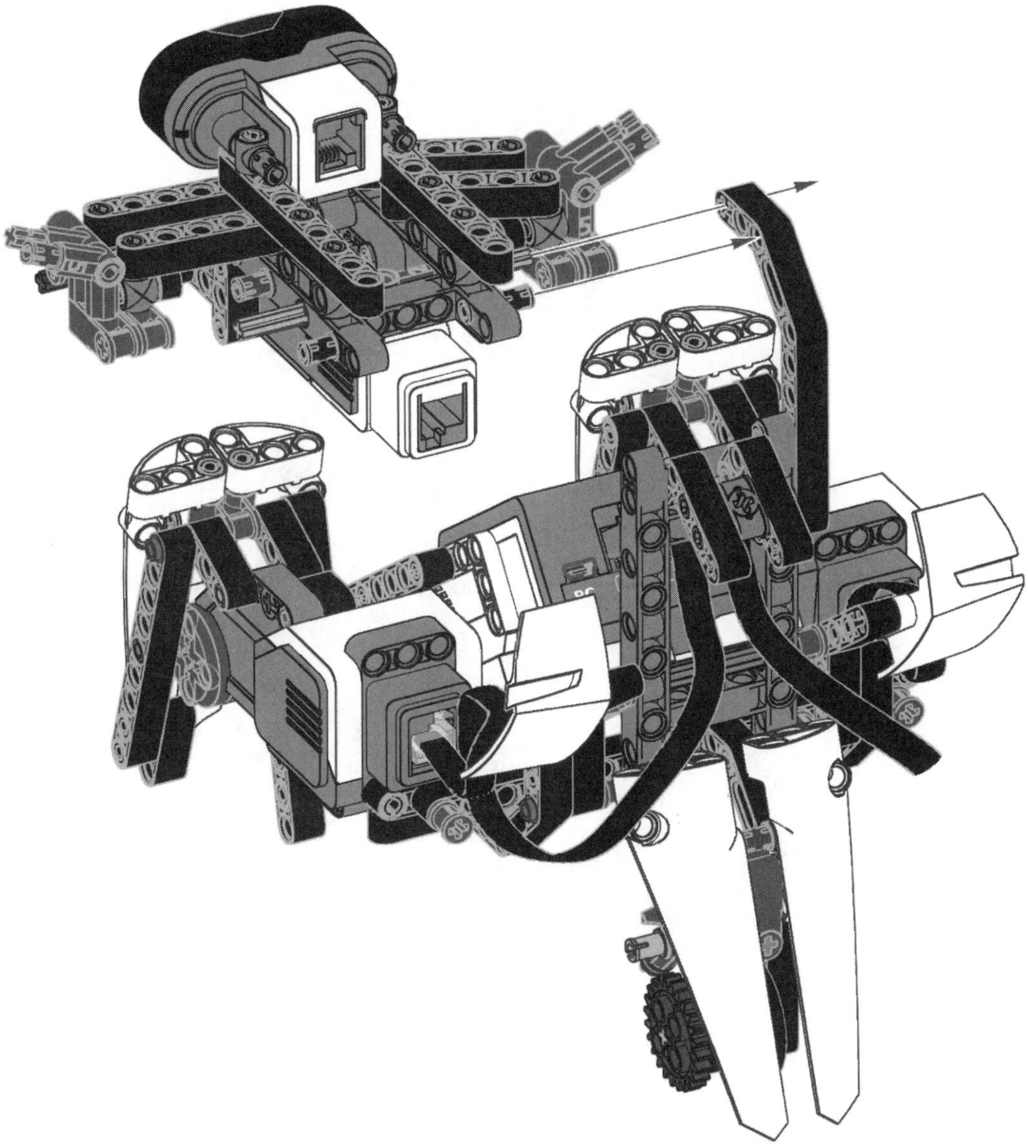

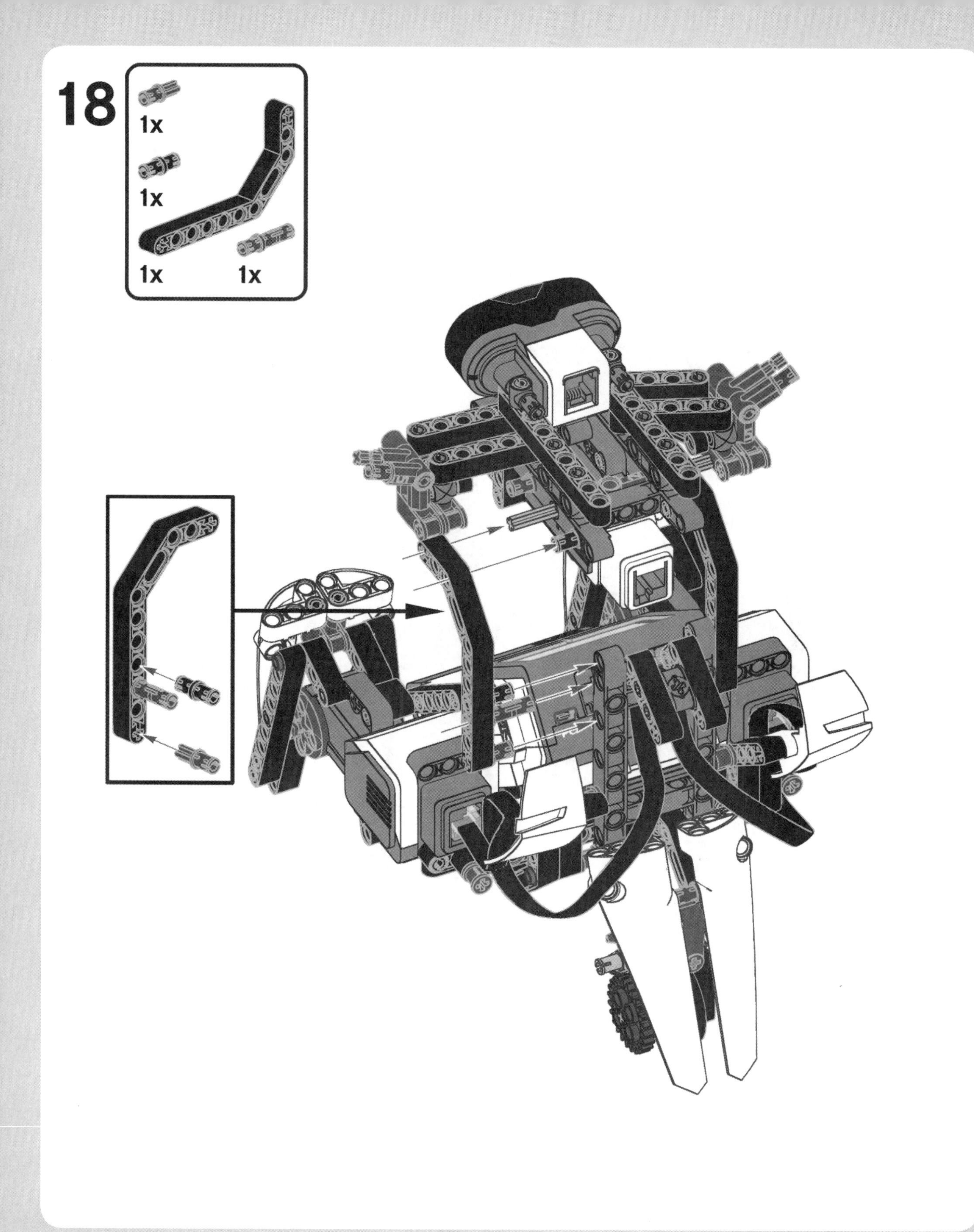
18
1x
1x
1x
1x

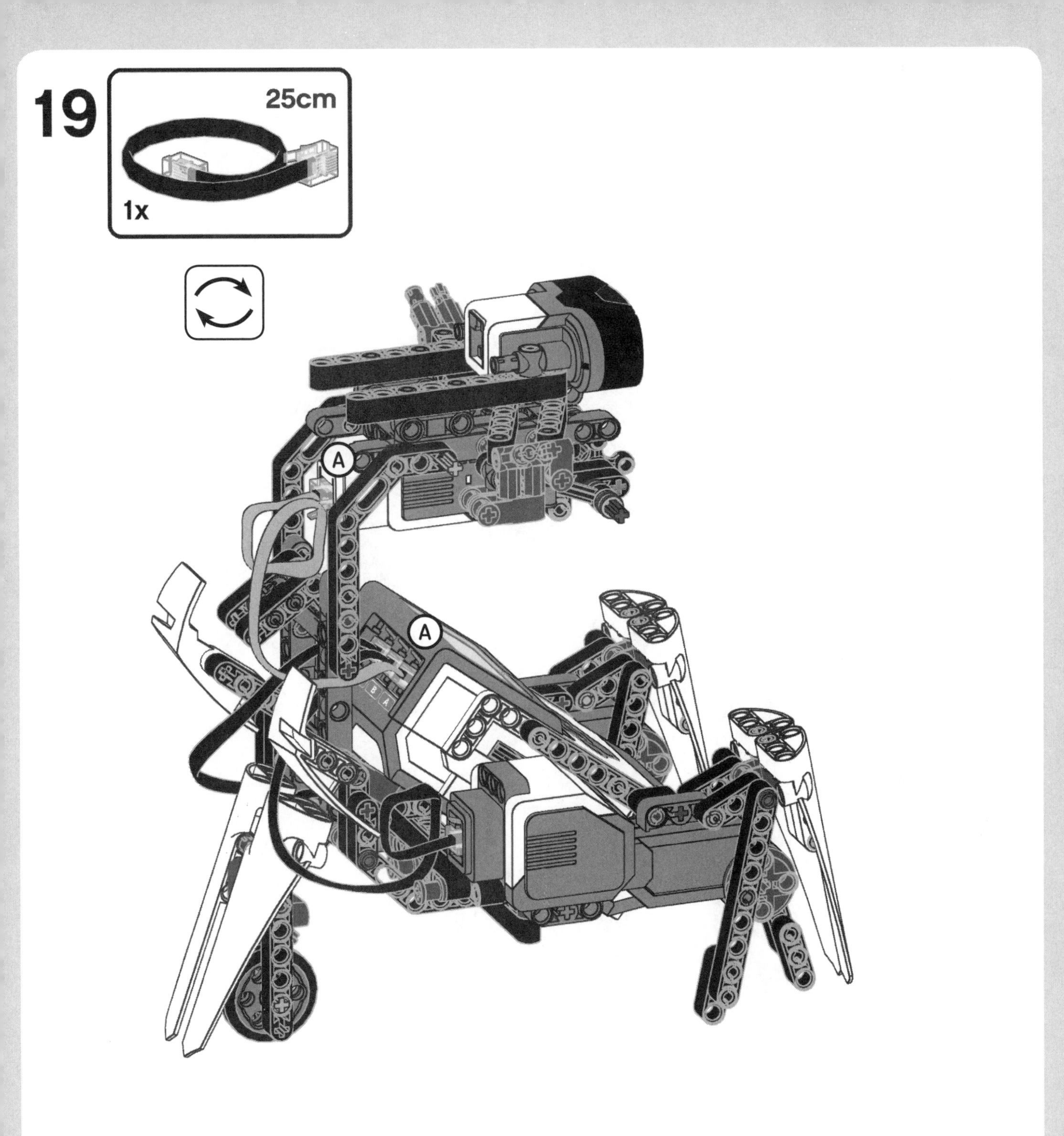
19
25cm
1x
A
A
B
A

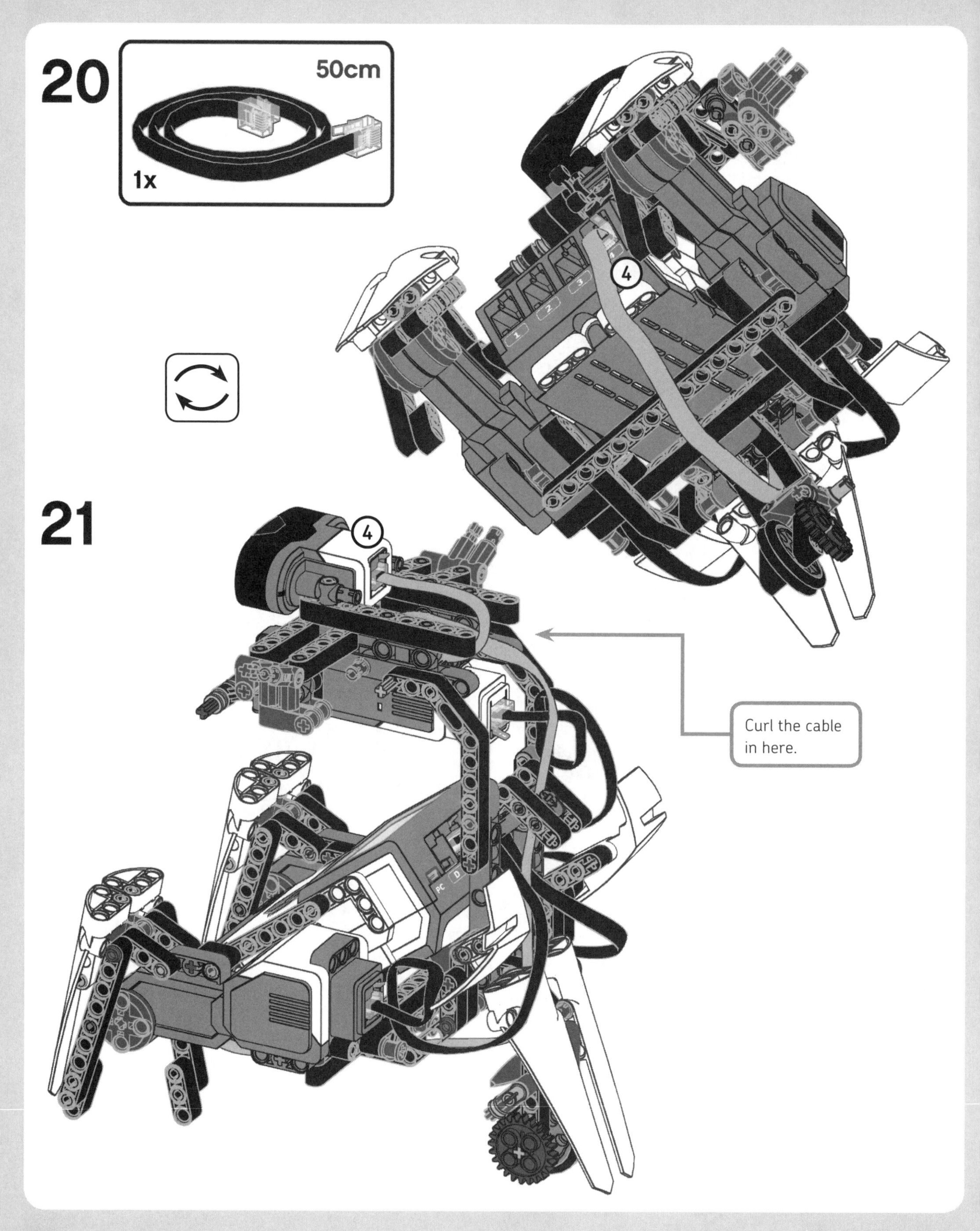
20
50cm
1x
4
21
4
Curl the cable
in here.

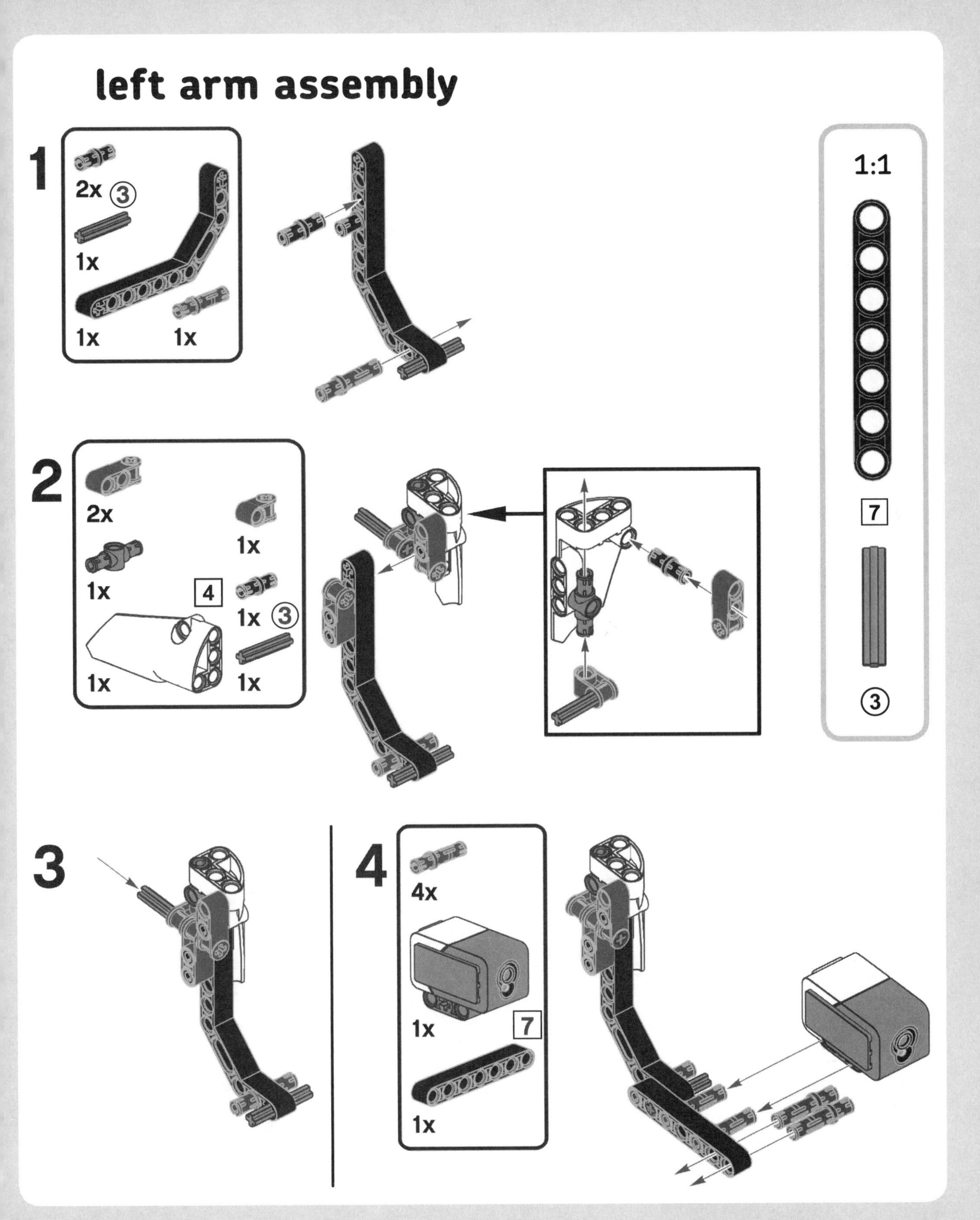
left arm assembly
1
2x ③
1x
1x
1x
2
2x
1x
1x
4
1x ③
1x
1x
3
4
4x
1x
7
1x
1:1
7
③

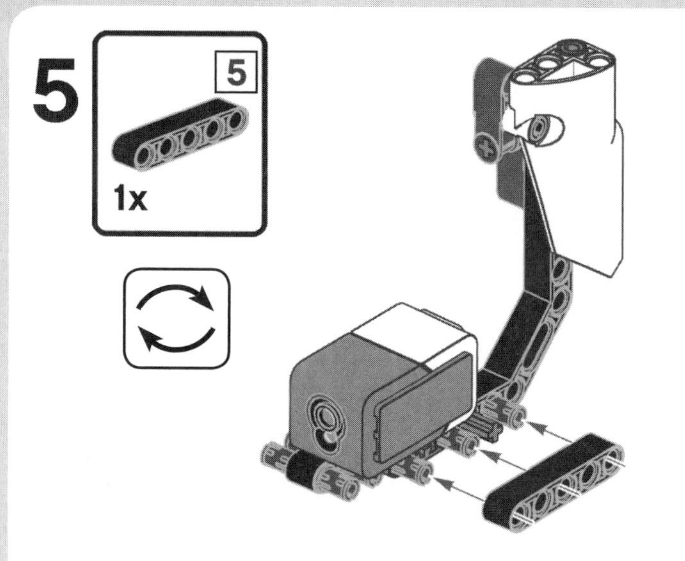

right arm assembly

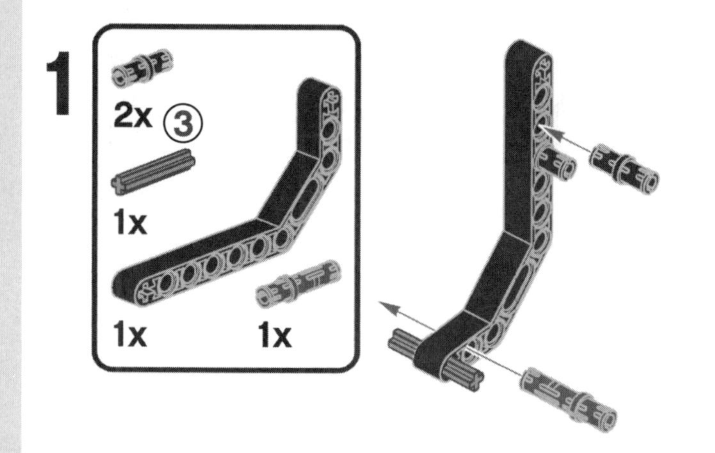

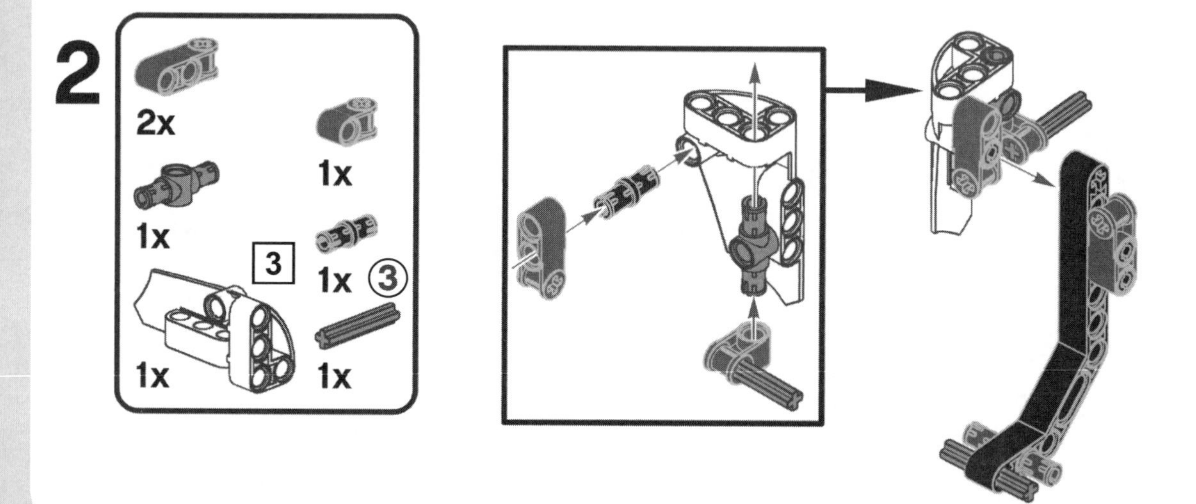

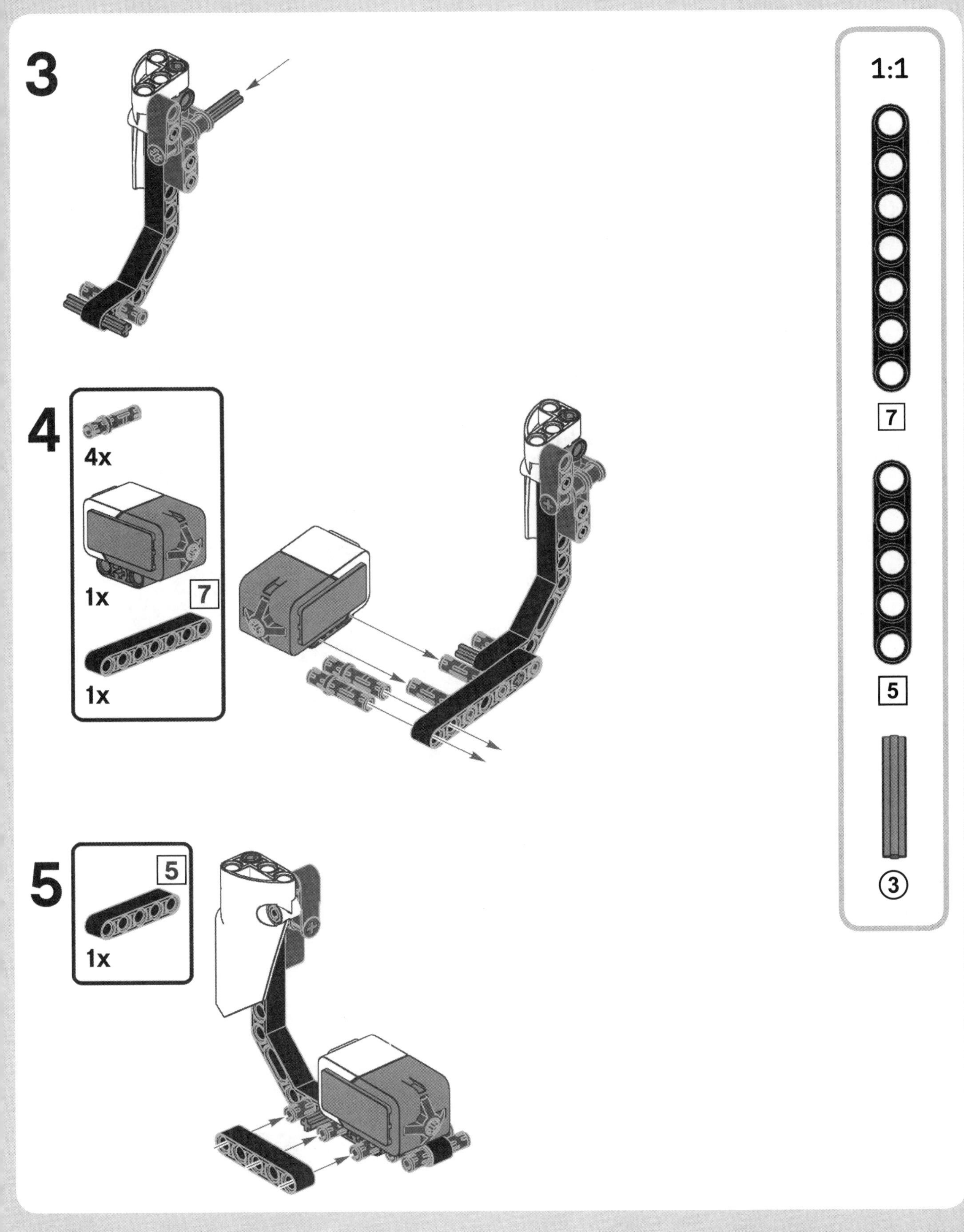
3
4
4x
1x
7
1x
5
5
1x
1:1
7
5
3

main assembly

22

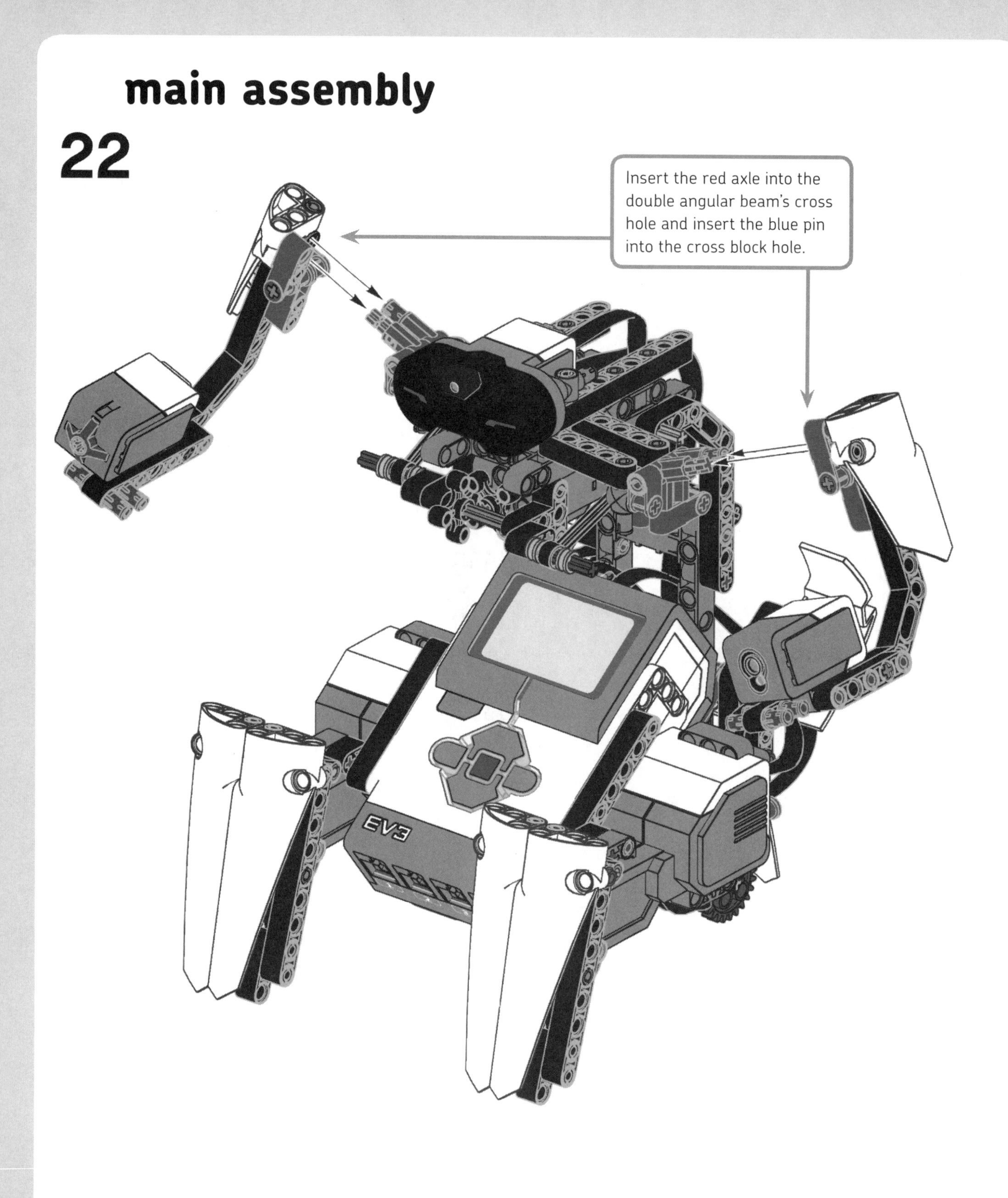

23

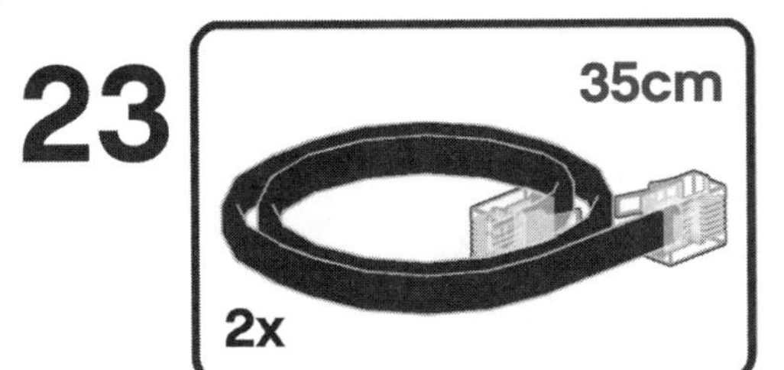

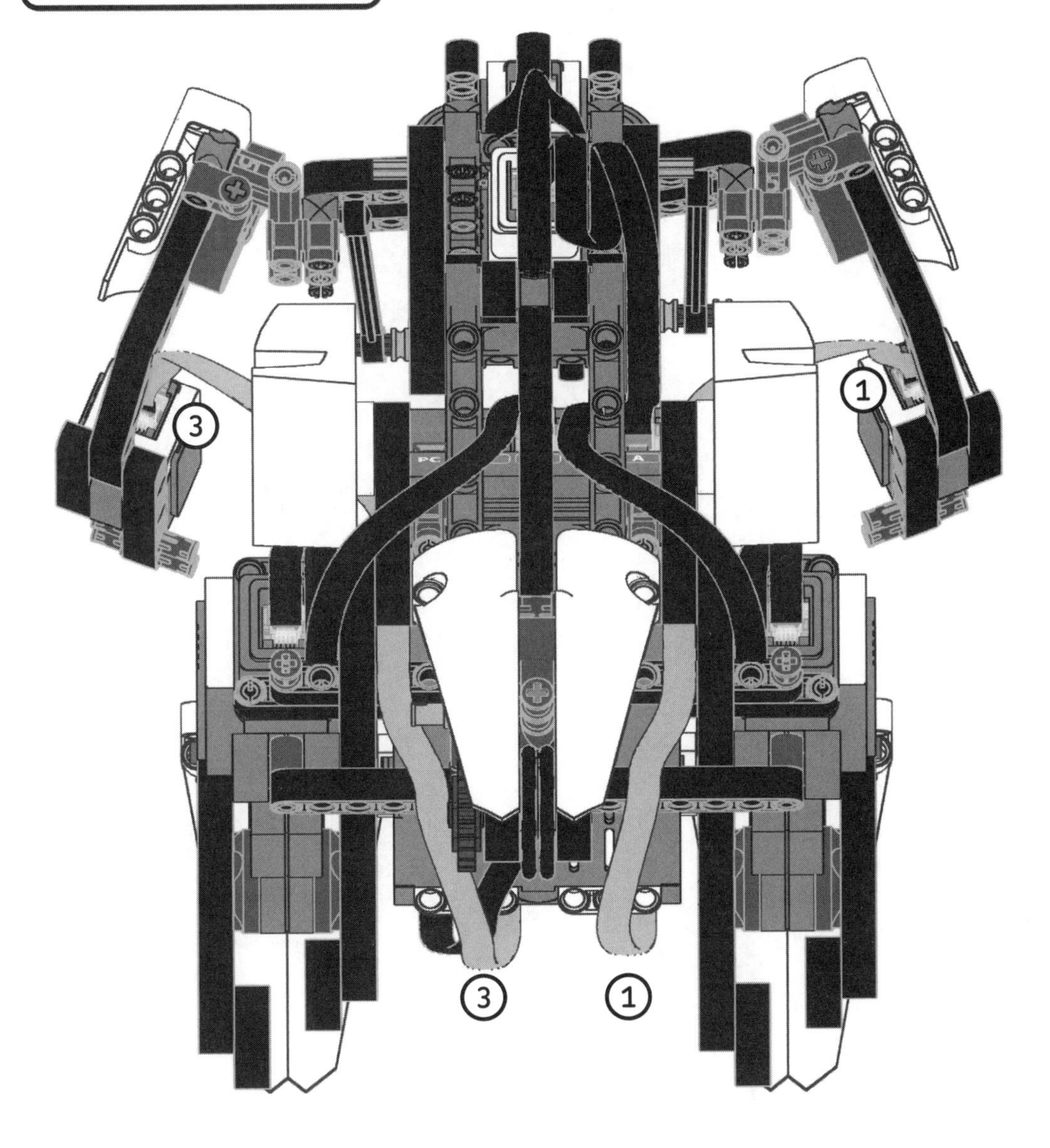

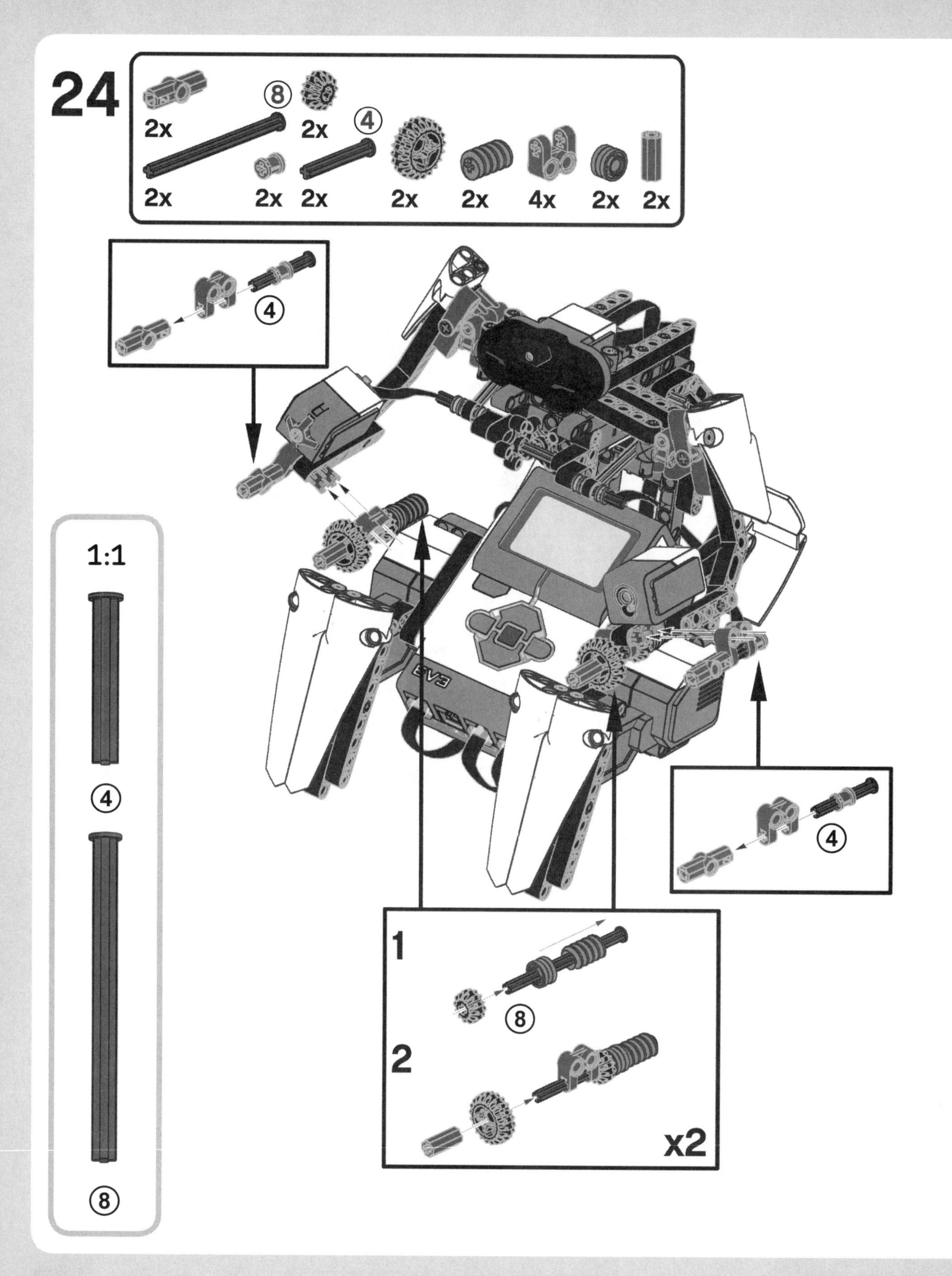
24
8
4
2x
2x
2x
2x
2x
2x
2x
4x
2x
2x
4
1:1
4
8
4
1
8
2
x2

back shield assembly

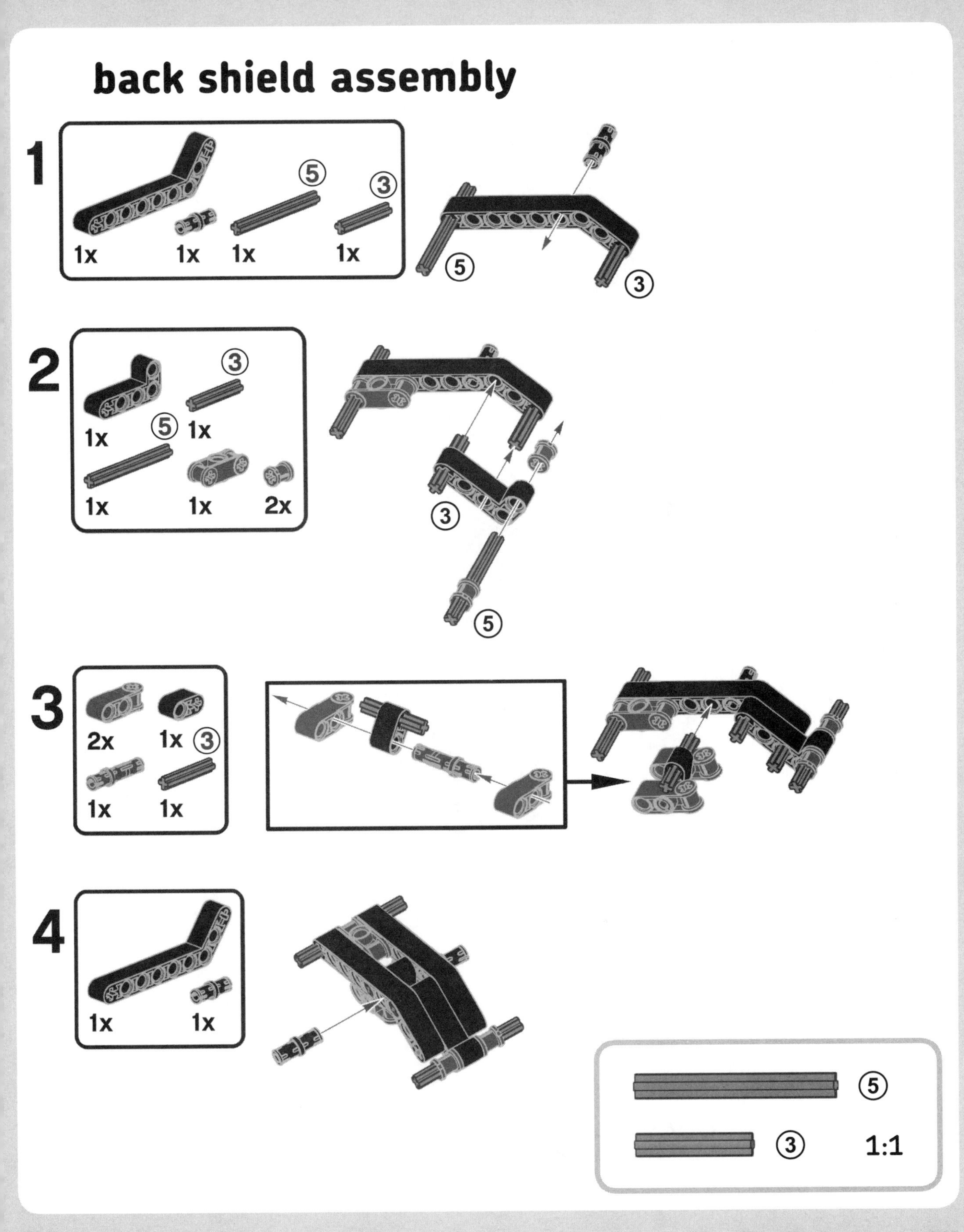

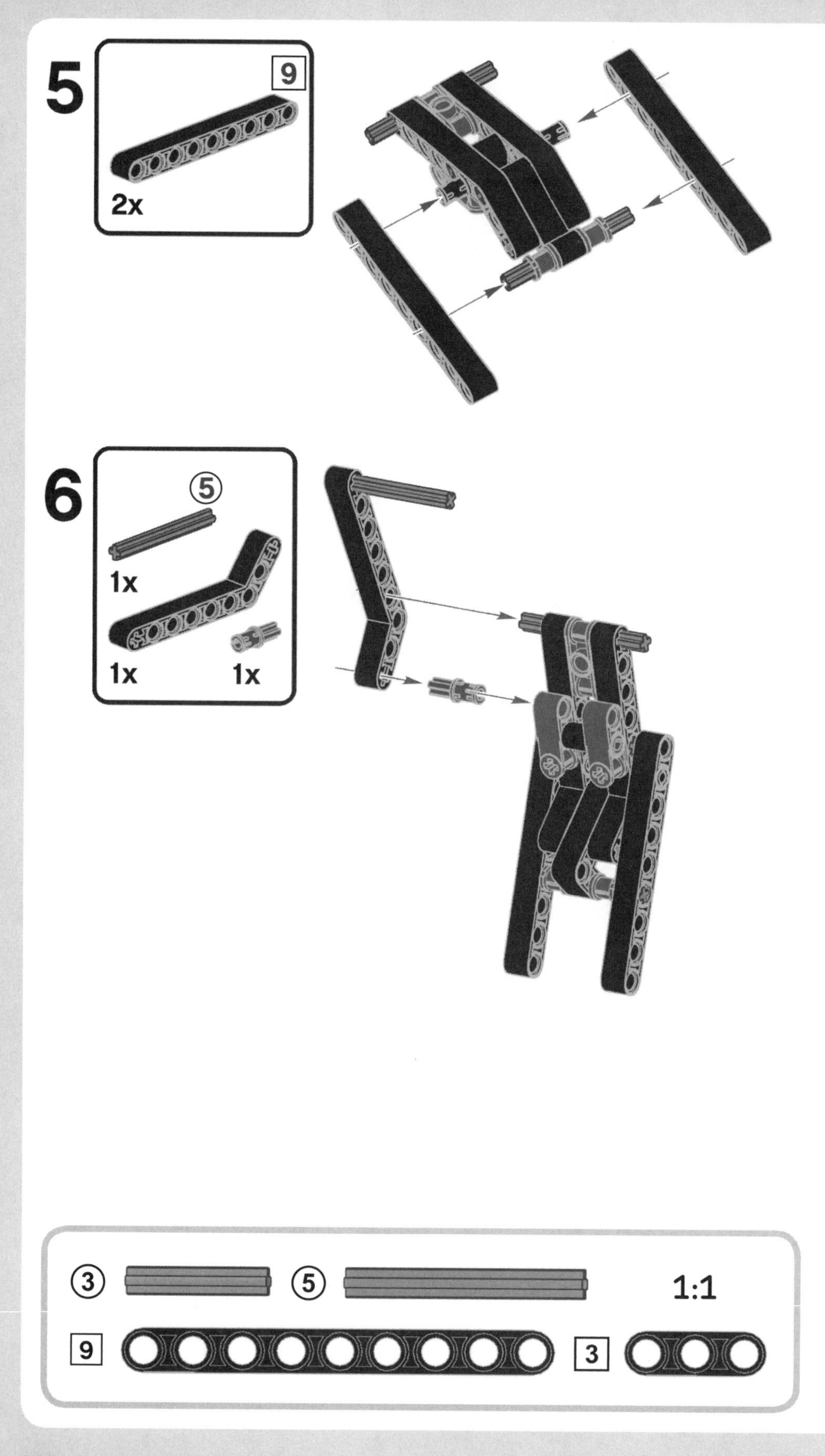
5
9
2x
6
5
1x
1x
1x
3
5
1:1
9
3

back shield middle subassembly

1

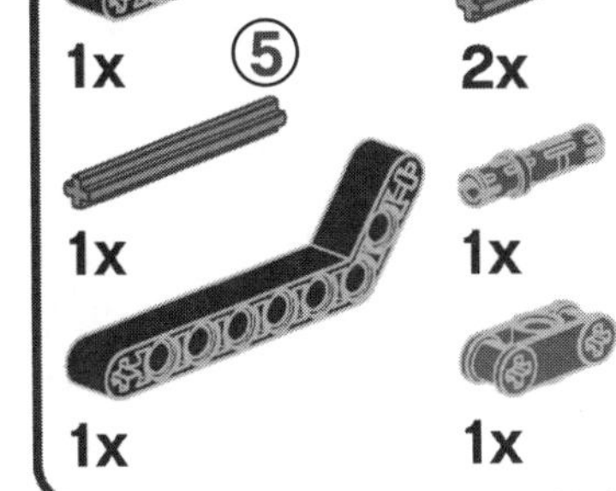

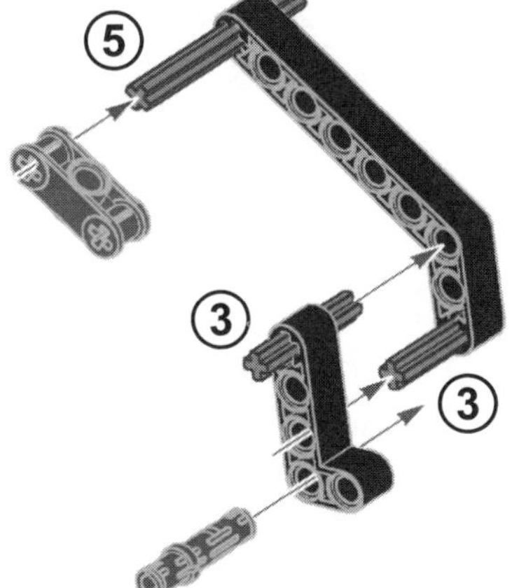

2

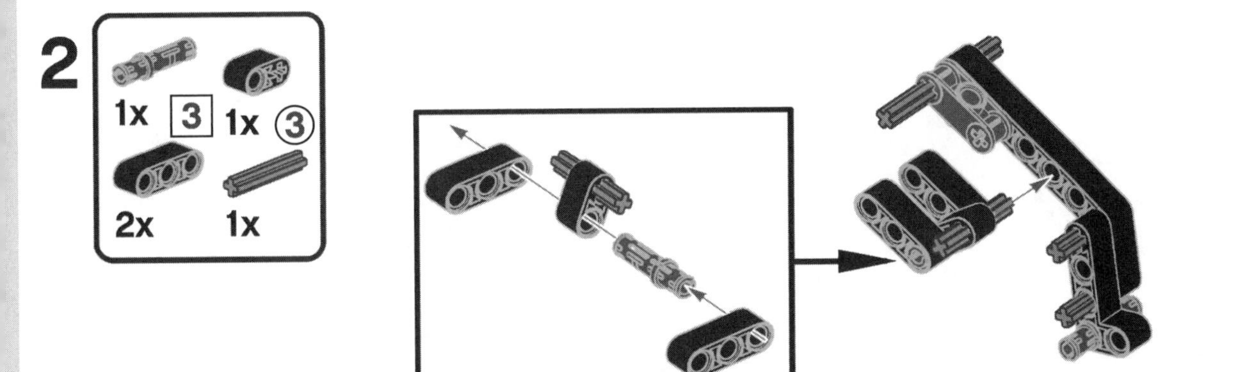

3

1x
5
1x
1x

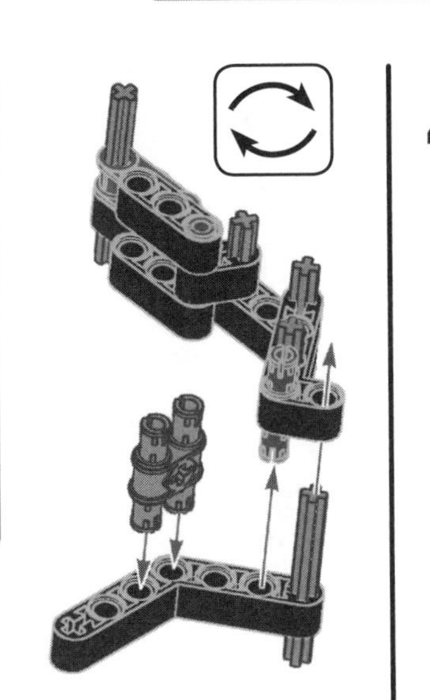

4

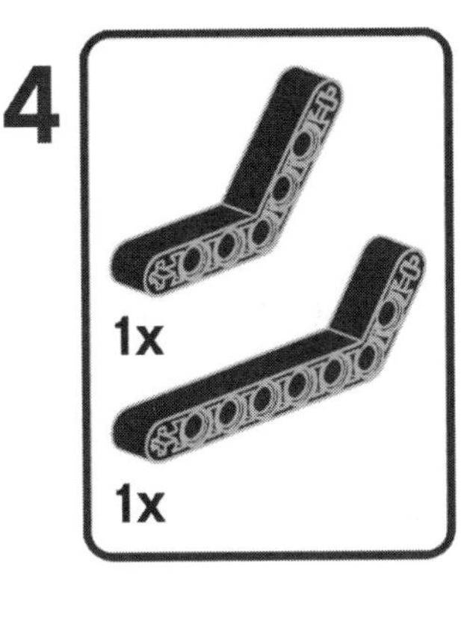

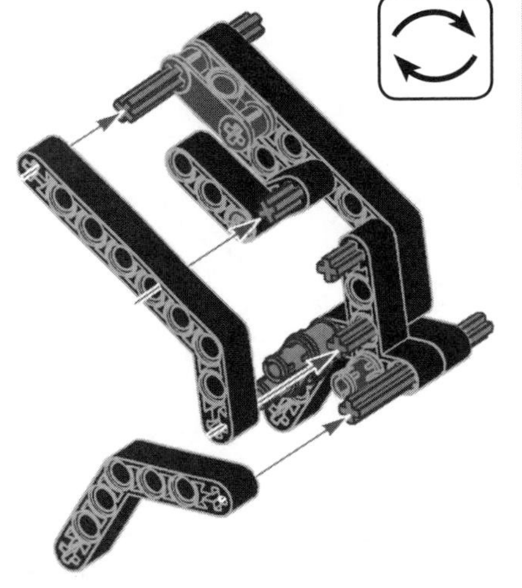

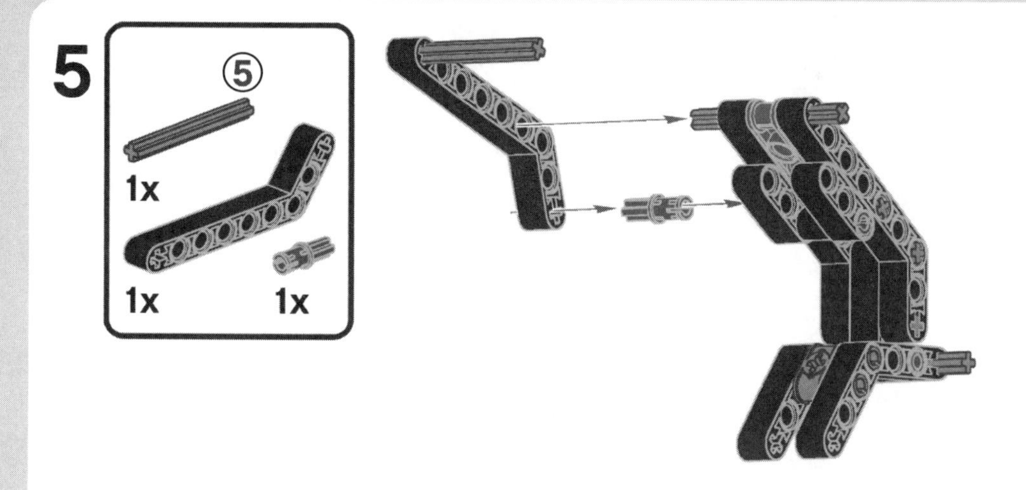

continuing the back shield assembly

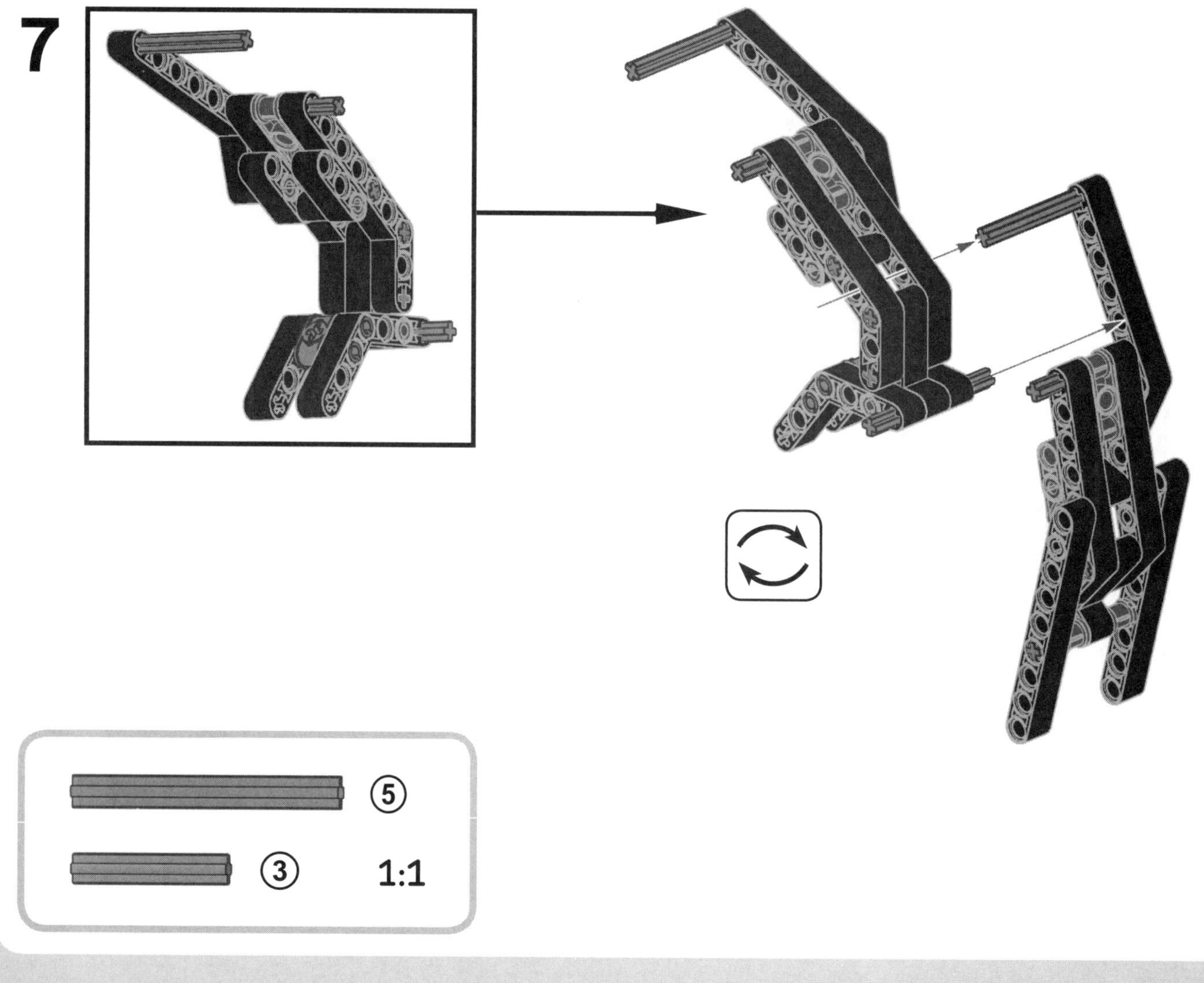

8

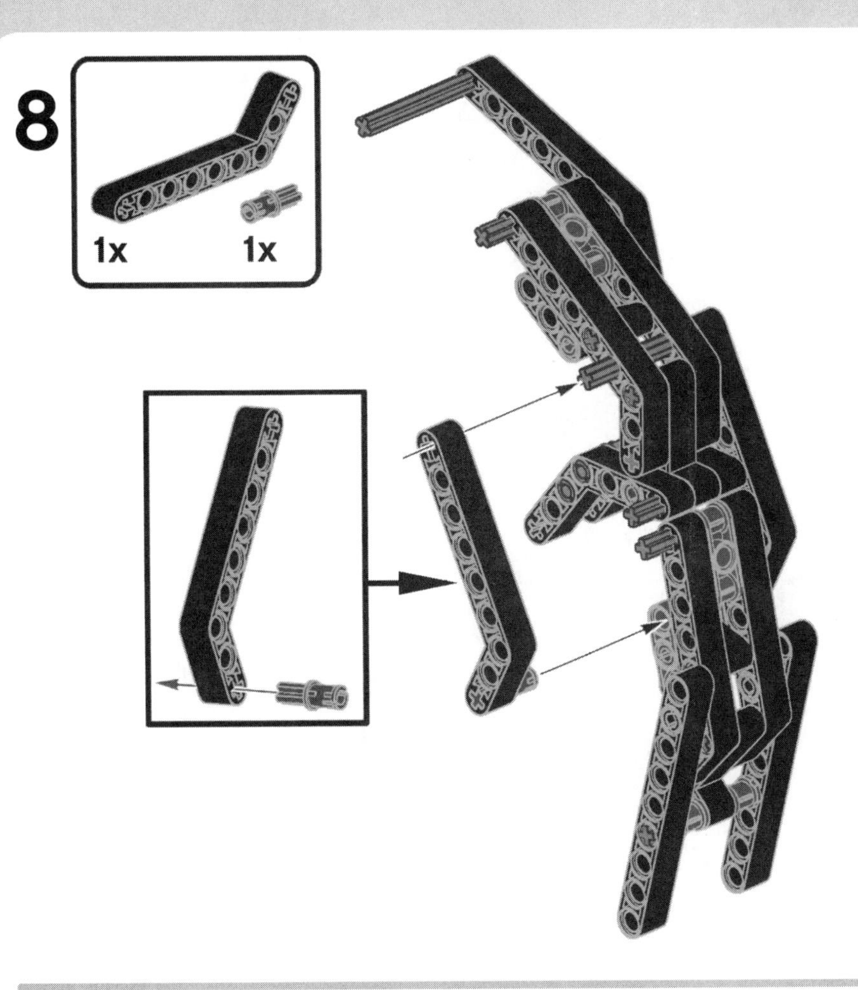

head subassembly

1

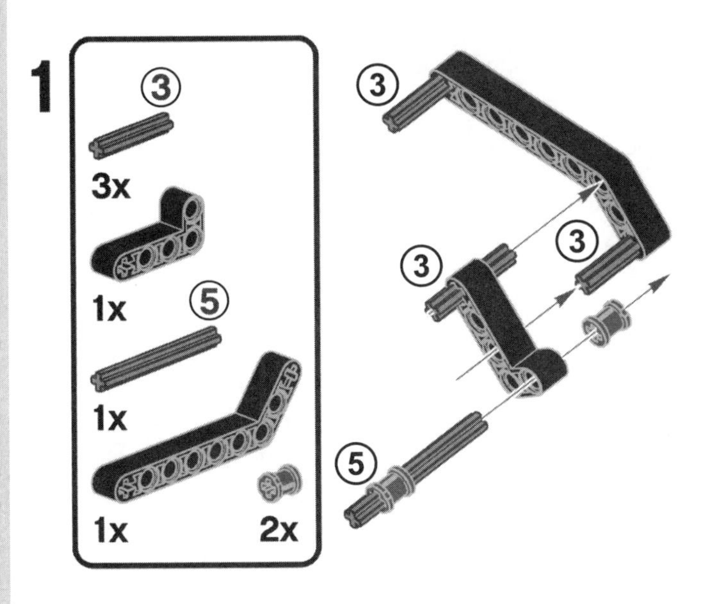

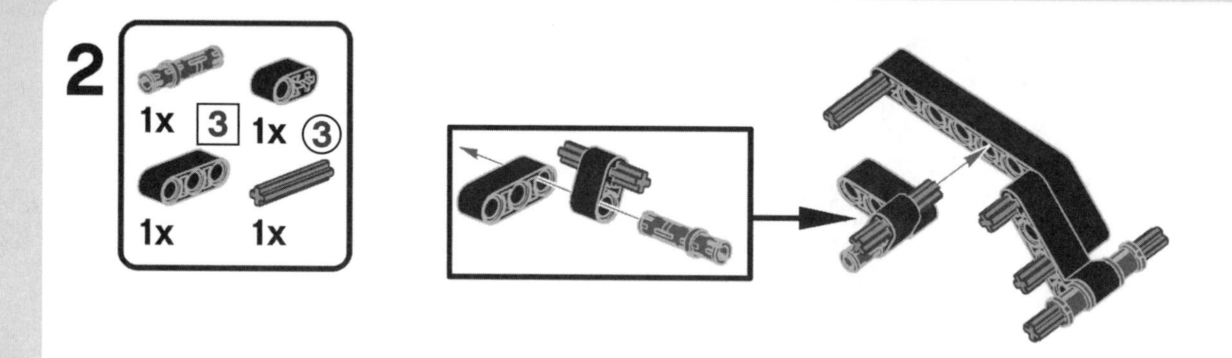
2
1x
3
1x
3
1x
1x

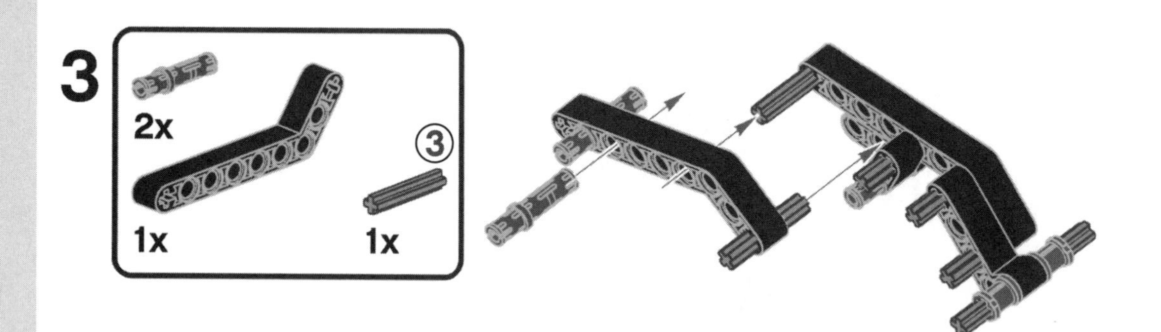
3
2x
3
1x
1x

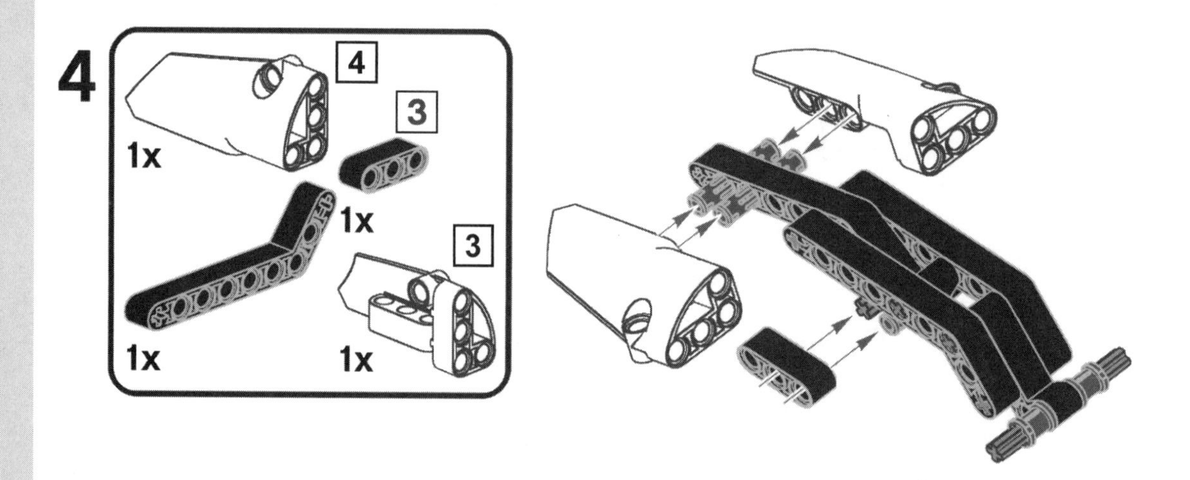
4
4
1x
3
1x
3
1x
1x

3
3
1:1

5

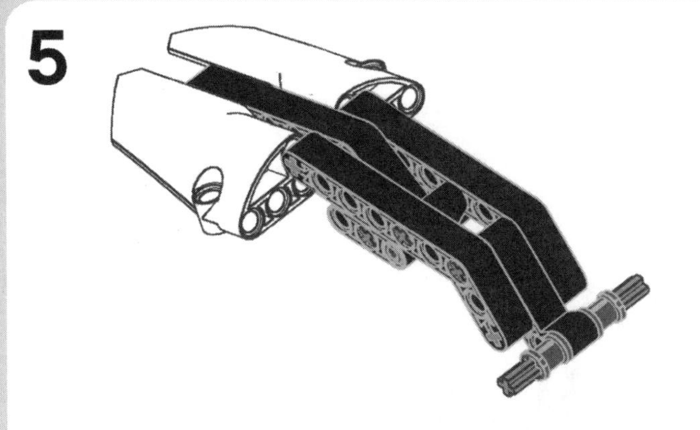

completing the back shield assembly

9

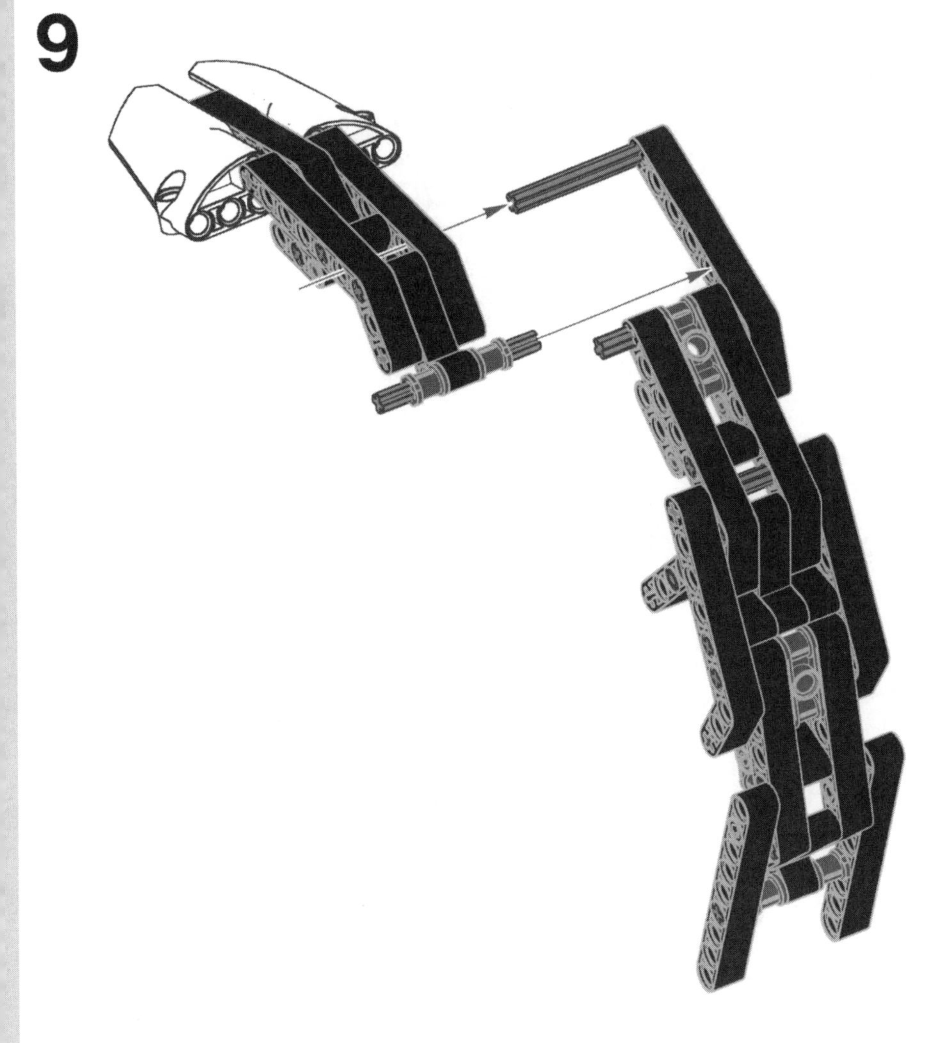

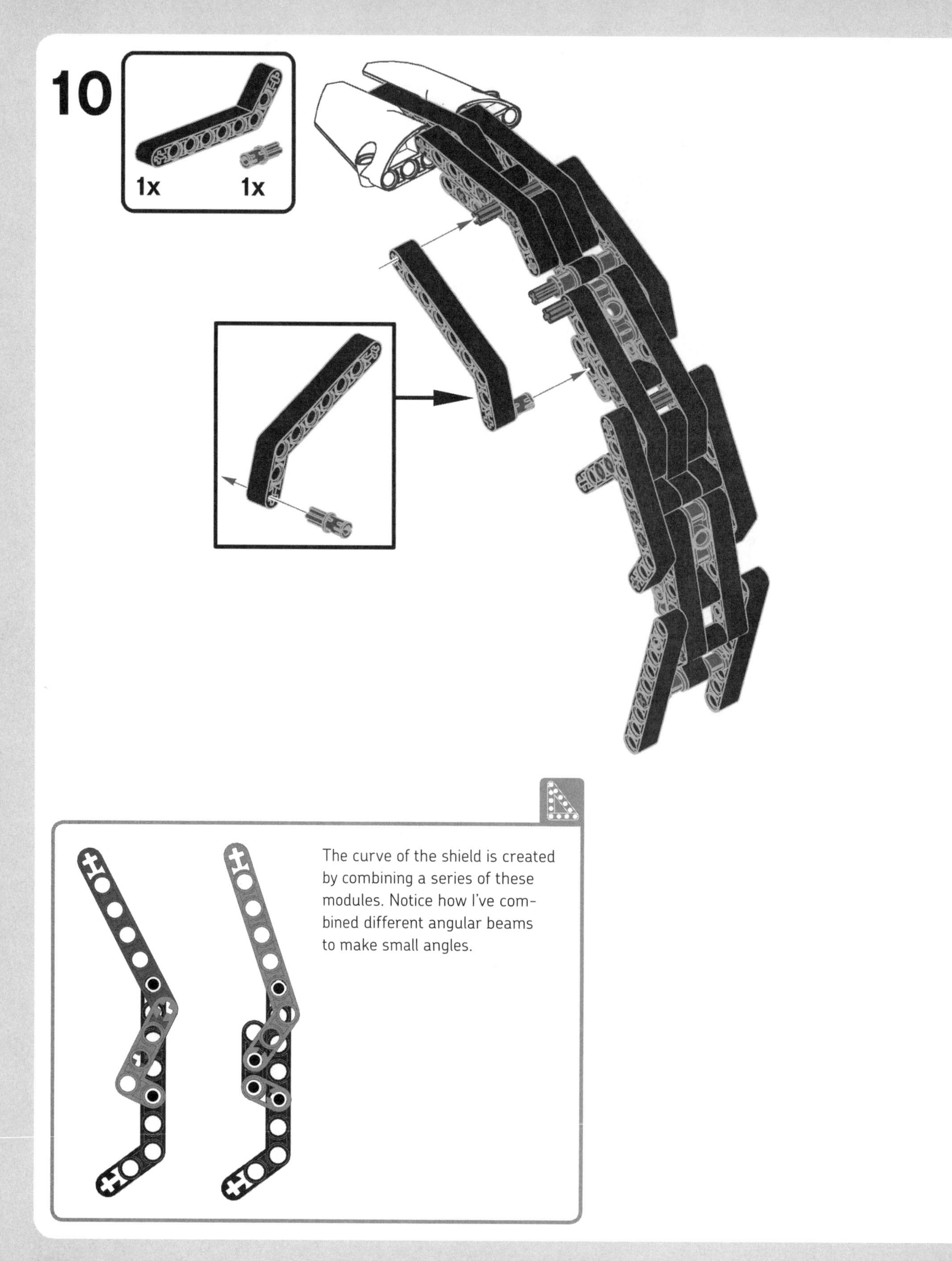

The curve of the shield is created by combining a series of these modules. Notice how I've combined different angular beams to make small angles.

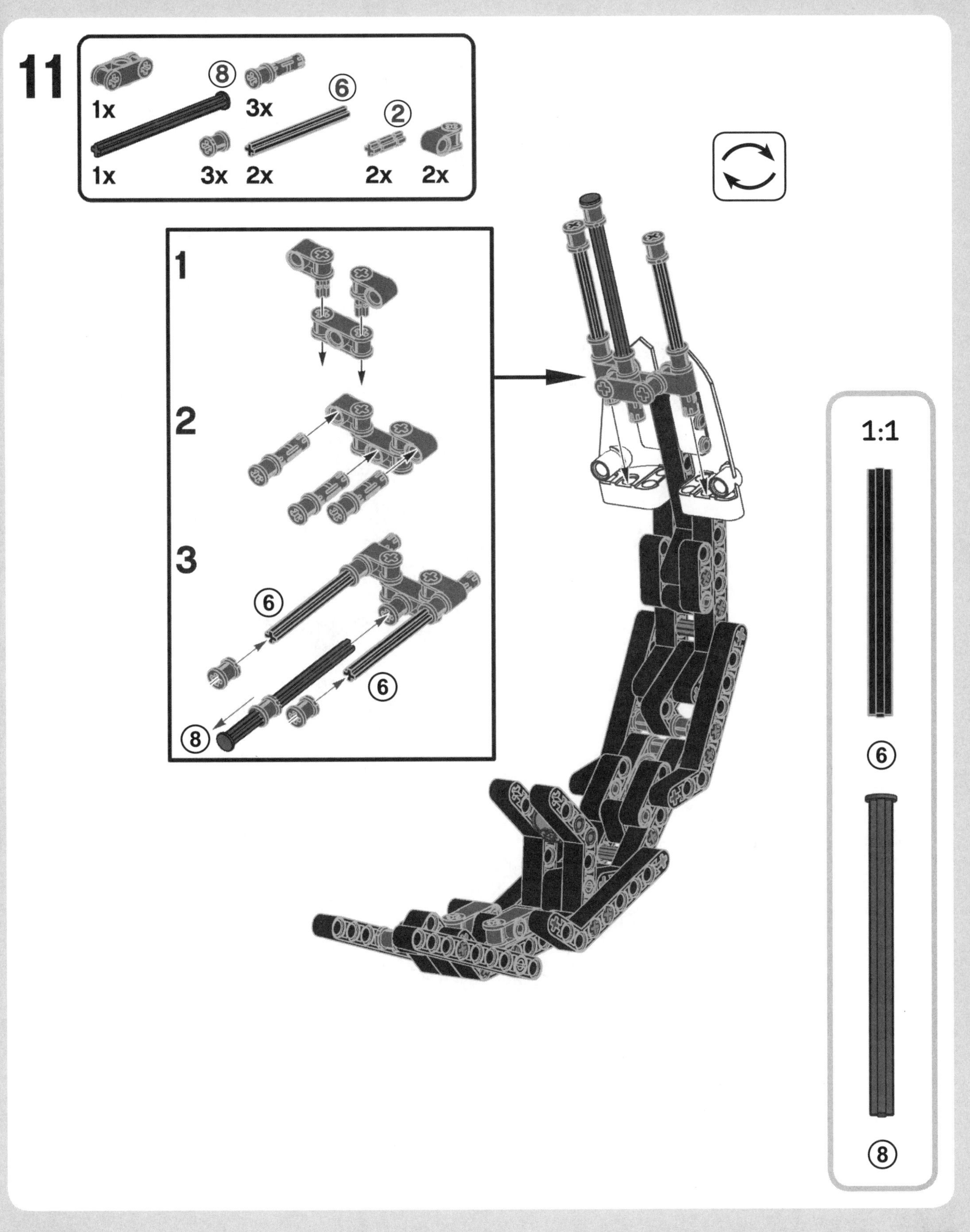
11
1x
8
3x
6
2
1x
3x
2x
2x
2x
1
2
3
6
6
8
1:1
6
8

main assembly

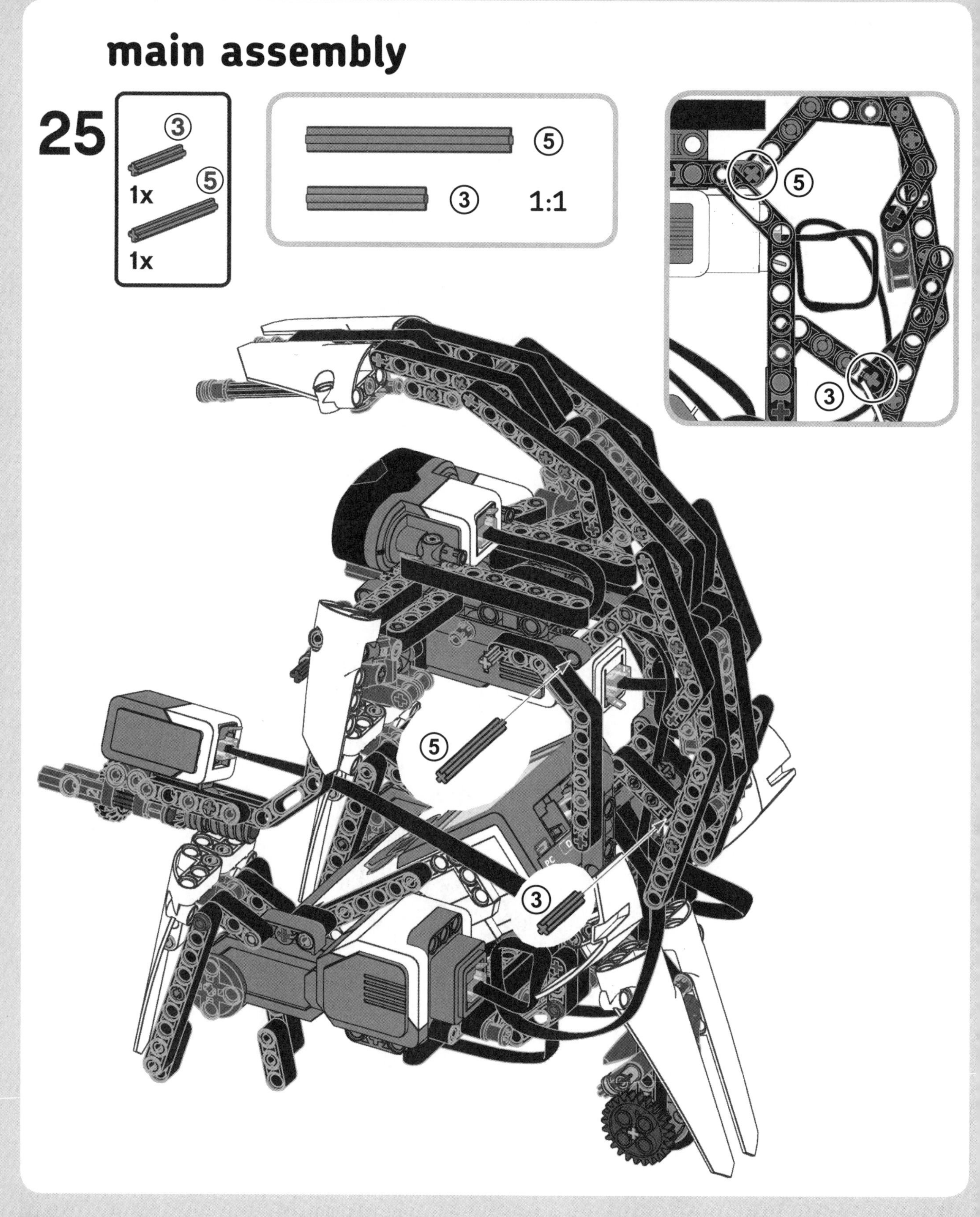

26

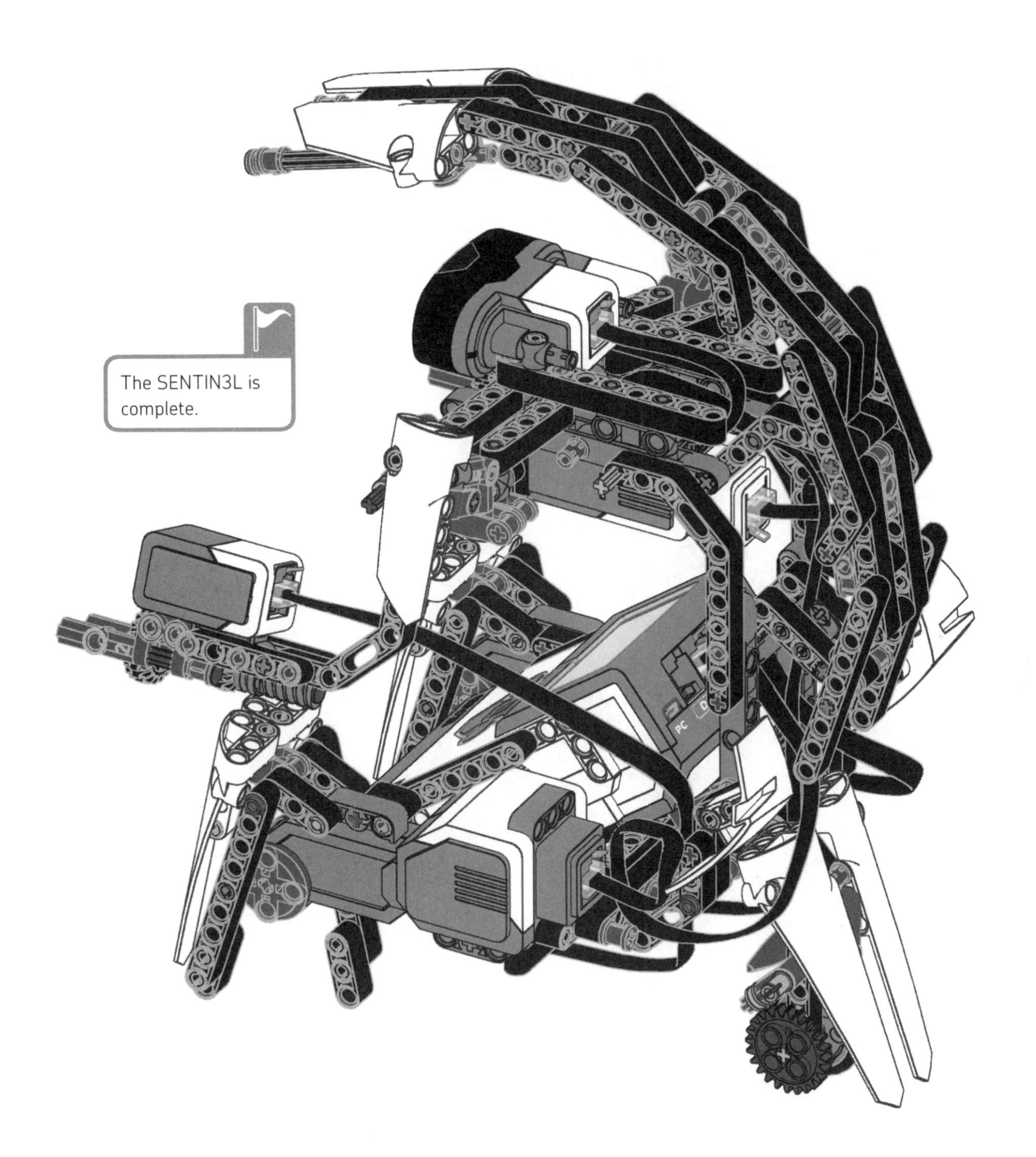

building the COLOR CUB3

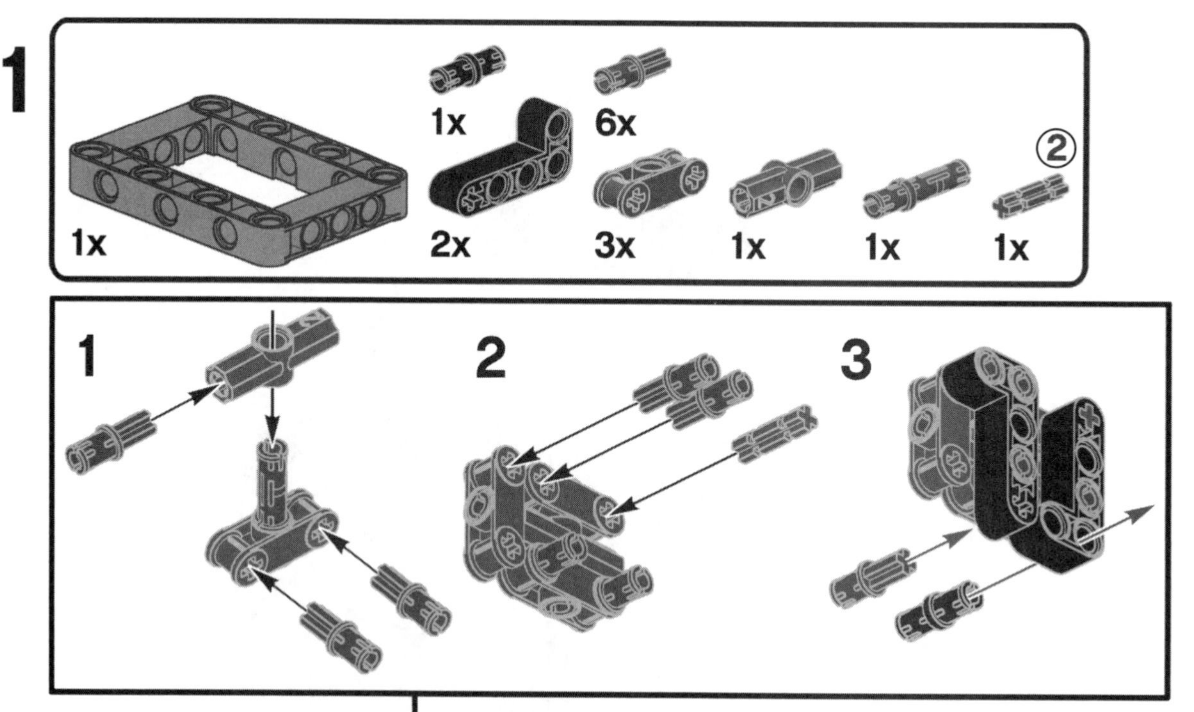

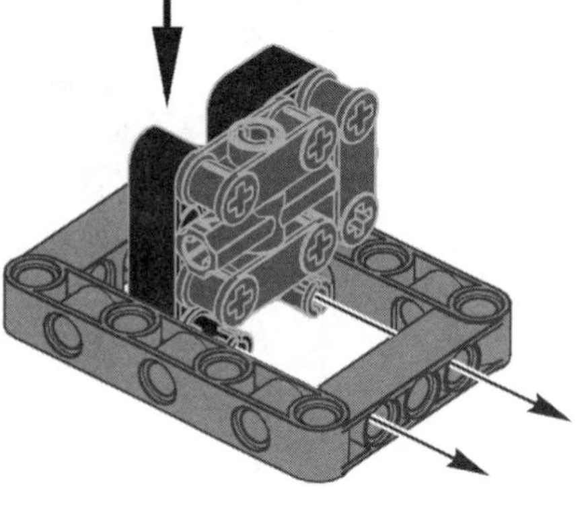

You will use the COLOR CUB3 to program sequences of actions by showing the colored sides to the SENTIN3L. This is the red face of the CUB3.

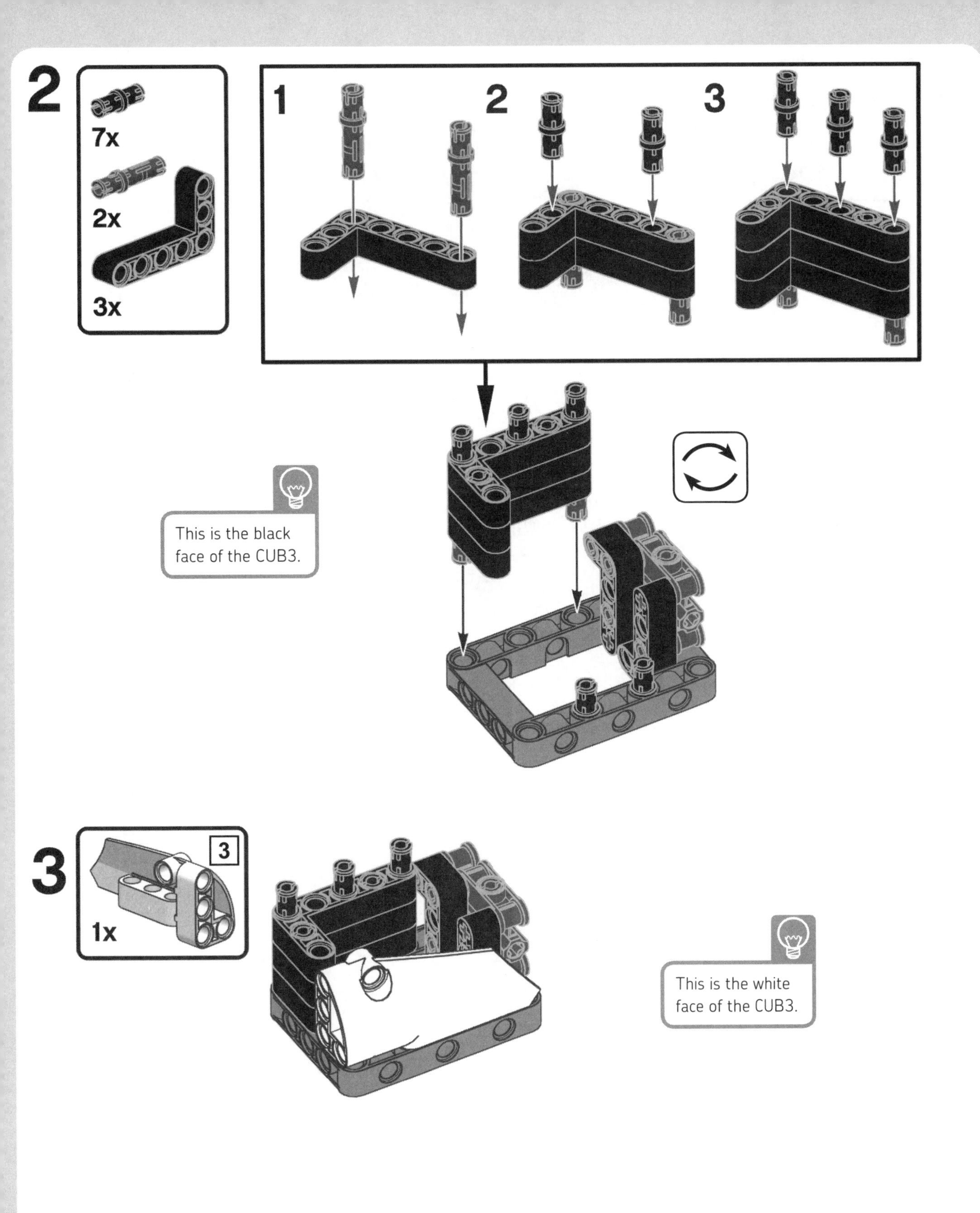
2
7x
2x
3x
1
2
3
This is the black face of the CUB3.
3
3
1x
This is the white face of the CUB3.

4

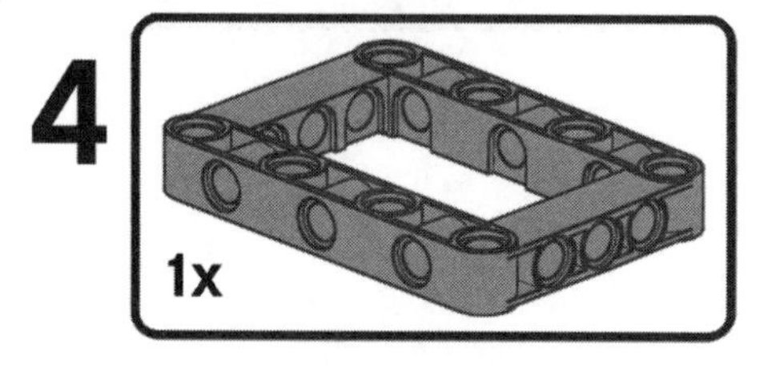

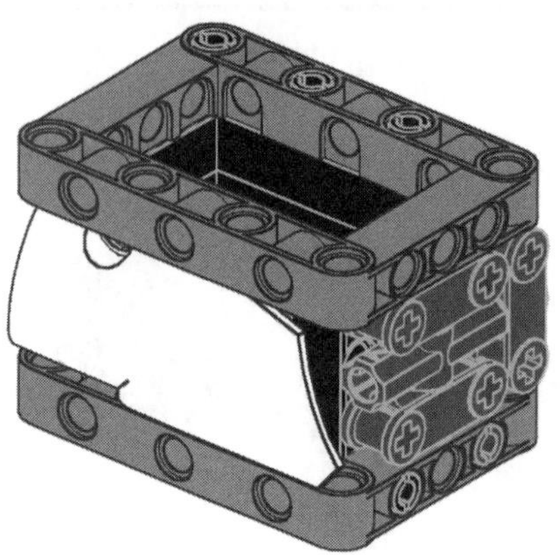

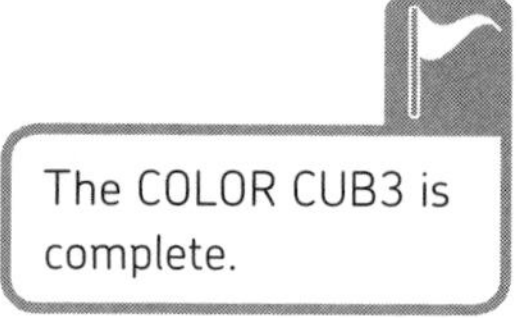

The COLOR CUB3 is complete.

conclusion

In this chapter, you built the SENTIN3L, a security robot that walks on three legs and is equipped with twin blaster cannons. (Well, actually it has four legs and a back wheel, but it *looks* like it's walking on three legs.) Since the motor's continuous rotation is transformed into walking motion, there's no need for special sequences in the program (as with WATCHGOOZ3), so you can easily program and remotely control the SENTIN3L as you would a wheeled robot.

In the next chapter, you'll learn how to make programs that allow you to record, save, and replay simple sequences of actions.

SO GOOD!
BURP!
BLESS YOU! GLAD YOU LIKED IT!
NOW THAT WE'VE HAD LUNCH TOGETHER, YOU CAN CALL ME DANNY— NO MORE "MISTER" OR "DOC."
THANKS, DOC ...ER...DANNY!
AND YOU CAN CALL ME SIMPLY "LORD DESTROYER OF VEGAMECHANIC COLONIES!"
WHAM!
SPLAT!
HA! HA! HA!
"DEXTER" WILL HAVE TO DO!
I'M SORRY FOR ALL THE CHORES. BUT I HAD TO FIND OUT IF I COULD TRUST YOU FOR THE NEXT MISSION!
WHAT WILL I BUILD?
ENOUGH WITH BUILDING FOR TODAY! *IT'S TIME TO DESTROY!* WHAT I'M TELLING YOU MUST REMAIN A SECRET BETWEEN US. PROMISE!
I PROMISE! SPIT IT OUT!
13X1

SOME TIME AGO, SOME PROTOTYPE DEFENSE ROBOTS WERE ACTIVATED...
...BUT WITHOUT ANY IMPRINTING!
THEY DON'T RECOGNIZE HUMANS AS THEIR MASTERS! AS A RESULT, A SECTOR OF THE EVOLUTION 3 LAB WAS PUT INTO TOTAL QUARANTINE, LOCKING THOSE ROBOTS INSIDE. WE MUST CLEAR IT!
$> Downloading imprinting pattern...
ERROR
YOUR FATHER DOESN'T KNOW, AND IT NEEDS TO STAY THAT WAY!
DOC...WHAT IF WE DON'T SUCCEED?
WE MUST SUCCEED! MY BROTHER ALEX IS TRAPPED INSIDE!
13X2

14

programming the SENTIN3L

In this chapter, you'll program the SENTIN3L, which you built in Chapter 13. The first program will simply make the robot patrol and shoot at objects that are in its way. Once you've mastered that, we'll make a program to record a sequence of actions at runtime by showing the robot the colored sides of the COLOR CUB3 from Chapter 13. The final program will also allow you to program the robot at runtime, but it will permanently save the recorded sequence to a file for later retrieval, even once the program has terminated or the EV3 Brick reboots.

the file access block

To create, write, read, and delete files in the EV3 Brick's memory, we'll use the File Access block, found in the Advanced palette (blue header). Figure 14-1 shows this block in its various modes.

creating and deleting a file and writing data

To create a file with a particular name, use the File Access block in **Write** mode and enter the filename in the File Name field (or select the Wired option to carry the File Name text with a Data Wire).

If the file with the name you specify does not exist yet, it will be created in the EV3 Brick's memory the first time you open the file for writing. If the file already exists, the data is appended at the end of it.

To be sure that you're writing to a new, empty file, first use the File Access block in **Delete** mode, specifying in the File Name field the filename you are going to write to.

For example, to write to a file named *log* (not appending data to it but writing data from the beginning), use the File Access block in Delete mode to delete *log* and then use another File Access block in Write mode to create and write to *log*.

To write data to a file, you can enter some text in the block Text input, or you can plug a Text or Numeric Data Wire into the Text input. If you plug in a Data Wire that carries a numeric value, that number will be converted into its textual representation. Each time you write to the file, the text string will be added on a new line. To verify this, use the Memory Browser tool to upload the file created by the File Access blocks from the EV3 Brick to your computer and then open the file with your favorite text editor.

NOTE When you've finished writing to a file, you must use the File Access block in *Close* mode to be able to read data back from that file.

WARNING Files (including programs) are temporarily kept in the RAM during a session and are saved to the flash memory only when the EV3 Brick is shut down. If the EV3 Brick is not shut down properly, you will lose the files created in that session. Likewise, if you update the EV3 Brick's firmware, you will lose any files from that session, so remember to back up your files on your computer using the Memory Browser tool before updating the firmware.

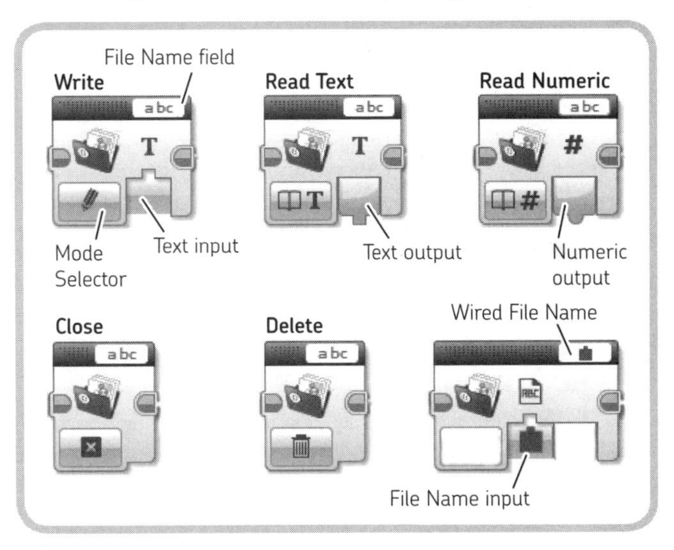

Figure 14-1: The File Access block allows you to create, delete, read, and write files to the EV3 Brick's memory.

reading data from a file

You can read data from a file using the **Read Text** and **Read Numeric** modes of the File Access block. In Read Numeric mode, the block tries to convert the text data into a numeric value: If the file contains text strings that do not represent a number, the value returned by the block is 0.

The block reads data from the file in sequence, one line at a time. In practice, this means that, for example, if you store three values to a file, you can read them back using three File Access blocks in Read mode, one after another.

To read data from the beginning of a file, use a File Access block in **Close** mode to close the file and then start reading again.

detecting the end of a file

If you keep reading from a file once you've reached the end, the block will return empty text strings or zero numeric values. In other words, you'll know that you've reached the *End Of File (EOF)* when the File Access block in Read Text mode reads an empty text string.

Alternatively, to mark the EOF when writing numbers to a file, you can choose to use a termination number (for example, –1 in a file that contains only positive numbers). When you read the numbers from that file (using Read Numeric mode), you'll know that you've reached the EOF when you read that termination number.

the random block

Later in this chapter, you'll learn to program the SENTIN3L to shoot a random number of times and then turn in a random direction when it detects an obstacle. To generate random numeric and logic values, you can use the Random block (Figure 14-2), found in the Data Operations palette (red header).

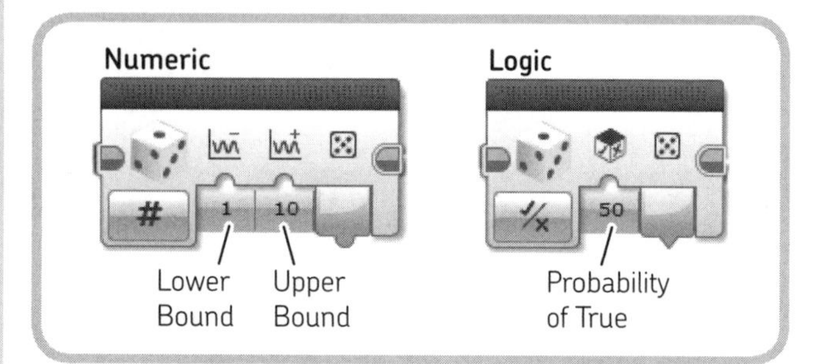

Figure 14-2: The Random block generates random numeric or logic values.

When generating random values in Numeric mode, the random value will be an integer in the range specified by the Lower and Upper Bound parameters. The Lower and Upper Bounds should be integers, too; if set to decimals, they will be truncated internally to the nearest integer. For example, if the bounds are set to 1 and 3, the generated values can be the integers 1, 2, and 3. If the bounds are set to 1.5 and 4.4, the generated values can be the integers 1, 2, 3, and 4.

The generated random numbers follow a *uniform distribution*. Each value is as equally likely to be drawn as another. For example, if you generate 6,000 random numbers within the bounds 1 and 6 (like rolling a die), each of the six possible values will be drawn about 1,000 times; that is, each value has a 1/6 or about 16 percent probability of being drawn.

When in Logic mode, the generated random value can be either True or False. You can use the input parameter (a percentage value ranging from 0 to 100) to set the probability that the True value will be generated. For example, when the *Probability of True* is set to 50, there is a 50 percent chance of getting the True value and a 50 percent chance of getting False (as when tossing a coin).

building the My Blocks

Before creating the programs that will bring the SENTIN3L to life, you'll need to prepare the My Blocks. As you build the My Blocks, I'll highlight noteworthy sequences and the ideas behind them. For each My Block, I'll show you the sequence of programming blocks and the image of the complete My Block so that you can check its icons, inputs, and outputs as well as the default values of the inputs.

the ResetLegs My Block

The building instructions in Chapter 13 show all of the robot's front legs touching the ground. For the robot to walk efficiently without wobbling, its front legs must all touch the ground at the beginning of the program and be kept in sync as the robot moves. The *ResetLegs* My Block shown in Figure 14-3 resets the position of the legs so they all touch the ground.

To determine whether a pair of legs is touching the ground, we run the motor that drives a pair of legs at low speed using the Unregulated Motor block. The point at which the motor encounters the most resistance (that is, when it is unable to lift the robot) is when both legs in a pair are on the ground. The motor will almost stop, and its *Current Power* will be nearly zero, as measured by the Motor Rotation block in Measure Current Power mode.

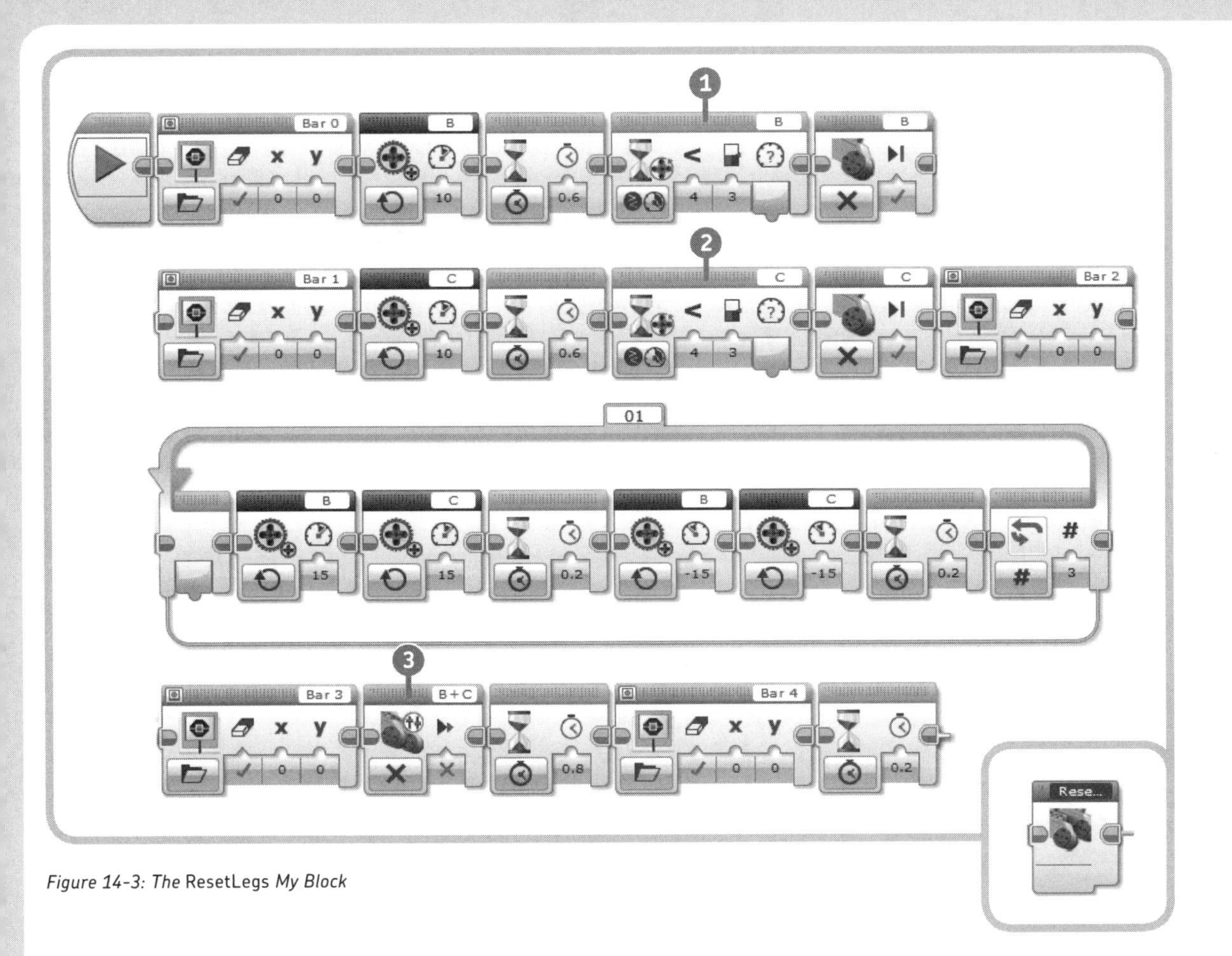

Figure 14-3: The ResetLegs *My Block*

DIGGING DEEPER: HOW "POWER" RELATES TO SPEED

The name of the **Measure Current Power** mode of the Motor Rotation block is misleading. In fact, the block does not measure the current flowing through the motor coils (in amperes) or the absorbed power (in watts); it measures the current speed of the motor in degrees per 100 milliseconds. (To convert the motor speed into degrees per second, multiply the Current Power value by 10. For example, a Current Power value of 10 means the motor is turning at 100 degrees per second, and a value of 40 means the speed of the motor is 400 degrees per second.)

Similarly, the Power parameter that you set for the Move and Motor blocks determines the speed of the motor and is expressed in degrees per 100 milliseconds. For example, setting a Power of 20 percent will make the motor turn at 200 degrees per second.

The Move and Motor blocks are regulated, and they drive the motors so as to keep their speed constant, even when you apply a load (a resisting torque) to the motor shaft. In contrast, the Unregulated Motor block sets the internal power level but does not guarantee that the corresponding motor's speed will remain constant under load. (For more on the feedback regulation of motors, see "Digging Deeper: Motor Speed Regulation" on page 188.)

In my tests, I found that the Large Motor cannot run faster than 850 degrees per second (rather than the 1,000 degrees per second that you would expect to measure based on multiplying a 100 percent power level by 10). The Medium Motor is faster and can reach 1,000 degrees per second, but its regulation is not very precise at low speeds.

In the *ResetLegs* My Block (Figure 14-3), each Large Motor is run at low speed until a Wait block in **Motor Rotation Compare Current Power** mode (1) and (2) detects a Current Power of less than 3 (meaning that the motor is almost still); at that point, the motor is stopped. Next, to be sure that the legs are touching the ground, the motors are moved back and forth a few times and then left floating by the Move Tank block in **Off** mode with *Brake at End* set to **False** (3). The motors can rotate freely when stopped without the electric brake, and the weight of the robot will level the legs on the ground.

During this entire reset process, a progress bar is displayed on screen using five Display blocks.

the WalkFWD My Block

This My Block (Figure 14-4) makes the robot walk forward until an object is detected by the IR Sensor at the proximity specified by the input. Notice that the Move Tank block inside the loop (in *On For Rotations* mode) turns the motors in steps of half a rotation each to keep the legs synchronized and to make sure the legs will touch the ground at the end of each step. Build the My Block with a numeric input called **Proximity**, Parameter Style set to **Vertical Slider**, Min set to **10**, Max set to **80**, and Default value set to **30**.

the Laser My Block

The *Laser* My Block (Figure 14-5) activates the Medium Motor that moves the robot's arms in sync with a laser audio effect. The alternating motion of the arms simulates the recoil of twin cannons firing. A Random block randomly chooses how many times the loop repeats.

The Color Sensor blocks are used to change the Color Sensor's RGB LED color from blue to red when the robot is shooting. The RGB LED can emit different colors when set in the various measurement modes: In Reflected Light Intensity mode, the LED is red; in Ambient Light Intensity mode, it's blue; and in Color mode, it uses red, green, and blue to produce a kind of pale purple light.

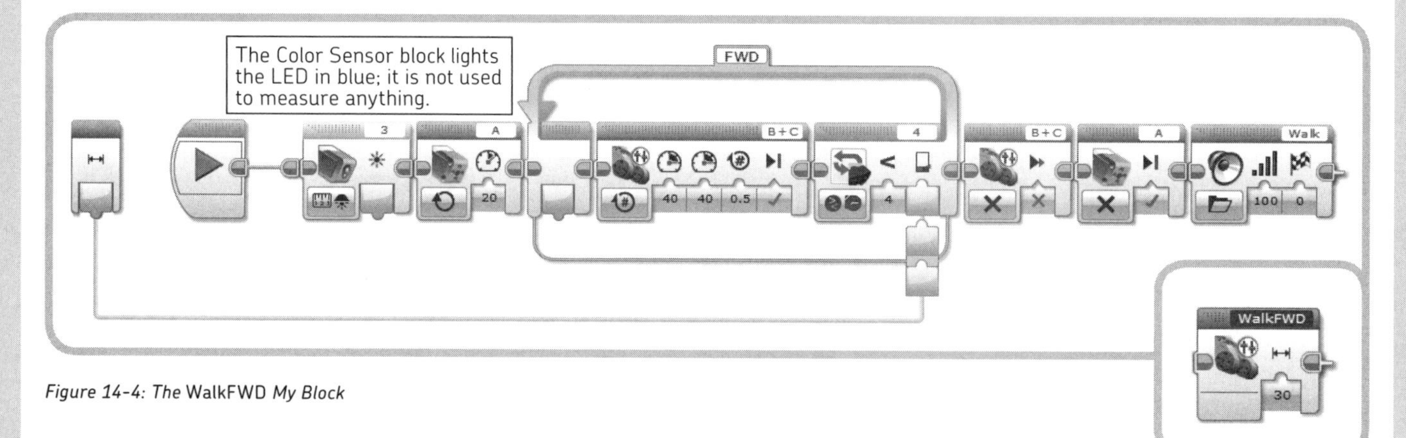

Figure 14-4: The WalkFWD My Block

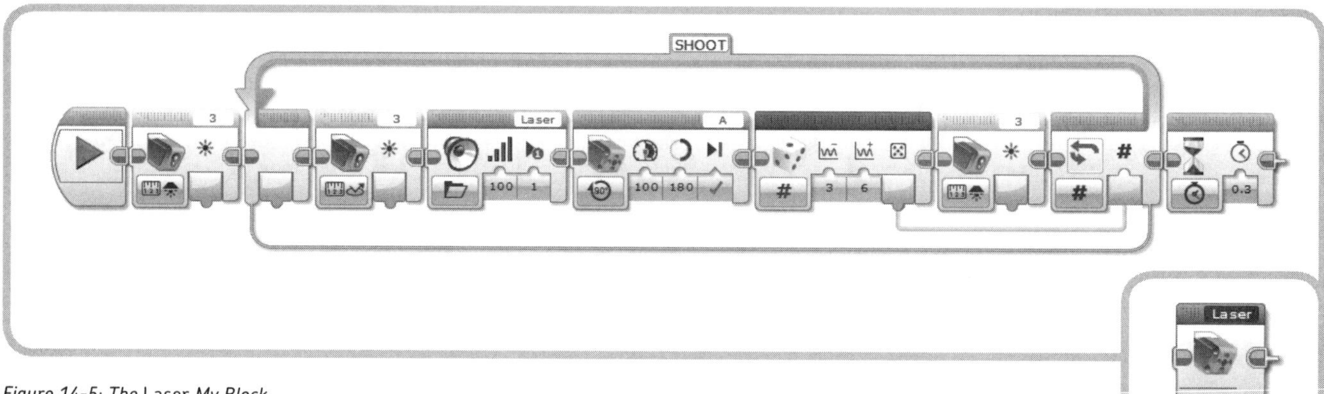

Figure 14-5: The Laser My Block

the Turn My Block

The *Turn* My Block (Figure 14-6) makes the robot turn in place in a random direction. The robot turns until it sees an obstacle nearer than the *Proximity* specified by the input and until at least the number of seconds specified by the *Time* input has elapsed.

First, a Random block in Logic mode generates a logic value with a 50 percent chance of being True. A Math block in Advanced mode transforms the logic values into Steering values for the Move Steering block using the formula **a*(1-2*b)**, with **a** set to 100 as constant input. When **b** is True, the result of the Math block is -100; when **b** is False, the result is 100.

The Move Steering block turns the motors in opposite directions by a half rotation for each iteration of the loop; this makes the robot turn in place. (The Steering value is 100 or -100.) The loop ends (and the robot stops turning) when the Proximity value measured by the IR Sensor block is greater than the *Proximity* input of the My Block and the time elapsed since the reset of Timer 1 is more than the *Time* input of the My Block. This method makes the robot steer for a minimum number of seconds even if an obstacle is removed at once, and it is similar to the one used in the *StepAdv* My Block for WATCHGOOZ3 (Figure 10-18 on page 186).

Create the My Block with two numeric inputs: **Proximity** (Vertical Slider, Min 20, Max 80, Default 30) and **Time** (Numeric input, Default 2).

the PowerDownFX My Block

This My Block plays a *Speed Down* sound while displaying a progress bar that falls to zero to simulate the robot losing power and shutting down (Figure 14-7).

the WaitButton My Block

The *WaitButton* My Block waits until the Enter button of the EV3 Brick or the Touch Sensor is pressed, as shown in Figure 14-8.

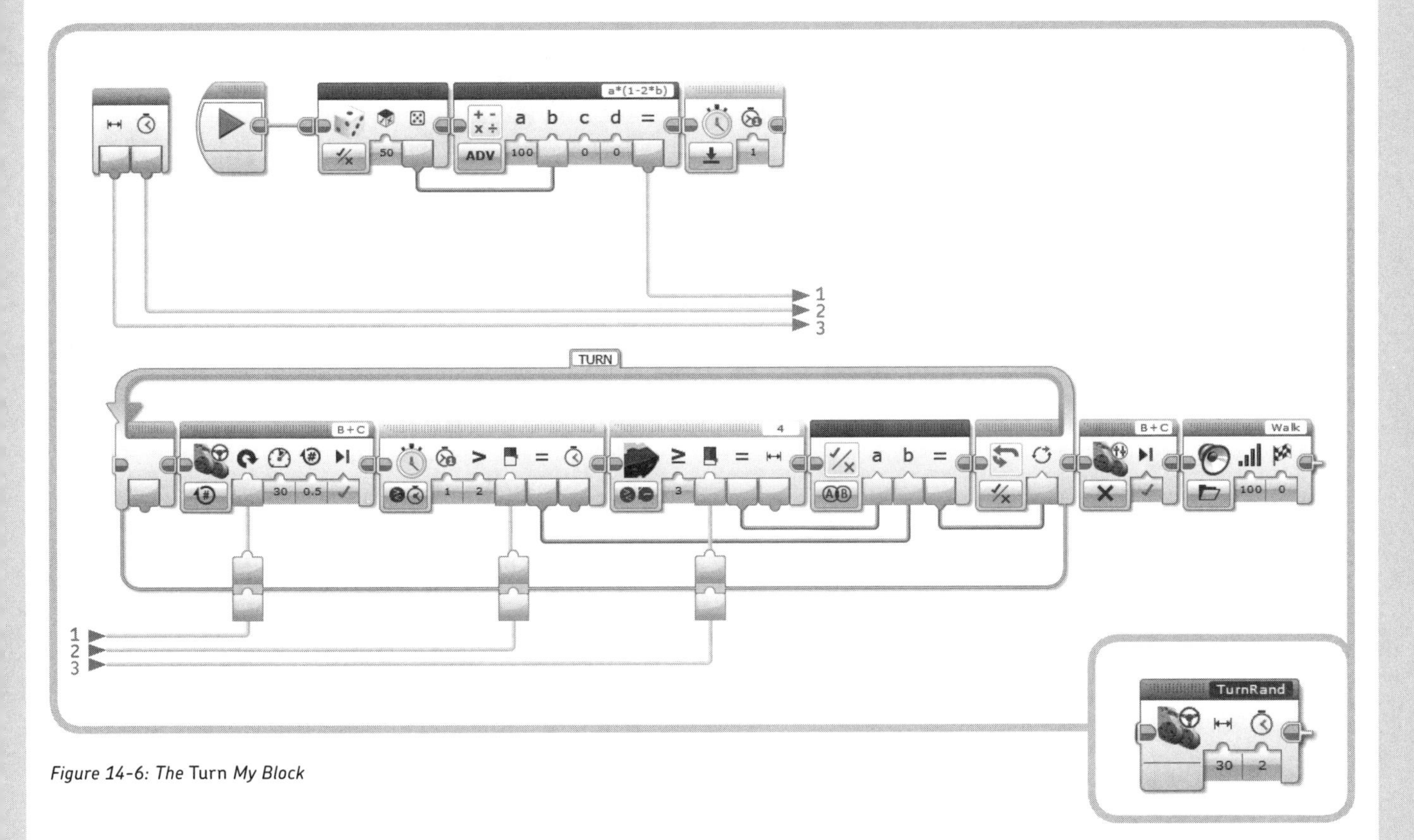

Figure 14-6: The Turn *My Block*

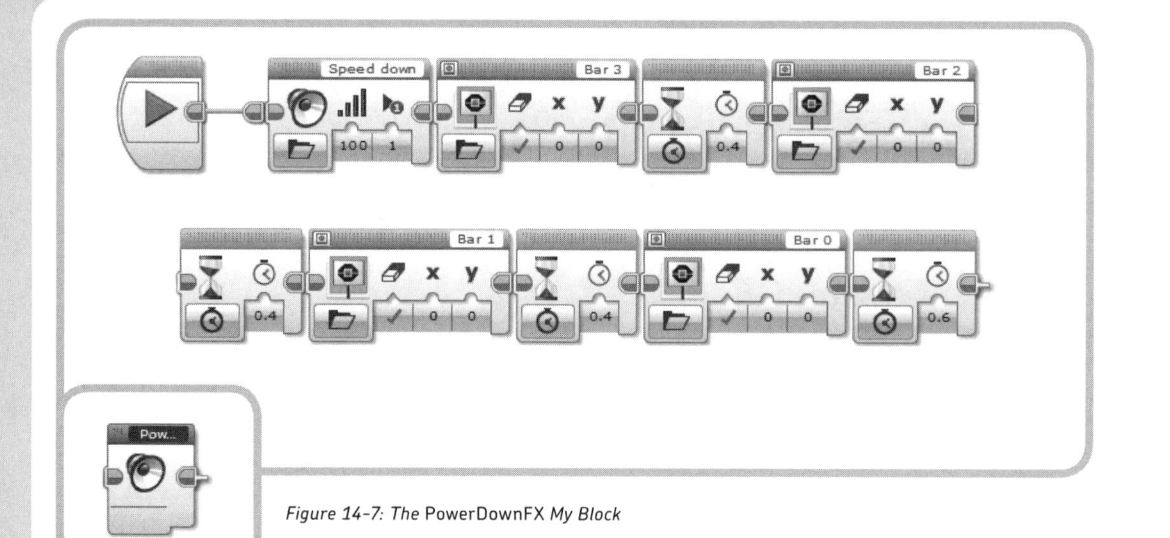

Figure 14-7: The PowerDownFX *My Block*

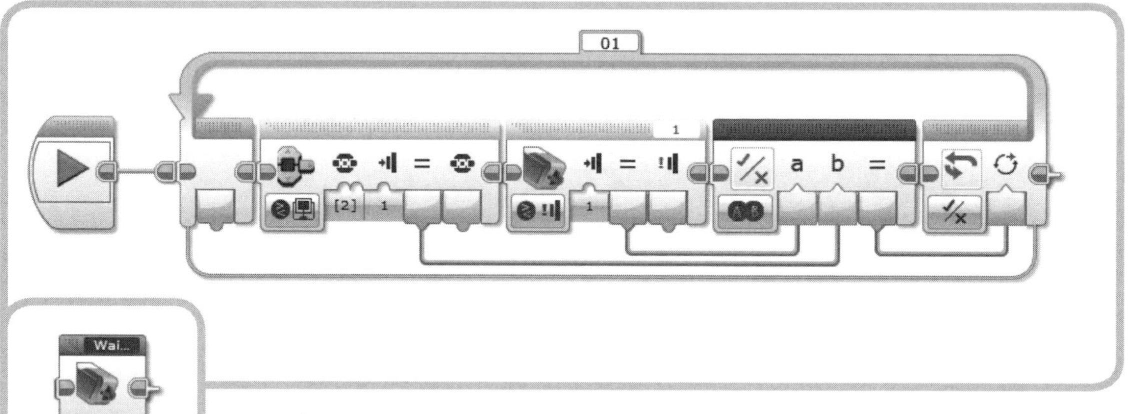

Figure 14-8: The WaitButton *My Block*

the SayColor My Block

The *SayColor* My Block (Figure 14-9) has a numeric input called *Color* that selects which case of the Switch block (in Numeric mode) to execute. The cases are numbered according to the Color Sensor codes (0 for none, 1 for black, 5 for red, and 6 for white). When the case corresponds to the color present on the COLOR CUB3, a Sound block plays the corresponding audio files: *Error*, *Black*, *Red*, or *White*.

the ExeCode My Block

The *ExeCode* My Block (Figure 14-10) executes an action according to the value of the numeric input *Code*. This value is used to select which case of the Switch block (in Numeric mode) to execute.

The cases are numbered according to the Color Sensor codes (0 for none, 1 for black, 5 for red, and 6 for white). When the case corresponds to the color present on the COLOR CUB3, a My Block makes the robot perform certain actions. Specifically, case 1 (black) contains a *Turn* My Block with Proximity set to **30** and Time set to **2**, case 5 (red) contains a *Laser* My Block, and case 6 (white) contains a *WalkFWD* My Block with Proximity parameter set to **30**. Other cases are empty.

the MakeProgram My Block

The *MakeProgram* My Block (Figure 14-11) allows you to record a program for the robot at runtime by showing the three colored sides of the COLOR CUB3 to the Color Sensor and entering the color codes by pressing the Touch Sensor. You can add actions to the program corresponding to the colors white (6), red (5), and black (1).

The program is saved as a sequence of numbers in the numeric array A, which is cleared with a Variable block in Write Numeric Array mode by specifying no elements in the drop-down menu ([]). Color codes are appended to the array in a loop

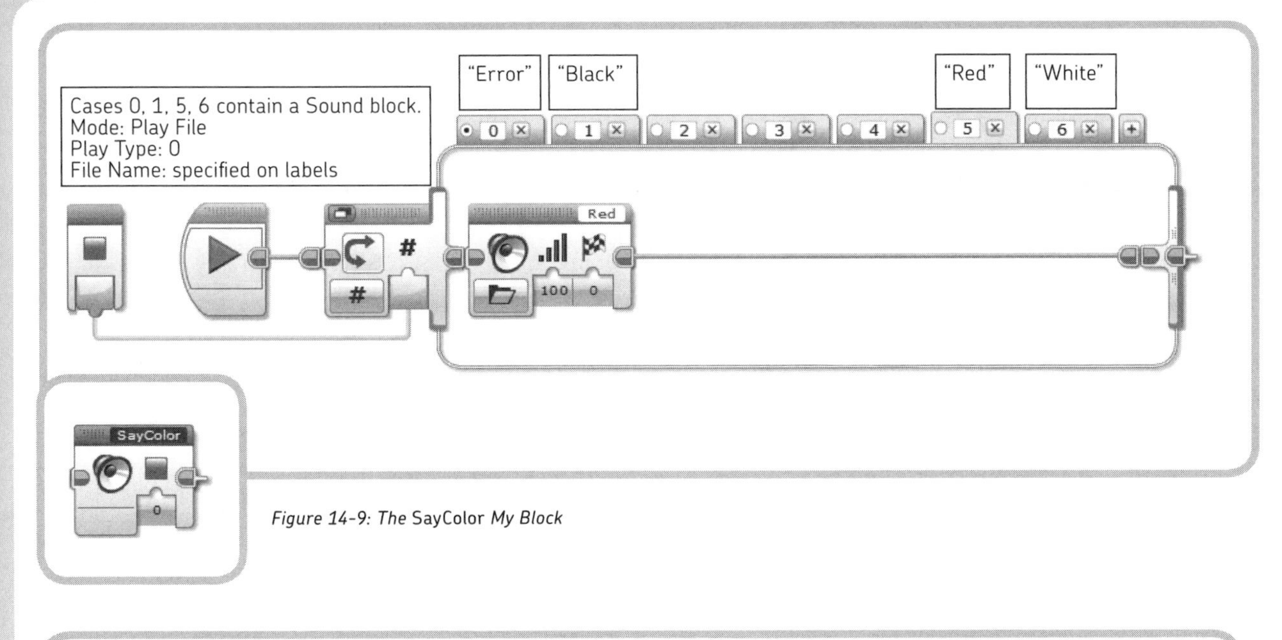

Figure 14-9: The SayColor *My Block*

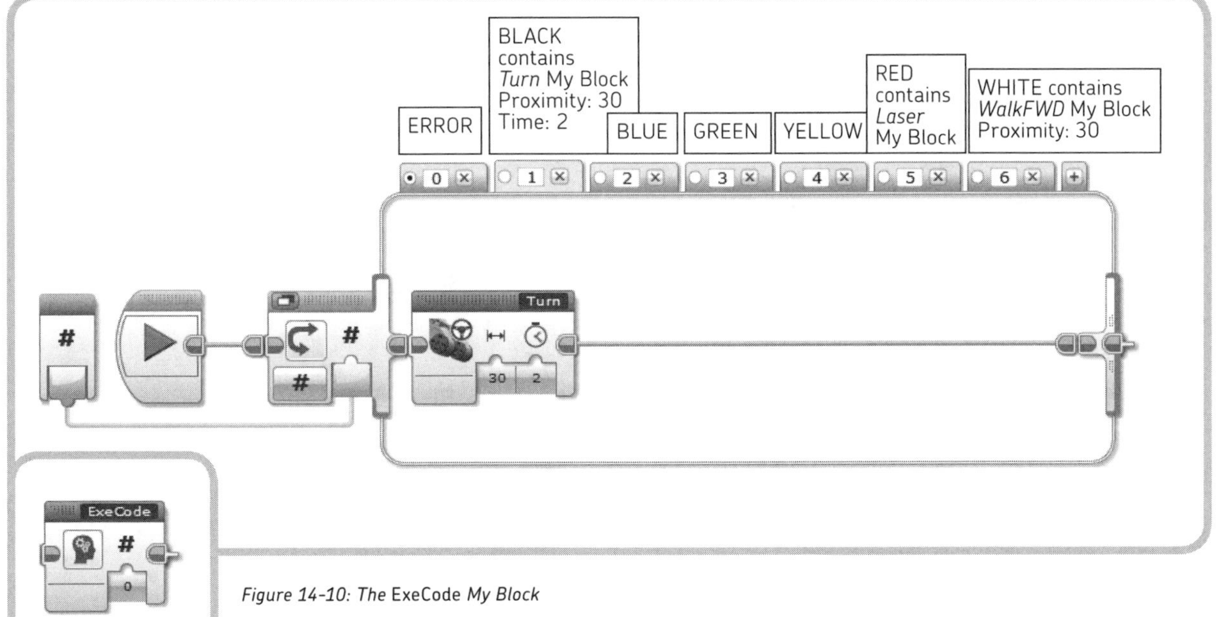

Figure 14-10: The ExeCode *My Block*

that ends when the EV3 Brick's Enter button is pressed. (To review how to use arrays, check Chapter 12.)

The *WaitButton* My Block waits for either the Touch Sensor or the Enter button to be pressed. If the Touch Sensor is pressed, the color shown to the Color Sensor is read, spoken by the *SayColor* My Block, and if it's different from 0 (no color), its value is stored in array A using an Array Operations block in Append mode. The Wait block in Touch Sensor Compare State mode waits for the Touch Sensor to be released in order to avoid adding the color code to the array multiple times (which would otherwise happen if you kept the Touch Sensor pressed).

the RunProgram My Block

The *RunProgram* My Block (Figure 14-12) reads the contents of array A, which contains the sequence of actions programmed with the COLOR CUB3. The Loop Index is used to read from the array. But before reading the array, we must check whether the current index value is less than the array length because if you access an element of the array outside its boundaries, the program will abort. The array is read in the *RUN* Loop block until the index reaches the array length or the EV3 Brick's Enter button is pressed.

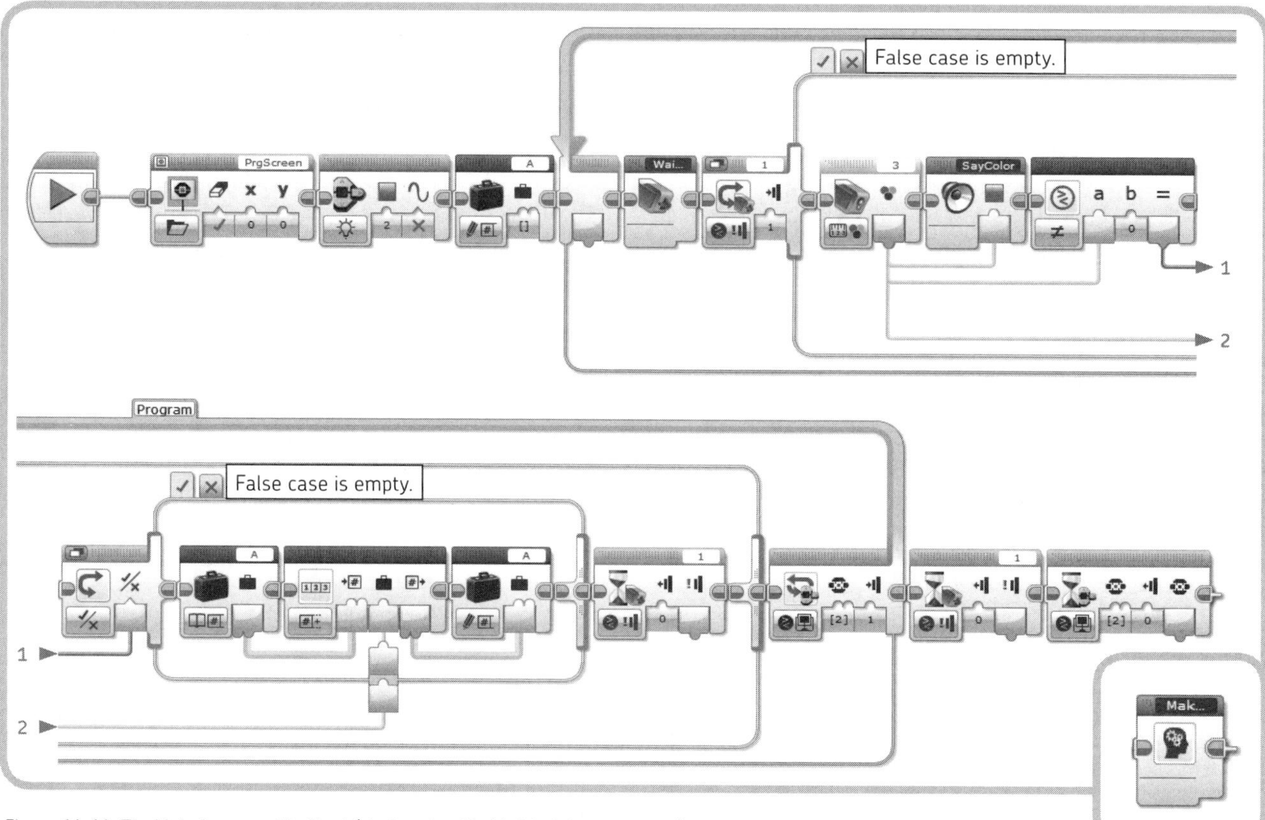

Figure 14-11: The MakeProgram *My Block (the icon for this My Block is a grey head)*

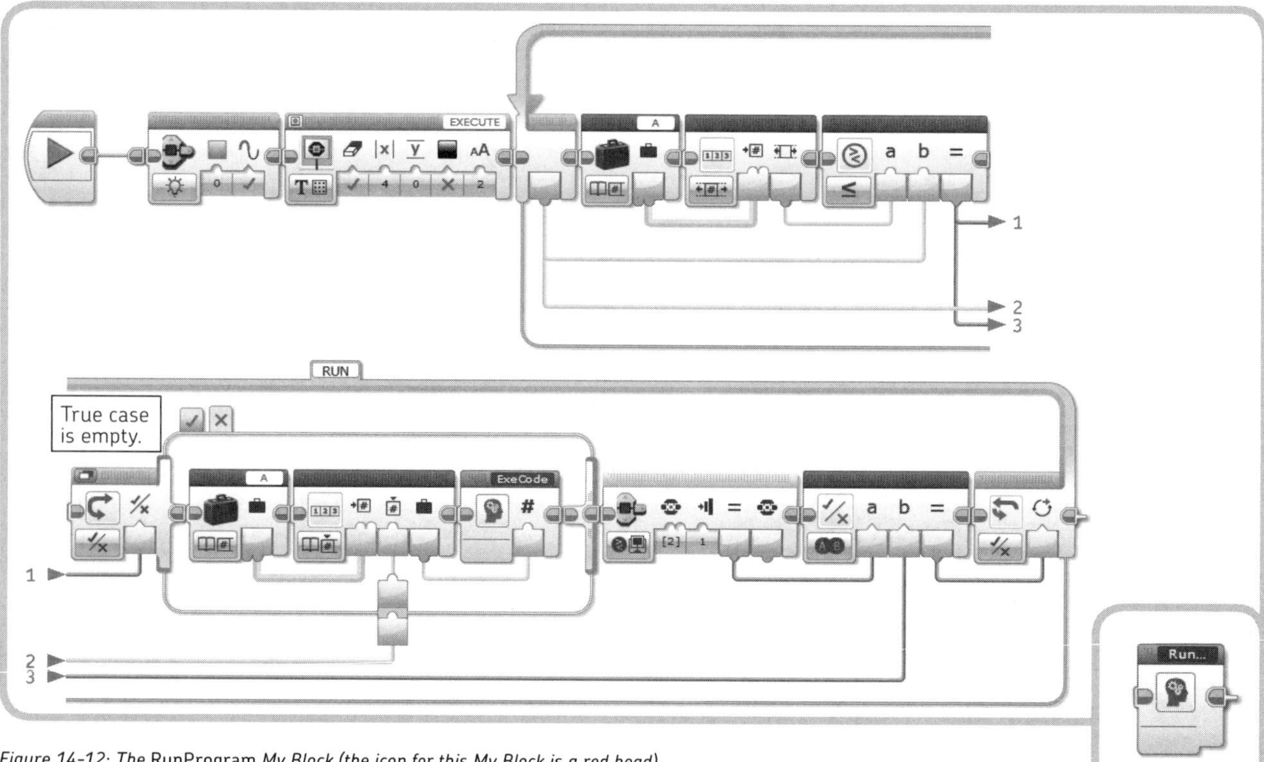

Figure 14-12: The RunProgram *My Block (the icon for this My Block is a red head)*

the MakePrgFile My Block

The *MakePrgFile* My Block (Figure 14-13) is like the *Make-Program* My Block, except that it stores the program you enter with the COLOR CUB3 into a file instead of an array. Storing the program in a file saves the color-programmed sequence in the EV3 Brick's memory; the sequence can be read back from the file even after the program stops or the EV3 Brick is rebooted.

To begin writing to a new file named *PRG*, set the first File Access block to **Delete** mode. Next, append the color codes to the file inside the Loop by using the File Access block in **Write** mode. When the Loop ends, a File Access block in **Close** mode closes the file *PRG* so that it can be read later.

the ParseFile My Block

The *ParseFile* My Block (Figure 14-14) converts the text strings read from the file to numeric values, with each string representing one of the possible numeric color codes (0 to 7). Usually, numbers should be read from a file using the File Access block in Read Numeric mode, but in this case, I read the data as text to detect the end of file, found when the read data is an empty string. The text string read from the file is fed into the input of a Switch block set in **Text** mode.

Initially, the **EOF** logic variable in this My Block is set to **False**, and the **_OUT** numeric variable is set to **0**. Each label of the Switch block contains the text representation of the numbers, and in each case, the _OUT variable is set to the numeric value corresponding to the text label. For example, in the case with the "2" label, the value 2 is written into the _OUT variable. In the default case of the Switch with an empty string ("") in its label, the EOF variable is set to True.

Create this My Block with a text input called *Text*, a numeric output named *Value*, and a logic output named *EOF*.

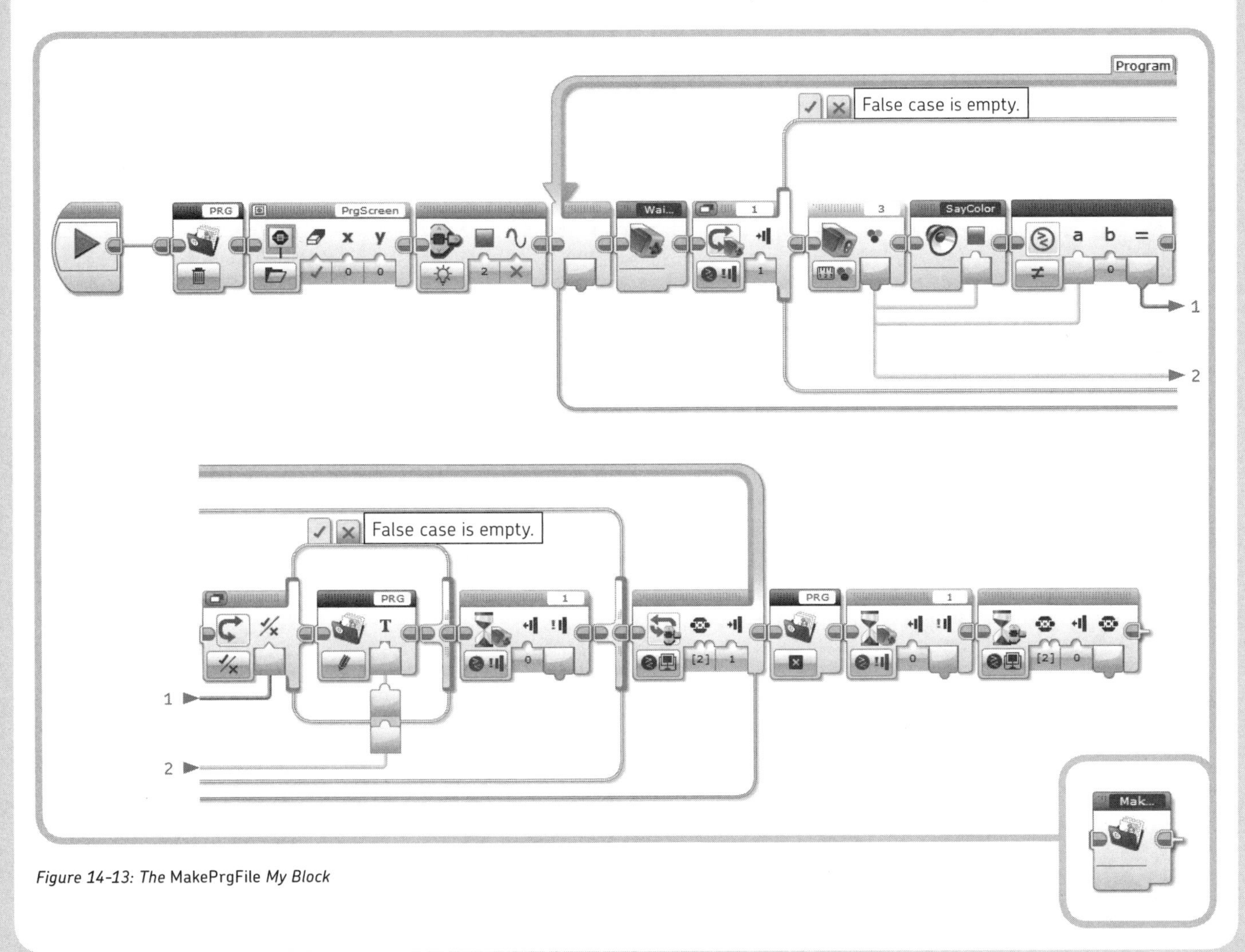

Figure 14-13: The MakePrgFile *My Block*

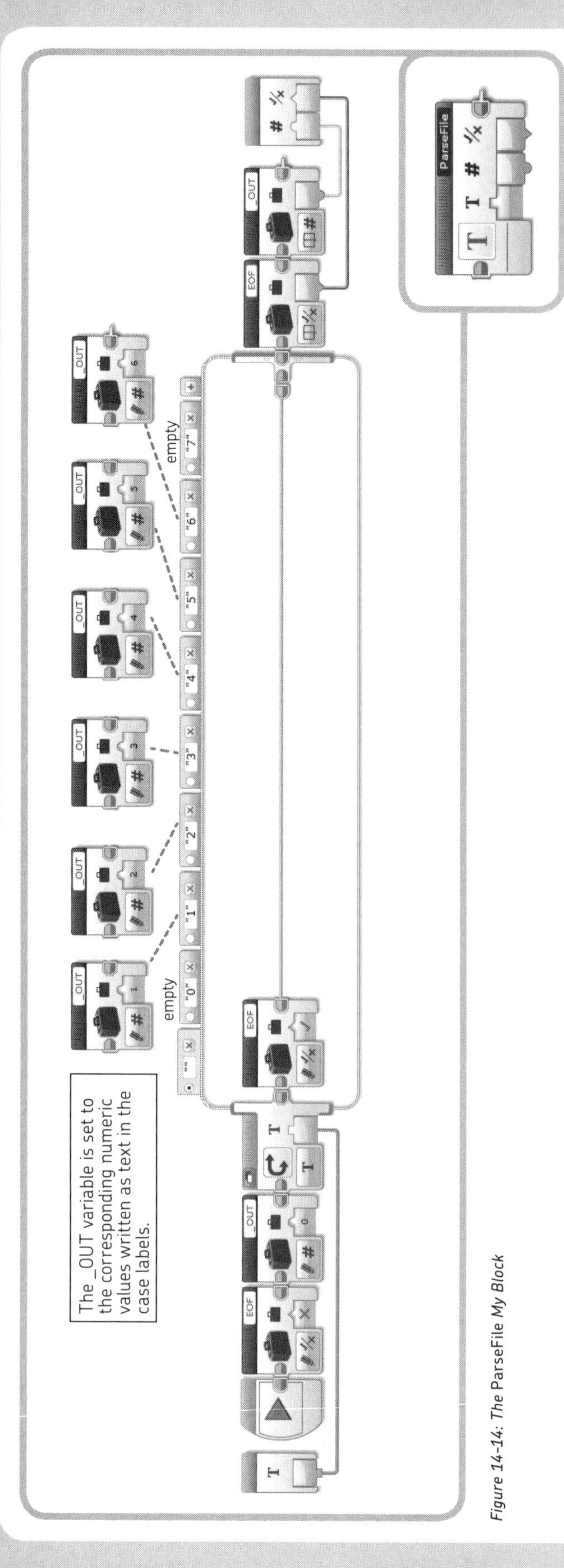

Figure 14-14: The ParseFile My Block

the RunPrgFile My Block

The *RunPrgFile* My Block (Figure 14-15) is similar to the *Run-Program* My Block, except that it reads the program you entered with the COLOR CUB3 from the *PRG* file instead of the array. The *RUN* Loop ends when its logic condition is True. The logic condition is True when the EOF flag coming from the *ParseFile* My Block is True (when the EOF is reached) or when the EV3 Brick's Enter button is pressed. Once the Loop ends, the file *PRG* is closed.

programming the SENTIN3L to patrol

The program that makes the SENTIN3L go on patrol is shown in Figure 14-16. First the legs are reset to make them all touch the ground (the *ResetLegs* My Block). Next, in an unlimited Loop called *MAIN*, the robot walks forward until it detects an obstacle, at which point it "shoots" the object with its twin laser blaster cannons. It then turns in a random direction until the obstacle is farther away than the specified threshold and at least two seconds elapse.

Press the Touch Sensor at any time to shut down the program. In the parallel sequence, a Wait block waits for the Touch Sensor to be pressed, at which point a Loop Interrupt block stops the *MAIN* Loop and all motors, and executes the *PowerDownFX* My Block. Finally, the program ends.

color-programming the SENTIN3L at runtime

The program shown in Figure 14-17 allows you to record and play back sequences of actions at runtime. To enter commands, show one of the colored sides of the COLOR CUB3 to the Color Sensor and press the Touch Sensor to confirm the action:

- White will make the robot walk forward until it detects an obstacle.
- Black makes it turn in a random direction until an obstacle moves past a set threshold.
- Red makes the robot "shoot" its cannons.

Press the EV3 Brick's Enter button to execute the sequence. Once the sequence has been played, you can enter a new sequence.

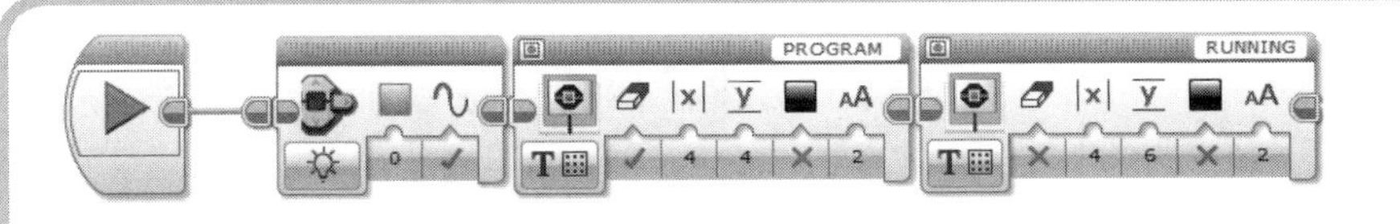

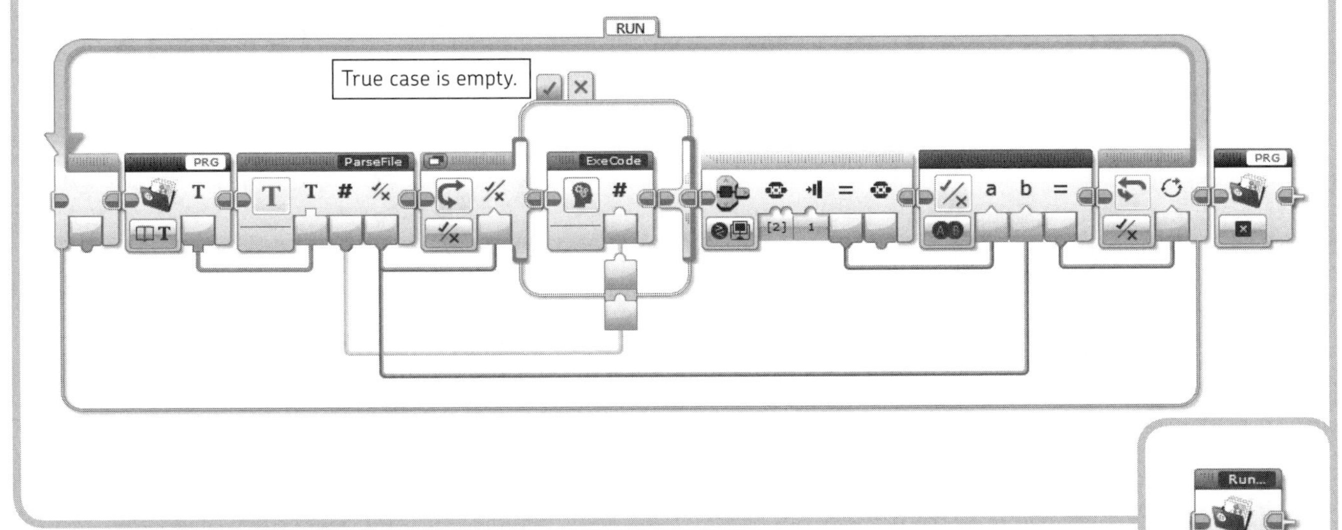

Figure 14-15: The RunPrgFile *My Block*

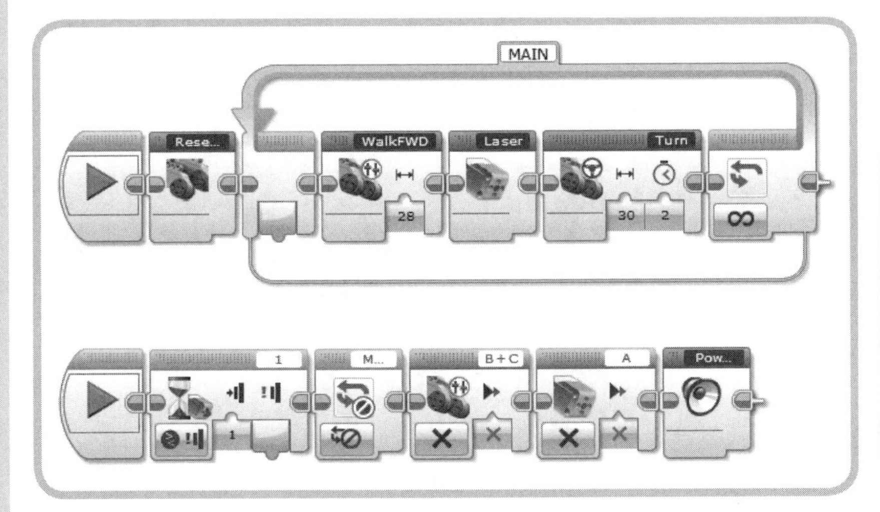

Figure 14-16: The Patrol *program*

EXPERIMENT 14-1

Make a program that allows you to command the robot directly by showing it the COLOR CUB3. Make the robot react to the colors according to your color-programming program: white for walking forward, black for turning, and red for "shooting" the lasers.

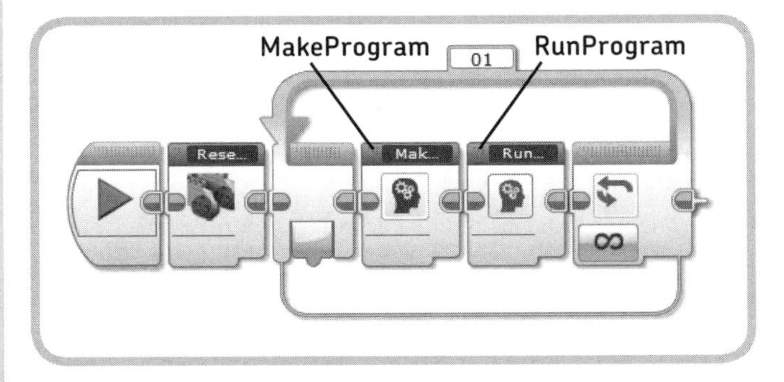

Figure 14-17: The ColorProgram *program*

EXPERIMENT 14-2

Extend the *ExeCode* My Block by filling the other cases for yellow, blue, and green to color-program the robot at runtime with more actions. You'll need to build a more colorful COLOR CUB3 with other Technic parts or use some colored cards.

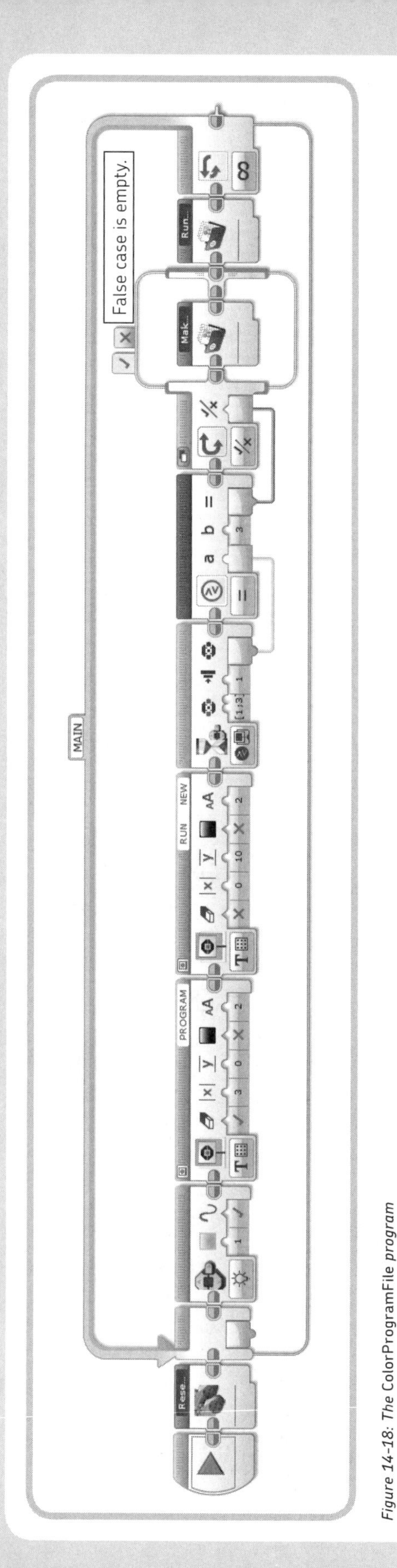

Figure 14-18: The ColorProgramFile *program*

making permanent runtime color programs

The program shown in Figure 14-18 saves the recorded sequence of actions to a file, which can be read back even after the program has stopped and restarted or the EV3 Brick has rebooted. A menu displayed on the EV3 Brick screen lets you choose whether to read and replay the sequence stored in the file or record and save a new one.

WARNING If you choose to run a sequence before having recorded one, the program won't find the file and will abort, displaying an error message on the EV3 Brick screen.

EXPERIMENT 14-3

Can you make a program that lets you control the SENTIN3L with the Remote IR Beacon? Take your inspiration from the SUP3R CAR remote control program *RC_switch* (see Figure 12-13 on page 240) and adapt the content of the *WalkFWD* and *Turn* My Blocks. Program the SENTIN3L to shut down when the Touch Sensor is pressed, as in the *Patrol* program. Modify the robot so that an identical robot can hit the Touch Sensor with the other cannon to shut it down. Then challenge a friend to a fight with two SENTIN3Ls! The first player who hits the other robot's Touch Sensor wins.

EXPERIMENT 14-4

Want more action? Build a working ball shooter to replace the SENTIN3L's arms. You can use the building instructions that come with the official EV3 models as inspiration.

conclusion

In this chapter, you learned how to make programs for the SENTIN3L that will send it on patrol and how to program it at runtime by showing it colors. You also learned how to generate random numbers and how to store and read back data using the File Access block as well as how to convert text values into numeric ones.

BUT WHY DID YOU WAIT FOR ME BEFORE TRYING TO FREE YOUR BROTHER?
DO YOU SEE ANYONE ELSE? WE ARE ALONE! IF THIS BUSINESS EVER GOT OUT, I WOULD HAVE TO SHUT DOWN!
MY BROTHER HAS BEEN LOCKED IN THERE FOR TWO WEEKS. HE ONLY BROKE THE QUARANTINE TO RETRIEVE SOME PROJECTS!
BUT ALEX WAS PREPARED FOR IT... HE REACHED HIS LAB, WHERE HE HAS ENOUGH SUPPLIES TO SURVIVE FOR A MONTH.
BLEAH!
THE QUARANTINE ZONE IS SHIELDED. THAT'S THE ONLY WAY TO COMMUNICATE FROM INSIDE!
Campbell's
CONDENSED
SOUP
HELLO? ALEX, ARE YOU THERE?
HEY! DO YOU HEAR ME?
14X1

ALEX, WHAT'S UP?
I'VE HAD BETTER DAYS!
DO YOU REMEMBER THE EMP SYSTEM? WE'RE GOING TO GET YOU OUT OF THERE TODAY!
I'VE READ THE EMERGENCY MANUAL SO MANY TIMES, I KNOW IT BY HEART!
DID YOU ENABLE THE EMP SYSTEM!? I'LL ACTIVATE IT THEN!
NO! FOOL! DO YOU WANT TO REMAIN TRAPPED IN THERE FOREVER? WE MUST UNLOCK THE DOORS FROM OUTSIDE, FIRST!
DON'T ACTIVATE THE EMP UNTIL YOU SEE THE WHITES OF OUR EYES!
GOT IT! SEE YOU SOON IN THE AUDITORIUM!
14X2

15 building the T-R3X

The T-R3X is an ominous surveillance robot in the shape of a *Tyrannosaurus rex* (see Figure 15-1)! It walks on two strong legs, wags its sharp tail, and bites its prey with powerful jaws.

As you build this robot, you'll learn another approach to building a weight-shifting biped walker. But be warned: Keep your fingers far away from its teeth!

Figure 15-1: T-R3X!

main assembly

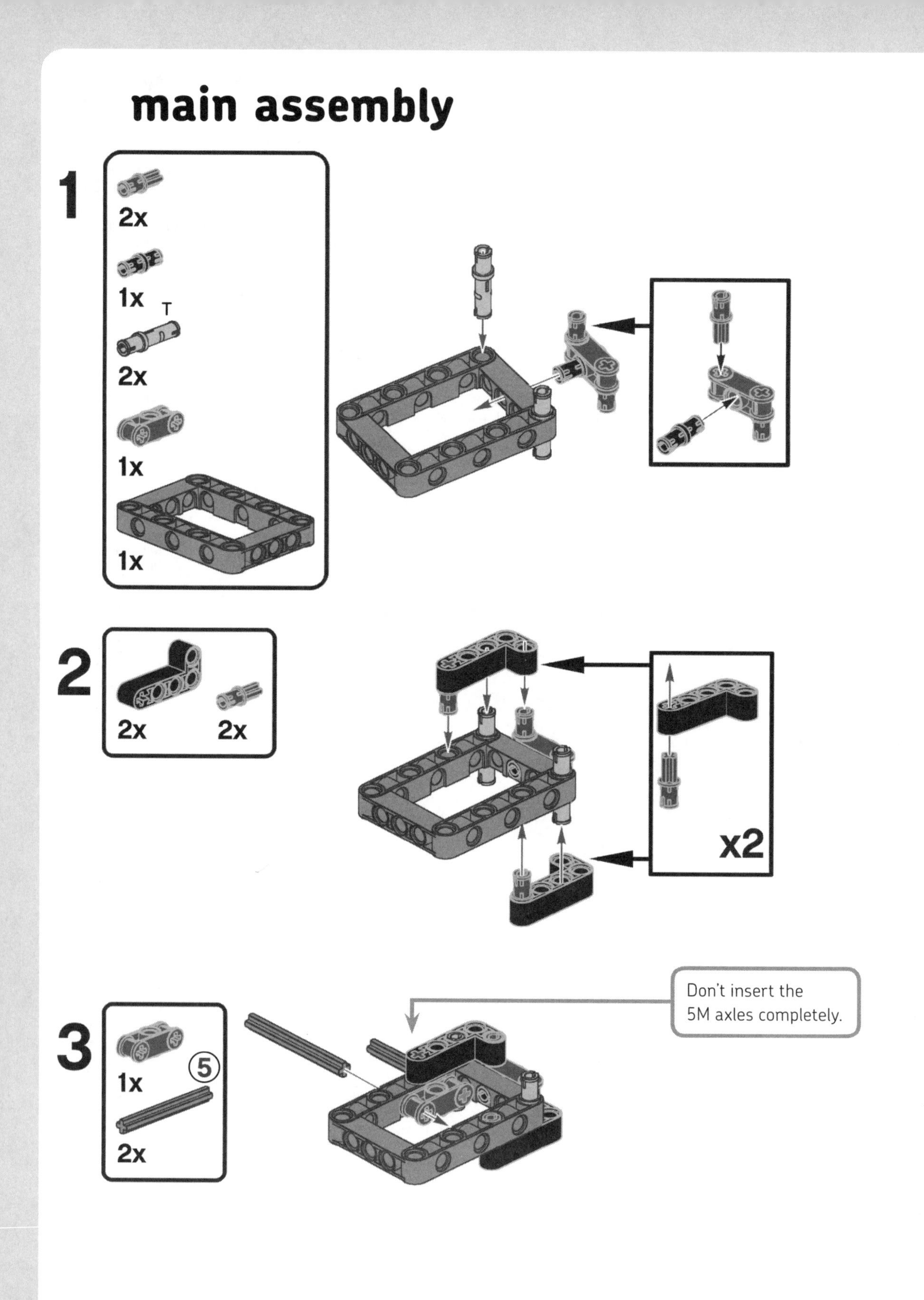

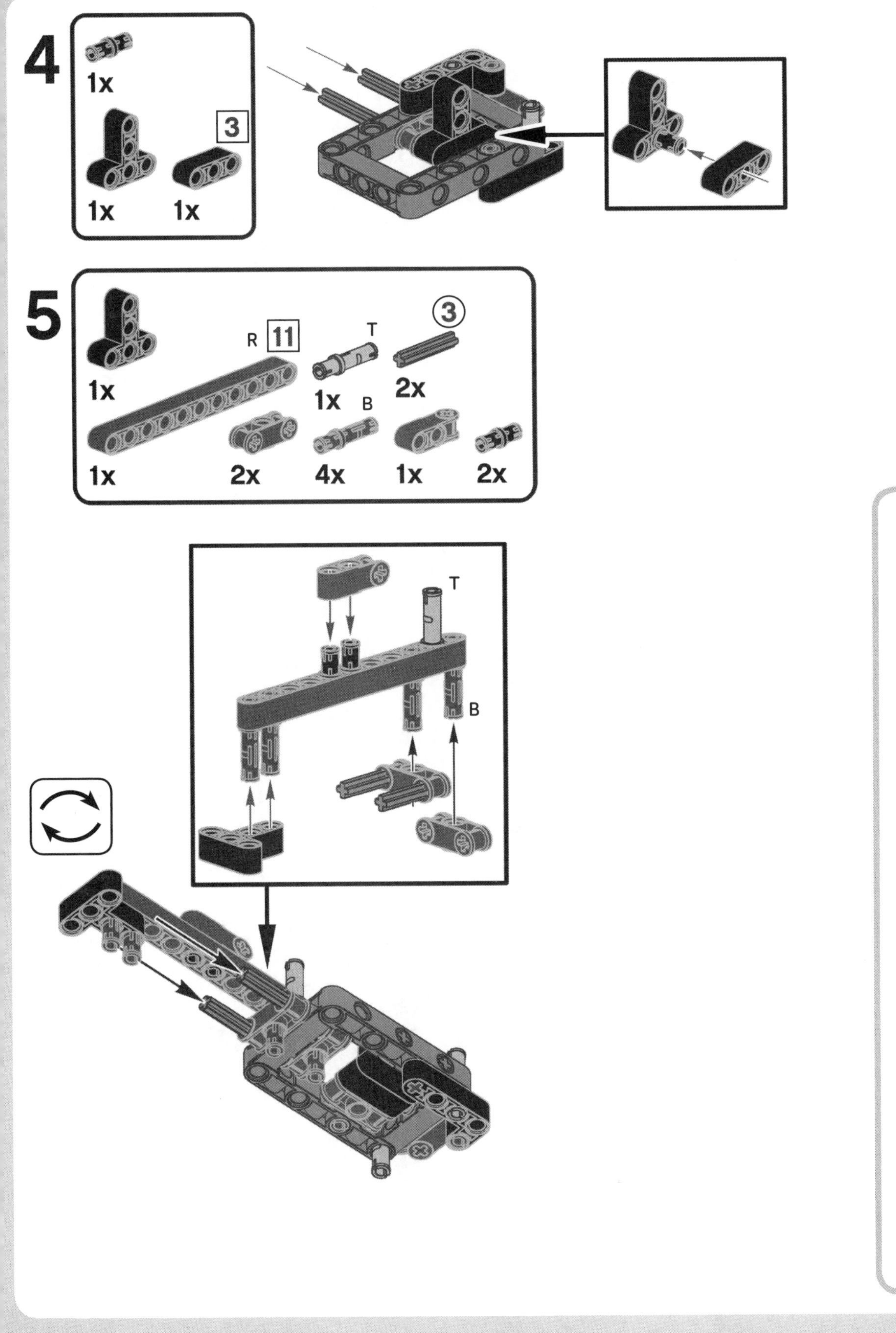
4
1x
3
1x
1x
5
R 11
T
3
1x
1x
2x
B
1x
2x
4x
1x
2x
T
B

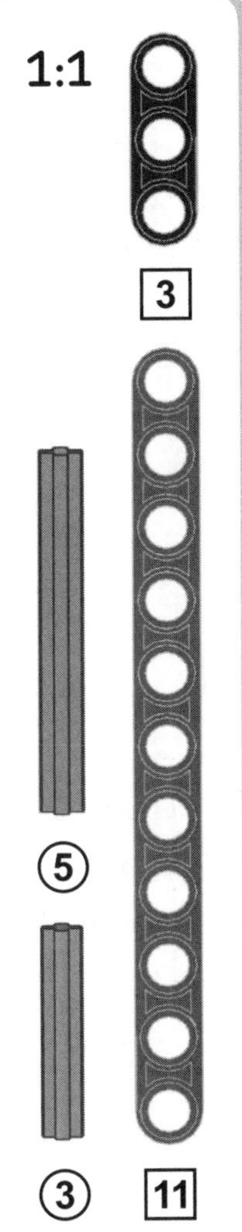
1:1
3
5
3
11

6

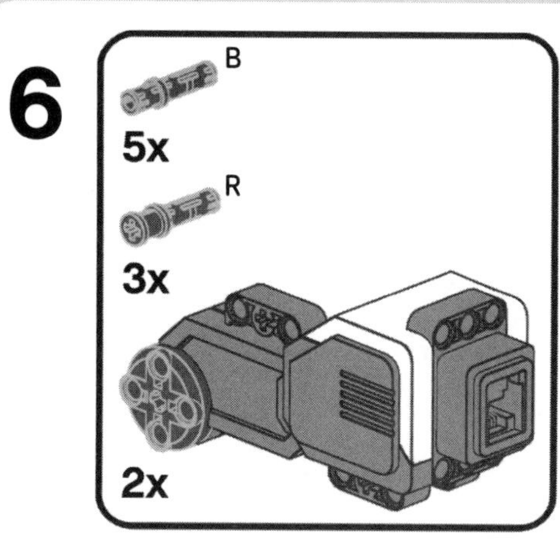

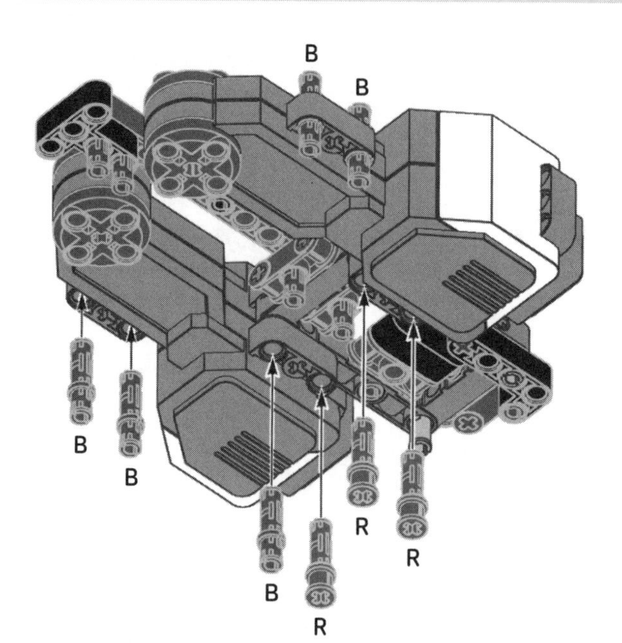

7

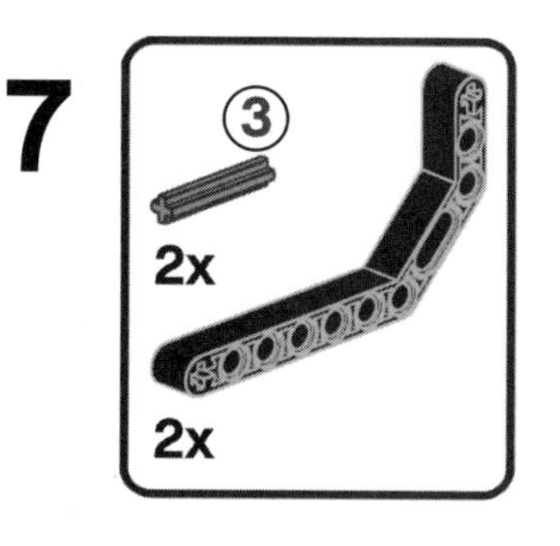

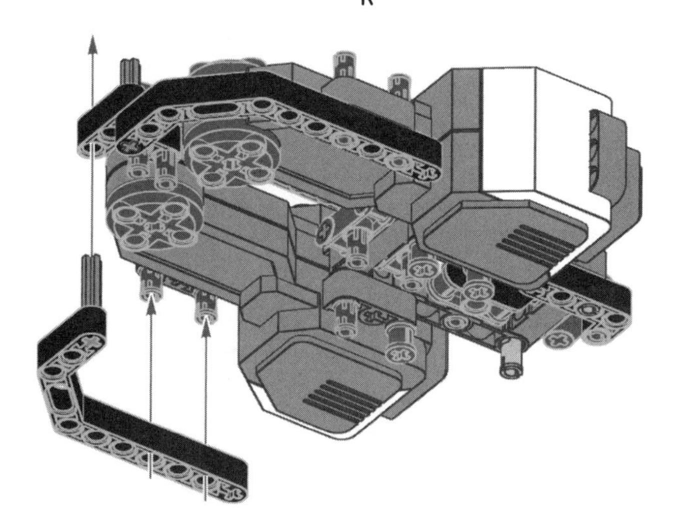

8

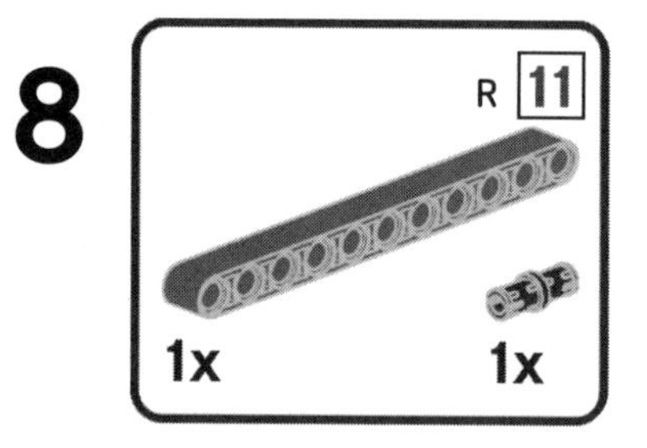

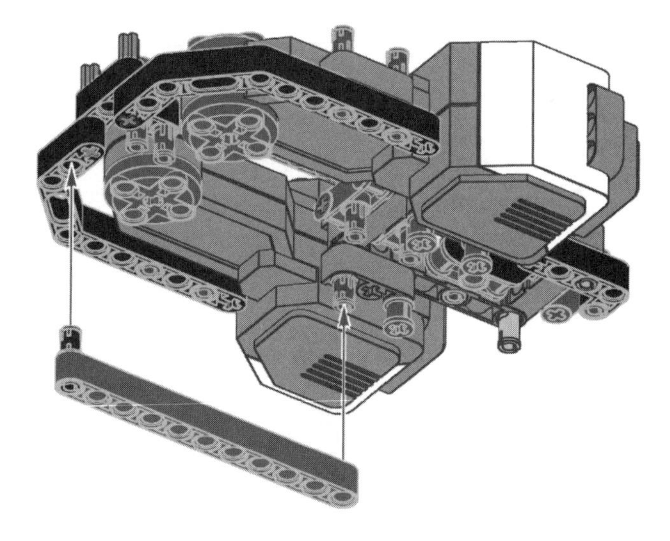

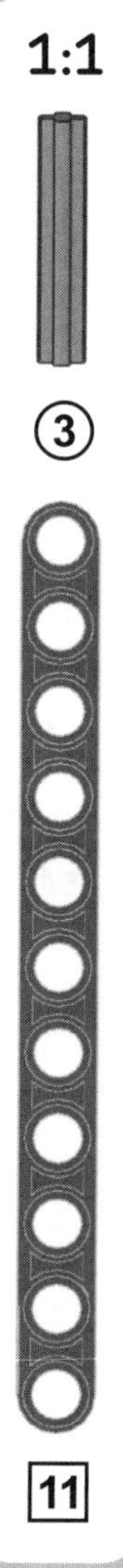

legs frame assembly

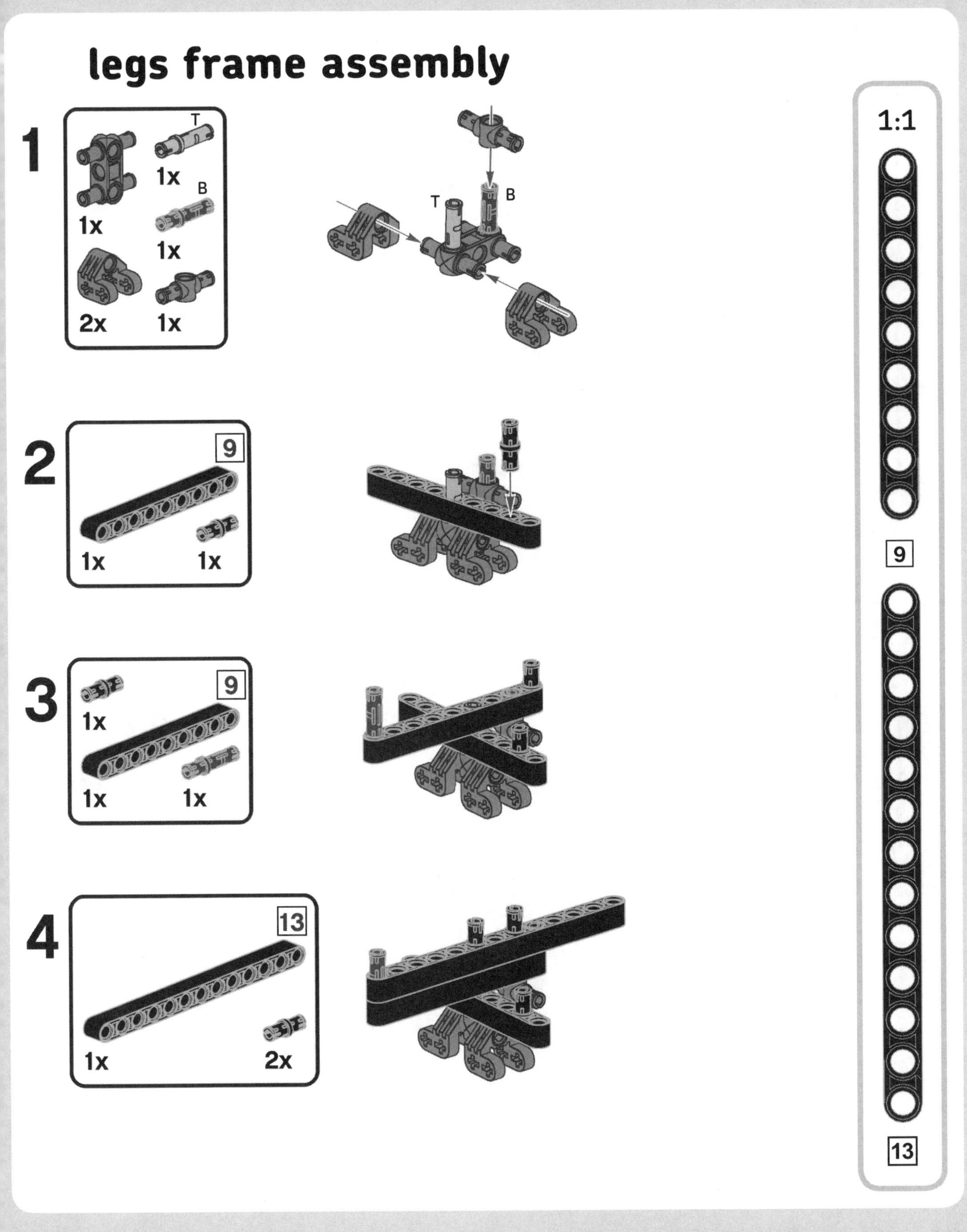

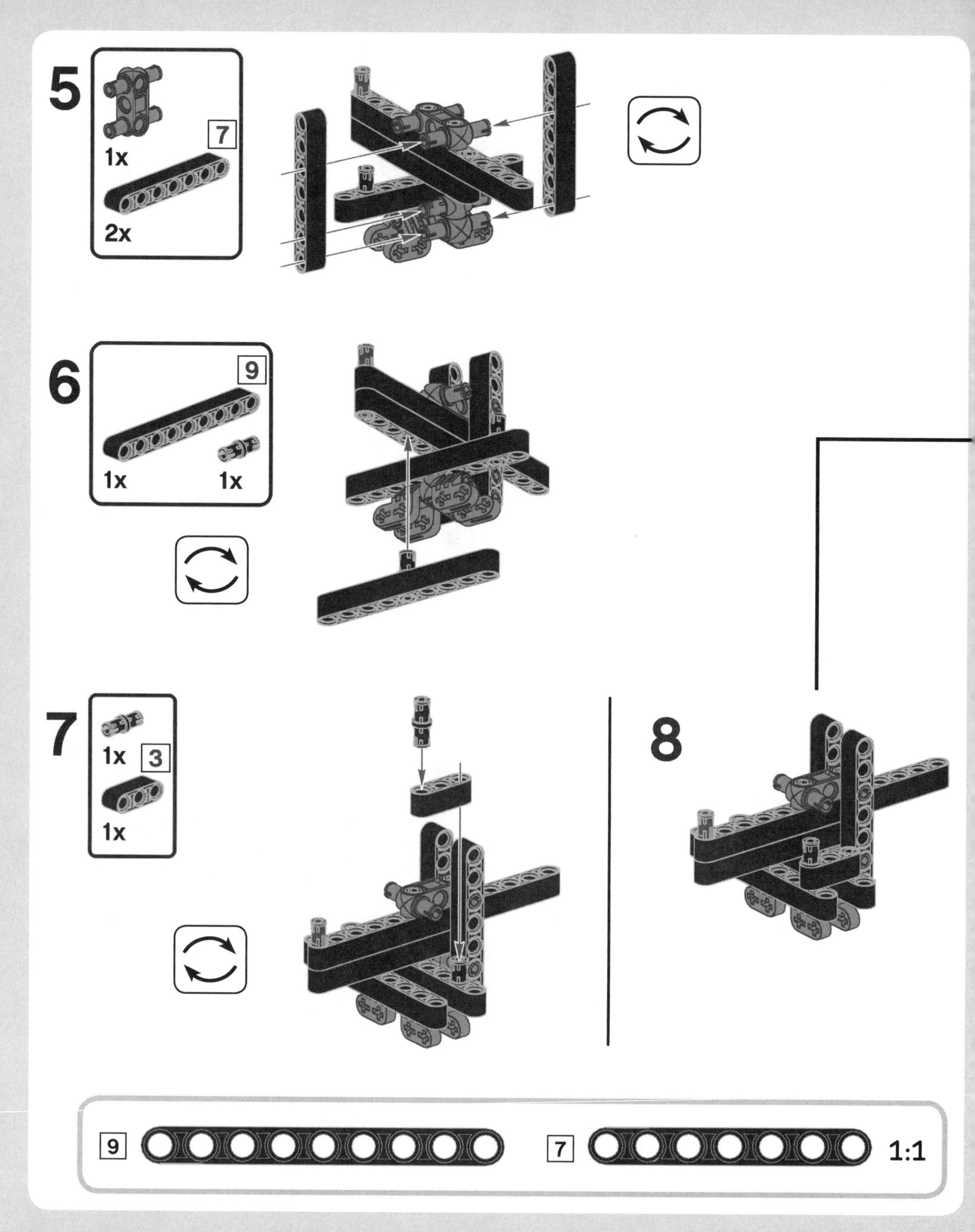
5
1x
7
2x
6
9
1x
1x
7
1x
3
1x
8
9
7
1:1

main assembly

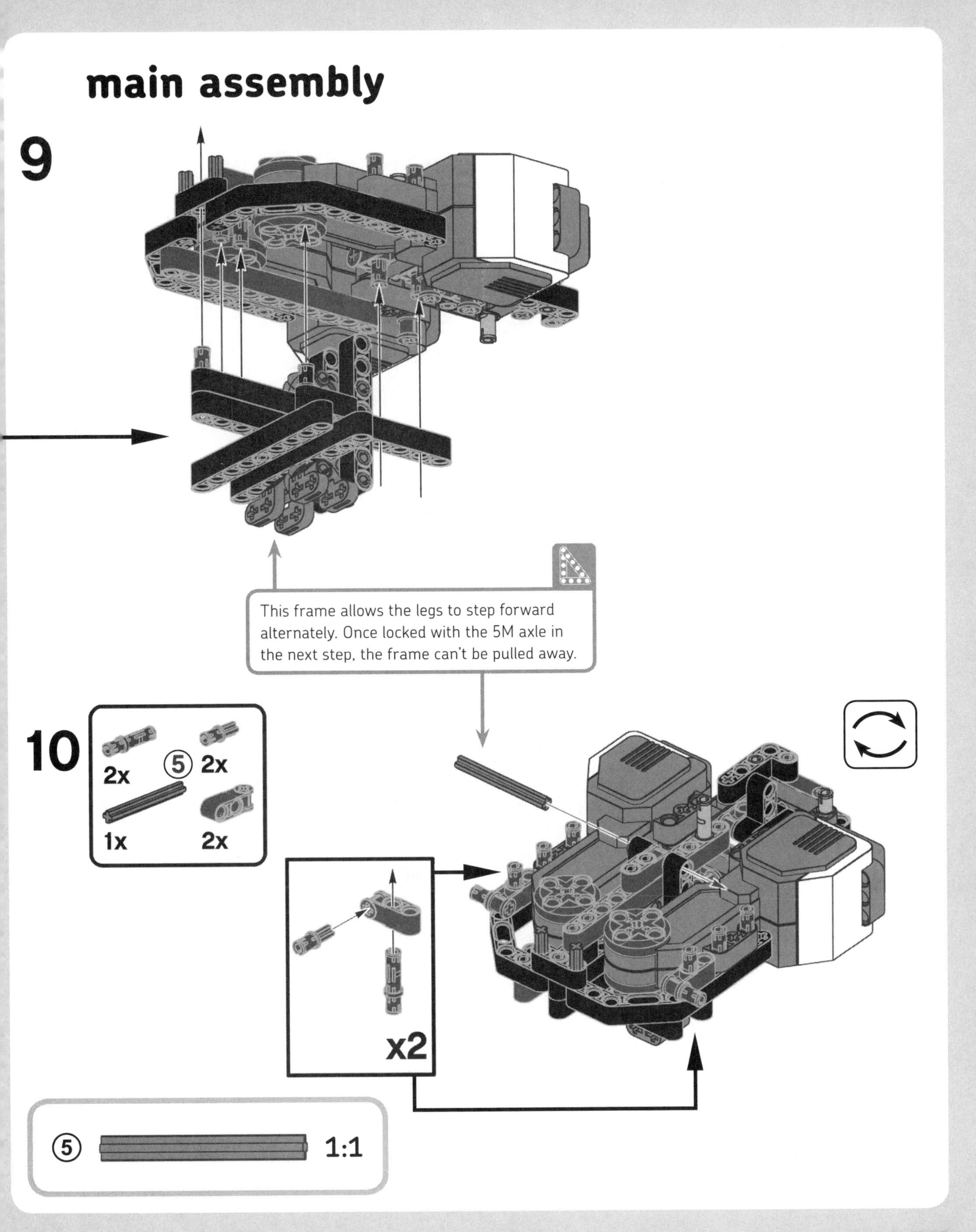

11

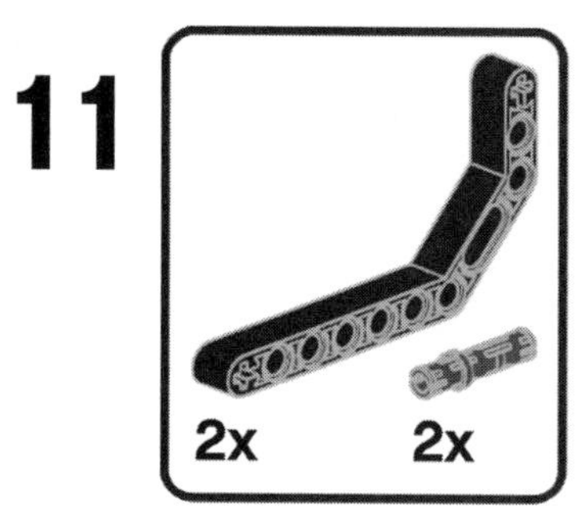

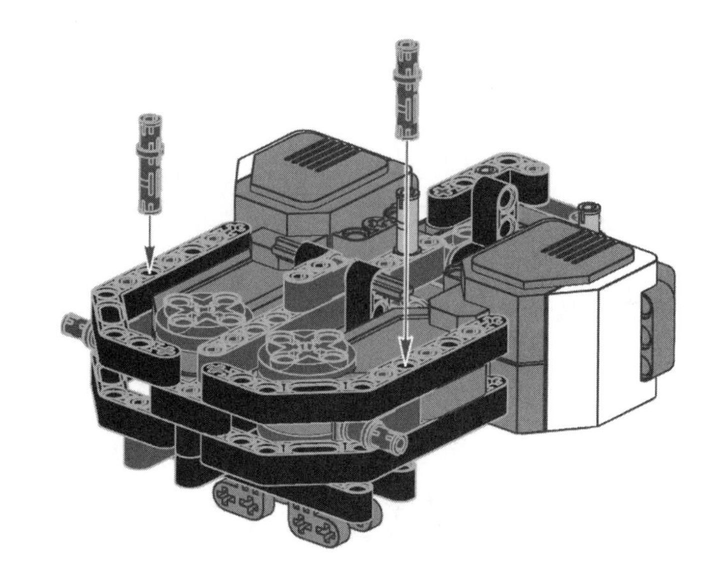

12

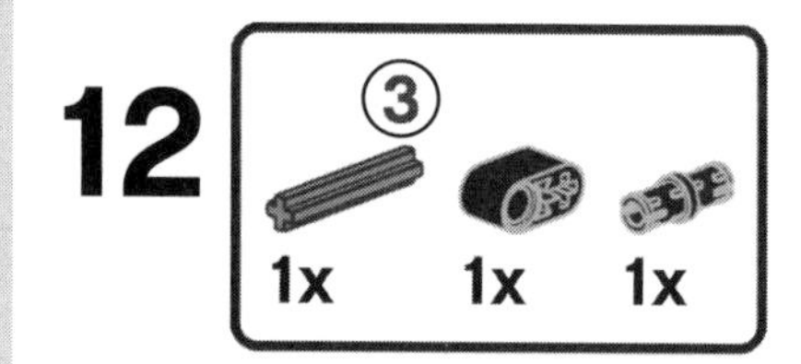

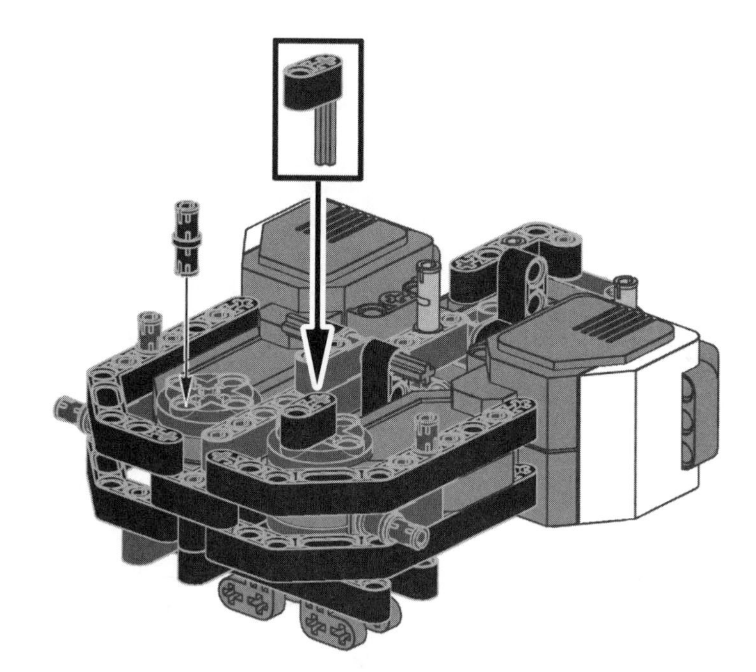

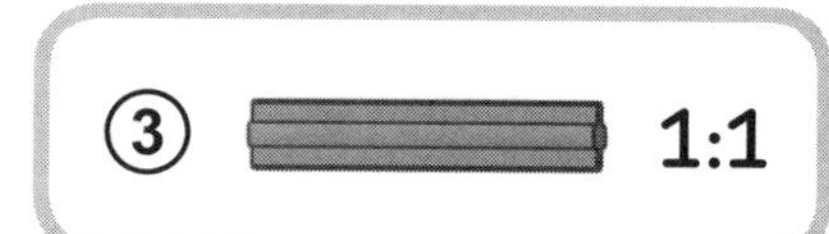

13

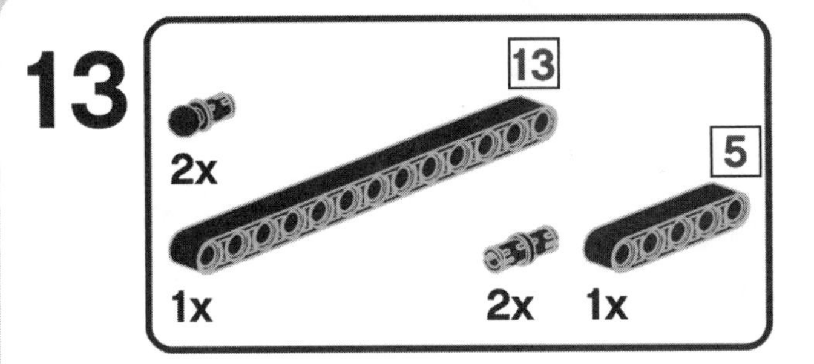

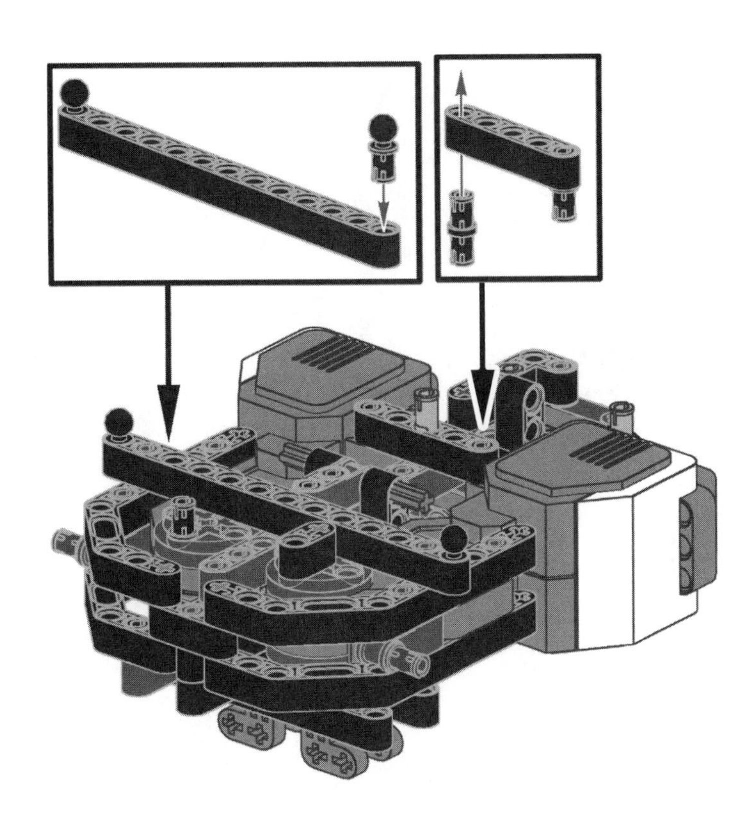

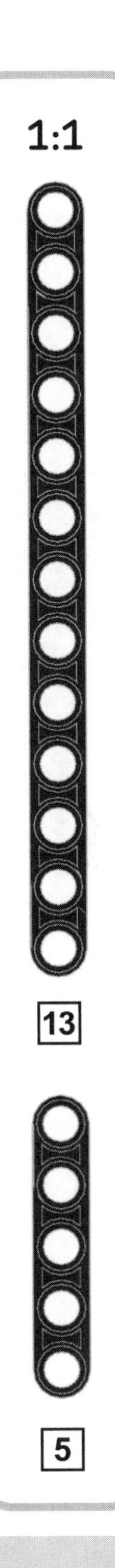

left leg

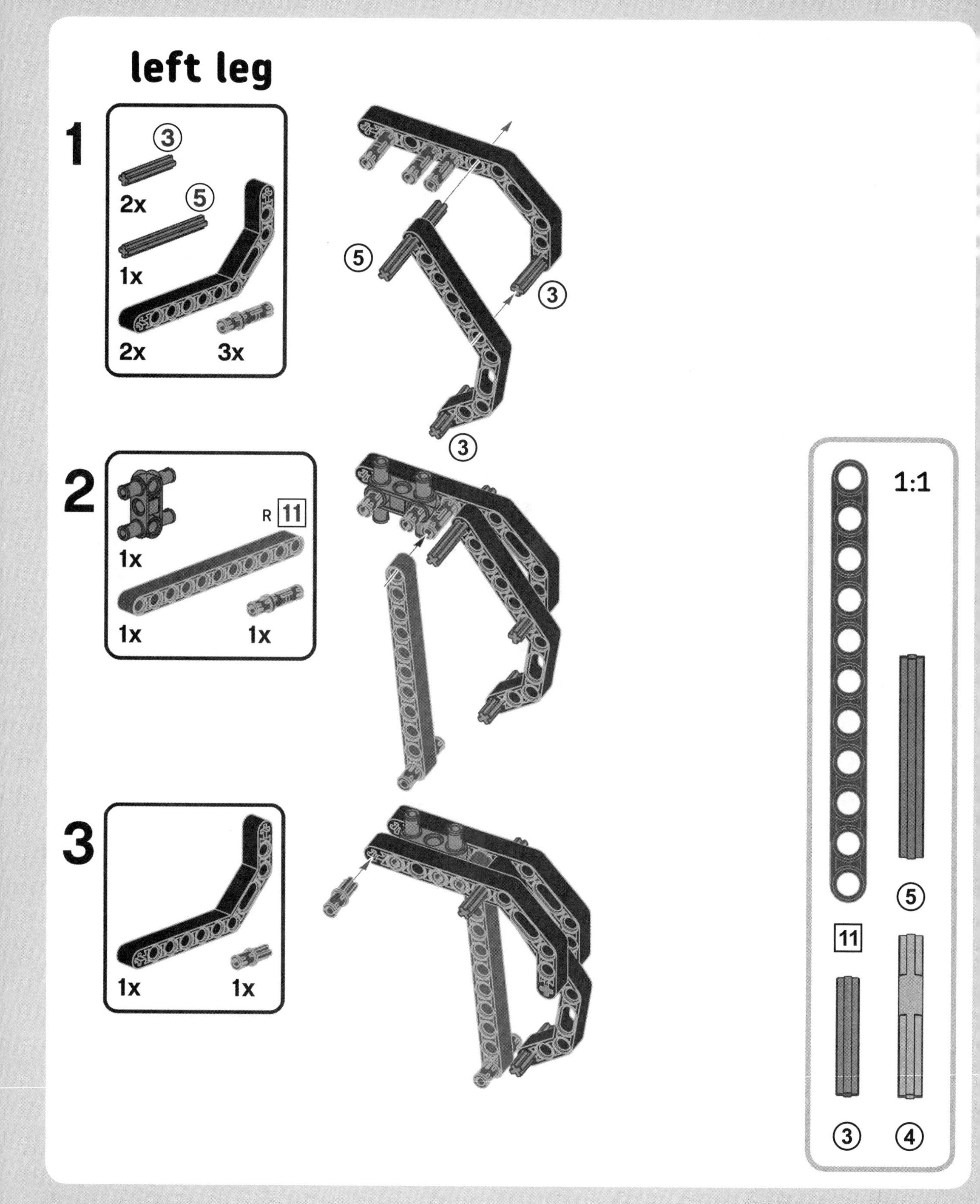

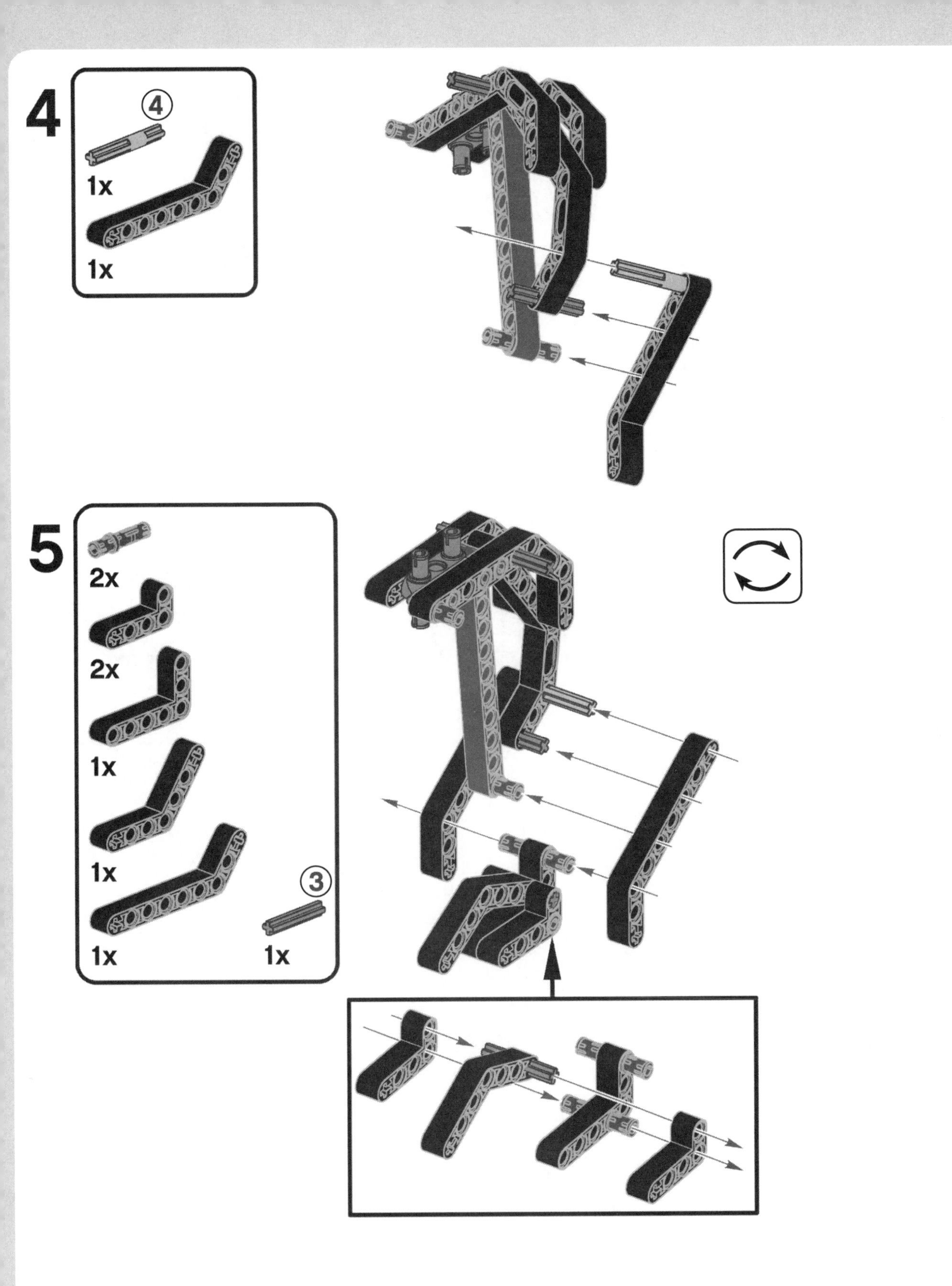
4
4
1x
1x
5
2x
2x
1x
1x
3
1x
1x

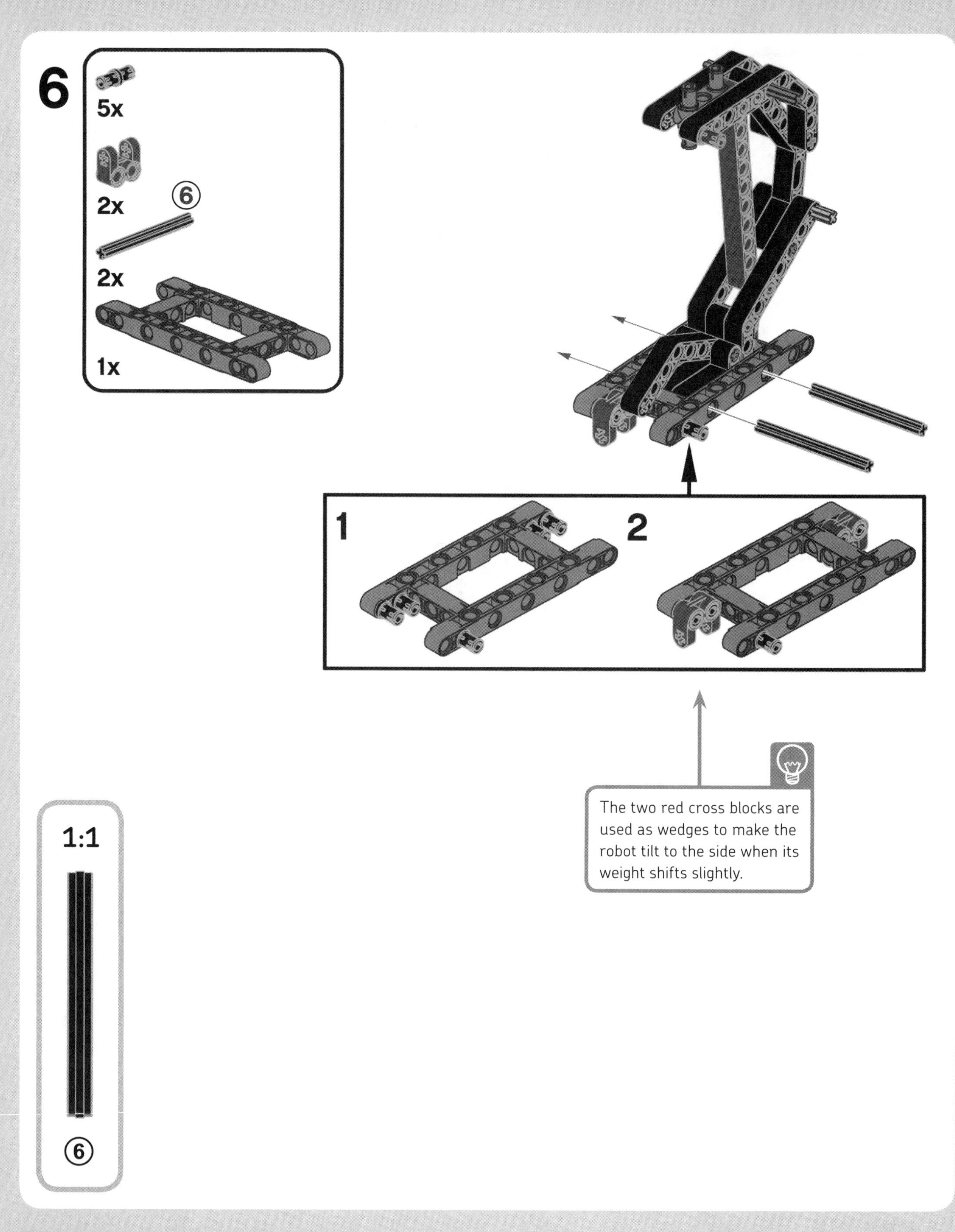
6
5x
2x
6
2x
1x
1
2
The two red cross blocks are used as wedges to make the robot tilt to the side when its weight shifts slightly.
1:1
6

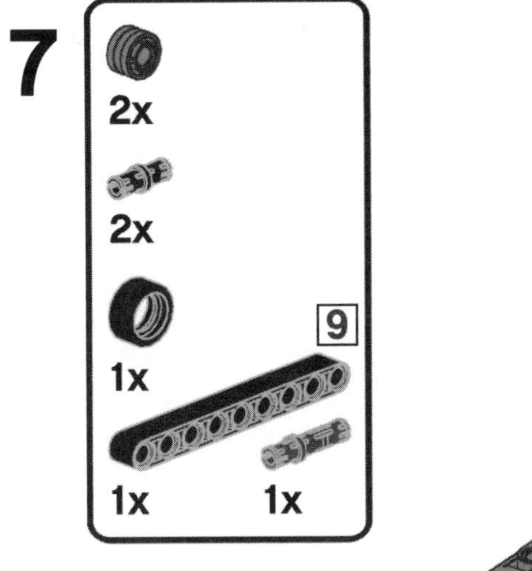

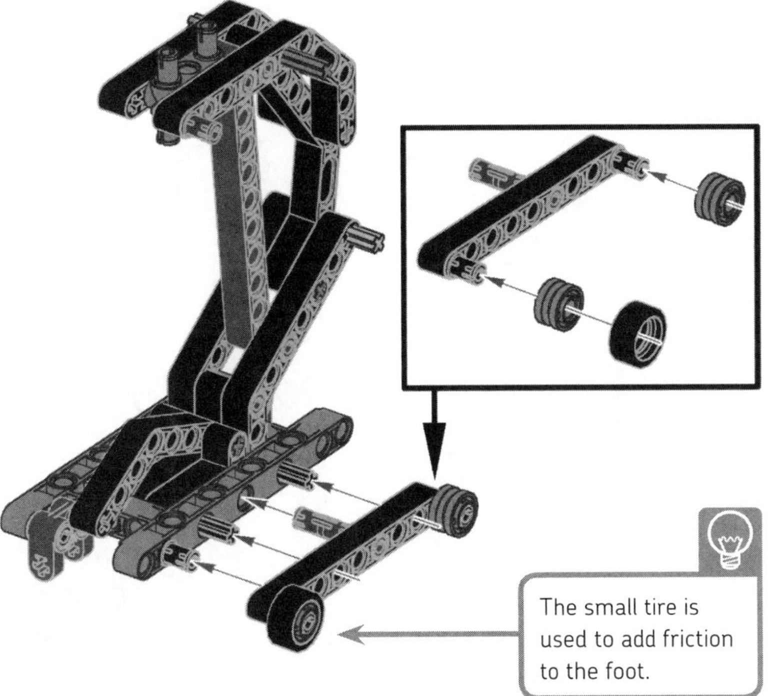

The small tire is used to add friction to the foot.

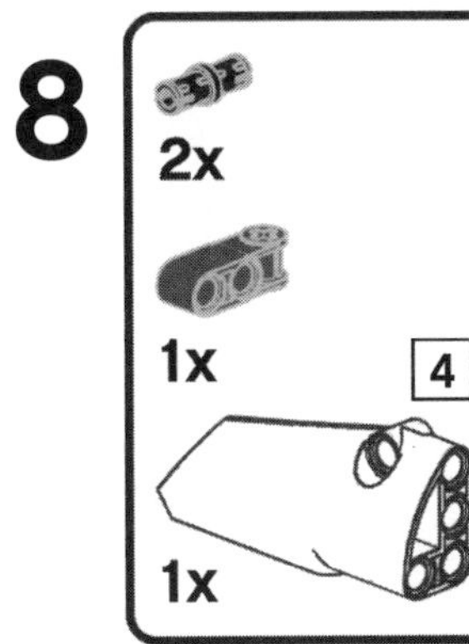

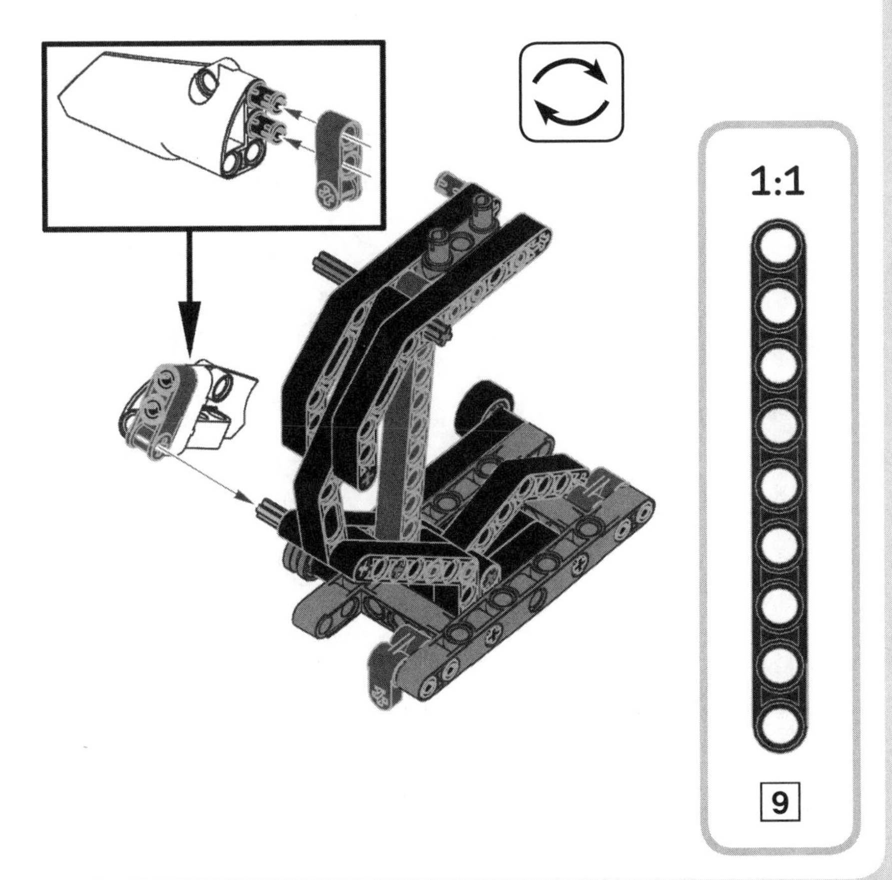

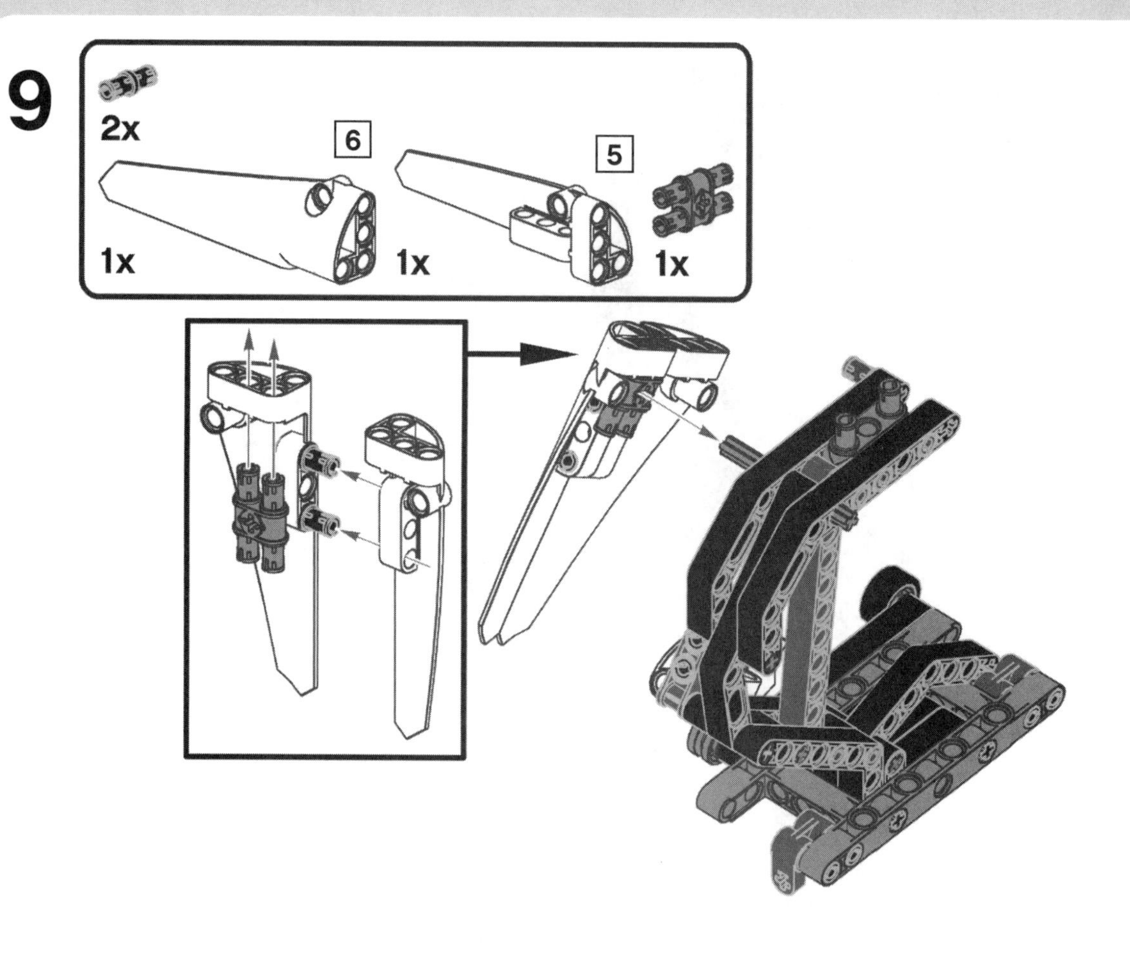
9
2x
6
5
1x
1x
1x

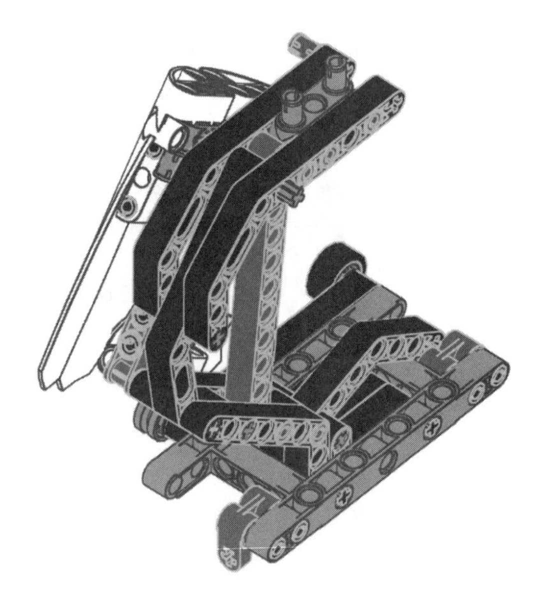

right leg assembly

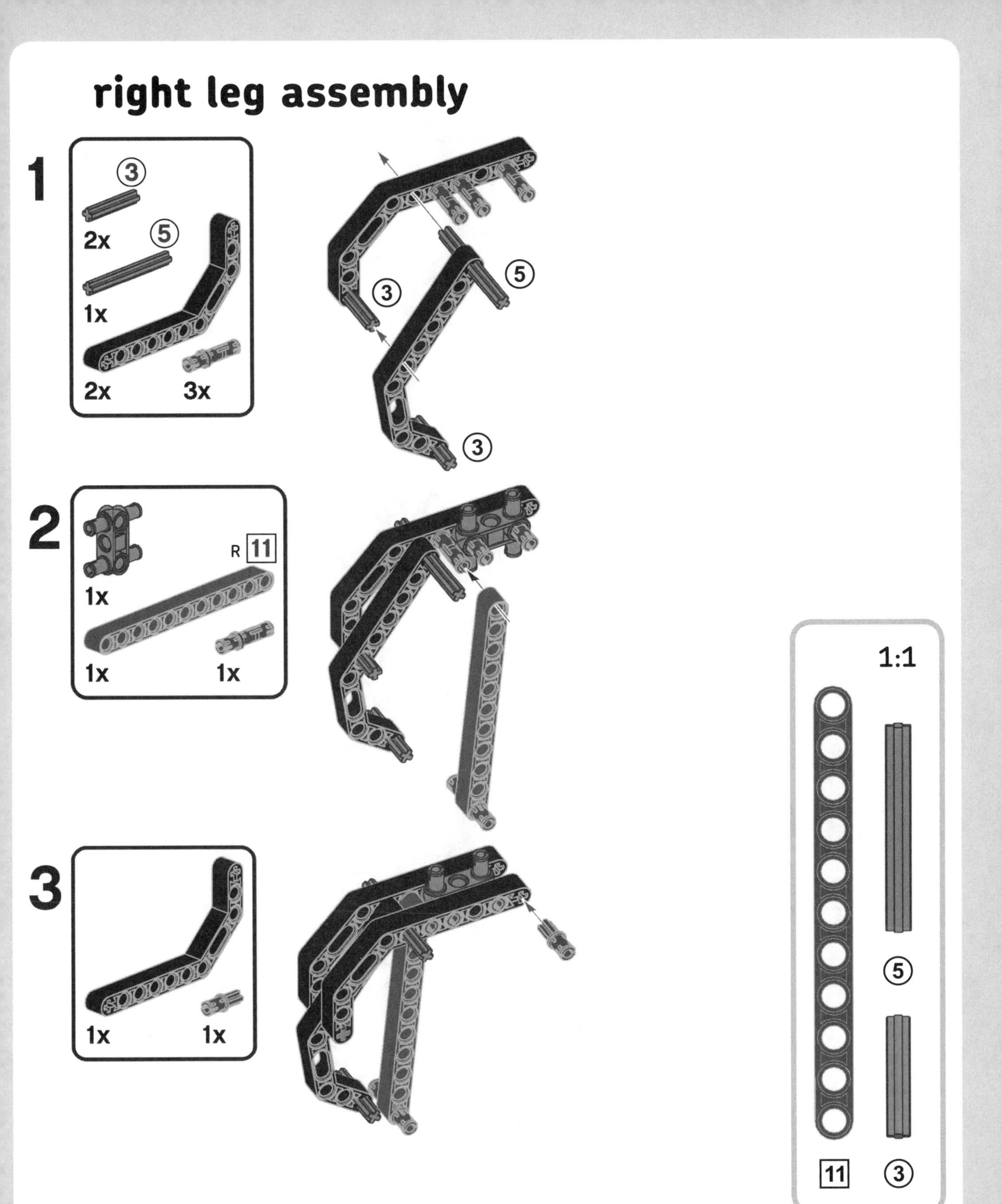

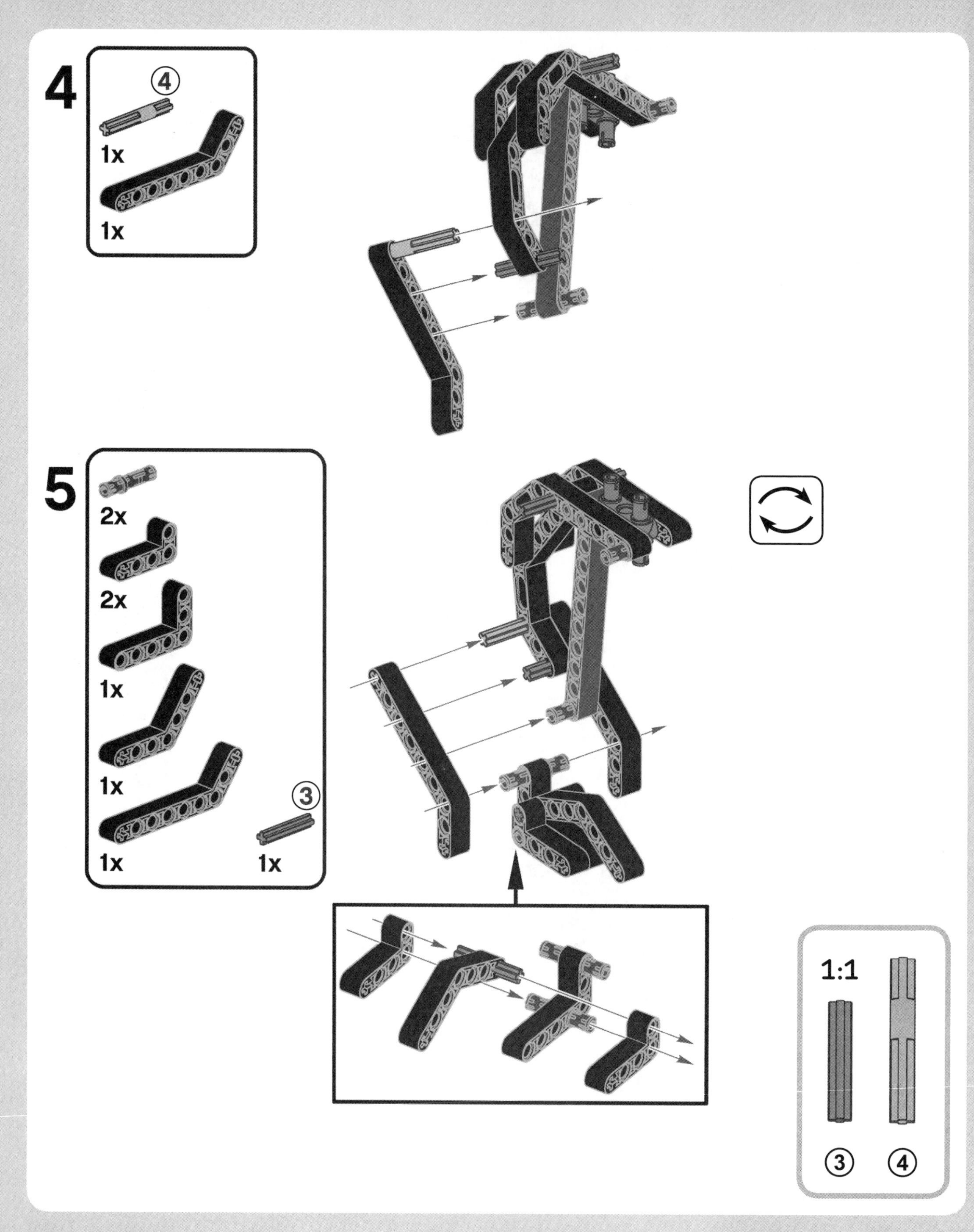
4
4
1x
1x
5
2x
2x
1x
1x
1x
3
1x
1:1
3
4

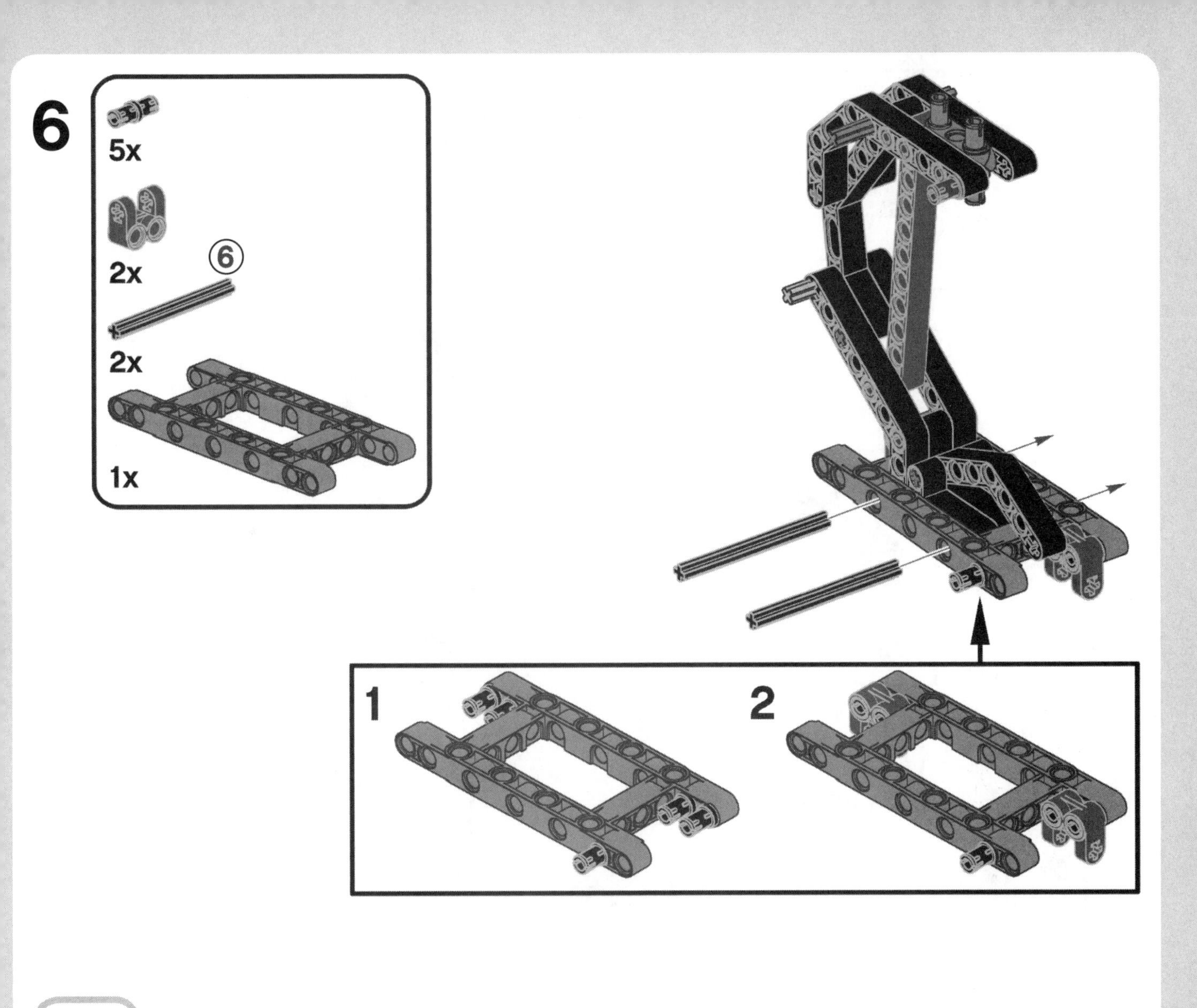
6
5x
2x
⑥
2x
1x
1
2

1:1
⑥

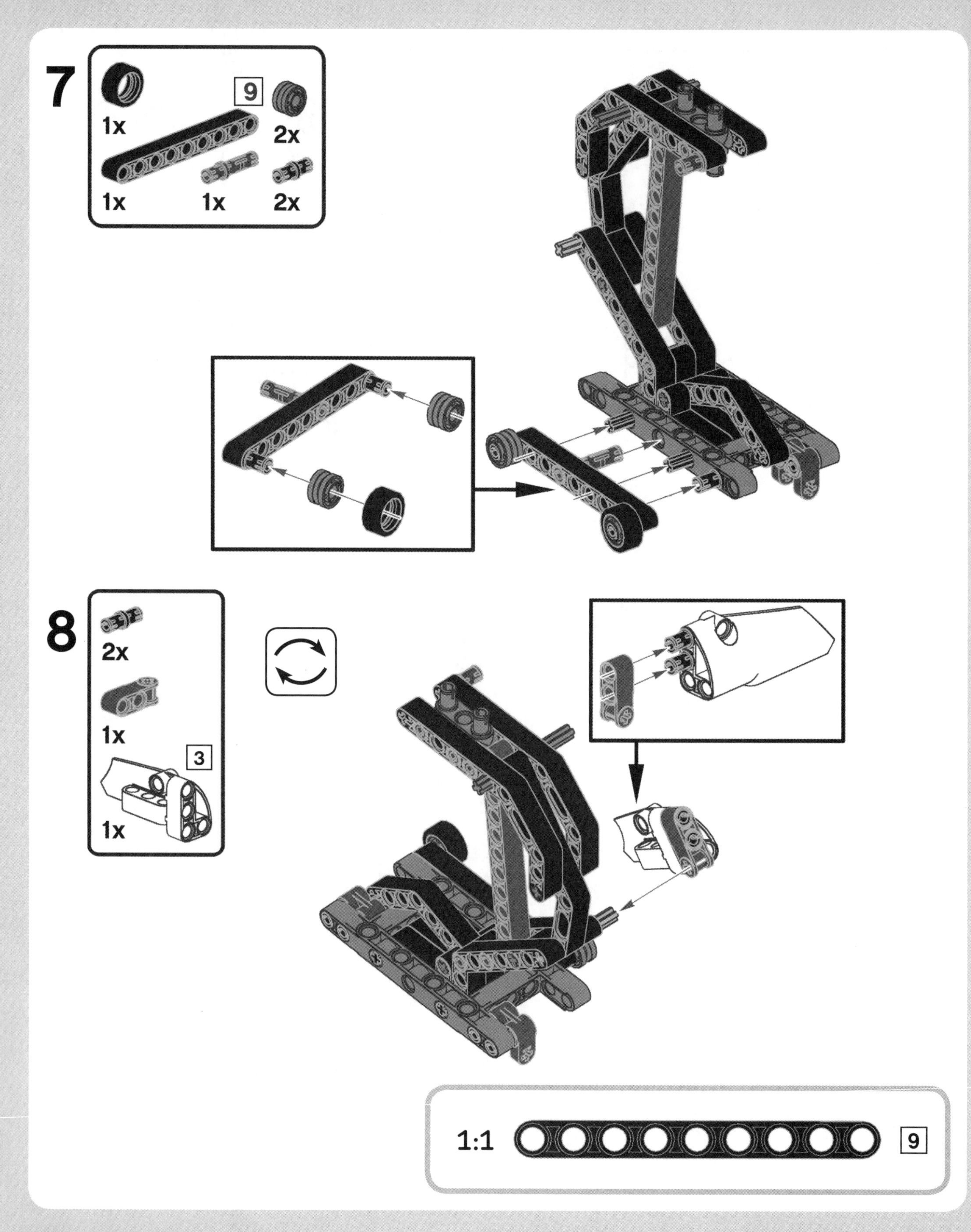
7
1x
9
2x
1x
1x
2x
8
2x
1x
3
1x
1:1
9

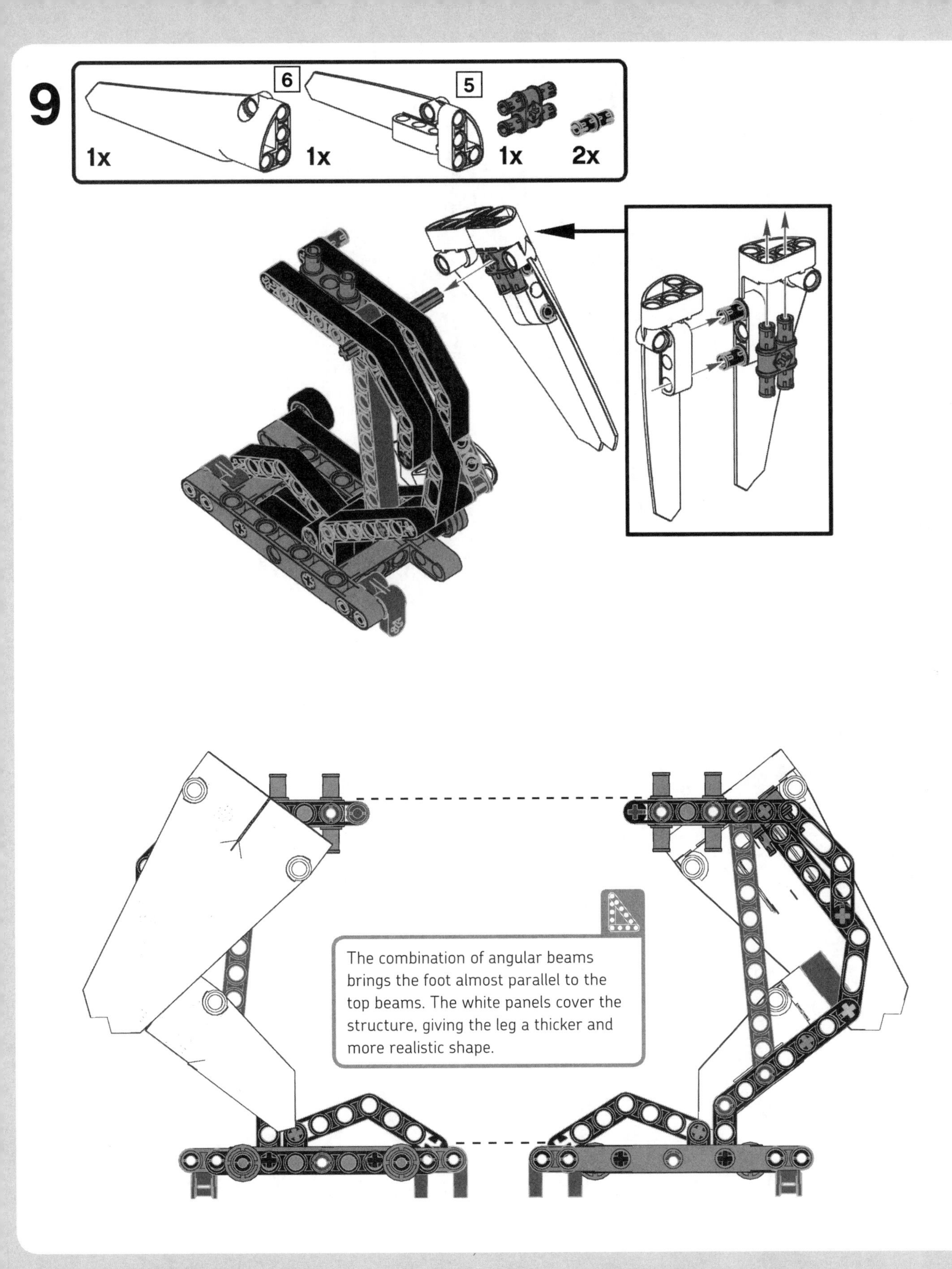
9
6
5
1x
1x
1x
2x
The combination of angular beams brings the foot almost parallel to the top beams. The white panels cover the structure, giving the leg a thicker and more realistic shape.

main assembly

14

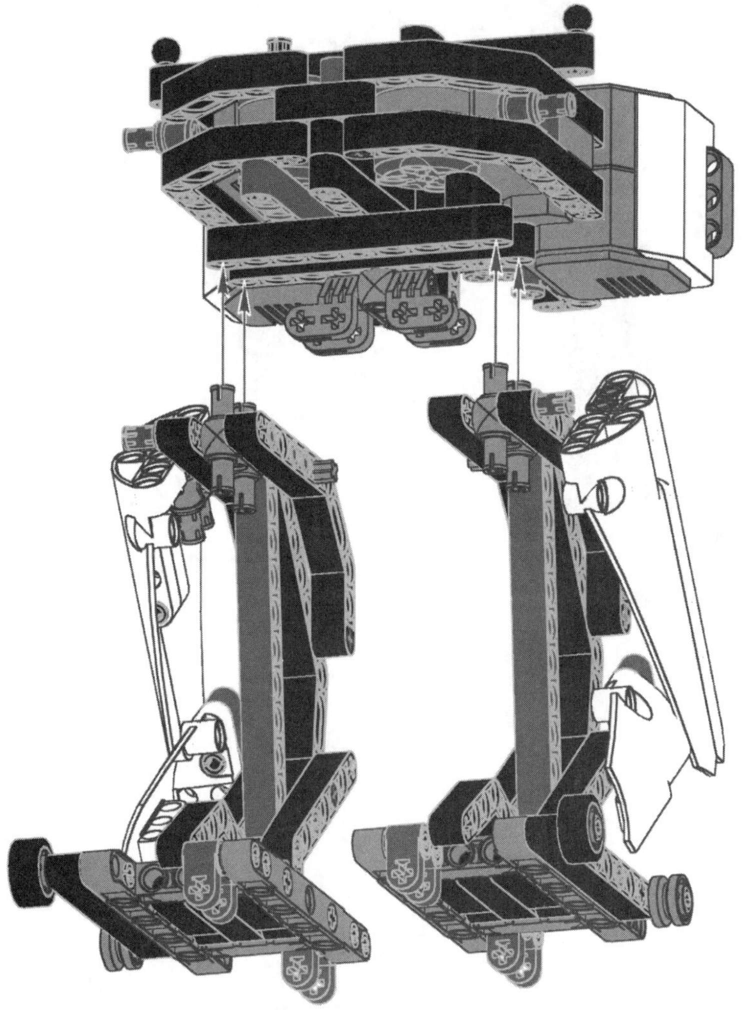

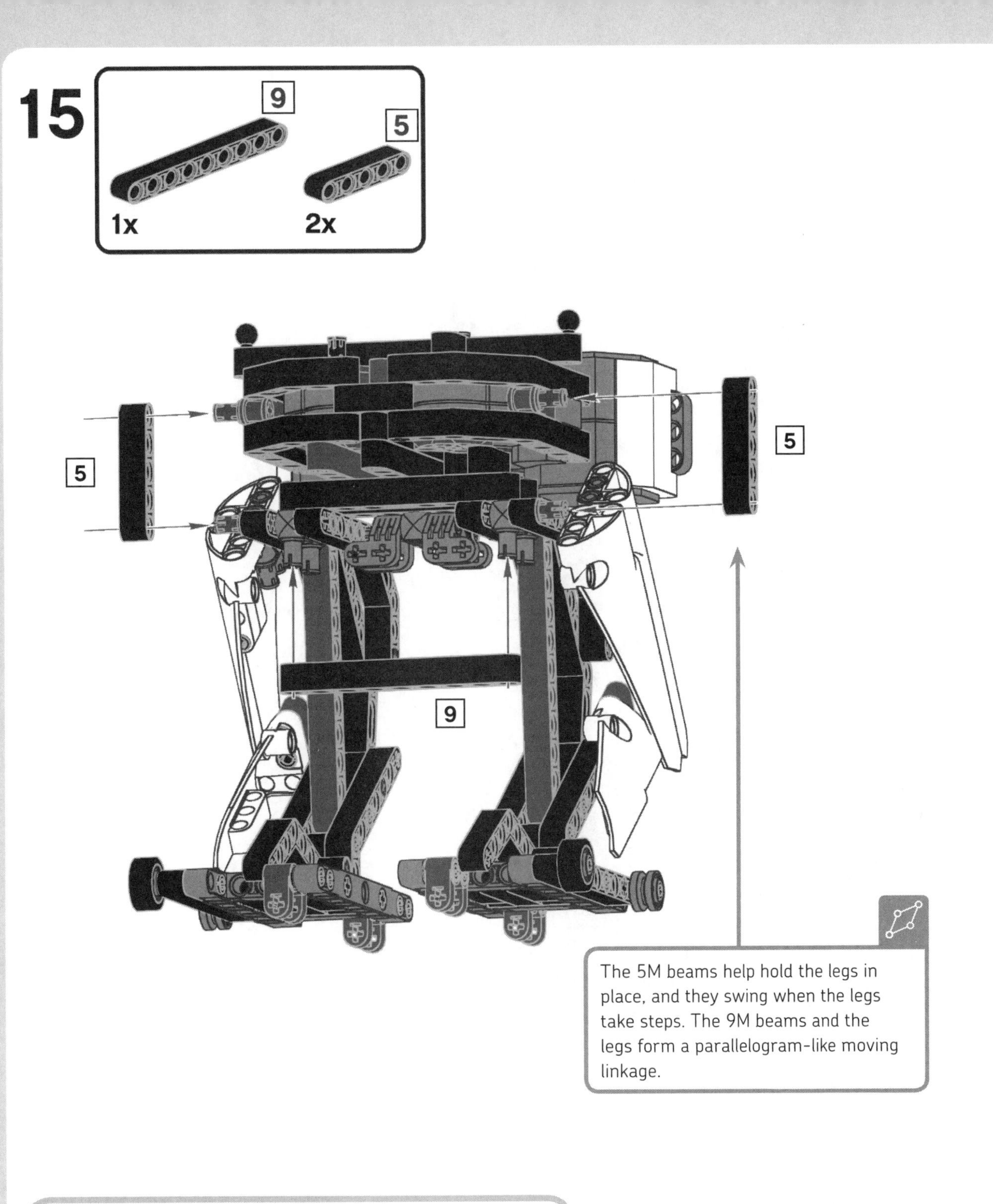
15
9
5
1x
2x
5
5
9
The 5M beams help hold the legs in place, and they swing when the legs take steps. The 9M beams and the legs form a parallelogram-like moving linkage.

5
1:1
9

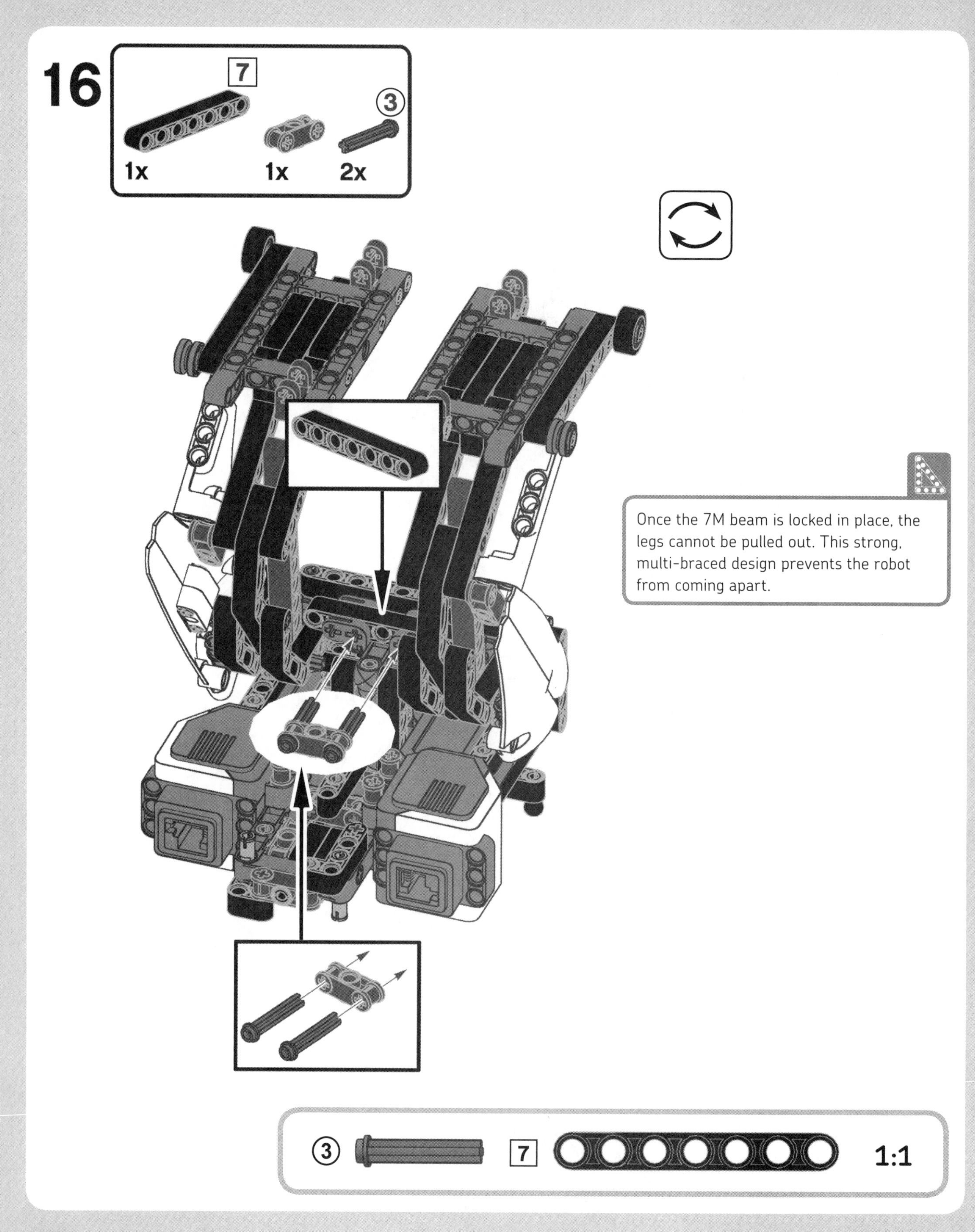
16
7
3
1x
1x
2x
Once the 7M beam is locked in place, the legs cannot be pulled out. This strong, multi-braced design prevents the robot from coming apart.
3
7
1:1

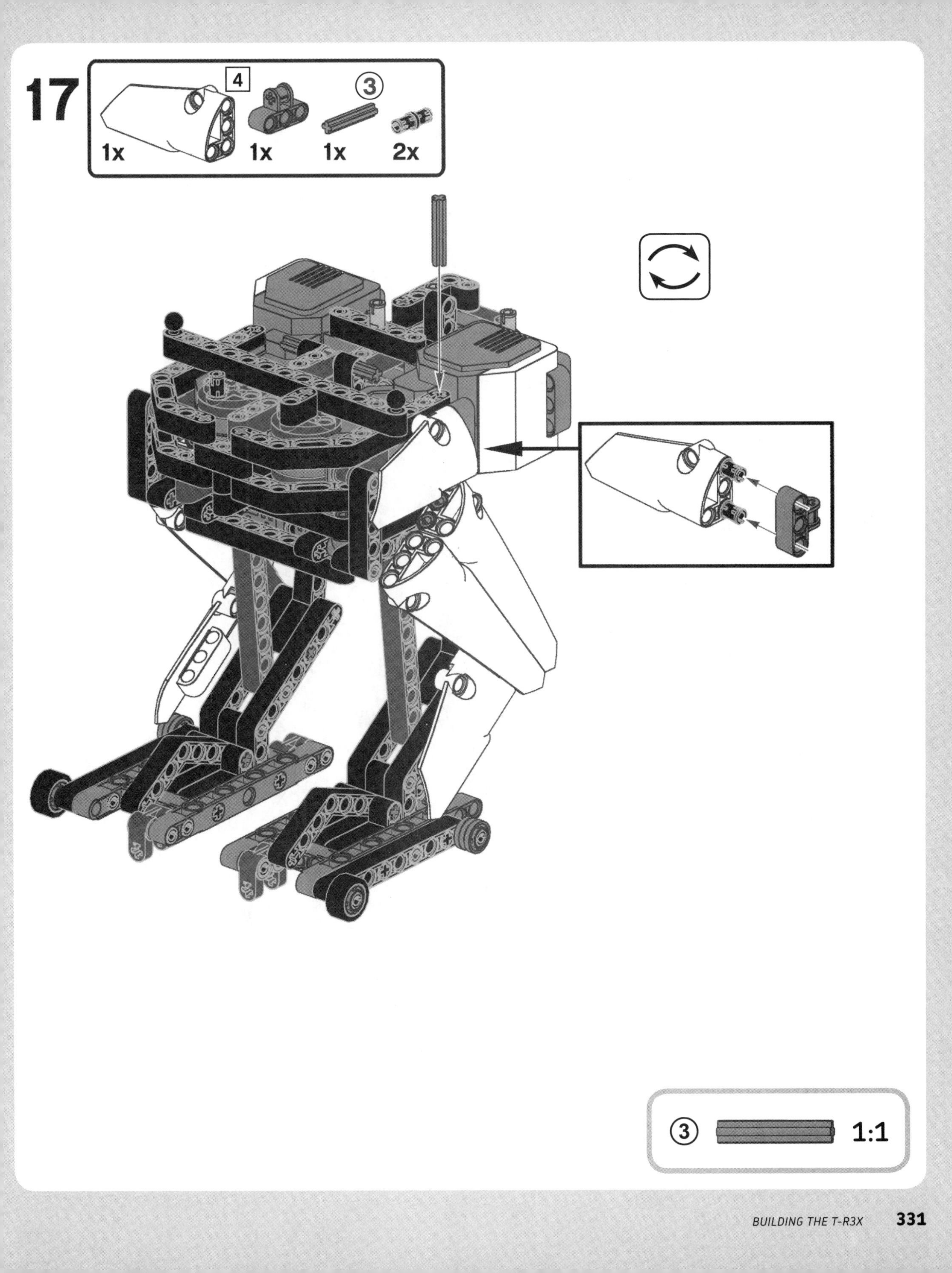
17
4
3
1x
1x
1x
2x
3
1:1

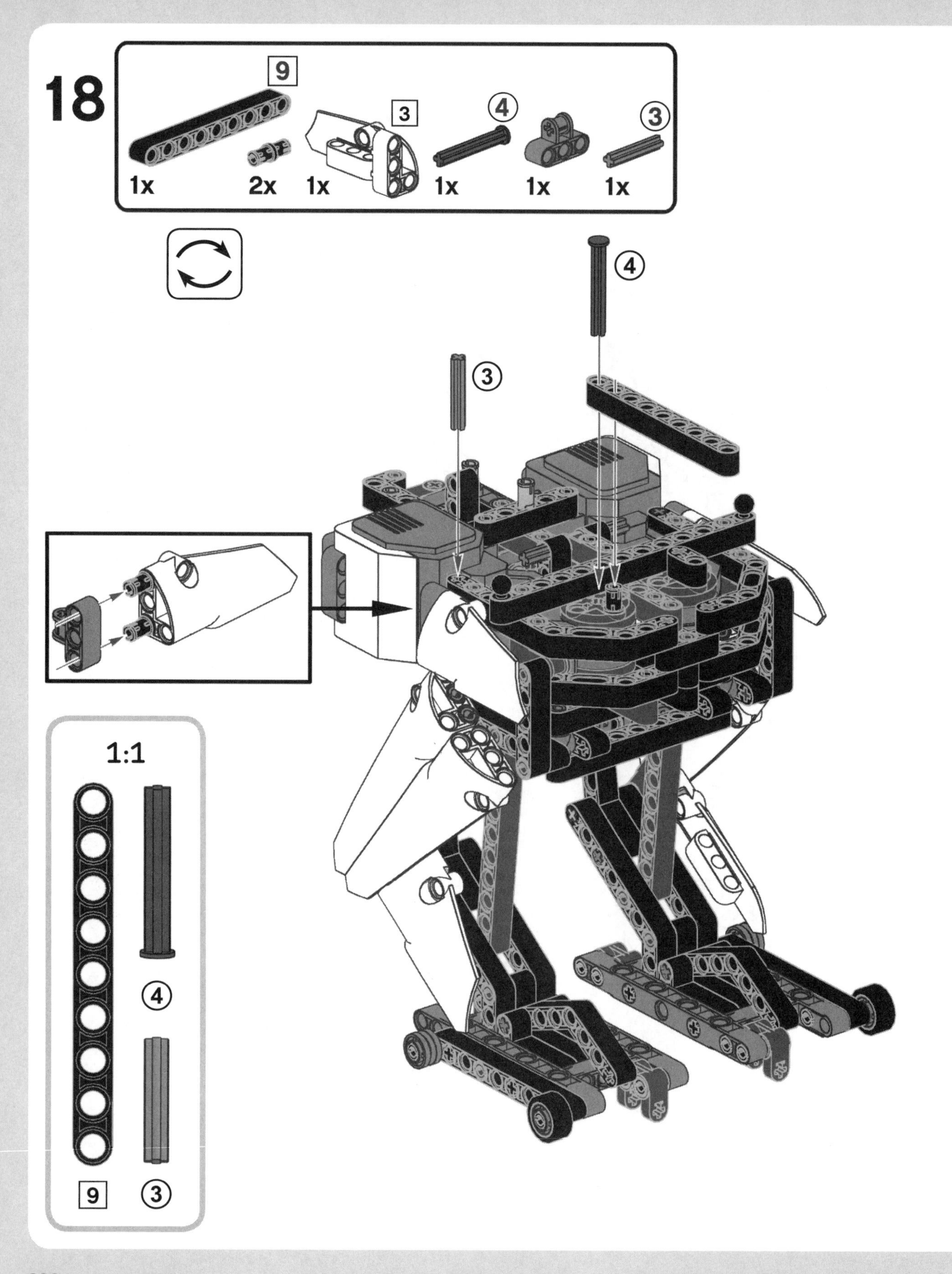
18
9
3
4
3
1x
2x
1x
1x
1x
1x
4
3
1:1
4
9
3

EV3 brick assembly

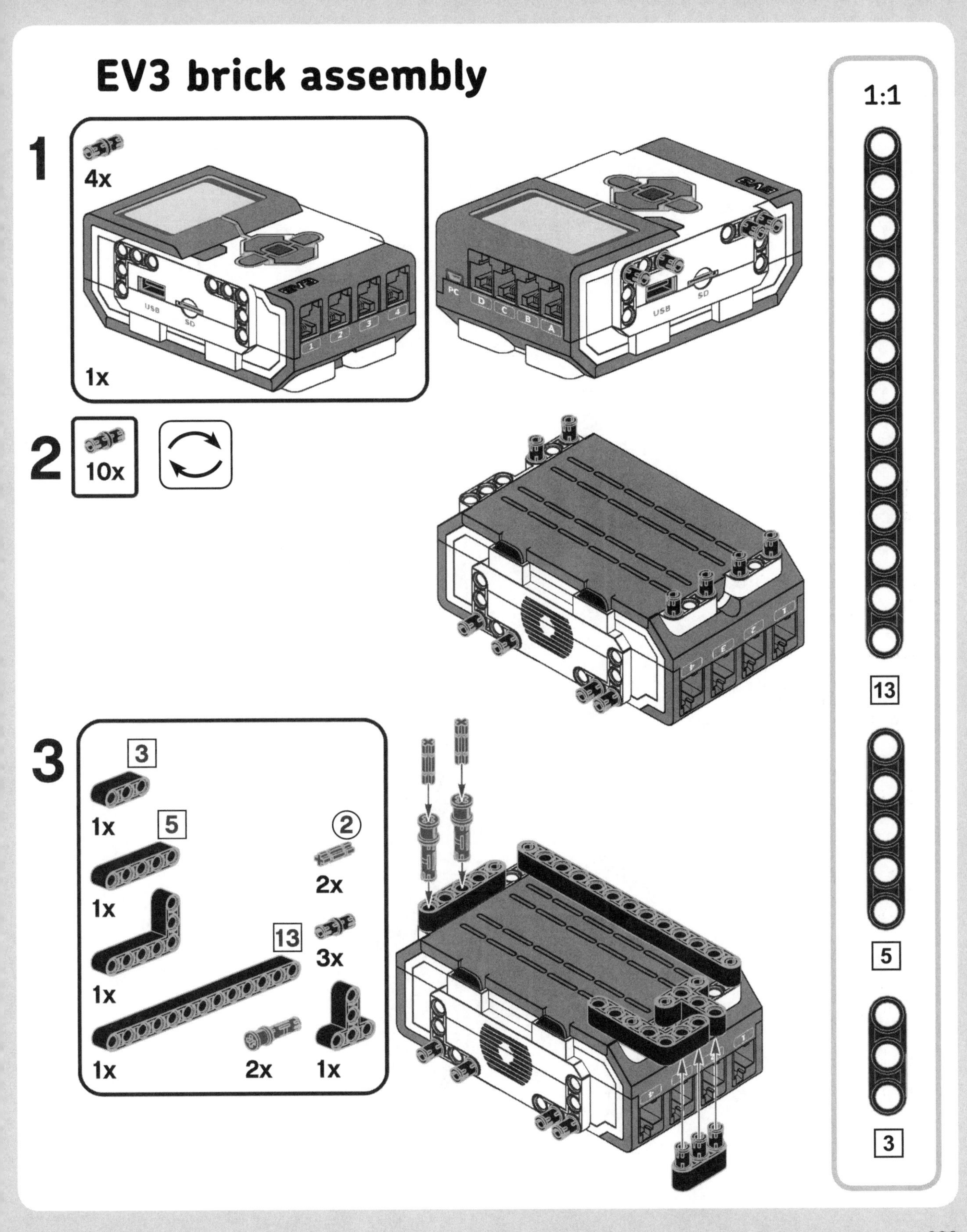

main assembly

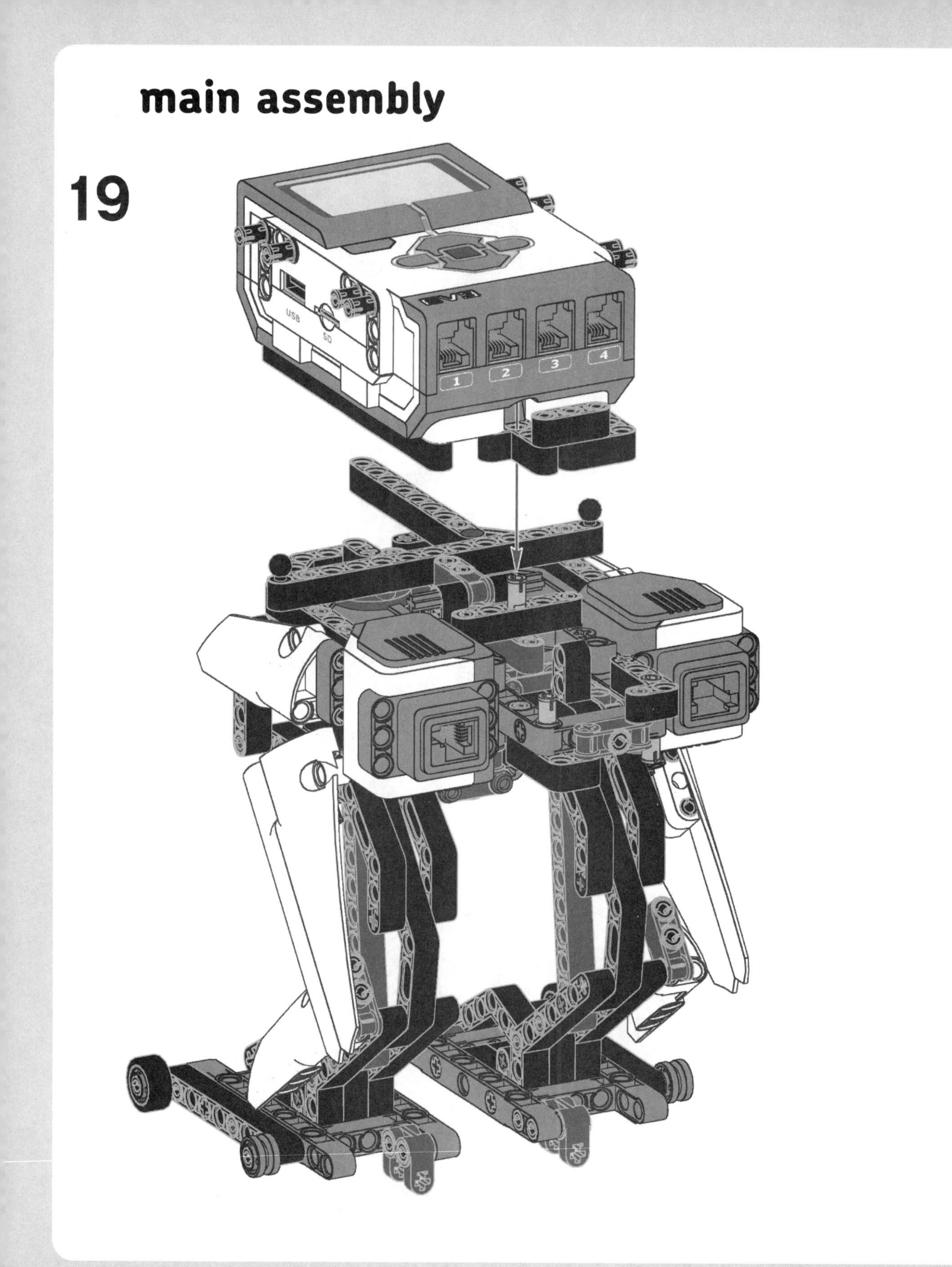

20

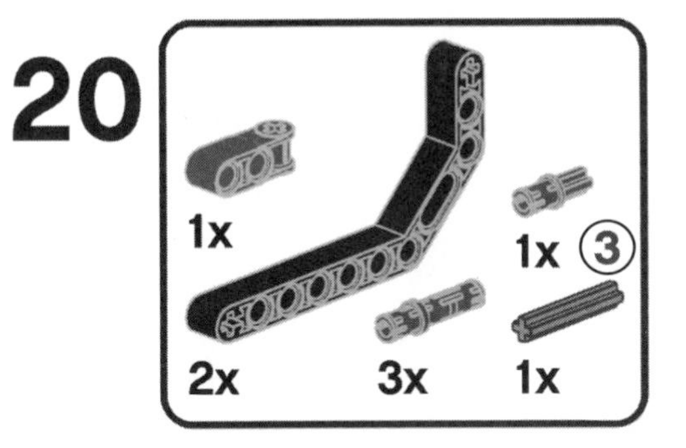

This is another example of bracing. First, place the double angular beam on the right, then lock it by attaching the subassembly on the left. These beams will hold the EV3 Brick in place. (You'll finish locking it later.)

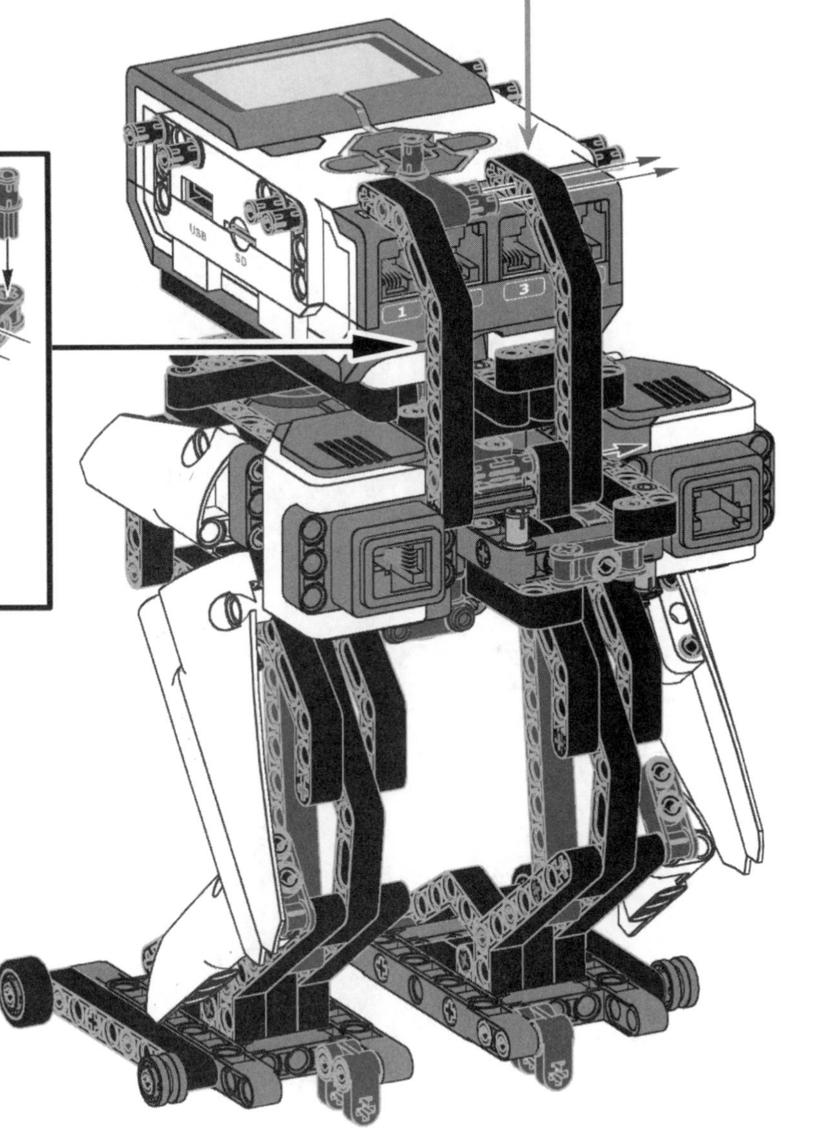

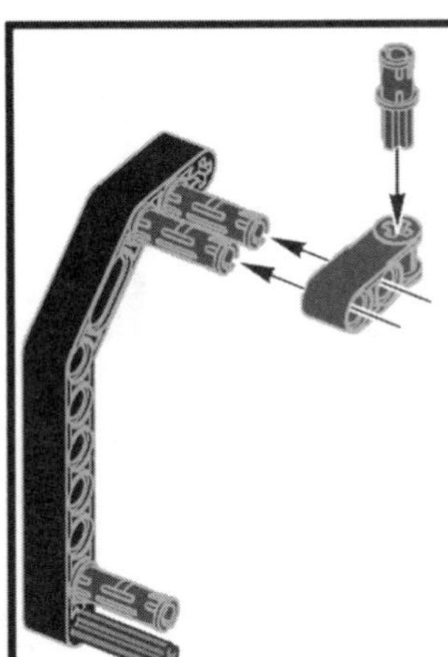

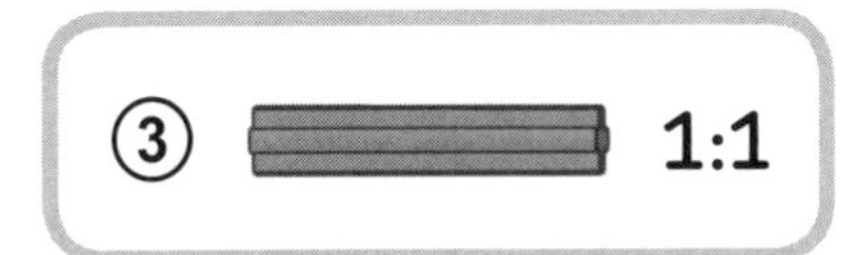

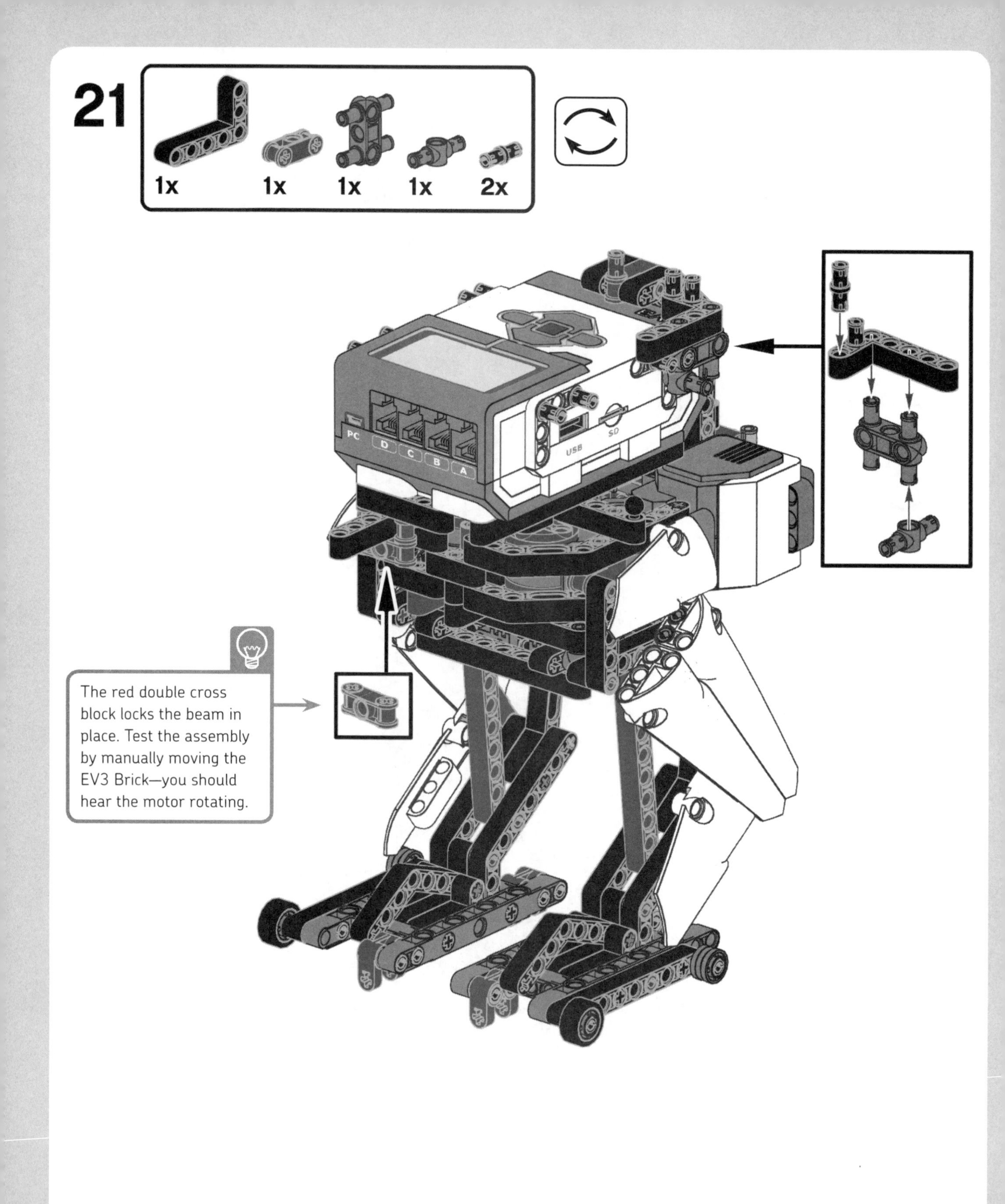
21
1x
1x
1x
1x
2x
PC
D
C
B
A
USB
SD
The red double cross block locks the beam in place. Test the assembly by manually moving the EV3 Brick—you should hear the motor rotating.

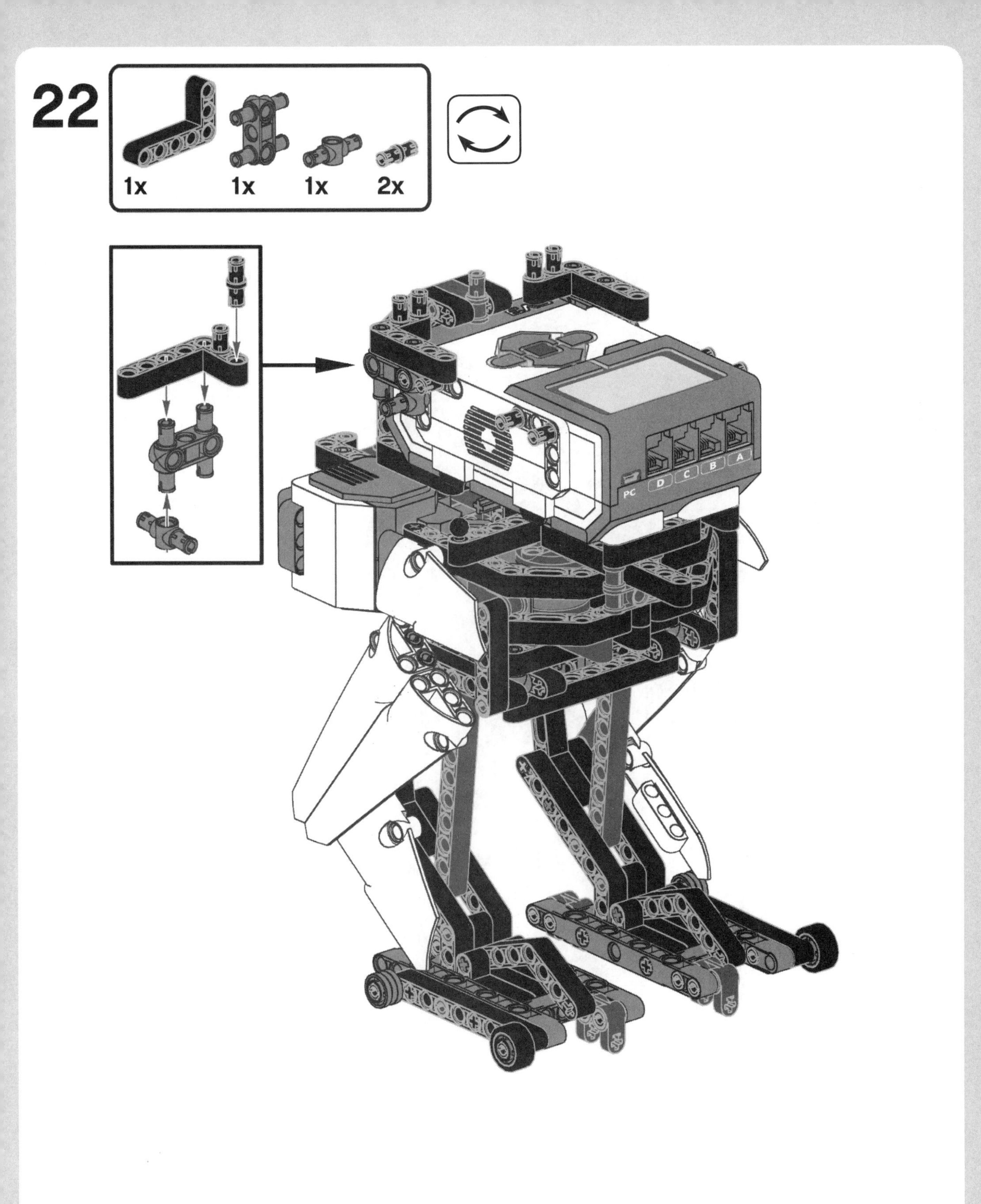
22
1x
1x
1x
2x
PC
D
C
B
A

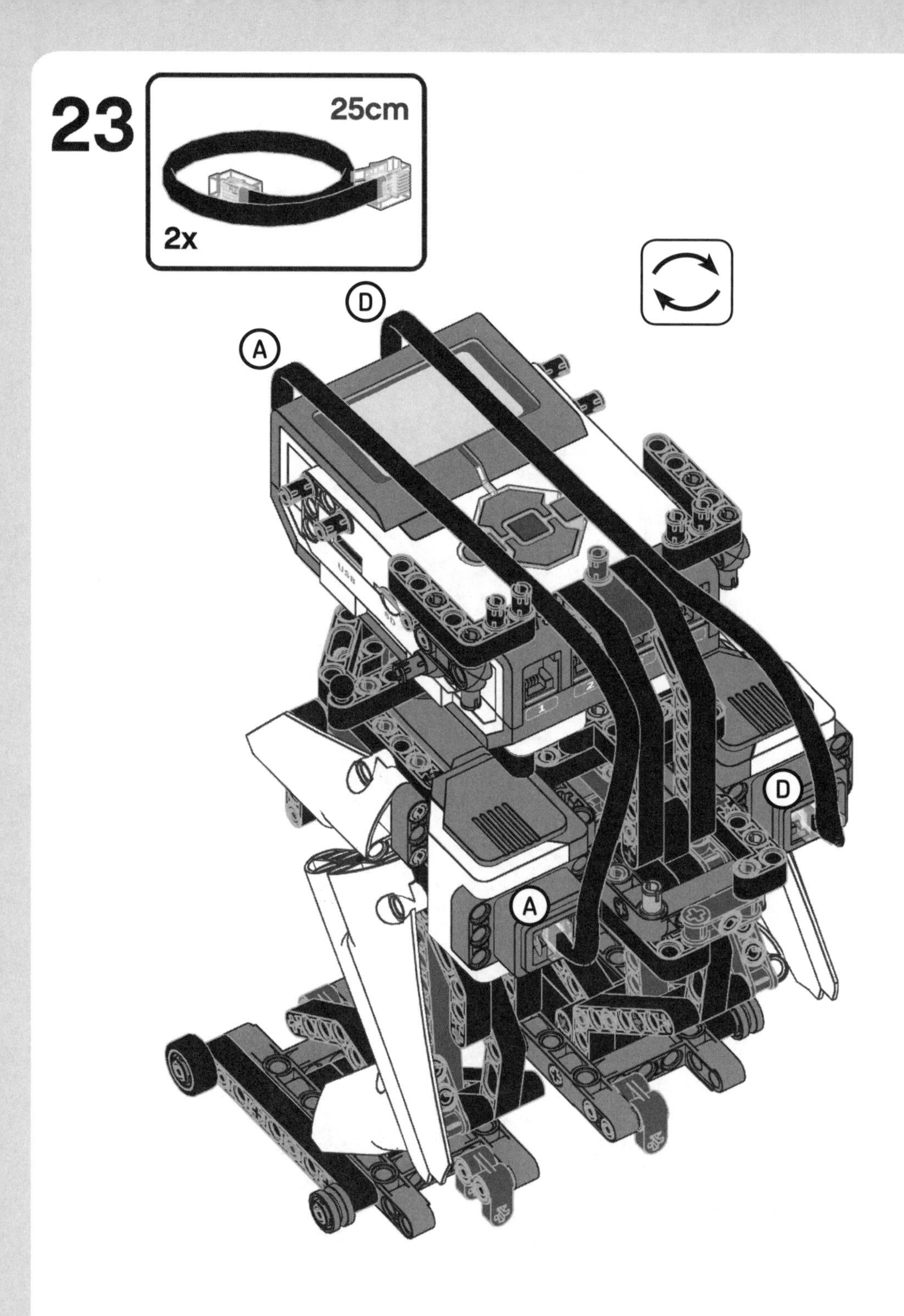
23
25cm
2x
D
A
D
A

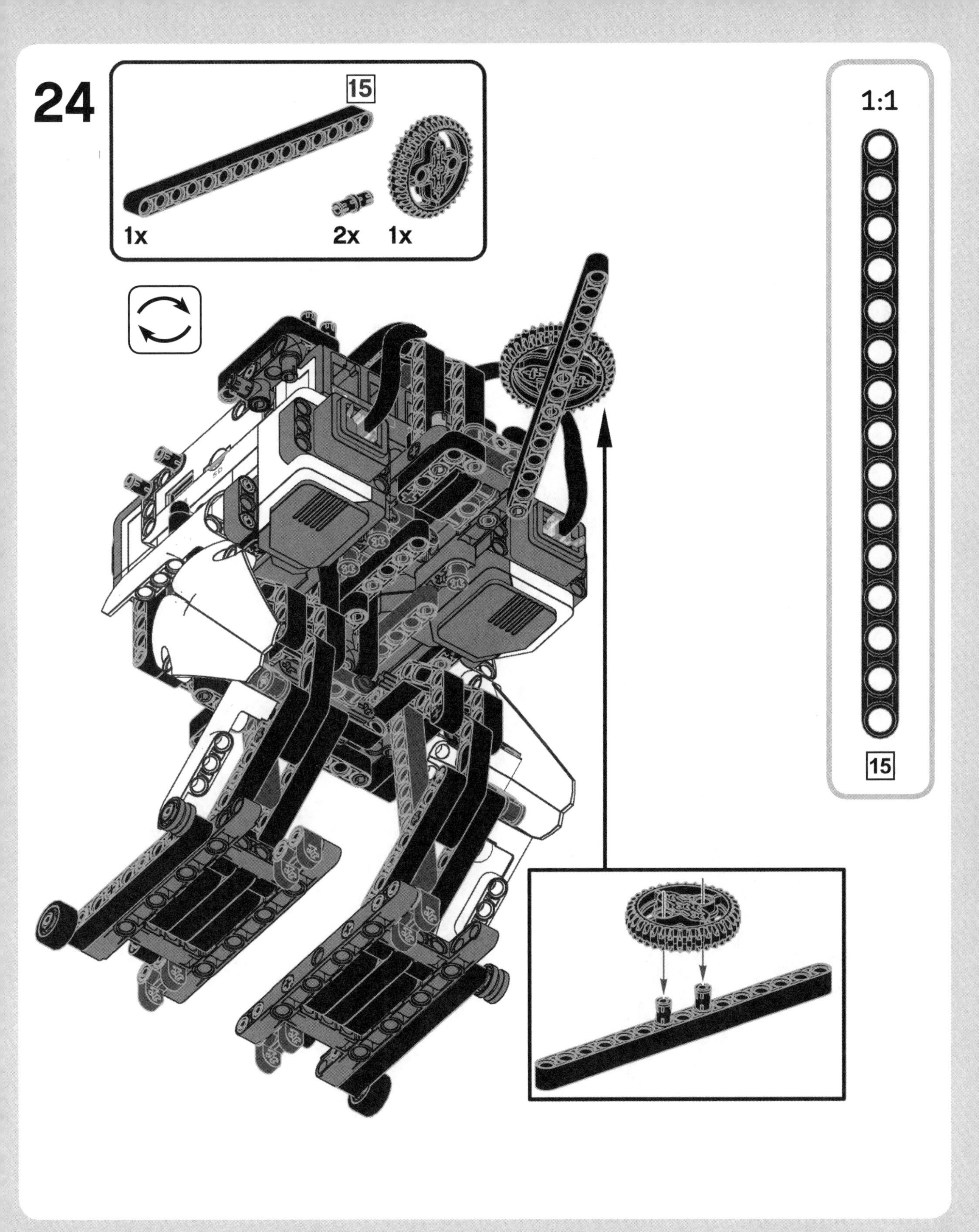
24
15
1x
2x
1x
1:1
15

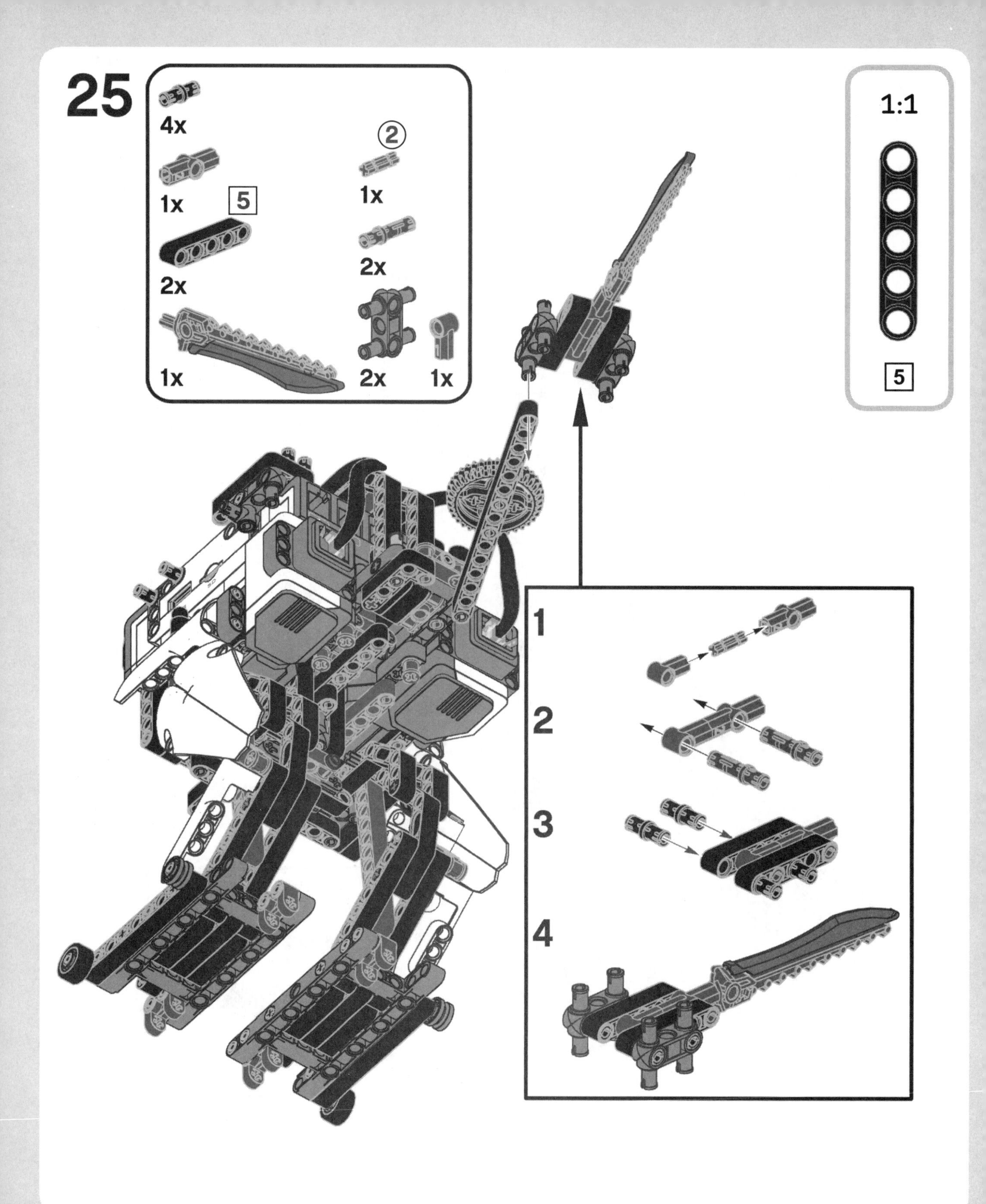
25
4x
1x
5
2x
1x
2
1x
2x
2x
1x
1:1
5
1
2
3
4

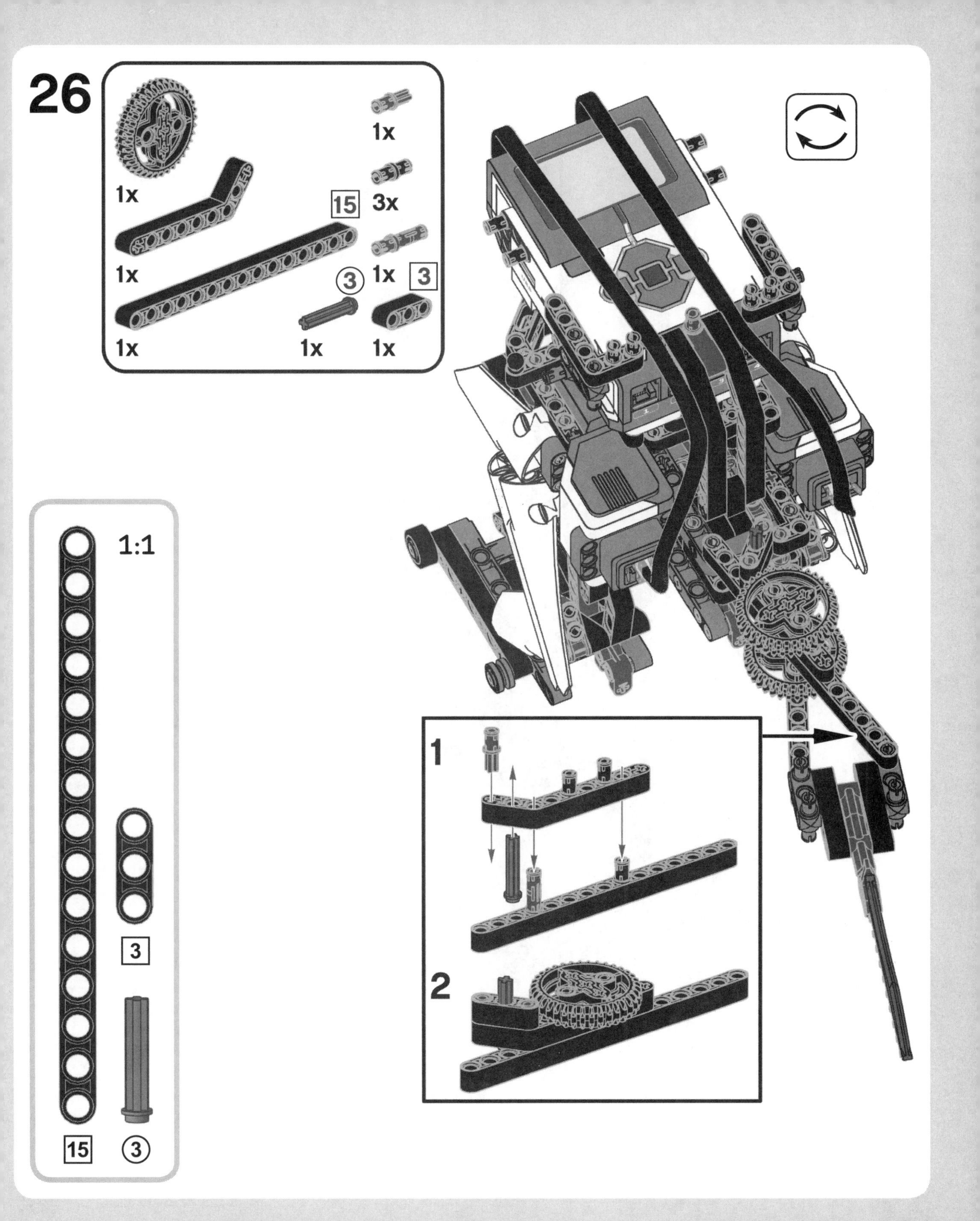
26
1x
3x
1x
15
1x
3
3
1x
1x
1x
1x
1x
1x
1:1
3
15
3
1
2

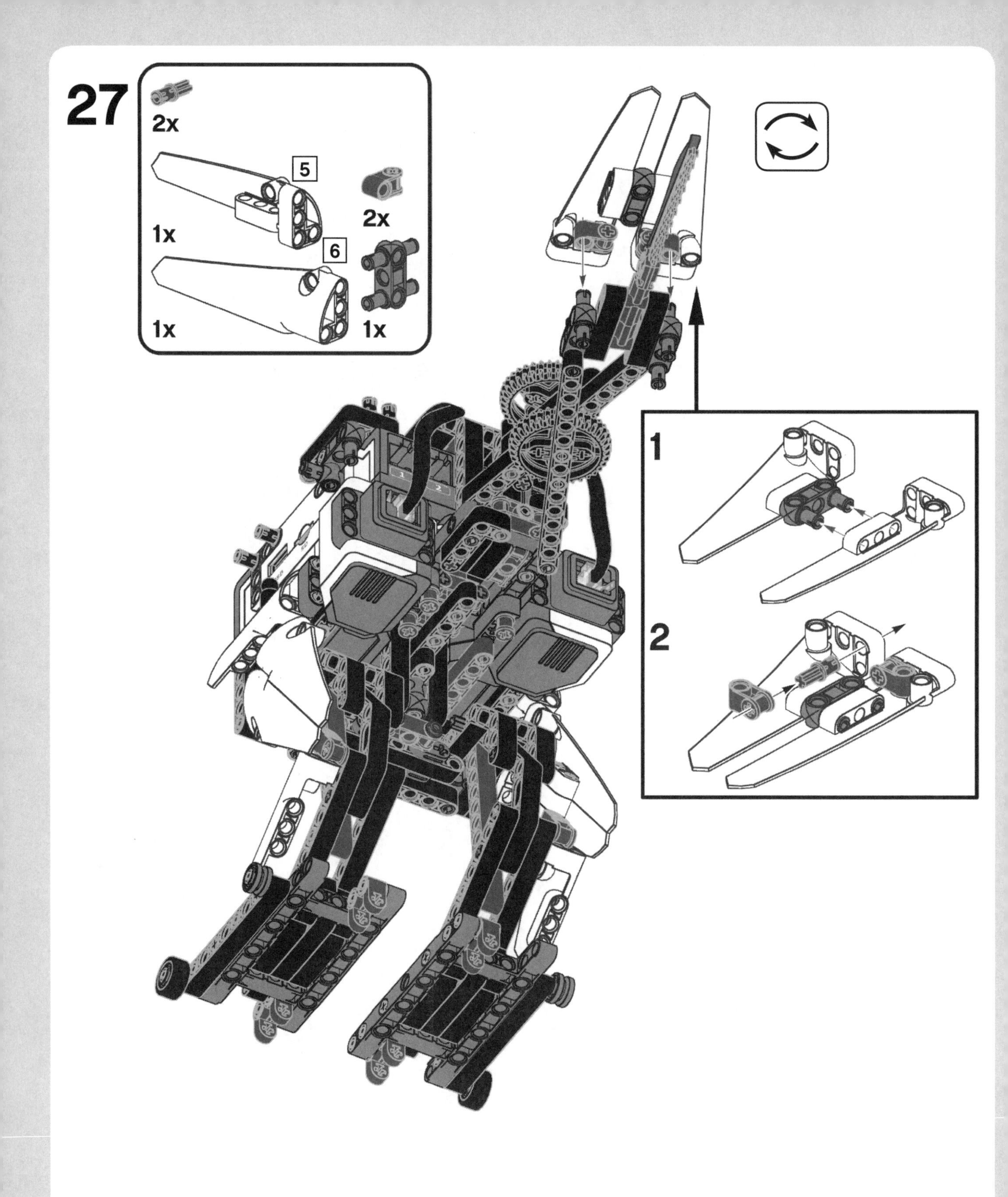
27
2x
5
1x
2x
6
1x
1x
1
2

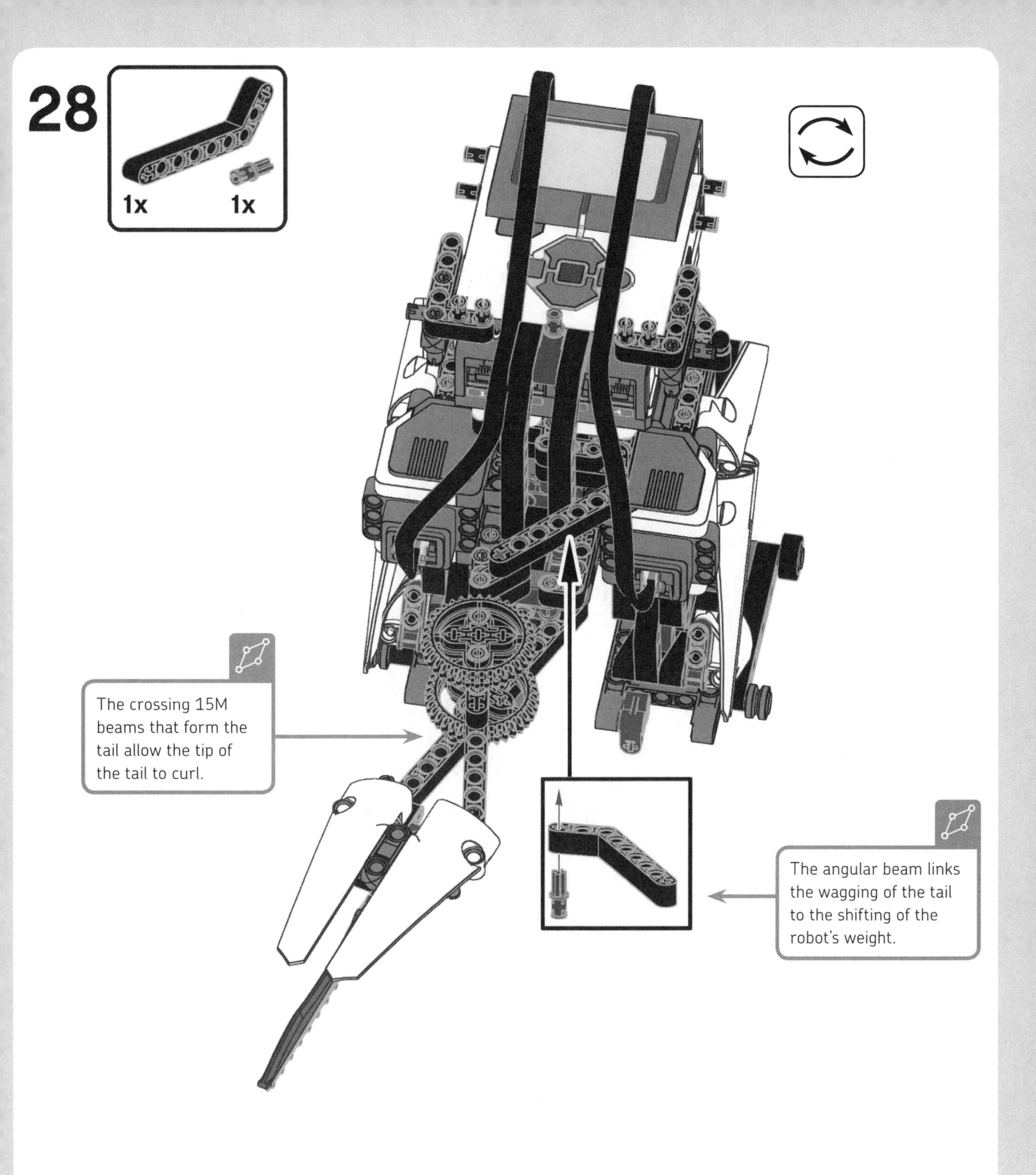
28
1x
1x
The crossing 15M beams that form the tail allow the tip of the tail to curl.
The angular beam links the wagging of the tail to the shifting of the robot's weight.

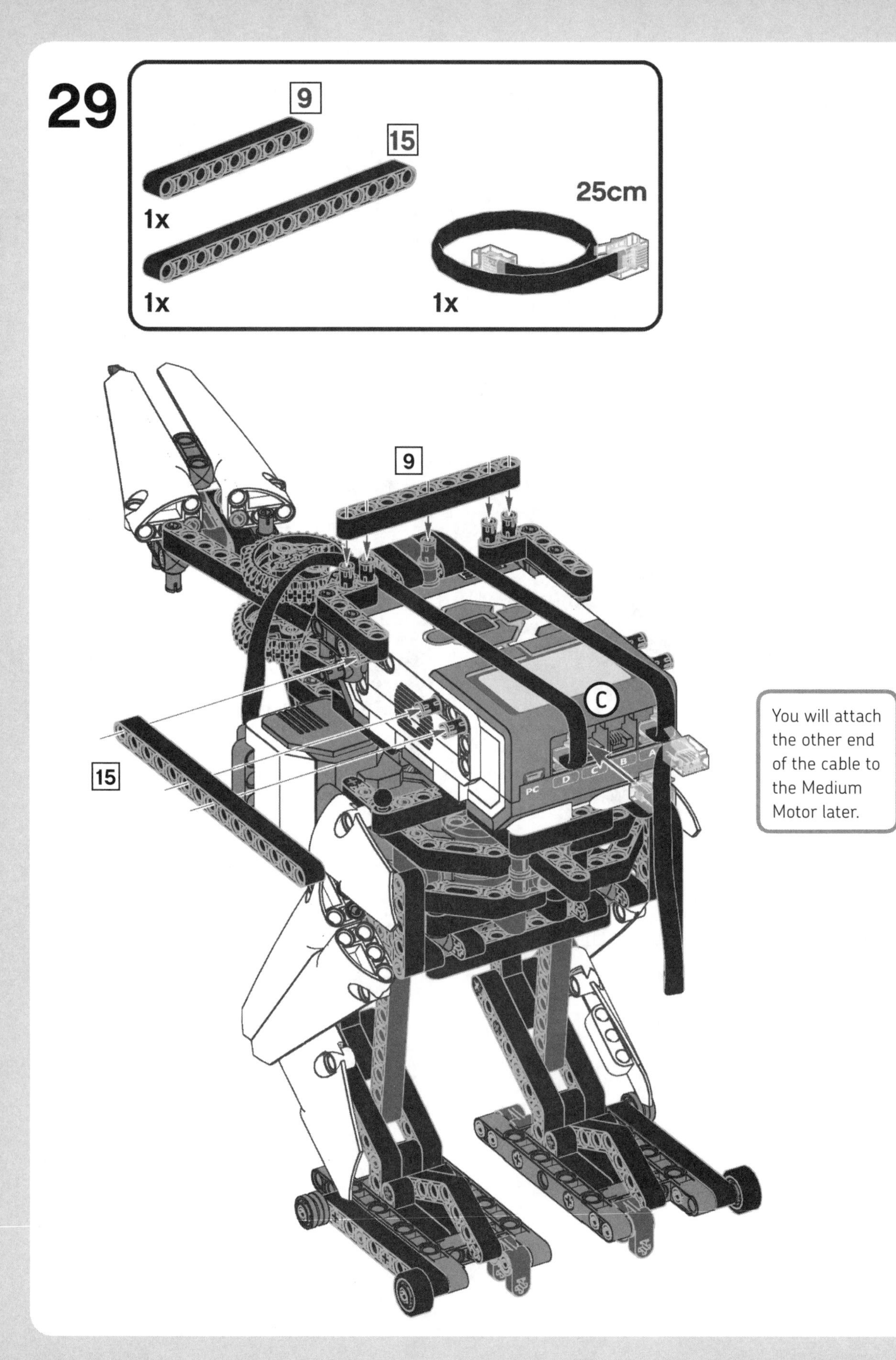

You will attach the other end of the cable to the Medium Motor later.

head and torso assembly

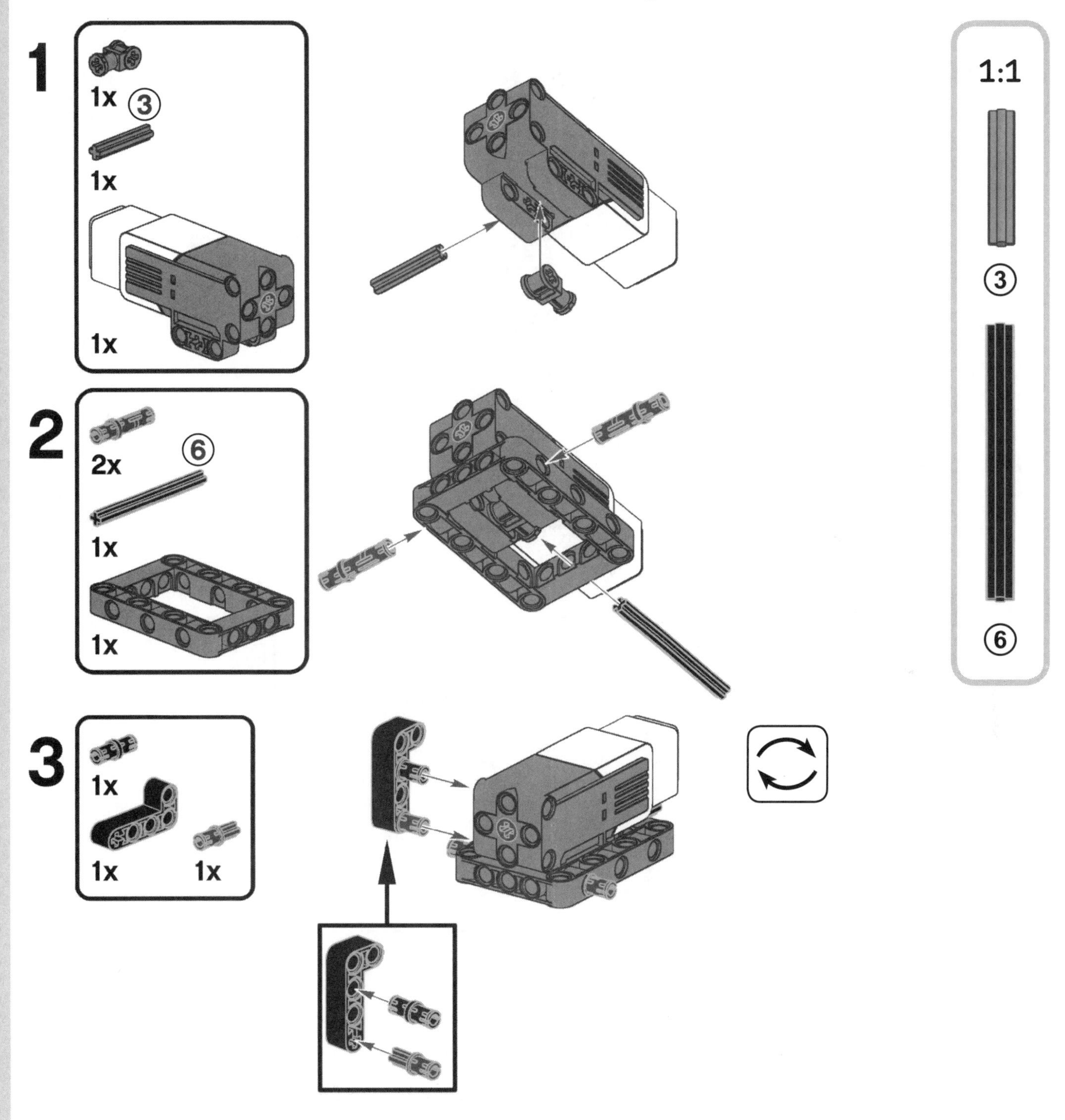

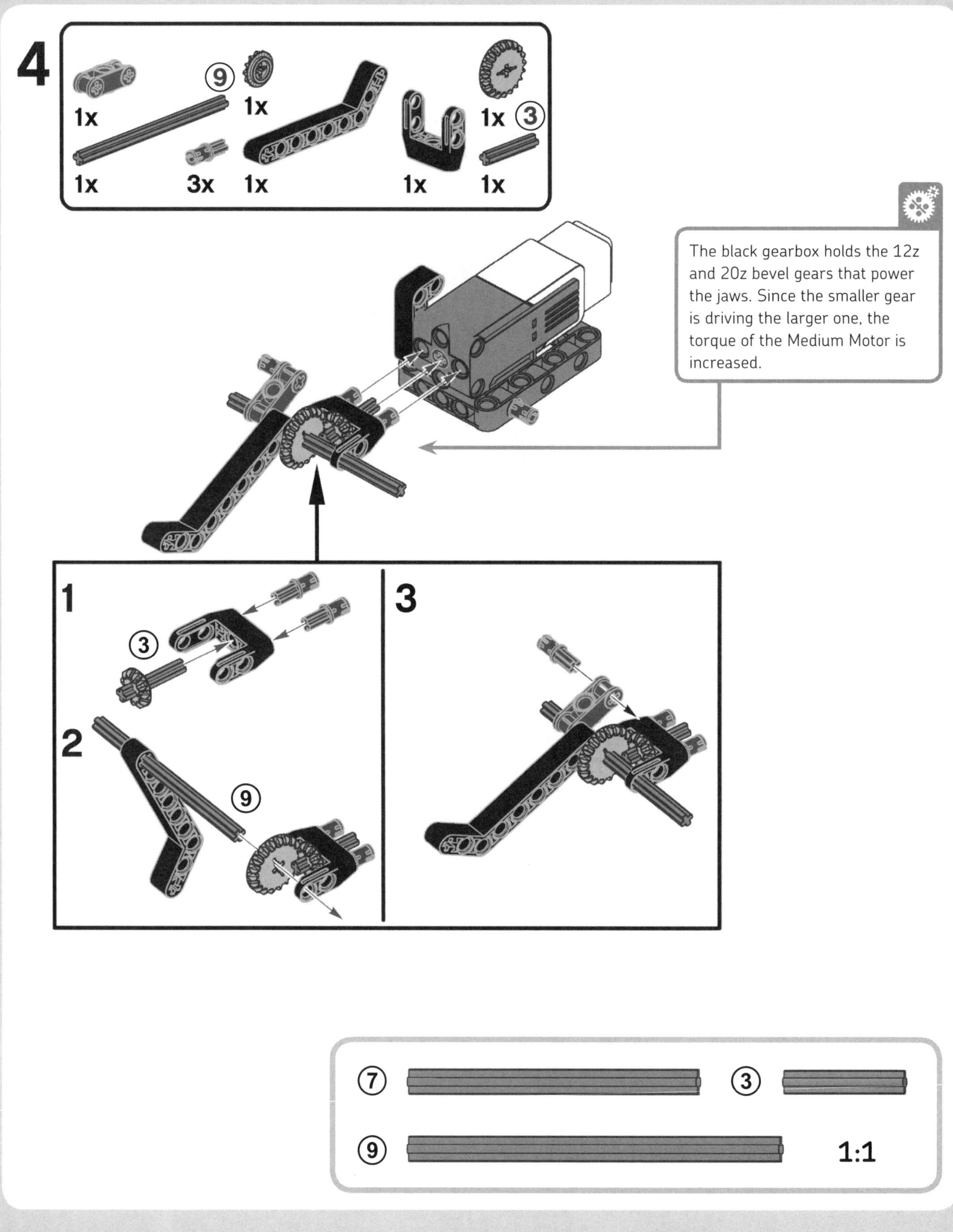
4
1x
⑨
1x
1x
1x ③
1x
3x
1x
1x
1x
The black gearbox holds the 12z and 20z bevel gears that power the jaws. Since the smaller gear is driving the larger one, the torque of the Medium Motor is increased.
1
③
2
⑨
3
⑦
③
⑨
1:1

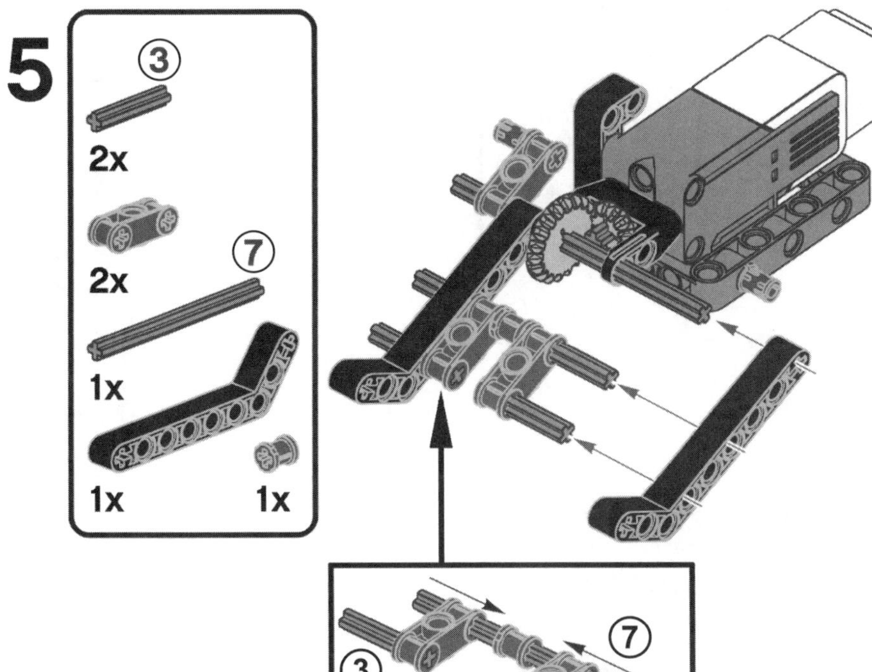
5
3
2x
2x
7
1x
1x
1x

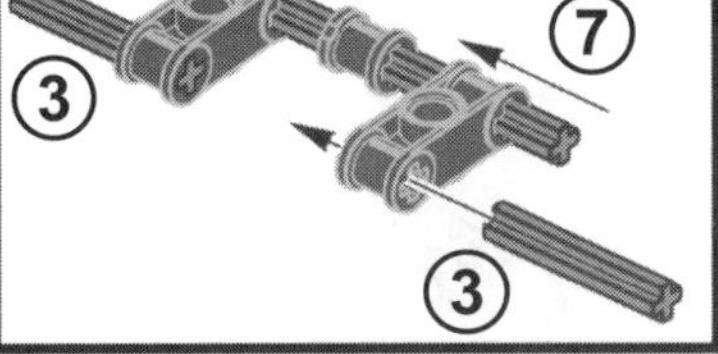
7
3
3

6

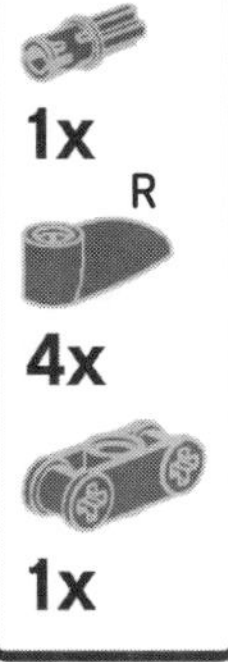
1x
R
4x
1x

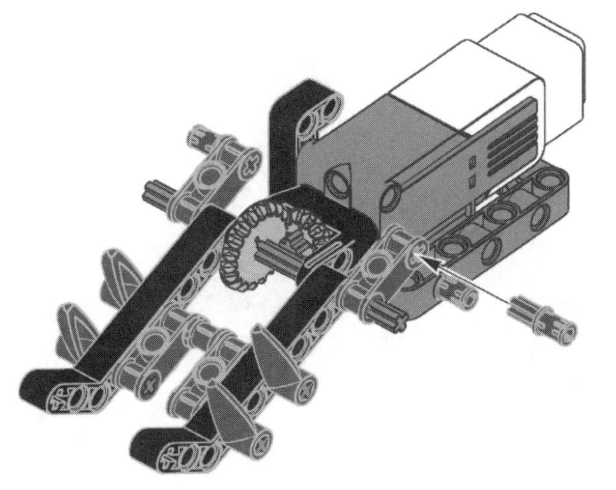

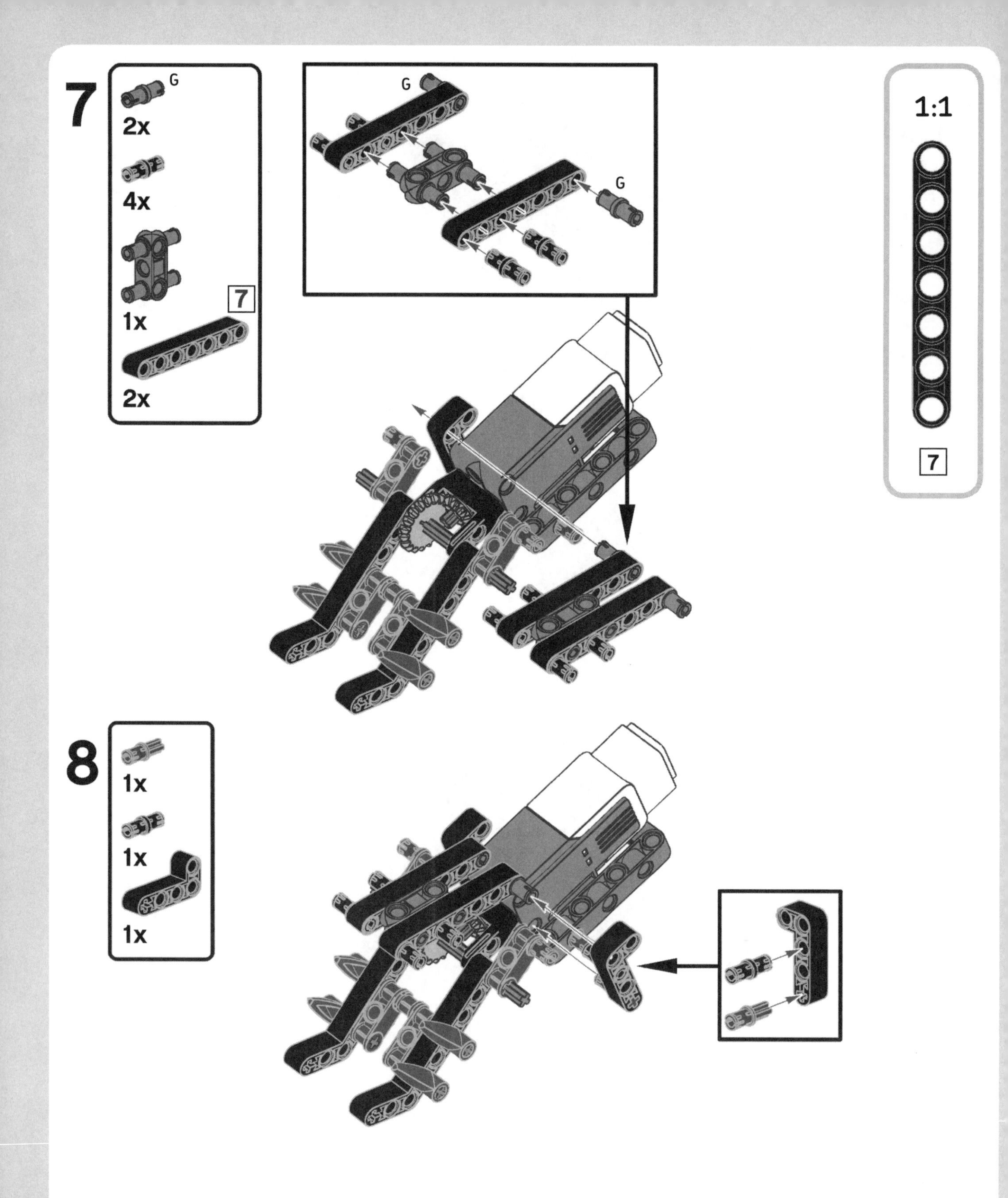
7
G
2x
4x
1x
7
2x
G
G
1:1
7
8
1x
1x
1x

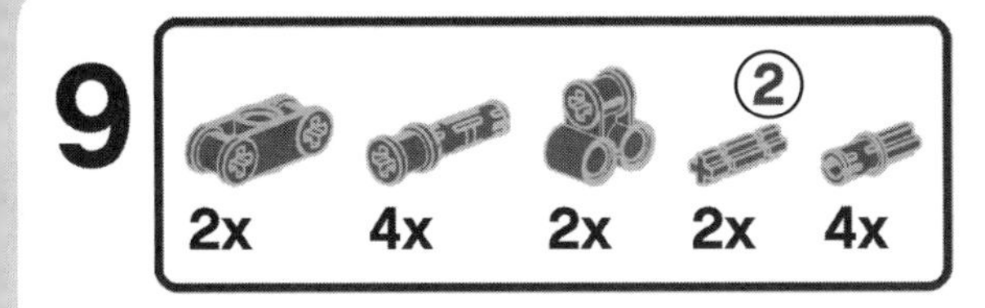
9
2
2x 4x 2x 2x 4x

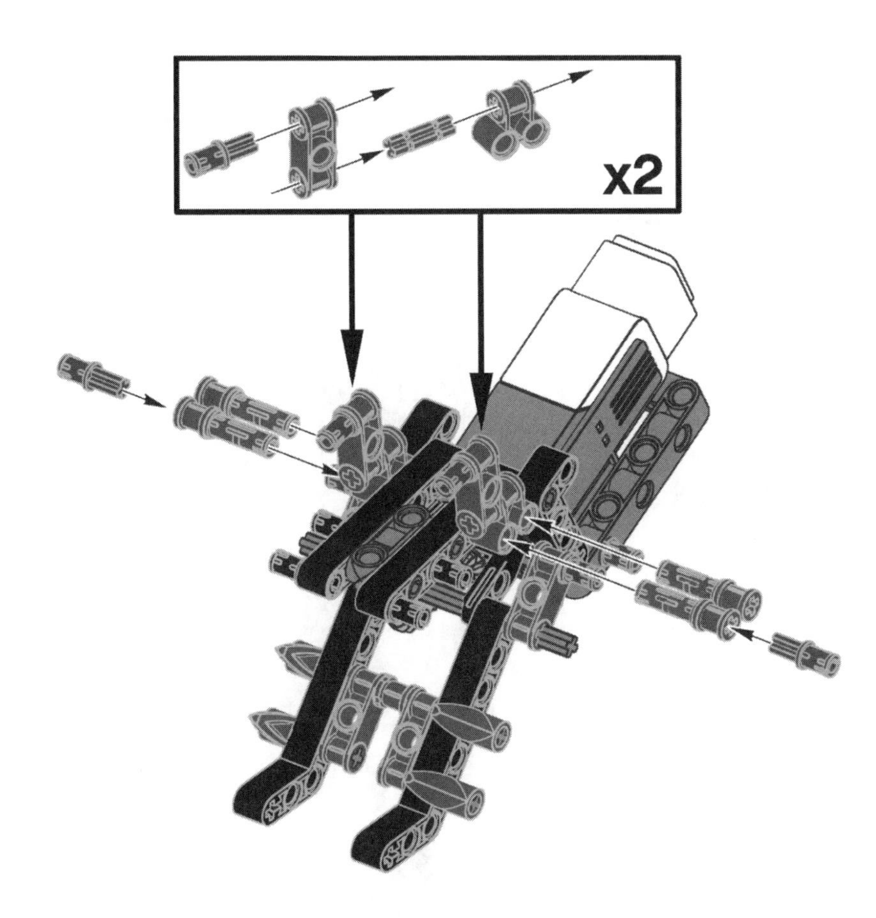
x2

10

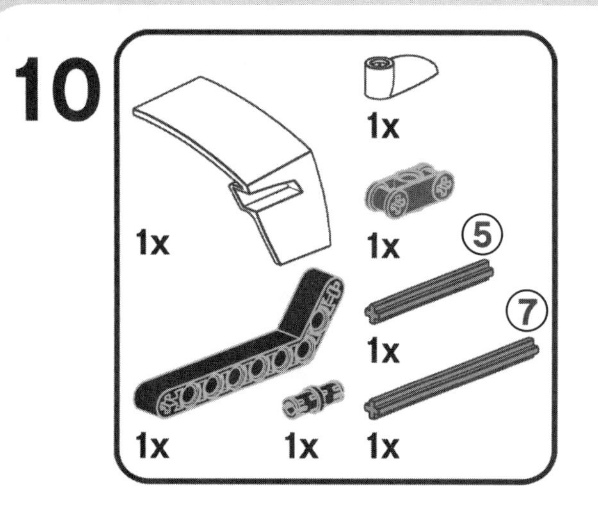

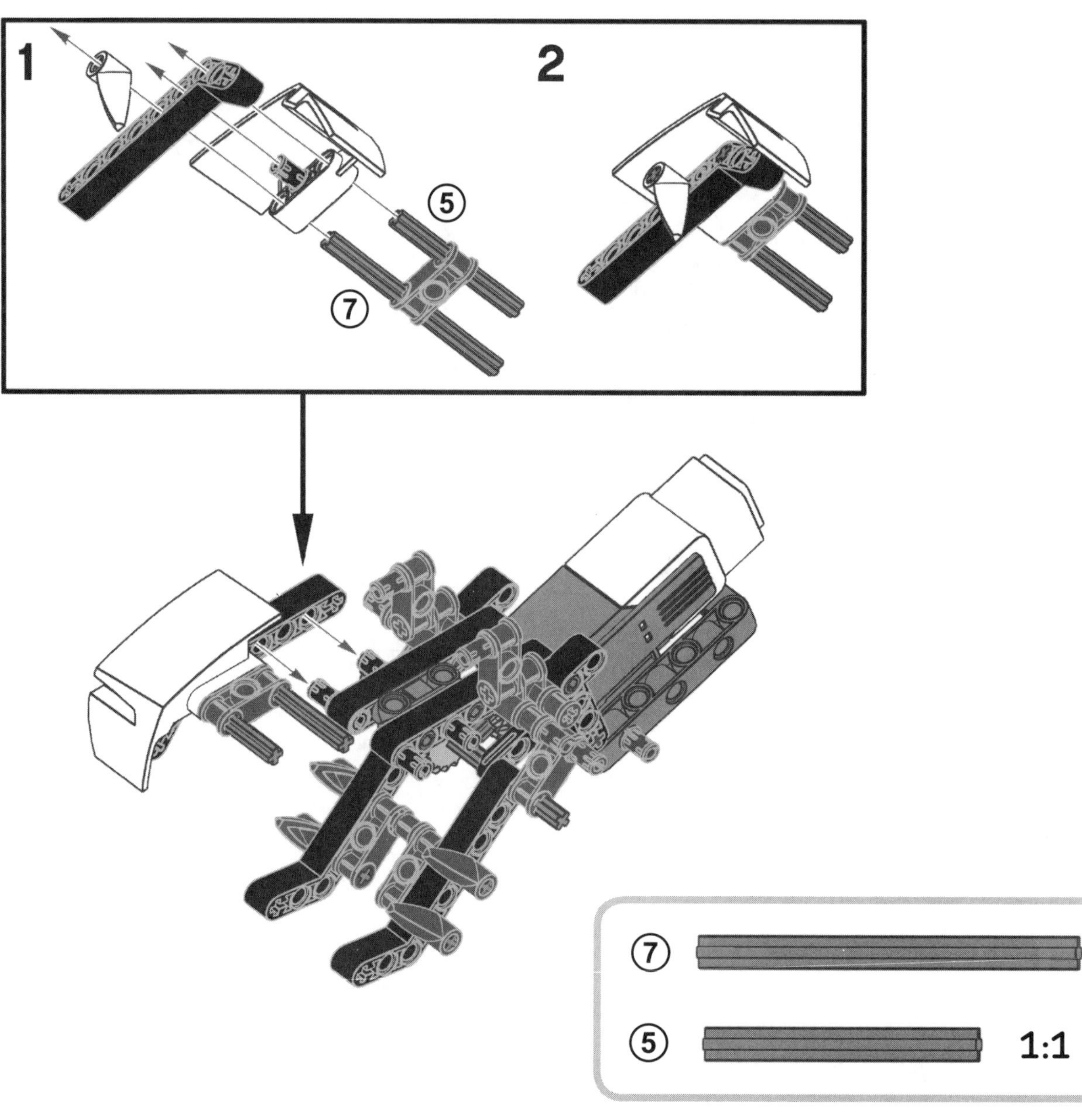

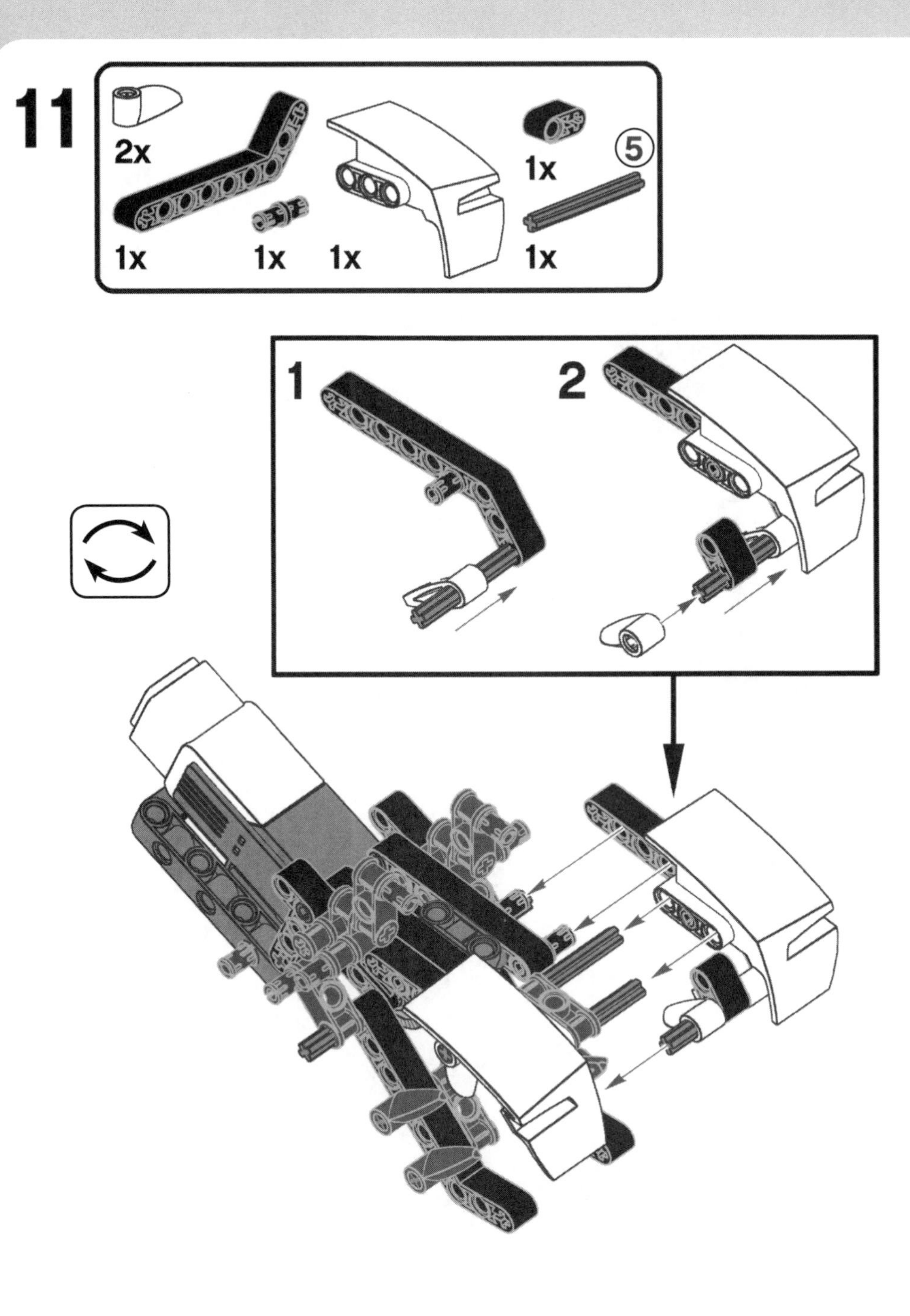

11
2x
1x
1x
1x
1x
5
1x
1
2

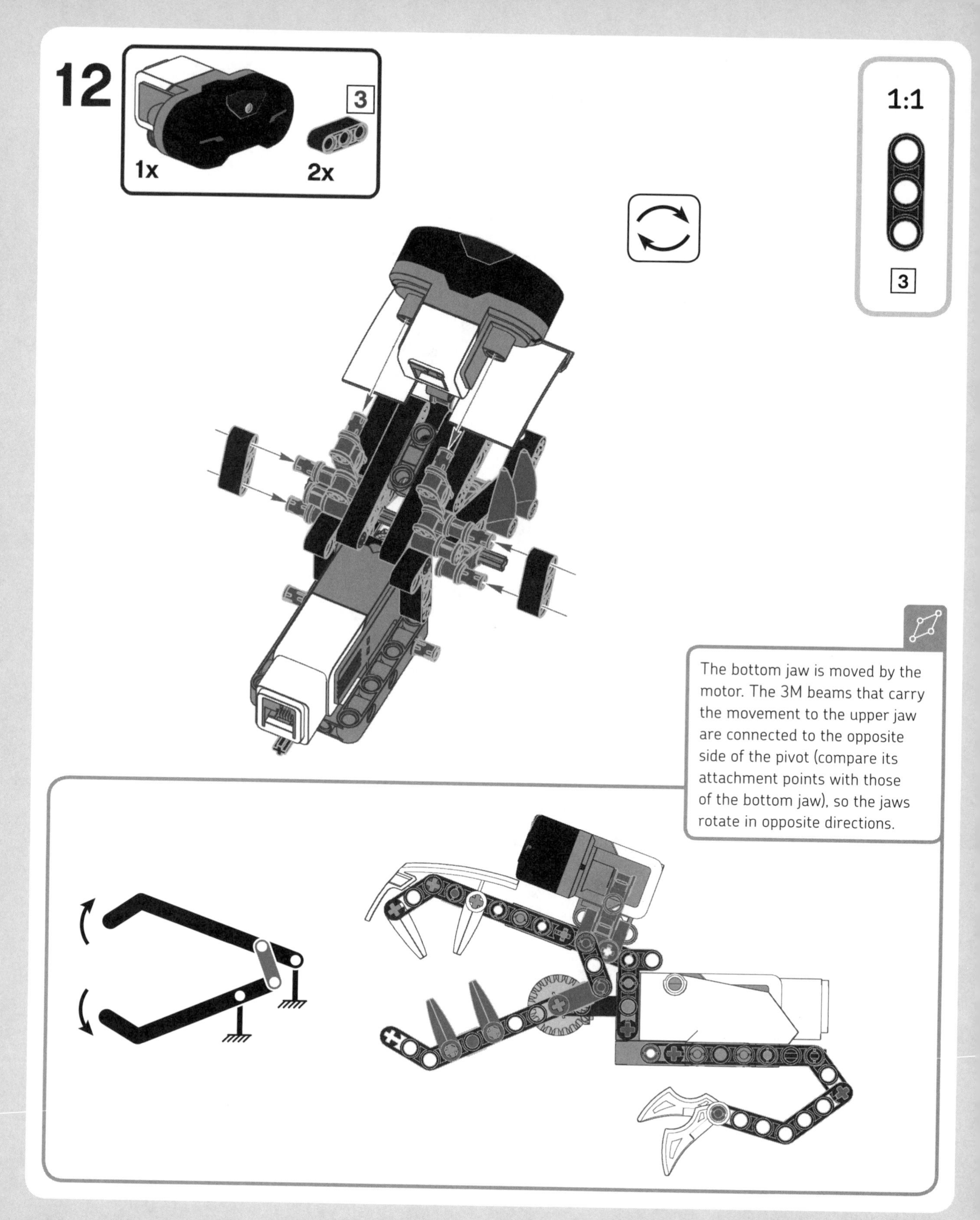

The bottom jaw is moved by the motor. The 3M beams that carry the movement to the upper jaw are connected to the opposite side of the pivot (compare its attachment points with those of the bottom jaw), so the jaws rotate in opposite directions.

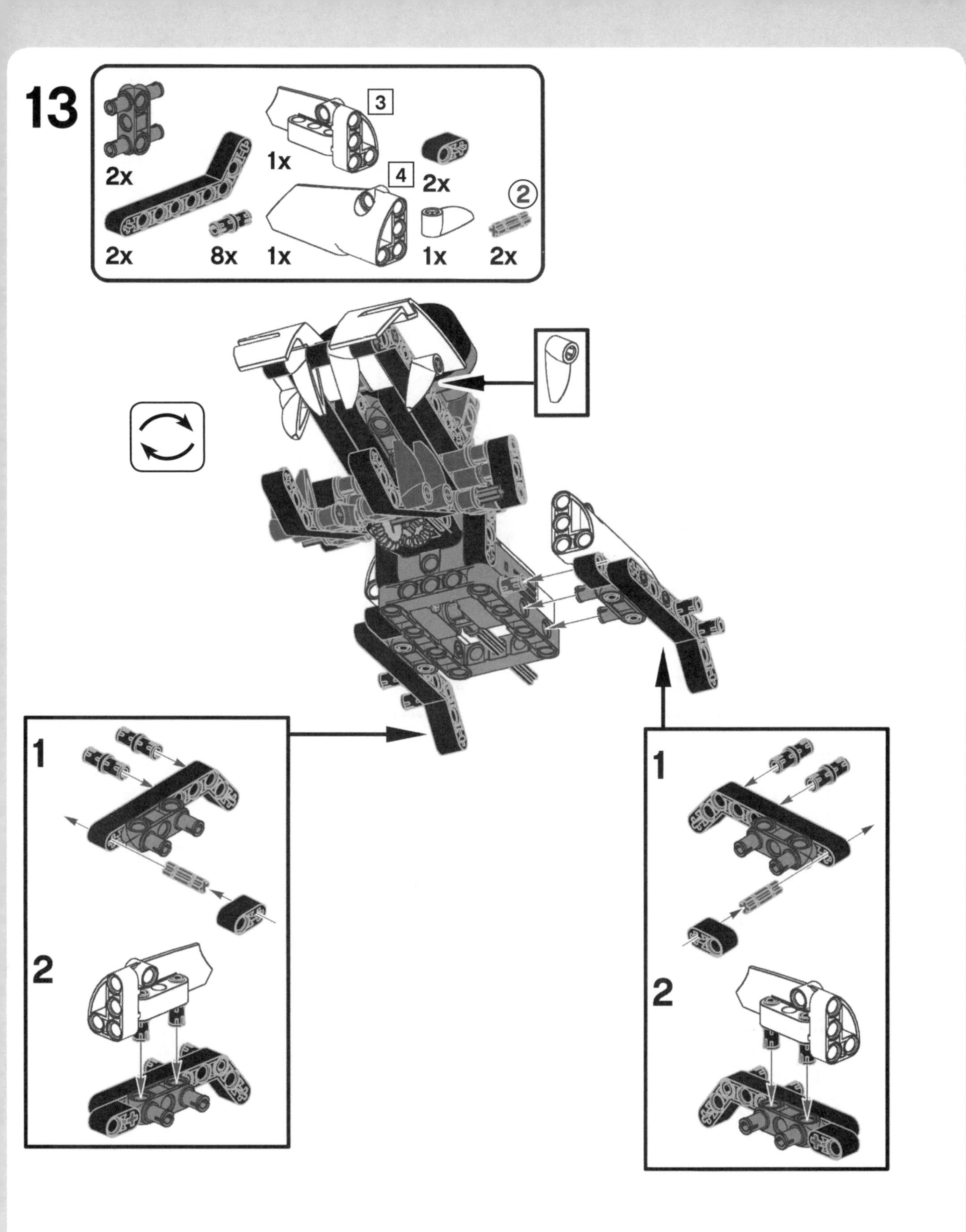
13
2x
2x
8x
1x
3
1x
4
2x
1x
2
2x
1
2
1
2

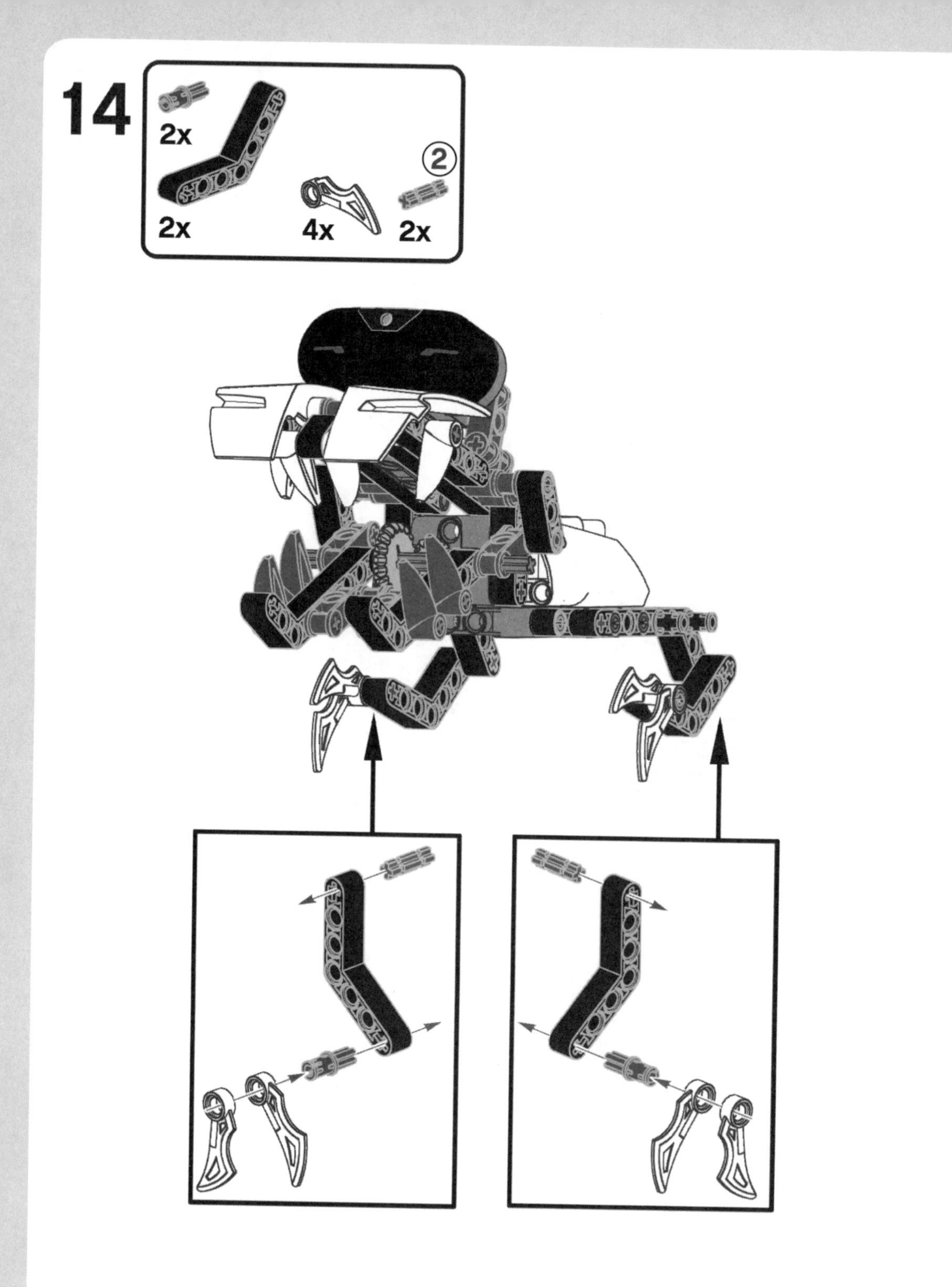
14
2x
2x
4x
2
2x

main assembly

30

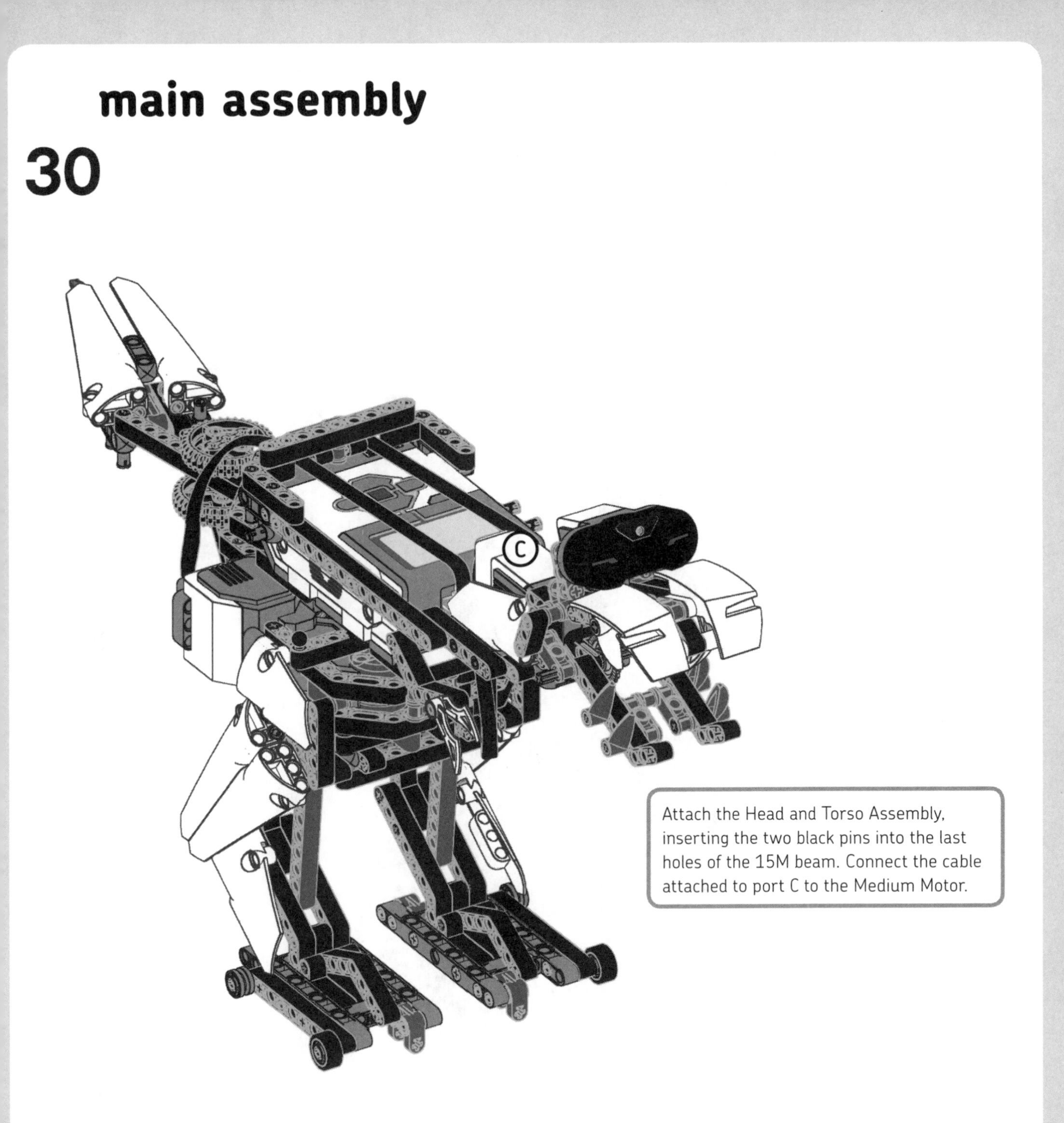

Attach the Head and Torso Assembly, inserting the two black pins into the last holes of the 15M beam. Connect the cable attached to port C to the Medium Motor.

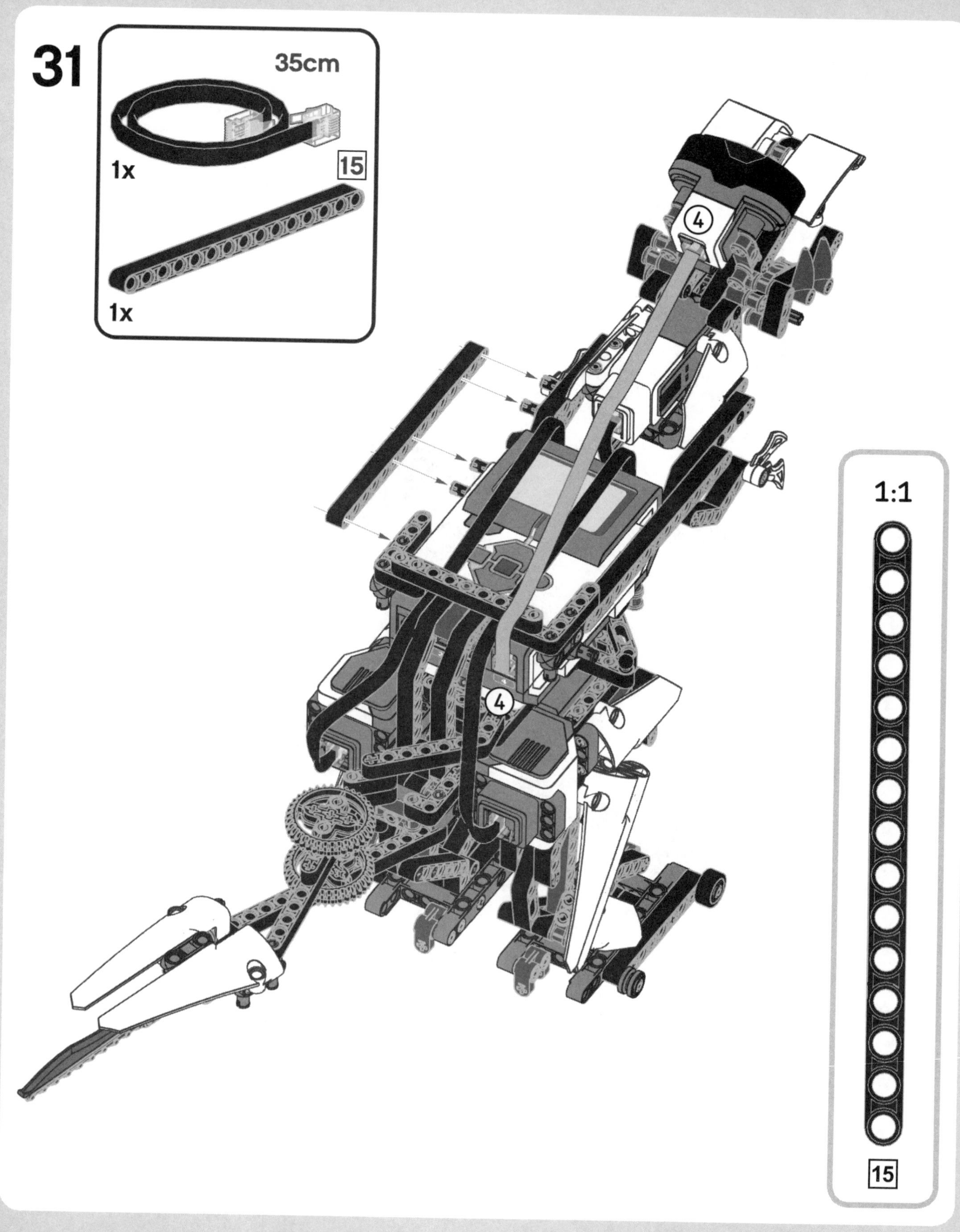
31
35cm
1x
15
1x
4
4
1:1
15

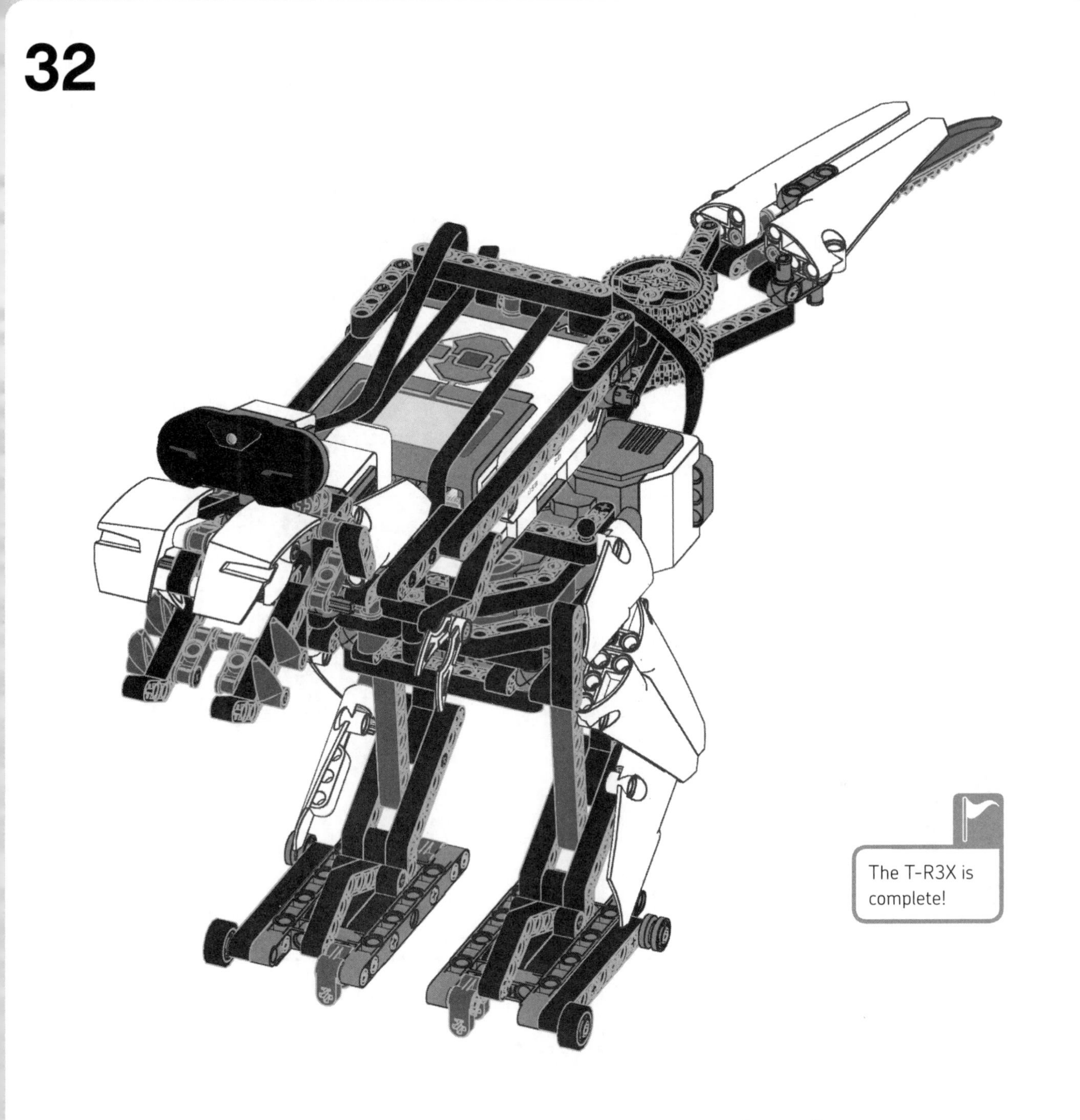

conclusion

In this chapter, you built the fierce T-R3X and discovered some new building techniques in the process. In the next chapter, you'll program the T-R3X to roam around and explore its surroundings. You'll also learn how to give such a creature autonomous behavior, allowing it to perform different actions based on its "mood," sensor readings, and timers.

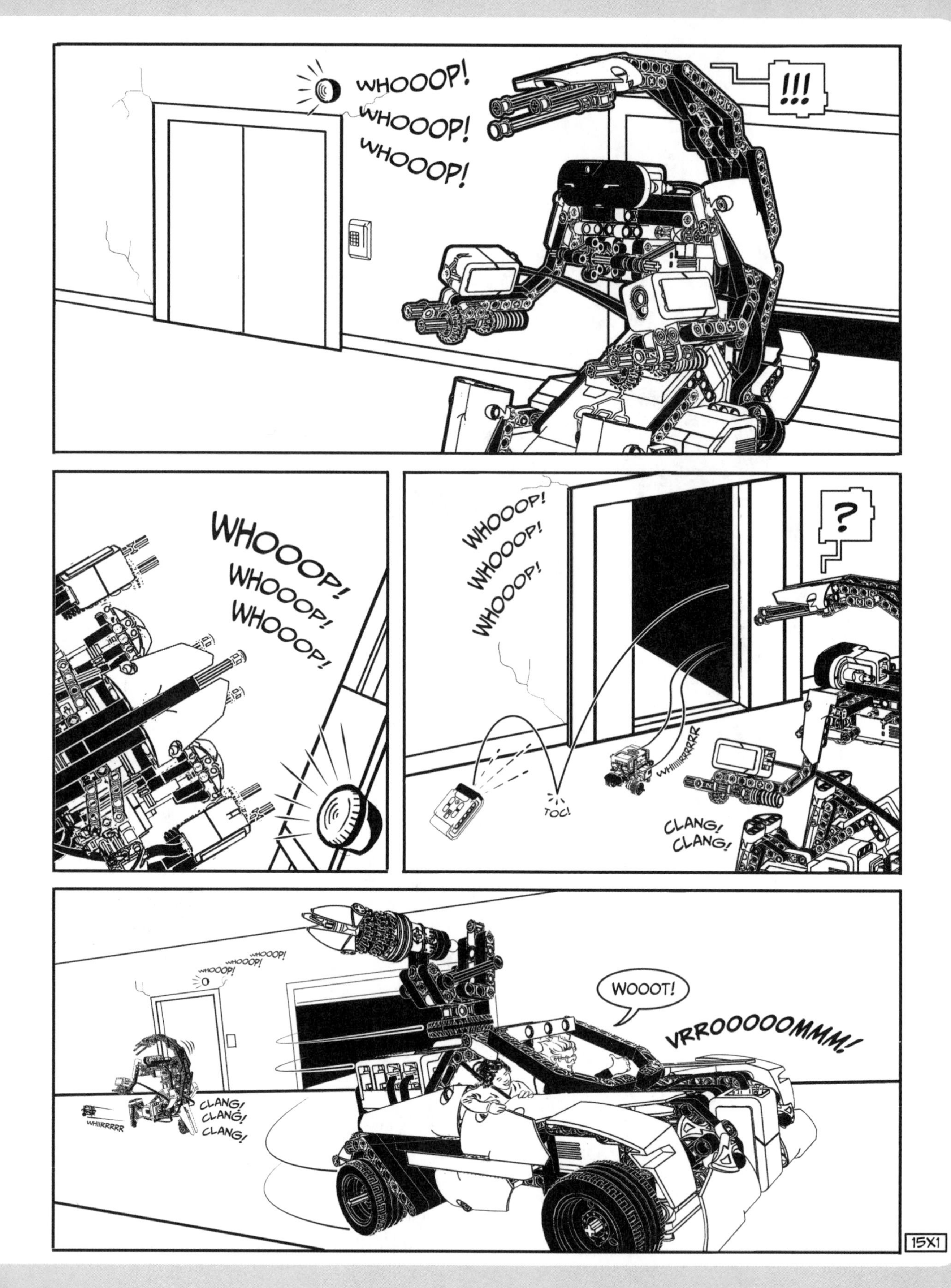
WHOOOP!
WHOOOP!
WHOOOP!
!!!
WHOOOP!
WHOOOP!
WHOOOP!
WHOOOP!
WHOOOP!
WHOOOP!
?
TOC!
WHIIRRRRR
CLANG!
CLANG!
WHOOOP!
WHOOOP!
WHOOOP!
WOOOT!
VRROOOOOMMM!
CLANG!
CLANG!
CLANG!
WHIIRRRR
15X1

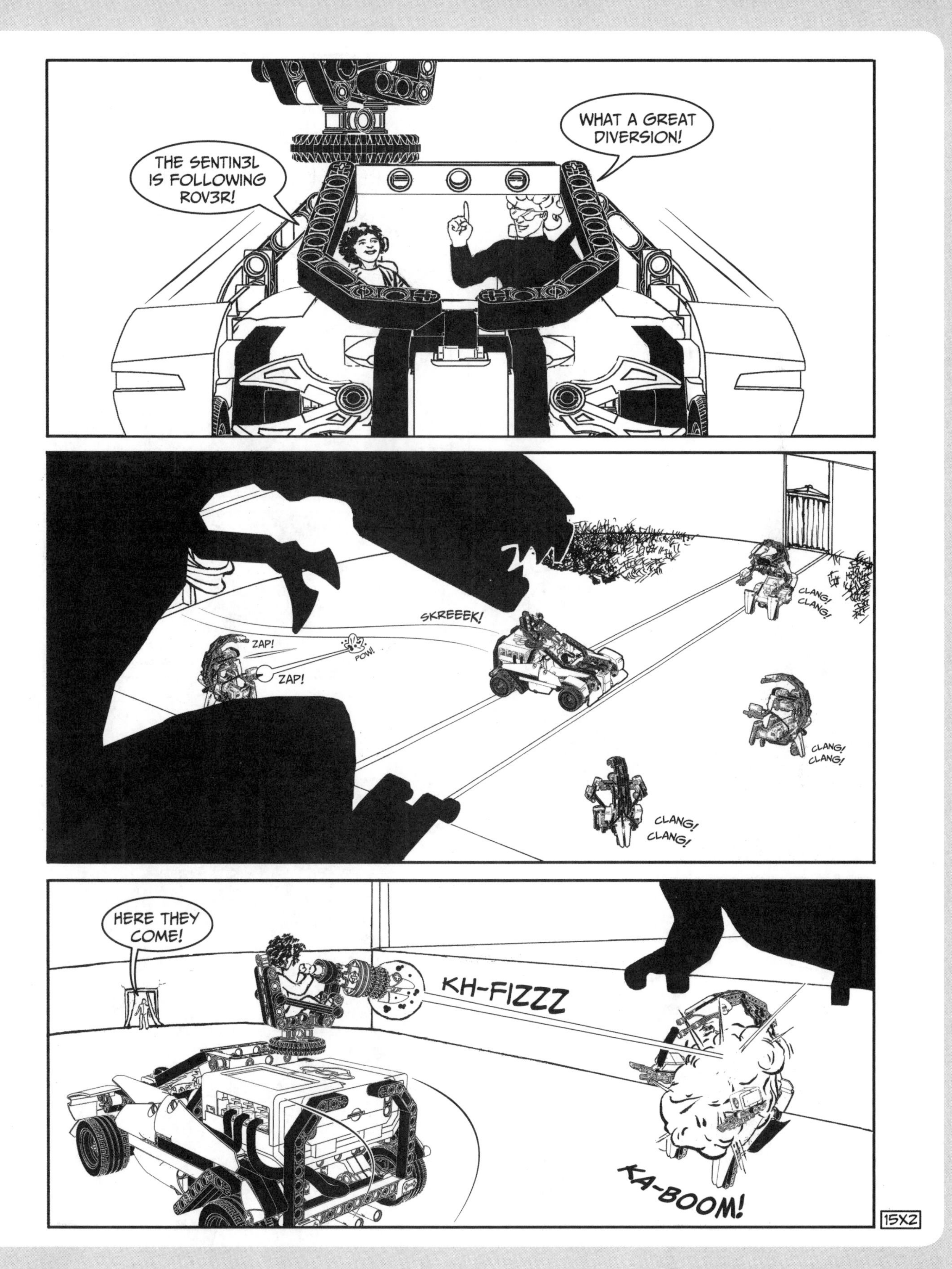
THE SENTIN3L IS FOLLOWING ROV3R!
WHAT A GREAT DIVERSION!
SKREEEK!
ZAP!
POW!
ZAP!
CLANG! CLANG!
CLANG! CLANG!
CLANG! CLANG!
HERE THEY COME!
KH-FIZZZ
KA-BOOM!
15X2

...UNTIL YOU SEE THE WHITES OF OUR EYES...
THEY HIT ME! THE CANNON IS GONE!
TIME TO ACTIVATE THE *EMP*!
LEAVE IT, GET BACK IN!
BUT WHY? THE SENTIN3LS ARE RETREAT...
ROAARR!
...ING!
SLAMM!
15X3

ROAARR!
GOSH!
CLANG! CLANG!
Magnetic
Pulse
FIIIIIIIIIIIIIIIIIIIIZZ
TLACK!
HOLD ON, DEXTER!
SSHBLAM!
CRASH!
HEY! ARE YOU OK?
THANKS, BRO!
15X4

programming the T-R3X

In this chapter, you'll program the fearsome T-R3X! The first program will make the dinosaur explore its environment by walking, detecting obstacles, looking around, and turning away from obstacles. Then, we'll model more complex autonomous behavior to make the T-R3X look around, react to "prey" (the IR Beacon), and hunt it down. You'll also learn the basics of using state machines to model behavior for a robot and how to compute complex logic expressions using mathematical formulas.

building the My Blocks for the Wander program

Before making the first program, *Wander*, for the T-R3X, we need to prepare the My Blocks. As you build the My Blocks, I'll point out noteworthy techniques and ideas. The My Blocks include the basic sequences to reset the legs at startup and to make the robot step forward, turn, roar, and chew. For each My Block, I'll list the sequence of blocks and show the final My Block showing the default input values and the icons used. Create each My Block using the figures as references.

the Reset My Block

The *Reset* My Block (Figure 16-1) is needed at the beginning of each program to reset all the mechanisms of the T-R3X to their initial positions. The body-shifting motor rotates until it reaches a mechanical stop, then moves the upper body to the center. Then, in a similar way, the motor that moves the legs aligns the legs. In a parallel sequence, the Medium Motor rotates to close the mouth. All the motor rotation counts are reset to zero.

the MoveAbsolute and MoveAbsolute2 My Blocks

The *MoveAbsolute* My Block (Figure 16-2) wraps around a Large Motor block and makes the specified motor rotate by an absolute number of degrees relative to the position in which its rotation count was reset, regardless of its actual position. (Compare this My Block with the *Steer* My Block for the SUP3R CAR shown in Figure 12-7 on page 236.)

The inputs are named *Port*, *Power*, *Angle*, and *Brake at End*. All but the last one are numeric inputs. *Brake at End* is a logic input like the Motor block. The Port parameter can be 1 (for output port A), 2 (B), 3 (C), or 4 (D).

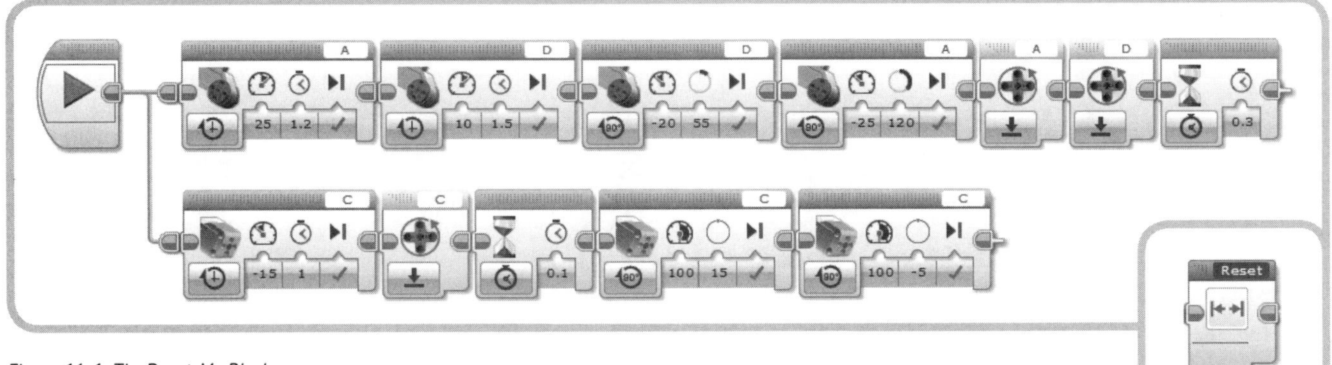

Figure 16-1: The Reset *My Block*

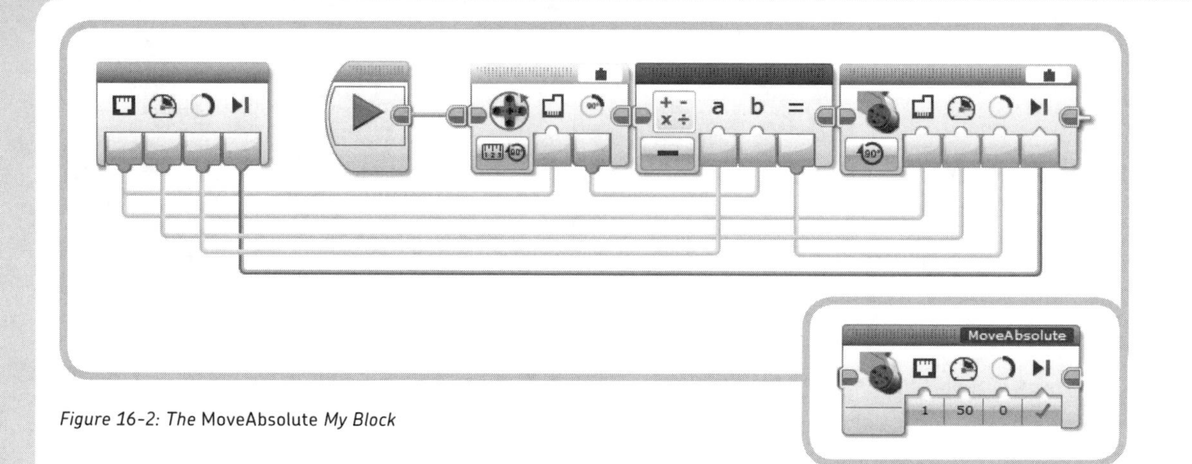

Figure 16-2: The MoveAbsolute My Block

To make the *MoveAbsolute2* My Block, use the My Blocks tab in Project Properties to copy and paste the *MoveAbsolute* My Block. The name *MoveAbsolute2* is assigned automatically to the copy when you paste the original My Block.

I need to make an identical copy of the *MoveAbsolute* My Block because My Blocks **are not re-entrant**, meaning that two instances of the same My Block **cannot be executed in parallel**. If you tried to do this, one of the My Blocks would have to wait for the other to complete before starting.

the Step My Block

The *Step* My Block (Figure 16-3) includes a basic sequence to make the T-R3X step forward by shifting the weight of its body left and right and moving its legs forward and back. The parameters for each *MoveAbsolute* My Block and Wait block are chosen precisely to make the robot take steps at a good pace.

the Roar My Block

The *Roar* My Block (Figure 16-4) plays the *T-rex roar* sound while opening the robot's mouth, and then closes the mouth.

the Chew My Block

The *Chew* My Block (Figure 16-5) makes the robot chew its prey. The mouth opens and closes in sync with the *Crunching* sound. Notice that the *Play Type* is set to **Play Once (1)** in the Sound block to make the program flow continue while the sound is played. (If we had selected the Wait for Completion Play Type, the block would pause the program until the sound finished.)

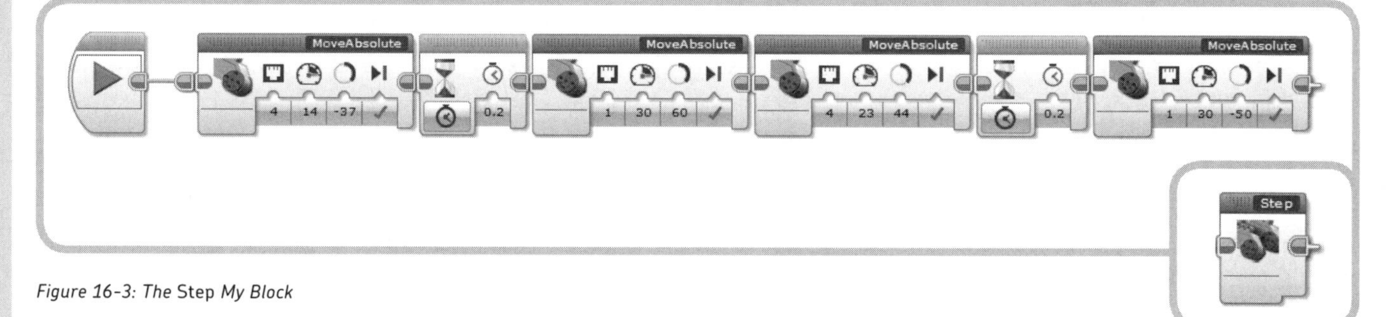

Figure 16-3: The Step My Block

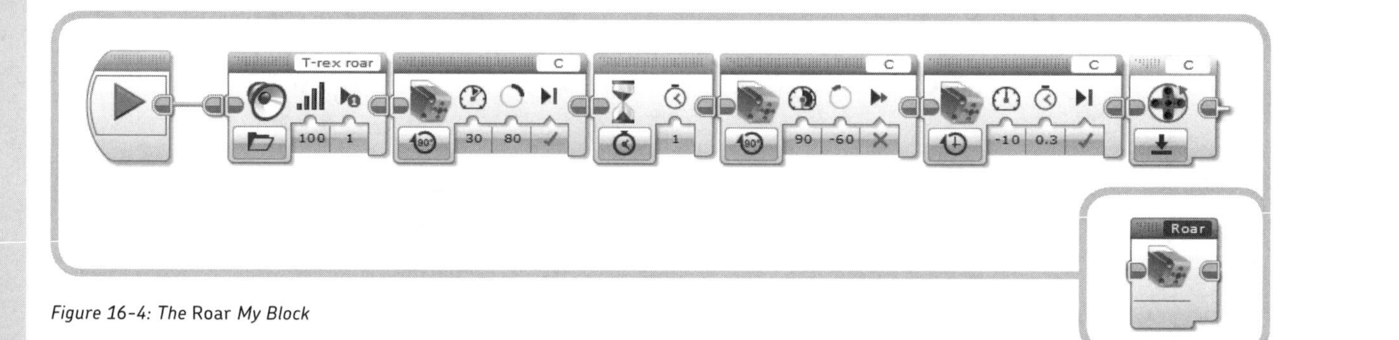

Figure 16-4: The Roar My Block

the Look My Block

The *Look* My Block (Figure 16-6) makes the robot turn its body (and the IR Sensor mounted on its head) left and right as it measures the proximity of the objects around it. This My Block outputs a logic value called *Clearest*, which is set according to the direction with the clearest line of sight, corresponding to the highest proximity value. When the *Clearest* output is True, the clearest direction is to the right; False means it's to the left. You can use this output to make the robot turn in either the direction deemed to be most free of obstacles at eye level or the direction where the nearest object is detected.

the Right My Block

The *Right* My Block (Figure 16-7) includes a basic sequence to make the robot turn right. The key to making the robot turn in place lies in exploiting the *conservation of angular momentum*. The heaviest part of the robot (its upper body) is shifted suddenly by the third *MoveAbsolute* My Block while the *MoveAbsolute2* My Block moves the robot's legs. In this way, the lightest part of the robot (the legs) is forced to turn in the direction counter to the body's rotation, while the upper body remains still with respect to ground.

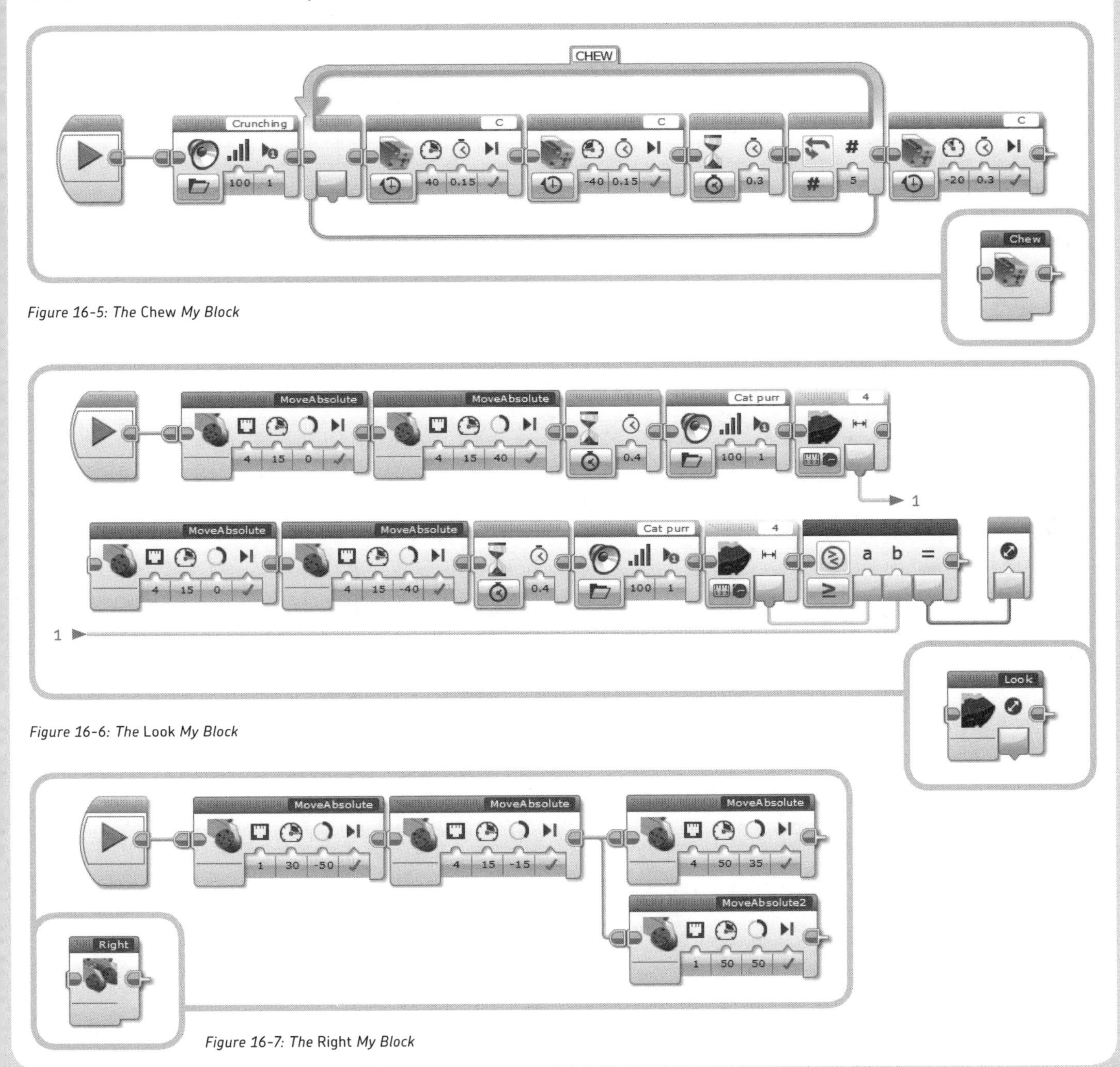

Figure 16-5: The Chew *My Block*

Figure 16-6: The Look *My Block*

Figure 16-7: The Right *My Block*

the Left My Block

The *Left* My Block (Figure 16-8) includes the basic sequence to make the robot turn left. As with the *Right* My Block, the parallel sequence executes a *MoveAbsolute2* My Block.

the TurnUntil My Block

The *TurnUntil* My Block (Figure 16-9) repeats the turning sequence to the *Right* or *Left*, according to the direction specified by the *Dir* logic input. The sequence continues until the number of times specified by the *Count* input is reached and the IR Sensor measures a proximity above the threshold specified by the *Prox* numeric input. The first Switch block shifts the weight to the correct side before the turning sequence begins.

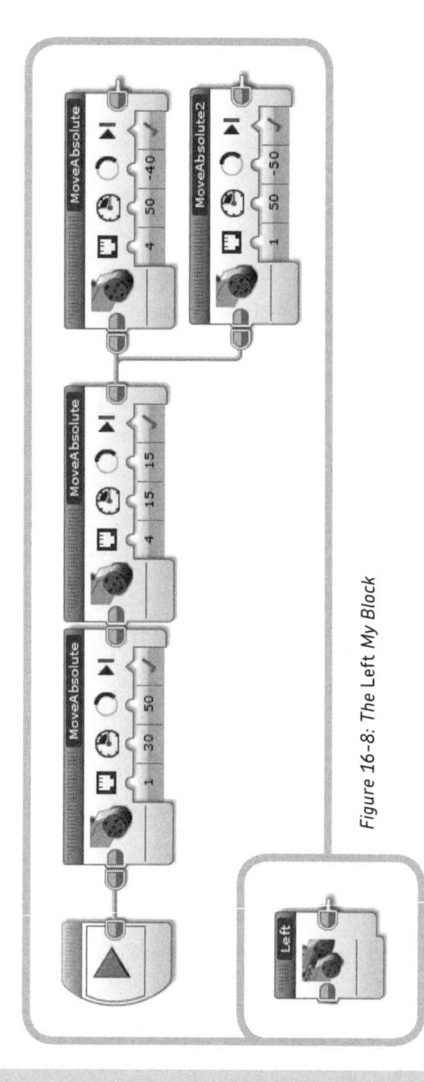

Figure 16-8: The Left My Block

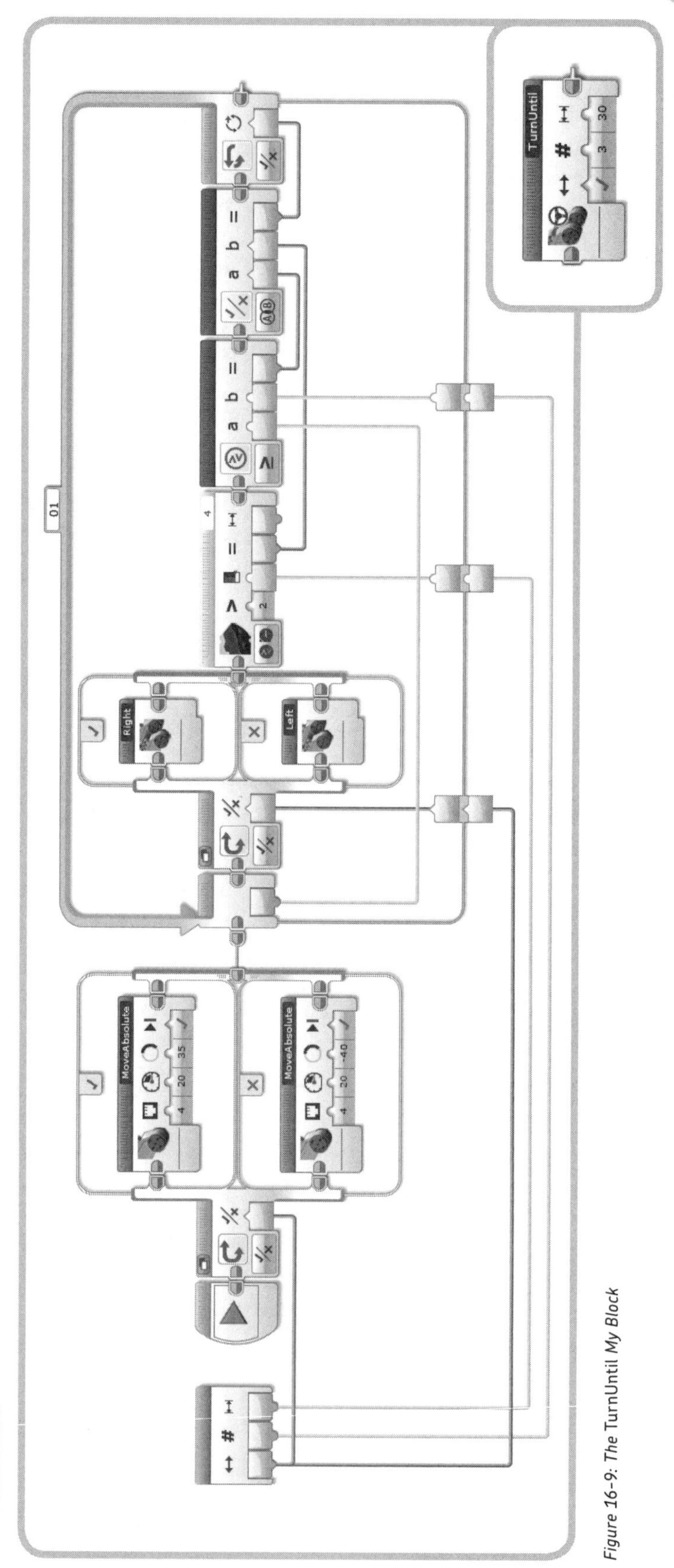

Figure 16-9: The TurnUntil My Block

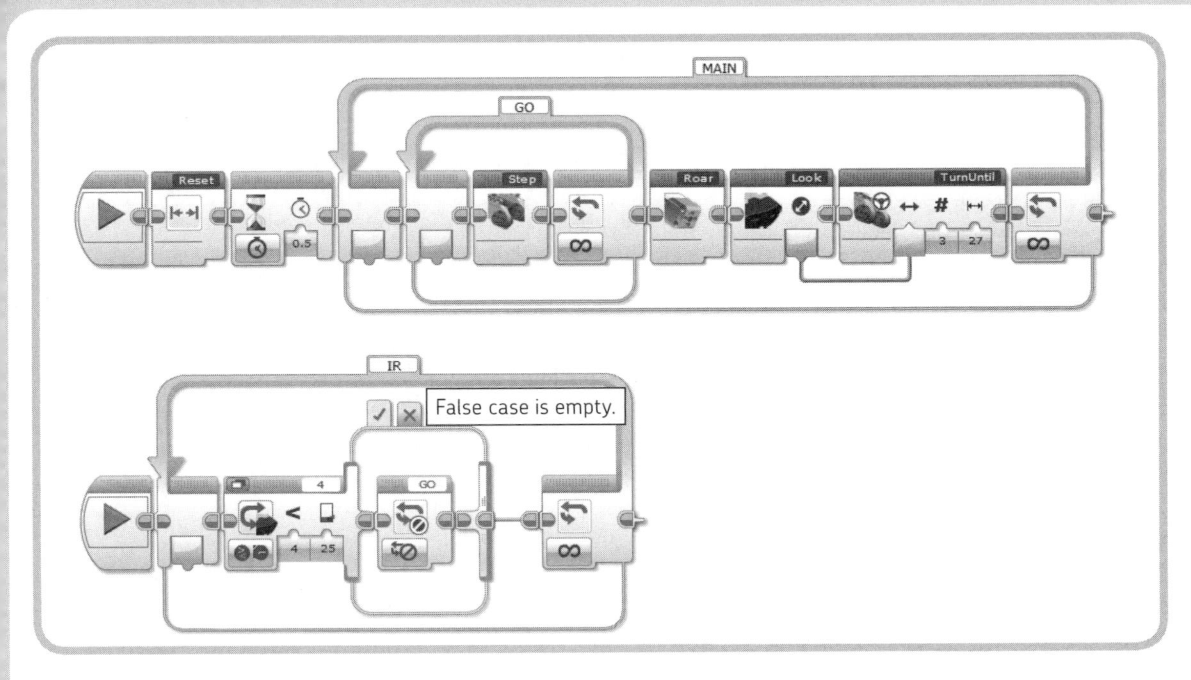

Figure 16-10: The Wander *program*

programming the T-R3X to wander

Having prepared your My Blocks, you can now make the *Wander* program for the T-R3X, which will make it walk and turn away from obstacles. The program is shown in Figure 16-10.

The *Step* My Block is repeated in the loop labeled *GO*. The IR Sensor is checked continuously in the loop named *IR* that is running in the parallel sequence. When an obstacle is detected, the Loop Interrupt block ends the *GO* loop so that the blocks after it in the *MAIN* loop can be executed: The T-R3X roars, looks around, and turns in the direction deemed to be the least cluttered.

NOTE The T-R3X walks well on flat, smooth surfaces but may fall over if asked to walk over rough surfaces like carpet.

designing the behavior of the T-R3X

When designing the behavior of the T-R3X, I first considered the fierce nature of this prehistoric predator. To have the robot display appropriate behavior, I decided to make it hunt using the Remote IR Beacon as its prey. The state diagram for the T-R3X is shown in Figure 16-11 (see "Digging Deeper: Behavior Modeling Using State Machines" on page 368 for more information on state diagrams). Notice that some transitions occur spontaneously after the actions of a certain state are performed, as indicated by dashed arcs. For example, after eating, the robot goes into the IDLE state.

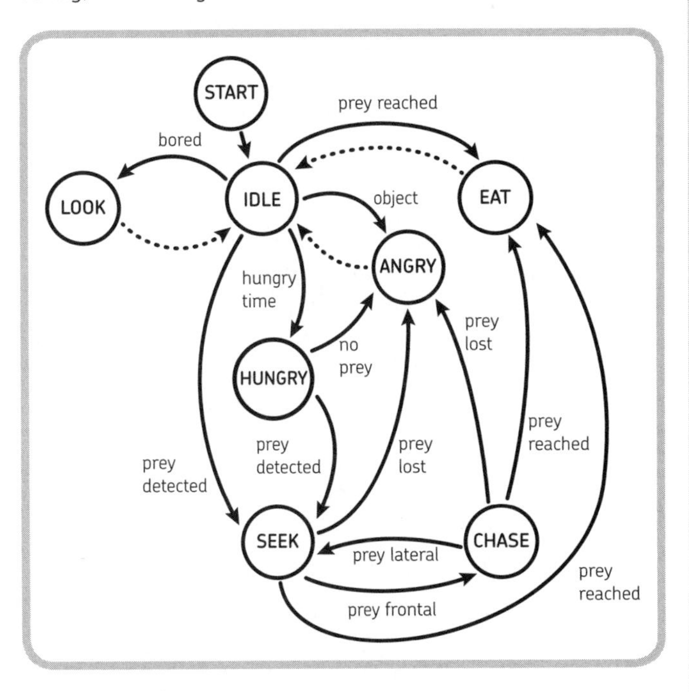

Figure 16-11: The state diagram of the T-R3X's behavior

To keep things simple, the nodes report only the name of each state. The actions performed in each state are as follows:

* START: In the START state, we initialize all of the robot's mechanisms, such as its mouth, its body, and its legs. We also reset the timers that will trigger the spontaneous stimuli of boredom and hunger, and we set their timeouts using random numbers.
* IDLE: Following the START state, the robot goes into the IDLE state. While in this state, the robot performs no visible actions, but the IR Sensor and timers are continuously checked because different events may occur: The IR Sensor can measure a nearby object or detect the Remote IR Beacon (the prey!), or timers for states like hunger or boredom may elapse at randomly chosen intervals.

DIGGING DEEPER: BEHAVIOR MODELING USING STATE MACHINES

Many robots don't exhibit very interesting behavior, usually because they execute actions in a loop. For example, the robot will walk until it sees an obstacle, turn, and then start walking forward again. How can you give your creatures a spark of life, so that their behavior makes them look like they have their own will?

The answer is to use finite-state machines, also known as simply state machines or SMs. State machines are a programming technique that you can use to model behavior in robots.

A *state machine* (or state automaton) models behavior based on a finite number of states, the transitions between these states, the set of events that can occur, and possible actions.

* The *state* describes the machine's situation, based on past or present events, thus reflecting the history of what happened from system startup until now. In most cases, all the things we should know about the machine's past are condensed into just the machine's current state.
* State *transitions* are not something you can see: When an event occurs in a certain state, a transition that brings the machine from one state to another is triggered. When events cause state changes, our state machine is called *event driven*.
* *Events* can be represented by, for example, input from a sensor, a timer running out, or an internal counter reaching a certain value. Events cause transitions between states.
* *Actions* are the outputs of the machine and are its visible behavior.

State machines can be represented using a *state diagram*, as shown in Figure 16-12. The circles (nodes) with letters inside them represent the states (S_0, S_i, S_j, and S_k). The arrows going from one state to another represent the transitions between these states. Every arrow is marked with the event E_{ij} that caused the state transition from state *i* to *j*. A_i is the action performed in the state S_i.

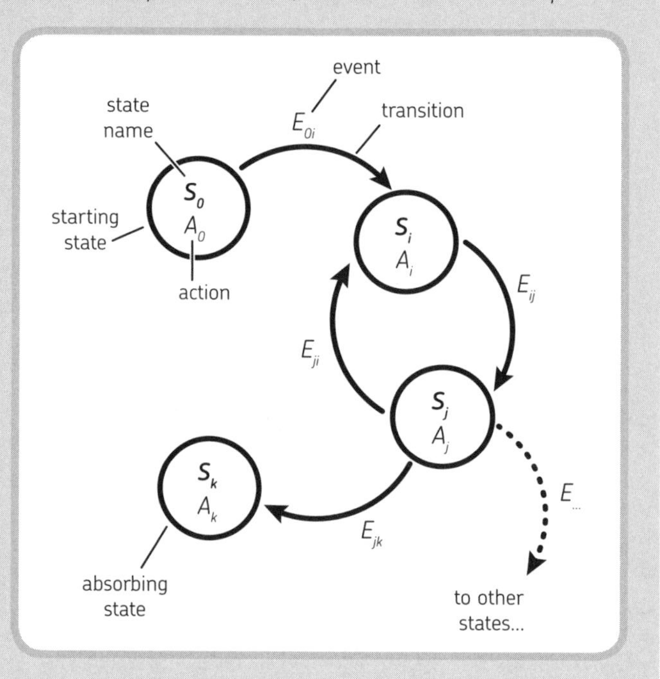

Figure 16-12: A state diagram—a representation of a state machine

As you can see in Figure 16-12, any state machine has a starting state (S_0) that does not have any arrows pointing to it. Once the robot has left this state, it cannot return to it. The actions of this state are executed only once, so usually this state initializes the variables and moving parts.

NOTE When designing a state machine, be sure to avoid so-called *absorbing states*, which lack exiting arcs; once such a state is reached, it cannot be escaped, and the machine stalls. To avoid getting stuck in an absorbing state, be sure each state has a transition to another state.

* LOOK: When the timer associated with boredom elapses, the robot looks right and left and turns in place in the direction deemed to be most free of obstacles. The boredom timer is then reset, a new interval is randomly chosen, and the robot returns to the IDLE state.
* ANGRY: The robot roars in fury if its prey escapes, if it's hungry and no prey is in sight, or if an object other than its prey approaches. Then it returns to the IDLE state.
* HUNGRY: When the hunger timer elapses, the robot becomes hungry and checks for the presence of prey. If no prey is detected, the T-R3X goes into the ANGRY state; otherwise, it goes into the SEEK state to begin the hunt.
* SEEK: When the prey is detected, this is the initial state that begins the hunt. The robot checks whether the prey is in front of it. If it is, the robot goes into the CHASE state to reach it. If the prey is to the left or right, the robot turns toward it. If the prey escapes (disappears from sight), the robot goes into the ANGRY state. If the T-R3X reaches its prey, it goes into the EAT state.
* CHASE: When the robot is in the SEEK state and spots its prey almost straight ahead, it reaches this state and walks straight toward the prey. If the prey is detected to the left or right, however, the robot returns to the SEEK state to turn toward it. If the prey escapes, the robot goes into the ANGRY state, but if the robot reaches the prey, it goes into the EAT state.
* EAT: When the prey is seen nearby, the robot tears the prey into pieces and chews it up! Then it returns to the IDLE state, the hunger timer is reset, and a new interval is randomly chosen.

As you can see, you can describe the behavior of a robotic creature using natural language.

implementing a state machine

In this section, you'll learn how to implement a *state machine (SM)* using the EV3 software. Each abstract component of the state machine (states, transitions, events, and actions) will be represented by a sequence of programming blocks. (The program shown in Figure 16-13 is just a generic structure; you'll see how to implement the SM for the T-R3X in detail later on.)

general structure

A state machine can be implemented in EV3 language by following the generic structure of Figure 16-13. Before the main loop, you can reset the robot's mechanics, reset event timers, and initialize the *state variable* **S** by assigning the first state that the state machine should execute (such as the default state named IDLE).

In the main loop, the state variable is read and used to switch among the various cases of a Switch block. Each case represents a state, and in each case you can perform actions, check for events, and transition to other states by assigning new values to the state variable. (Transitions can be either spontaneous or based on events.)

To keep the program tidy, group the actions, event checks, and transitions into a single My Block for each state.

starting state

The initial START state can be implemented by placing one or more My Blocks before the main program loop. In the generic program shown in Figure 16-13, the *Reset* My Block resets the robot's mechanisms, while the *INIT* My Block sets the initial value for the state variable, resets the timers, and chooses random timeouts for them.

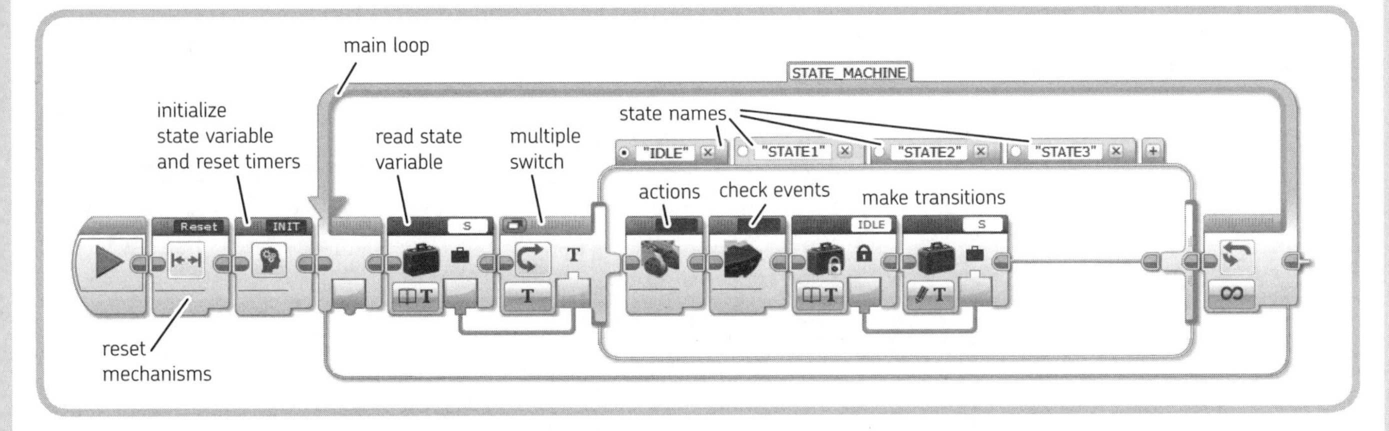

Figure 16-13: The generic implementation of a state machine using EV3 language

state variable

We represent the state of the machine with a single variable **S**. The state variable is global, which means that it can be accessed from anywhere in the program. If you choose a numeric variable, you must associate an integer value with each of the states. For example, IDLE would be 0, LOOK would be 1, and so on. The problem with using numeric state variables is that you might forget the codes for the various states and get confused.

Alternatively, you can make the state variable a text variable, so that you can just write the state name into it, such as IDLE, LOOK, and so on. If you use text variables when making a transition, be sure to write the exact name of the state into the state variable, or the machine will not behave as expected. It will be trying to reach a state that does not exist and will execute the default state (in this case, the IDLE state). The state variable in the generic program shown in Figure 16-13 is a text variable.

WARNING The Switch block in Text mode is case sensitive, which means that *IDLE* is different from *Idle* or *idle*. Be sure to write state names with all capitals to avoid errors.

sensor events

In each state, you can check for the occurrence of particular events and transition to another state when one of those events occurs. For example, you might compare the value of a sensor against a threshold, and if the comparison is true, assign a new value to the state variable, as shown in Figure 16-14.

You can also group all blocks that read sensors into one monitor-like My Block and then place that My Block in the main Loop, outside the Tabbed Switch block. This My Block will always execute, no matter the machine's state.

timer events

To make it seem as if your robot behaves spontaneously, you can generate events that trigger when timers exceed randomly chosen thresholds, as shown in Figure 16-15. Setting timer events is kind of like setting an alarm clock. You need one timer for each behavior that you want to simulate: one for hunger (Timer 1), one for boredom (Timer 2), one for sleep, and so on.

Once the timer-generated transition ends, reset the timer and choose a new random value for its threshold, as shown in Figure 16-16.

transitions

Transitions can be triggered by an event, or they can be programmed to occur automatically once all of a state's actions have been performed. To make a transition, you assign a new value to the state variable **S**, either in the main program or in any My Block. Once the blocks of the current state finish running, the state variable is read again in the main loop, and the value assigned to the state variable will determine the next state to be executed.

If you don't change the state variable, the machine will remain in the current state, and the main loop will continue to execute the current state's actions.

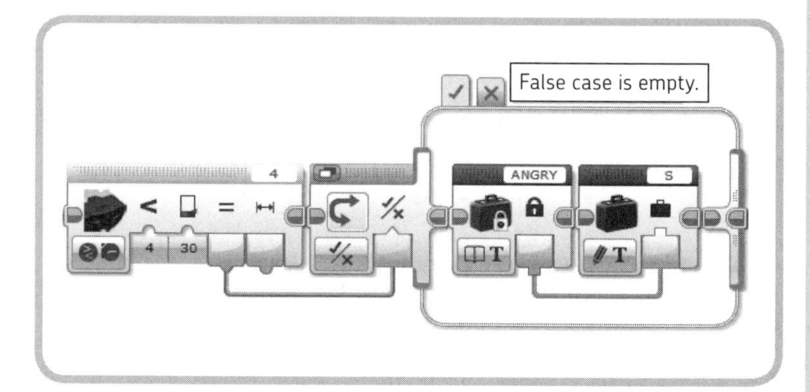

Figure 16-14: An event can be generated by checking the value of a sensor.

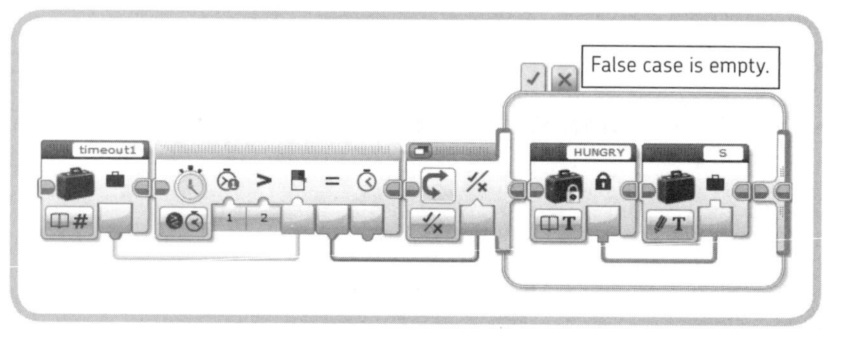

Figure 16-15: A timer that exceeds its threshold generates a spontaneous transition to another state.

Figure 16-16: A timer is reset, and its threshold is set to a random number to generate a spontaneous future event.

Instead of using a timer, you can use a variable to count how many times an event has occurred—for example, how many times the Touch Sensor has been pressed. If the count surpasses a certain threshold, the state machine will transition to another state.

timer-filtered events

Say you want an event to occur (and trigger a transition) when a certain logic condition holds for a specified amount of time, while filtering out conditions that last only a fraction of a second. You can do this as shown in Figure 16-17: The state variable is set to ANGRY if the logic value carried by the Data Wire is False for more than 1 second (the Switch block is in Timer Compare Time mode, Threshold Value set to 1). The timer is reset every time the logic condition is True (the opposite of the desired value). If the timer is not reset for a certain amount of time, its Elapsed Time will exceed the Threshold Value, and the True case of the Switch block will execute, setting the state variable to ANGRY.

actions

Your robots can perform certain actions—such as moving motors, playing sounds, and flashing lights—depending on the current state. The sequence of actions should last only a few seconds (so that the robot can respond promptly to new events) and must not contain infinite loops or the machine will stall. The main loop handles repeated actions.

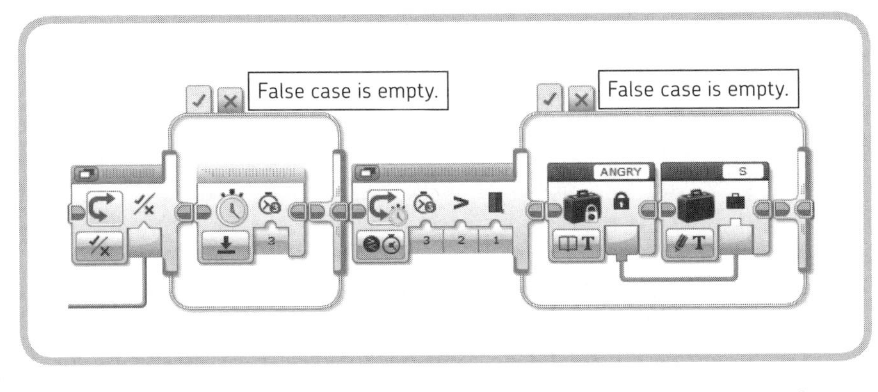

Figure 16-17: You can use a timer to trigger a state transition only if a condition remains constant for a certain period.

DIGGING DEEPER: COMPUTING COMPLEX LOGIC OPERATIONS USING THE MATH BLOCK

It's easy enough to program complex logic expressions with EV3 blocks, but sometimes the resulting sequence of blocks can become cumbersome. For example, combining the logic variables *A*, *B*, and *C* to compute this expression would take four Logic Operations blocks and a lot of Data Wires, as shown in Figure 16-18(a).

```
Result = ( A AND B ) OR ( C AND NOT(B) )
```

But there's a more elegant way. In Chapter 6, you learned that input logic values are converted automatically to numeric values (True = 1, False = 0) when the Data Wires carrying logic values are plugged into numeric inputs. Here's how to compute any logic expression using the Math block, without creating a mess of blocks and wires in the process:

1. In a Math block set in Advanced mode, enter a formula containing the corresponding algebraic representation of each logic operation, as listed in Table 16-1.
2. Convert the numeric result of the Math block to a logic value using the Compare block, as explained in "Converting Numeric Values to Logic Values" on page 98: Set mode to Not Equal To and input **B** to 0 and drag a Data Wire from the Math block to input **A**. In this way, any value other than 0 is seen as True.

For example, you can get the same result as in Figure 16-18(a) by using a Math block in Advanced mode with the formula `(A*B)+(C*(1-B))` and a Compare block to check whether the numeric result is different from 0, as shown in Figure 16-18(b).

(continued)

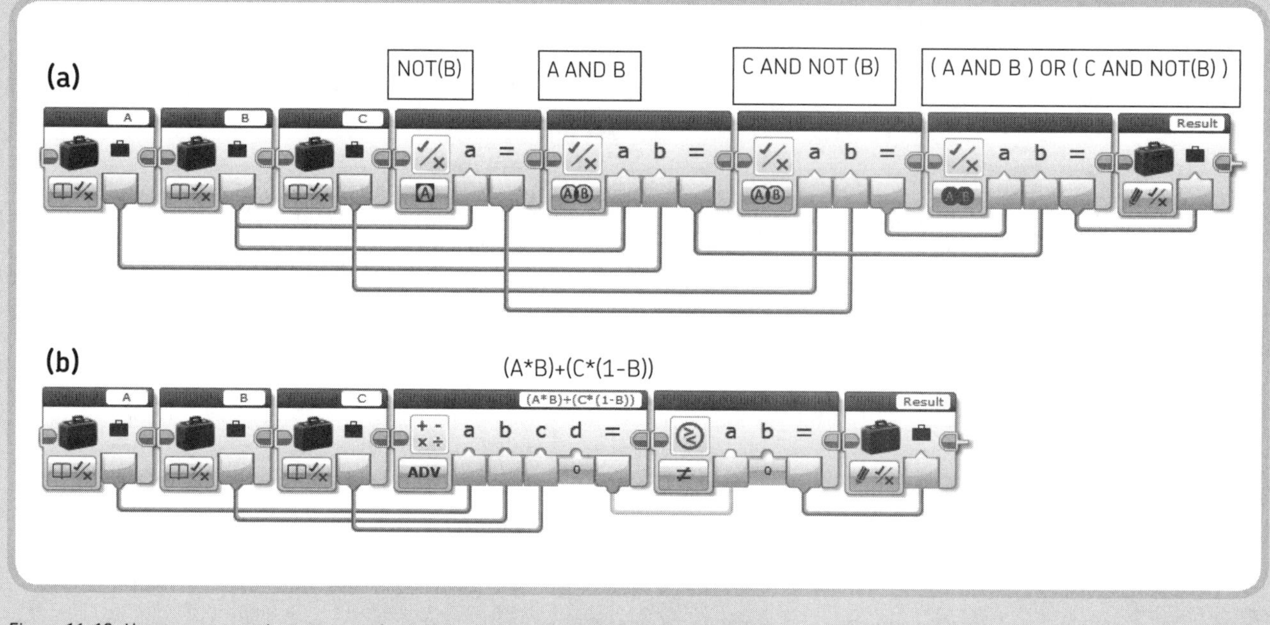

Figure 16-18: How to compute the same complex logic expression using four Logic Operations blocks and many Data Wires (a) and the Math block in Advanced mode and a Compare block (b)

table 16-1: conversion of logic expressions into algebraic formulas

Logic Expression	Algebraic Formula
NOT(A)	(1 - A)
A OR B	(A + B)
A AND B	(A * B)
A XOR B	(A - B)^2
A NOR B	(1 - A) * (1 - B)
A NAND B	(1 - A * B)

NOTE The logic AND function and multiplication are equivalent for any value for the inputs A and B (0 × 0 = 0, 0 × 1 = 0, 1 × 1 = 1). On the other hand, the OR function and addition are equivalent only because we later convert any nonzero value to True. For example, when both A and B are equal to 1 (True), the result of A OR B would be A + B = 2, a result that is not a valid binary digit (that is, neither 0 nor 1). However, 2 is different from zero, and so the Compare block converts it to True.

WARNING A and B should be input logic values and should not be replaced by another algebraic formula, or you might get incorrect results. For example, you can't compute NOT(A OR B) as (1 - (A + B)), where you've taken the formula (1 - A) and substituted in (A + B) for the operand A. If you did, you would get (1 - A - B), which is wrong: If both A and B were True (equal to 1), the result of the algebraic formula would be -1 (incorrect) rather than 0 (correct).

Table 16-2 lists some examples of logic expressions converted into algebraic formulas.

table 16-2: examples of converting logic expressions to algebraic formulas

Logic Expression	Algebraic Formula
NOT(A) OR NOT(B) OR C	(1 - A) + (1 - B) + C
(A AND B) OR (A AND C) OR NOT(A)	A * B + A * C + (1 - A) = A * (B + C) + (1 - A)
A AND NOT(B) AND C OR NOT(A)	A * (1 - B) * C + (1 - A)

I'll use this conversion method to clean up some of the My Blocks needed by the T-R3X. I think you'll find that your programs will also benefit from this technique.

DIGGING DEEPER: DE MORGAN'S LAWS

To convert the logic operator NOR of Table 16-1, I applied *De Morgan's laws*, which state:

* NOT (A OR B) = NOT (A) AND NOT(B)
* NOT (A AND B) = NOT(A) OR NOT(B)

These laws can sometimes help simplify the logic expressions needed by your programs.

making the My Blocks for the final program

To implement the final *StateMachine* program to make the T-R3X behave autonomously and go hunting, we need to prepare some more My Blocks.

the Turn My Block

The *Turn* My Block (Figure 16-19) is like the *TurnUntil* My Block. It makes the robot turn based on the *Dir* logic input, where True means turn right and False means turn left. The turning sequence is repeated by the number of times specified by the *Count* numeric input.

the ReadBeacon My Block

The *ReadBeacon* My Block (Figure 16-20) wraps an IR Sensor Block in Measure Beacon mode. It has four logic outputs:

* *Detected* is True when the beacon is detected *and* its proximity is less than 100. This double check is necessary because the IR Sensor sometimes measures a proximity of 100 even when its *Detected* Output is True and the beacon is effectively out of sight.
* *Outside* is True when the Heading to the beacon is below –10 or above 10. When inside these boundaries, the beacon can be considered to be in front of the robot. This broad range of values allows for the wide swing of the robot's head as it walks.
* *Dir* is True when the Heading is greater than zero (beacon seen at the right side). This value can be used as input to the *Turn* My Block to make the robot turn toward the beacon.
* *Near* is True when the Proximity to the beacon is less than 6. This value tells the robot that the prey is within range of its jaws.

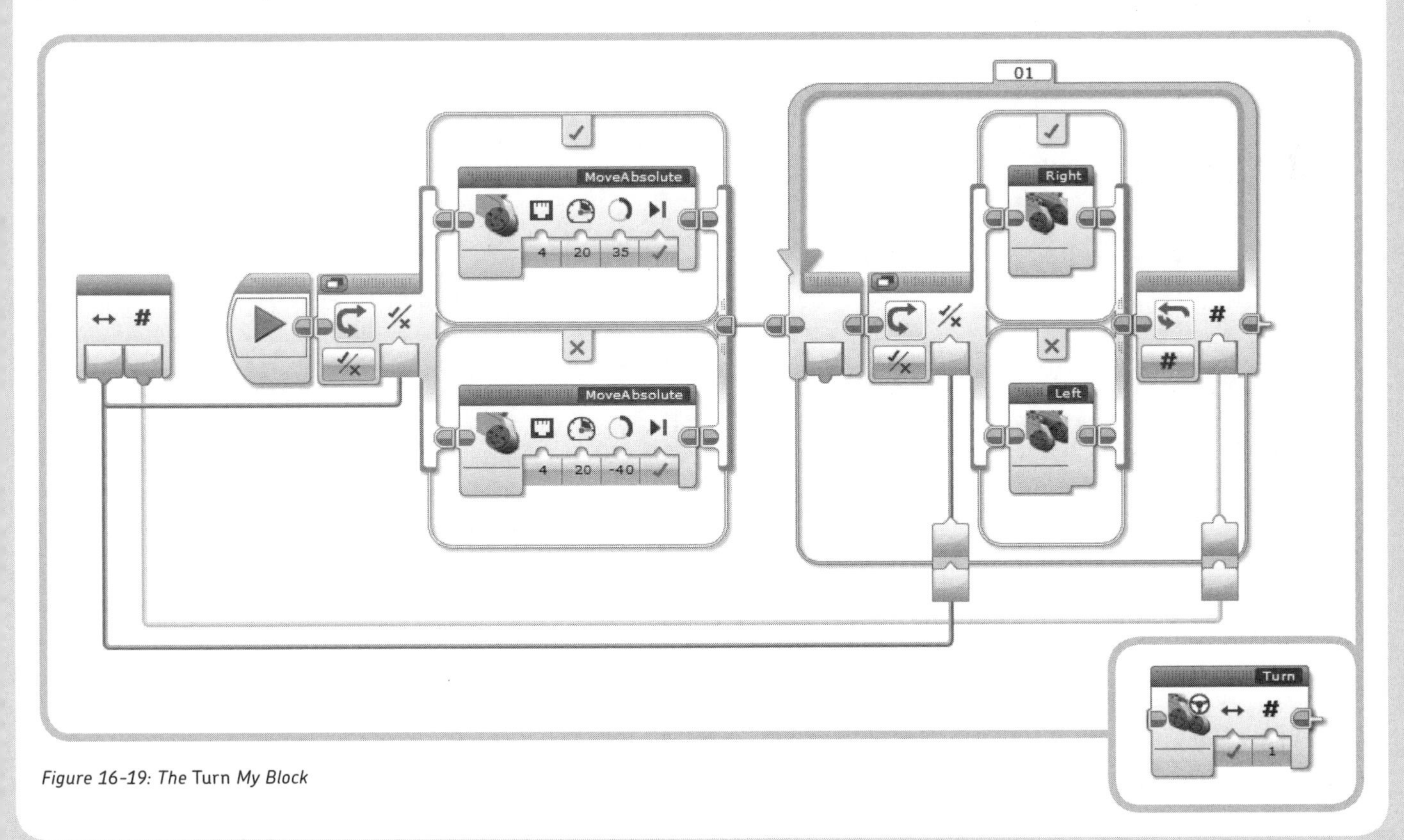

Figure 16-19: The Turn *My Block*

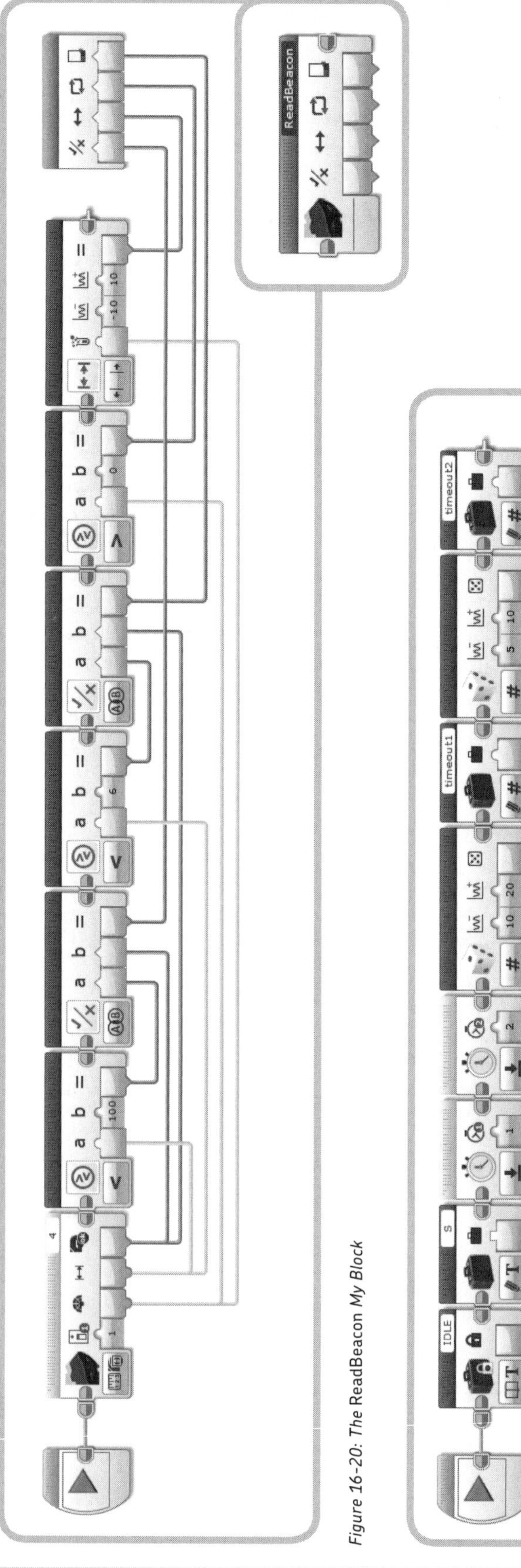

Figure 16-20: The ReadBeacon My Block

Figure 16-21: The INIT My Block (the icon for this My Block is a red head)

the INIT My Block

The *INIT* My Block (Figure 16-21) initializes the state machine by setting the text state variable **S** to IDLE, resetting Timers 1 (hunger) and 2 (boredom), and drawing random values for the corresponding time thresholds. The text state variable is set with the text data from a Constant block to make the content of the string more evident to whoever reads the program. (Remember, you can write text to a variable simply by entering the text in the input field.)

the IDLE My Block

The *IDLE* My Block (Figure 16-22) implements the IDLE state of the diagram shown in Figure 16-11. It performs no visible action; instead, it silently checks the IR Sensor for nearby objects, checks whether the timers have surpassed their thresholds, and checks whether the beacon is in sight.

Notice that the transition to the LOOK state is made if Timer 2 exceeds the threshold (condition A), which means that the robot is bored, and if there is no object in sight (condition B). The logic expression (A AND NOT(B)) is translated into the formula `A*(1-B)` in the Math block in Advanced mode. (If you don't remember how to produce such an expression, read "Digging Deeper: Computing Complex Logic Operations Using the Math Block" on page 371.)

the HUNGRY My Block

The *HUNGRY* My Block (Figure 16-23) implements the HUNGRY state of the diagram shown in Figure 16-11. It checks to see whether the prey is in sight, and if the beacon is detected, it makes the transition to the SEEK state. If not, it transitions the T-R3X to the ANGRY state. In the latter case, Timer 1 (for hunger) is reset, and a new random value is generated for its threshold.

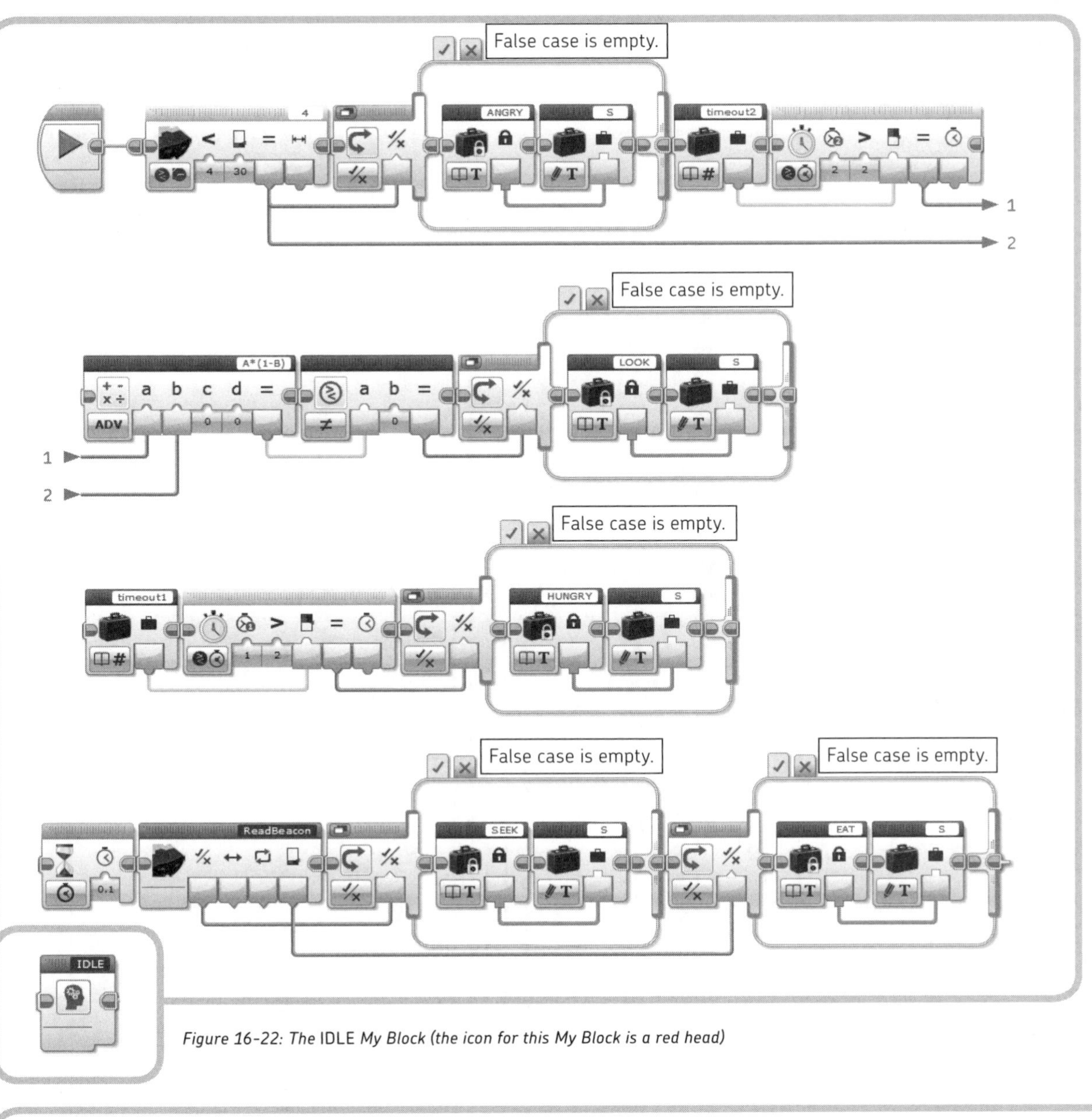

Figure 16-22: The IDLE *My Block (the icon for this My Block is a red head)*

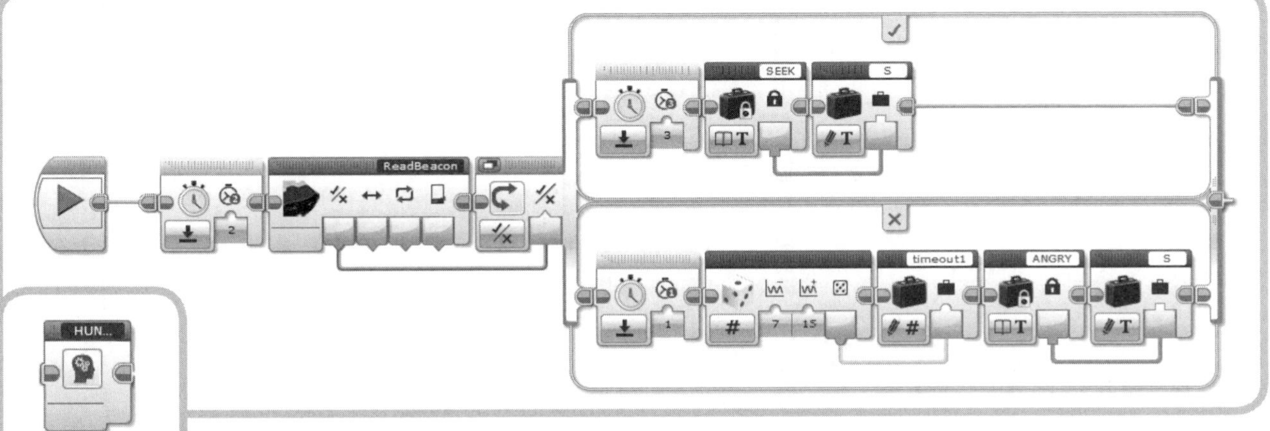

Figure 16-23: The HUNGRY *My Block (the icon for this My Block is a red head)*

the SEEK My Block

The *SEEK* My Block (Figure 16-24) implements the SEEK state of the diagram shown in Figure 16-11. The *ReadBeacon* My Block provides the logic values that regulate the operation of this state. If the beacon is detected to the side of the robot (the *Outside* output is True), the *Turn* My Block makes the robot turn in the direction specified by the *Dir* output. If the beacon is detected in front of the robot (the *Outside* output is False), the state variable is set to CHASE. If the beacon is not detected, the transition to another state is not instantaneous: The beacon must remain undetected for 1 second before Timer 3 triggers the state transition to ANGRY (see "Timer-Filtered Events" on page 371). If the beacon is seen nearby (*Near* output is True), the state variable is set to EAT.

the CHASE My Block

The *CHASE* My Block (Figure 16-25) implements the CHASE state of the diagram shown in Figure 16-11. It makes the transition to the SEEK state if the beacon is seen to the side of the robot (that is, if the *Outside* output of the *ReadBeacon* My Block is True). If the beacon is not detected for more than 1 second, the state is changed to ANGRY. Finally, if the beacon is nearby (the *Near* output is True), the state is changed to EAT.

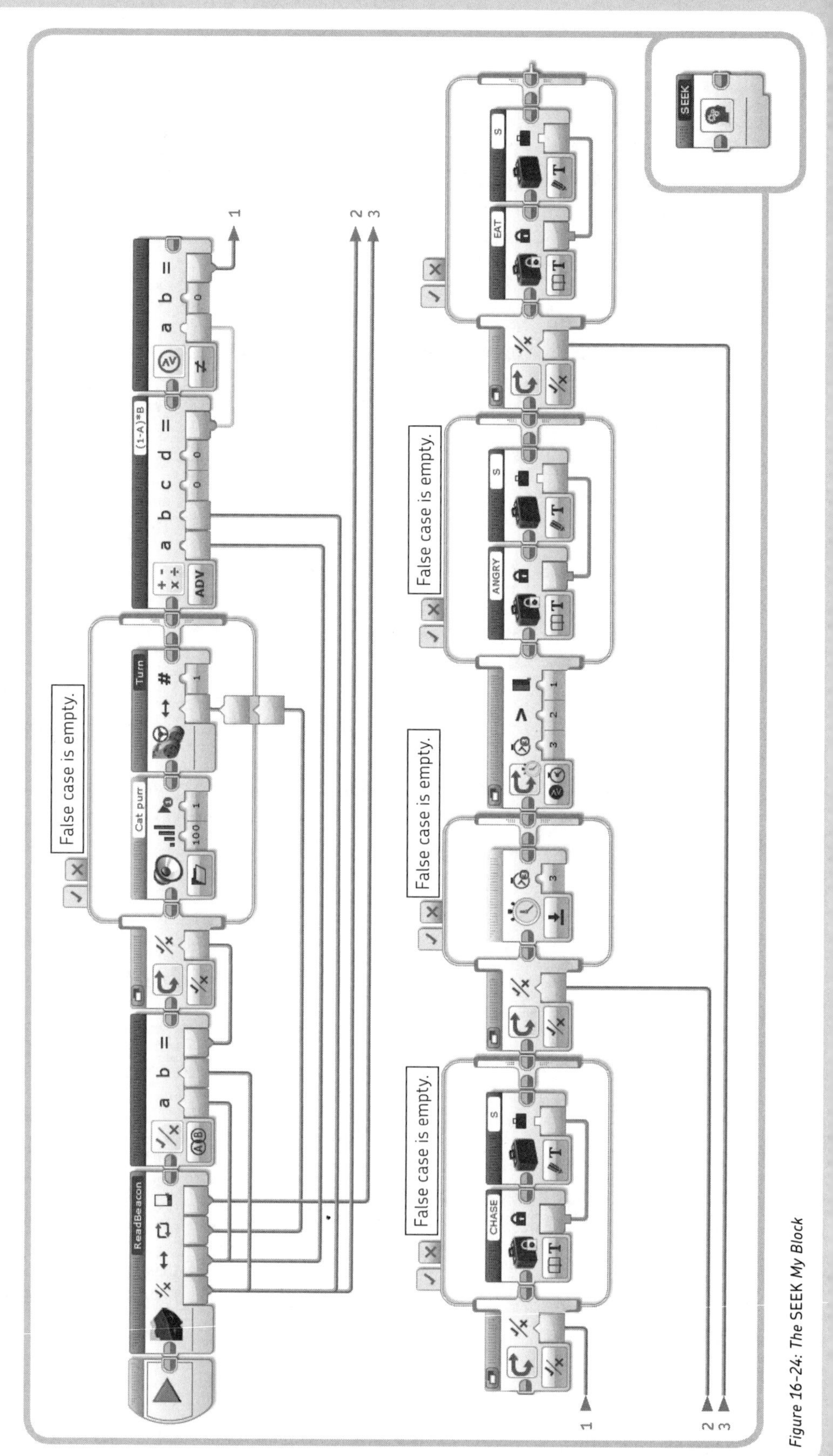

Figure 16-24: The SEEK My Block

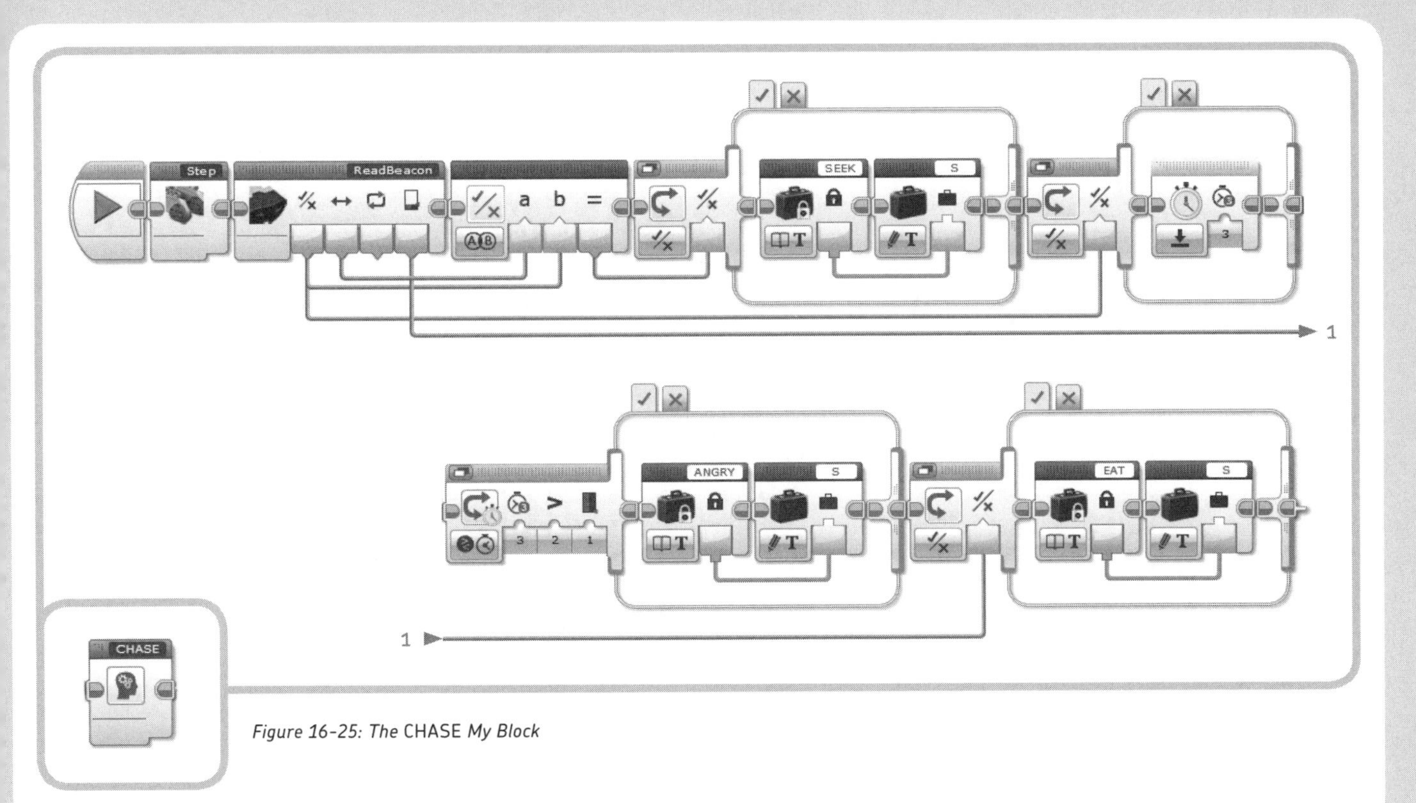

Figure 16-25: The CHASE *My Block*

ORDERING STATE TRANSITIONS BY PRIORITY

Each time the test of a Switch block succeeds, a new value is written to the state variable, overwriting the old value. Therefore, all possible transitions should be placed in order of increasing priority: The transition with the lowest priority should be first in the sequence of tests, and the transition to the state with the highest priority should be the last.

In the case of a hunting T-R3X, its highest priority is to attack its prey. Therefore, in the *IDLE* My Block, the transition to the LOOK state has a lower priority than the transition to the EAT state.

Even if Timer 2, corresponding to boredom, causes the Switch block to set the state variable to LOOK when the beacon is nearby (the *Near* output of the *ReadBeacon* My Block is True), the state variable is set to EAT, overwriting the previous value.

programming the T-R3X's behavior

Now that all the My Blocks are ready, you can finally build the program shown in Figure 16-26 that implements the state machine for the T-R3X shown in Figure 16-11. Some cases of the Tabbed Switch block contain My Blocks, while others contain sequences of blocks.

Now run the program to test the behavior of the T-R3X. Try tweaking the range of the random thresholds for the timers to make the robot more or less temperamental.

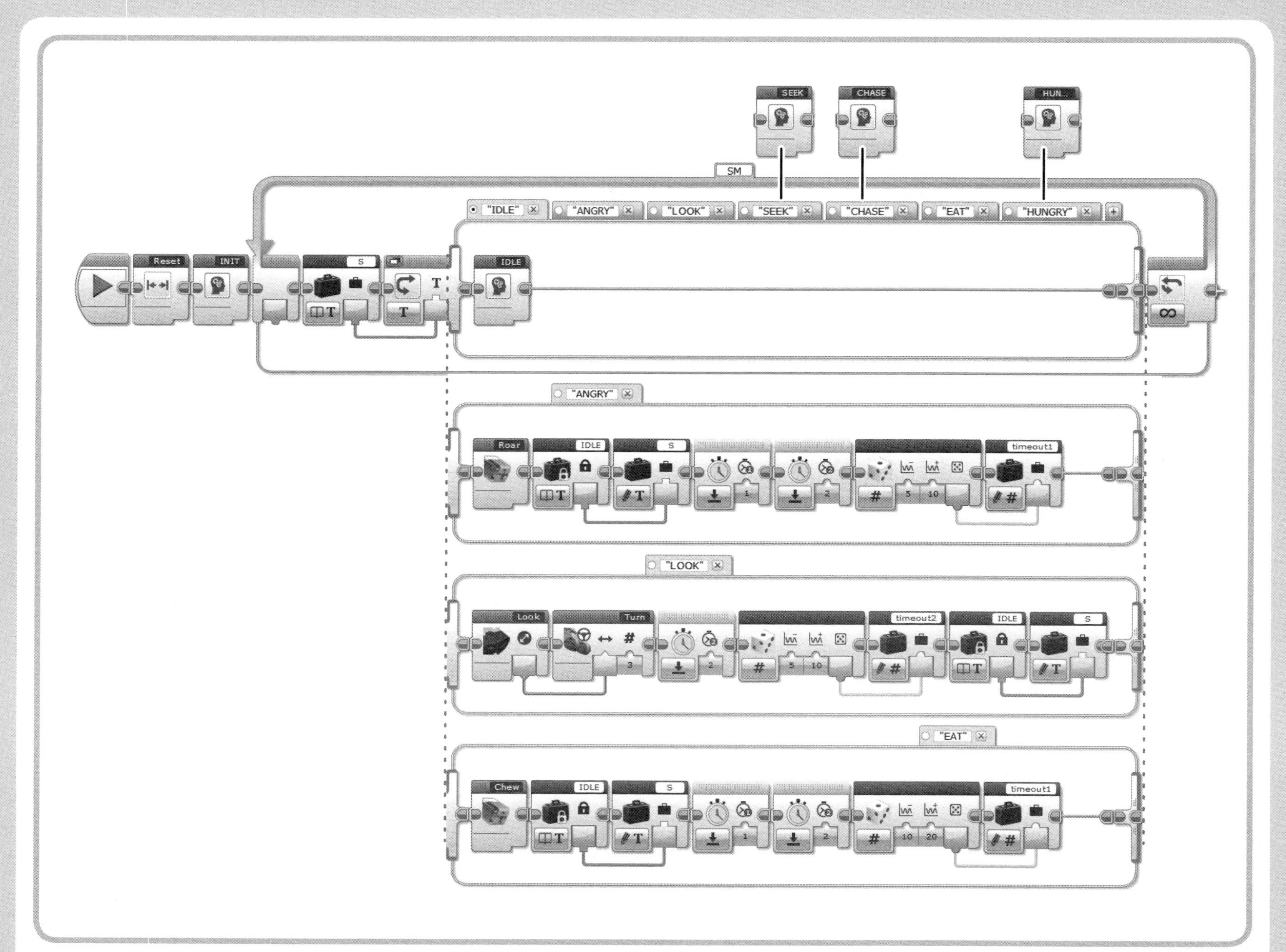

Figure 16-26: The StateMachine *program*

EXPERIMENT 16-1

Design a new behavior for the T-R3X, either beginning with the state machine discussed in this chapter or starting from scratch. For example, try "taming" the T-R3X by making it act like an electronic pet that's happy if you feed it and pet it frequently enough. You can use the Remote IR Beacon to feed and pet the T-R3X, or you can use the IR Sensor for detecting "food" and the EV3 Brick buttons for petting.

EXPERIMENT 16-2

Create a program to remotely control the T-R3X. Use Remote IR Beacon commands to perform different actions using a Switch block.

EXPERIMENT 16-3

Using the few LEGO elements remaining in the set, decorate the Remote IR Beacon to see if the T-R3X finds it tastier.

EXPERIMENT 16-4

If you stop building at step 23 (page 338) and add the 9M beam shown in step 29 (page 344), you have a walking bipedal robot that you can customize. Start from that base to make your own bipedal creature!

conclusion

In this chapter, you learned to program the T-R3X to make it walk, turn, and avoid obstacles. After an introduction to modeling advanced behaviors for robots using state machines, you programmed the T-R3X to show autonomous behavior, giving it some random attitude changes and a fierce hunting instinct at the expense of the beacon-prey. You also learned how to compute complex logic expressions using the Math block in Advanced mode instead of many Logic Operations blocks.

WE CAME, WE SAW, AND WE KICKED ITS...
I WANNA SHOW YOU SOMETHING!
WHILE I WAS LOCKED IN HERE, I FINALLY COMPLETED IT!
WOW! AN EL3CTRIC GUITAR!
COOL! I WANNA TRY IT!
THE GAZILLION-WATT AMPLIFIER!
I STILL NEED TO TWEAK IT. THERE'S A SLIGHT POSSIBILITY OF OVERLOAD!
DEXTER, DON'T...
...
NGH!
KH-NEEEEOOOOOOOWWW!
16X1

the EV3 31313 set bill of materials

This appendix lists the LEGO elements included in the EV3 Retail set. Each entry includes:

* An image of the piece, the quantity included in the set, and the color
* The design ID (used to identify the parts in the LDraw computer-aided design system), which you can also use to look for spare LEGO parts on BrickLink (*http://www.bricklink.com/*)
* The unique LEGO part ID (which takes color into account), which you can use to look for spare parts using LEGO Customer Service (*http://service.lego.com/en-us/replacementparts/*)
* The LEGO Group's internal name for the element
* A short, easy-to-remember name

LOOKING UP PIECES ON BRICKSET

To check the official LEGO name and see what a part looks like, add the LEGO ID at the end of this address: *http://www.brickset.com/parts/?part=*. For example, to see the first element in the following table, go to *http://www.brickset.com/parts/?part=6006140*. This takes you to a unique page for that part at Brickset, an online resource for LEGO collectors and hobbyists.

NOTE I've omitted the prefix *Technic* at the beginning of each name for brevity. When referring to axles and beams, *XM* is the abbreviation for *X modules long*. For example, 11M stands for 11 modules long. And a part like the 3×7 angular beam is read as "3 by 7 angular beam."

Image	Qty	Color	BrickLink LDraw ID	LEGO ID	LEGO Name	Easy Name
	10	Black	60483	6006140	Beam 1×2 with Cross and Hole	2M Beam with Cross Hole
	12	Black	32523	4142822	3M Beam	3M Beam
	10	Black	32316	4142135	Beam, 5M	5M Beam
	6	Black	32524	4495935	Beam, 7M	7M Beam
	8	Black	64289	4645732	Beam, 9M	9M Beam

Image	Qty	Color	BrickLink LDraw ID	LEGO ID	LEGO Name	Easy Name
	4	Red	64290	4562805	Beam, 11M	11M Beam
	4	Black	41239	4522933	Beam, 13M	13M Beam
	4	Black	64871	4542573	Beam, 15M	15M Beam
	2	Grey	64179	4539880	Beam Frame 5×7 Ø4.85	O Frame
	2	Grey	64178	4540797	Beam R. Frame 5×11 Ø4.85	H Frame
	4	Black	60484	4552347	T-Beam 3×3 w/ Hole Ø4.8	T Beam
	8	Black	32140	4120017	Angular Beam, 2×4, 90 deg	2×4 Angular Beam
	6	Black	32526	4142823	Angular beam, 3×5, 90 deg	3×5 Angular Beam
	4	Black	32348	4128593	Angular Beam, 4×4	4×4 Angular Beam
	12	Black	32271	4140327	Angular Beam, 3×7	3×7 Angular Beam
	12	Black	32009	4111998	Double Angular Beam 3×7 45°	Double Angular Beam

Image	Qty	Color	BrickLink LDraw ID	LEGO ID	LEGO Name	Easy Name
	1	Grey	6632	4211566	Lever 3M	3M Thin Beam
	2	Black	6575	4143187	Comb Wheel	Cam
	2	Black	32005	4629921	Track Rod 6M	6M Link
	4	Black	32293	4141300	LT Steering Gear	9M Link
	95	Black	2780	4121715	Connector Peg with Friction	Pin with Friction
	38	Blue	6558	4514553	Connector Peg with Friction 3M	3M Pin with Friction
	28	Blue	43093	4206482	Connector Peg with Friction/ Cross Axle	Axle Pin with Friction
	10	Red	32054	4140806	3M Fric. Snap w/ Cross Hole	3M Pin with Stop Bush
	4	Grey	3673	4211807	Connector Peg	Pin without Friction
	4	Tan	32556	4514554	3M Connector Peg	3M Pin without Friction
	6	Black	6628	4184169	Ball with Friction Snap	Pin with Towball
	6	Grey	2736	4211375	Ball with Cross Axle	Axle Pin with Towball
	9	Red	6590	4227155	Bush for Cross Axle	Bush
	11	Yellow	32123	4239601	Half Bush	Half Bush
	12	Red	32062	4142865	2M Cross Axle with Groove	2M Axle
	22	Grey	4519	4211815	Cross Axle 3M	3M Axle
	4	Dark Tan	6587	4566927	Cross Axle 3M with Knob	3M Axle with Stop
	4	Dark Grey	87083	4560177	Cross Axle 4M with End Stop	4M Axle with Stop
	3	Tan	99008	4666999	Cross Axle with Stop 4M	4M Axle with Middle Stop
	9	Grey	32073	4211639	Cross Axle 5M	5M Axle
	2	Dark Grey	59426	4508553	Cross Axle 5.5 with Stop 1M	5.5M Axle with Stop

Image	Qty	Color	BrickLink LDraw ID	LEGO ID	LEGO Name	Easy Name
	9	Black	3706	370626	Cross Axle 6M	6M Axle
	2	Grey	44294	4211805	Cross Axle 7M	7M Axle
	6	Dark Grey	55013	4499858	Cross Axle 8M with End Stop	8M Axle with Stop
	1	Grey	60485	4535768	Cross Axle 9M	9M Axle
	4	Red	32013	4254606	Angle Element 0 degrees [1]	Angle Connector #1
	6	Red	32034	4234429	Angle Element 180 degrees [2]	Angle Connector #2
	4	Red	32192	4189936	Angle Element 135 degrees [4]	Angle Connector #4
	1	Red	32014	4189131	Angle Element 90 degrees [6]	Angle Connector #6
	3	Red	59443	4513174	Cross Axle Extension 2M	Axle Connector
	1	Grey	57585	4502595	3-Branch Cross Axle with Cross Hole	Connector with 3 Axles
	2	Grey	32039	4211553	Catch with Cross Hole	Connector with Axle Hole
	2	Grey	62462	4526985	Tube with Double Hole Ø4.85	Pin Connector
	8	Red	6536	4188298	Cross Block 90°	2M Cross Block
	17	Red	32184	4128598	Double Cross Block	Double Cross Block
	14	Red	42003	4175442	Cross Block 3M	3M Cross Block
	2	Red	32291	4128594	Cross Block 2×1	2×1 Cross Block ("Mickey")
	4	Red	41678	4173975	Cross Block/Fork 2×2	2×2 Fork Cross Block ("Minnie")
	2	Grey	63869	4538007	Cross Block 3×2	3×2 Cross Block
	12	Grey	48989	4225033	Beam 3M with 4 Snaps	3M Cross Block with 4 Pins

Image	Qty	Color	BrickLink LDraw ID	LEGO ID	LEGO Name	Easy Name
	6	Grey	87082	4560175	Double Bush 3M Ø4.9	3M Pin with Hole
	4	Grey	32138	4211888	Module Bush	2M Beam with 4 Pins
	4	Grey	32068	6013936	Steering Gear 3M	3M Cross Block, Steering
	2	Grey	92907	4630114	Cross Block/Form 2×2×2	2×2×2 Fork Cross Block
	1	Black	87408	4558692	Beam 3M Ø4.85 with Fork	Gearbox Cross Block
	4	Black	32072	4248204	Angular Wheel	4z Knob Wheel
	1	Tan	6589	4565452	Conical Wheel z12	12z Bevel Gear
	1	Tan	32198	6031962	Bevel Gear z20	20z Bevel Gear
	2	Black	32270	4177431	Double Conical Wheel z12 1M	12z Double-Bevel Gear
	4	Black	32269	4177430	Double Conical Wheel z20 1M	20z Double-Bevel Gear
	5	Black	32498	4255563	Double Conical Wheel z36	36z Double-Bevel Gear
	2	Dark Grey	3648	4514558	Gear Wheel z24	24z Gear
	2	Grey	4716	4211510	Worm	Worm Gear
	4	Grey	42610	4211758	Hub Ø11.2×7.84	Small Wheel
	3	Grey	4185	4494222	Wedge-Belt Wheel Ø24	Medium Wheel
	4	Black	56145	4299389	Rim Wide with Cross 30/20	Large Wheel
	2	Black	50951	4246901	Tyre Low Narrow Ø14.58×6.24	Small Tire

Image	Qty	Color	BrickLink LDraw ID	LEGO ID	LEGO Name	Easy Name
	3	Black	2815	6028041	Tyre for Wedge-Belt Wheel	Medium Tire
	4	Black	44309	4184286	Tyre Normal Wide Ø43.2×22	Large Tire
	2	Black	53992	4502834	Caterpillar Track	Rubber Tread
	4	White	41669	4173941	Bionicle Eye	Tooth
	6	Red	41669	4185661	Bionicle Eye	Tooth
	1	White	61070	6015596	Right Screen Ø4.85 4×7×4	Right Mudguard
	1	White	61071	6015597	Left Screen Ø4.85 4×7×4	Left Mudguard
	3	White	64391	4547582	Right Panel 3×7	Medium Panel #4
	3	White	64683	4547581	Left Panel 3×7	Medium Panel #3
	3	White	64393	4558797	Right Panel 3×11	Long Panel #6
	3	White	64681	4558802	Left Panel 3×11	Long Panel #5
	4	White	98347	4656205	Blade with Technic Hole 1	Curved Blade

Image	Qty	Color	BrickLink LDraw ID	LEGO ID	LEGO Name	Easy Name
	6	Red/Grey	98568	4657296	Sword	Sword
	1	Red	85544	4544143	V-Belt Ø24	Rubber Band
	1	Black	53550	6024109	Magazine for Balls Ø16.5	Ball Magazine
	1	Black	54271	6024106	Shooter	Ball Shooter
	3	Red	54821	4545430	Ball Ø16.5	Ball
	1	Various	95646	6009996	MS-EV3, P-Brick	EV3 Brick
	2	Various	95658	6009430	MS-EV3, Large Motor	EV3 Large Motor
	1	Various	99455	6008577	MS-EV3, Medium Motor	EV3 Medium Motor
	1	Various	95648	6008472	MS-EV3, Touch Sensor	EV3 Touch Sensor
	1	Various	95650	6008919	MS-EV3, Color Sensor	EV3 Color Sensor

Image	Qty	Color	BrickLink LDraw ID	LEGO ID	LEGO Name	Easy Name
	1	Various	95654	6009811	MS-EV3, IR Sensor	EV3 IR Sensor
	1	Various	72156	6014051	MS-EV3, IR Beacon	EV3 Remote IR Beacon
	4	Black	11145	6024581	Cable 250 mm	Cable 25 cm / 10 in
	2	Black	11146	6024583	Cable 350 mm	Cable 35 cm / 14 in
	1	Black	11147	6036899	Cable 500 mm	Cable 50 cm / 20 in

differences between the education set and retail set

The tables in this appendix list the differences between the Retail set 31313 and the Education Core set 45544. The sets have a different assortment of elements and sensors. The Education Core set has fewer parts and is designed to be more of a tool kit, with a more balanced quantity of different parts. For example, the Education Core set lacks many of the decorative blades and swords, but it has more gears. Also, the Education Core set has a special steel ball that you can insert into a socket to make a smooth caster wheel for wheeled robots. This appendix also provides tables that list the parts you would need to turn Retail set 31313 into Education Core set 45544 and the parts you would need to create the 31313 set assortment using the 45544 set or Education Expansion set 45560.

electronic devices

The Education Core set lacks the IR Sensor and the Remote IR Beacon of the Retail set, but it does have a Gyroscopic (Gyro) Sensor (which can measure rotation speed along an axis) and an Ultrasonic (US) Sensor (which can measure distances in centimeters or inches). It also has two Touch Sensors. The Education set's assortment of cables is the same as that of the Retail set. Its EV3 Brick hardware is exactly the same, but the firmware is different: The Education version of the firmware includes an on-brick Data Logging App.

the EV3 software

Compared to the Home Edition that comes with the Retail set, the Education EV3 Software has a different Lobby and Activities, and it has a sophisticated Data Logging environment for experiments and data analysis. Its Content Editor has additional features designed for use by teachers in classrooms, and its Programming Palettes have extra blocks to control the different sensors (the US, Gyro, Temperature, and Power Meter) and to manage data logging.

The Home Edition has only a subset of the features of the Education Edition. The EV3 Software Home Edition lets you open a project created in the EV3 Software Education Edition. In fact, even if the project has blocks that are not found in the Home Edition, you can still open and download and run programs to the EV3 Brick. However, you won't be able to edit any of the extra blocks. (Extra sensor blocks for the Home Edition are available for download at *http://lego.com/mindstorms/*.)

turning the retail set into the education core set

The Retail set 31313 and the Education Core set 45544 share many elements, though some have different colors. The table below lists only the parts you'll need to build models designed with the LEGO MINDSTORMS Education EV3 Core set 45544. This table does not include parts that are simply a different color (like the bushes), but I do note the differences.

Although there are enough 3M beams in the 31313 set, they are all black, so I've added the 16 colored 3M beams included in the 45544 set because these can be used to perform various activities with the Color Sensor. For example, you could create a robot that sorts LEGO pieces by color or one that can react to colors in other ways.

Image	Qty	Color	BrickLink LDraw ID	LEGO ID	LEGO Name	Easy Name
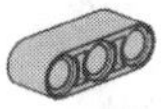	4	Yellow	32523	4153707	3M Beam	3M Beam
	4	Red	32523	4153718	3M Beam	3M Beam
	4	Blue	32523	4509376	3M Beam	3M Beam
	4	Green	32523	6007973	3M Beam	3M Beam
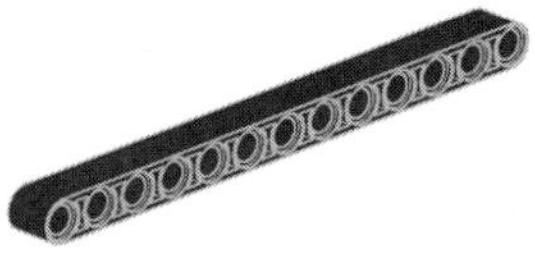	2	Black	41239	4522933	Beam, 13M	13M Beam
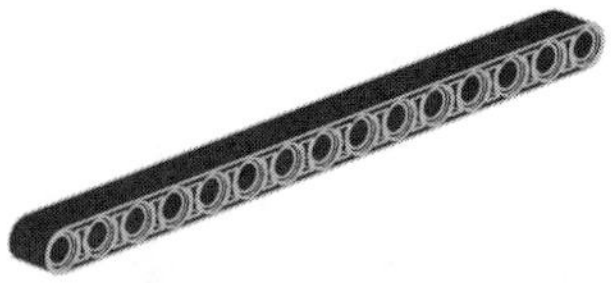	2	Black	64871	4542573	Beam, 15M	15M Beam
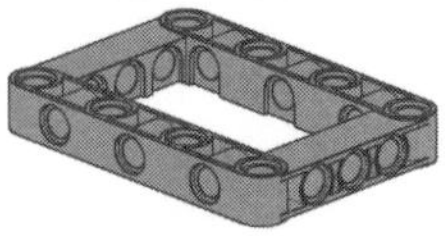	1	Grey	64179	4539880	Beam Frame 5×7 Ø4.85	O Frame
	2	Black	32348	4128593	Angular Beam, 4×4	4×4 Angular Beam
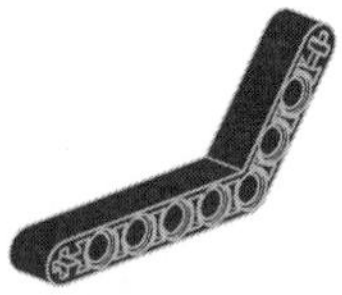	4	Black	6629	4112282	Angular Beam, 4×6	4×6 Angular Beam
	2	Black	32449	4142236	Half Beam 4M	4M Thin Beam
	2	Black	33299	4563044	Connector Peg with Handle	3M Beam with Pin
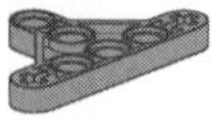	4	Grey	99773	6009019	Half Triangle Beam 5×3	Triangular Thin Beam
	8	Tan	3749	4666579	Connector Peg / Cross Axle	Axle Pin without Friction
	12	Red	32054	4140806	3M Fric. Snap w/ Cross Hole	3M Pin with Stop Bush
	6	Grey	3673	4211807	Connector Peg	Pin without Friction
	2	Tan	32556	4514554	3M Connector Peg	3M Pin without Friction
	1	Red	6590	4227155	Bush for Cross Axle	Bush

Image	Qty	Color	BrickLink LDraw ID	LEGO ID	LEGO Name	Easy Name
	4	Black	3705	370526	Cross Axle 4M	4M Axle
	3	Grey	44294	4211805	Cross Axle 7M	7M Axle
	2	Black	3707	370726	Cross Axle 8M	8M Axle
	1	Grey	60485	4535768	Cross Axle 9M	9M Axle
	2	Black	3737	373726	Cross Axle 10M	10M Axle
	2	Black	3708	370826	Cross Axle 12M	12M Axle
	1	Red	32014	4189131	Angle Element 90 degrees [6]	Angle Connector #6
	3	Red	59443	4513174	Cross Axle Extension 2M	Axle Connector
	1	Grey	57585	4502595	3-Branch Cross Axle with Cross Hole	Connector with 3 Axles
	2	Grey	62462	4526985	Tube with Double Hole Ø4.85	Pin Connector
	4	Black	45590	4198367	Rubber Beam with Cross Holes, 2M	2M Rubber Beam
	2	Red	32291	4128594	Cross Block 2×1	2×1 Cross Block ("Mickey")
	4	Grey	63869	4538007	Cross Block 3×2	3×2 Cross Block
	4	Grey	55615	4296059	Angular Connector Peg, 3×3	3×3 Cross Block with 4 Pins ("Puppy")
	2	Red	44809	6008527	Hto V Beam 90 Degr.	V Cross Block
	4	Dark Grey	10928	6012451	Gear Wheel z8	8z Gear
	1	Tan	6589	4565452	Conical Wheel z12	12z Bevel Gear
	4	Grey	94925	4640536	Gear Wheel z16	16z Gear

Image	Qty	Color	BrickLink LDraw ID	LEGO ID	LEGO Name	Easy Name
	2	Dark Grey	3648	4514558	Gear Wheel z24	24z Gear
	2	Grey	3649	4285634	Gear Wheel z40	40z Gear
	2	Grey	99009	4652235	Turntable Bottom, 28-Tooth	28z Small Turntable Bottom
	2	Black	99010	4652236	Turntable Top, 28-Tooth	28z Small Turntable Top
	1	Grey	4185	4494222	Wedge-Belt Wheel Ø24	Medium Wheel
	1	Black	2815	6028041	Tyre for Wedge-Belt Wheel	Medium Tire
	2	Grey	41896	4634091	Hub, 43.2×26	43.2×26 Wheel
	2	Black	41897	6035364	Low Profile Tyre Ø56×28	56×28 Large Tire
	1	Dark Grey	92911	4610380	Power Joint	Ball Socket
	1	Steel	99948	6023956	Steel Ball Ø36	Steel Ball
	54	Black	57518	6014648	Track Link 5×1.5	Large Track Link
	4	Black	57519	4582792	Sprocket 40.7×15	Sprocket

Image	Qty	Color	BrickLink LDraw ID	LEGO ID	LEGO Name	Easy Name
	1	Black	87080	4566251	Left Panel 3×5	Short Panel #1
	1	Black	87086	4566249	Right Panel 3×5	Short Panel #2
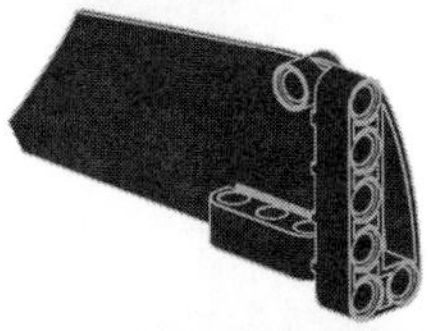	1	Black	64392	4541326	Left Panel 5×11	Long Wide Panel #17
	1	Black	64682	4543490	Right Panel 5×11	Long Wide Panel #18
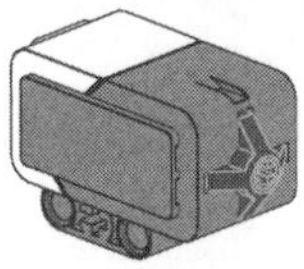	1	Various	95648	6008472	MS-EV3, Touch Sensor	EV3 Touch Sensor
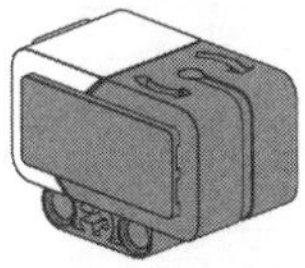	1	Various	99380	6008916	MS-EV3, Gyroscopic Sensor	EV3 Gyro Sensor
	1	Various	95652	6008924	MS-EV3, Ultrasonic Sensor	EV3 US Sensor
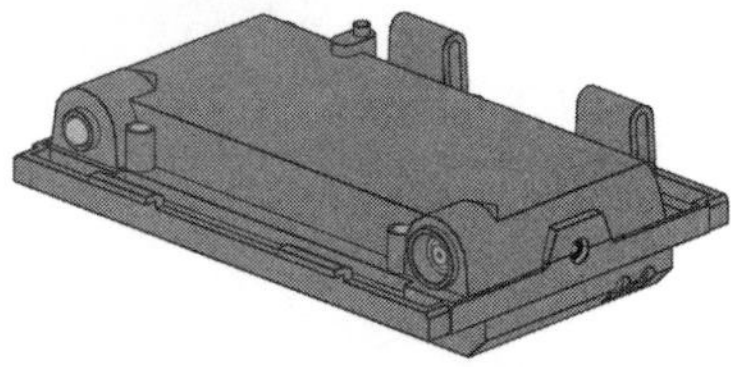	1	Grey	95656	6012820	MS-EV3, Rechargeable Battery	EV3 Rechargeable Battery

turning the education core set into the retail set

This table lists all the parts you would need to get if you have the Education Core set 45544 and you want to build all of the robots in this book using the parts found in Retail set 31313.

Image	Qty	Color	BrickLink LDraw ID	LEGO ID	LEGO Name	Easy Name
	6	Black	60483	6006140	Beam 1×2 with Cross and Hole	2M Beam with Cross Hole
	6	Black	32316	4142135	Beam, 5M	5M Beam
	2	Black	32524	4495935	Beam, 7M	7M Beam
	2	Black	40490	4645732	Beam, 9M	9M Beam
	1	Grey	64178	4540797	Beam R. Frame 5×11 Ø4.85	H Frame
	2	Black	32140	4120017	Angular Beam, 2×4, 90 deg	2×4 Angular Beam
	8	Black	32271	4140327	Angular Beam, 3×7	3×7 Angular Beam
	8	Black	32009	4111998	Double Angular Beam 3×7 45°	Double Angular Beam
	1	Grey	6632	4211566	Lever 3M	3M Thin Beam

Image	Qty	Color	BrickLink LDraw ID	LEGO ID	LEGO Name	Easy Name
	2	Black	6575	4143187	Comb Wheel	Cam
	2	Black	32005	4629921	Track Rod 6M	6M Link
	4	Black	32293	4141300	LT Steering Gear	9M Link
	35	Black	2780	4121715	Connector Peg with Friction	Pin with Friction
	8	Blue	6558	4514553	Connector Peg with Friction 3M	3M Pin with Friction
	8	Blue	43093	4206482	Connector Peg with Friction/ Cross Axle	Axle Pin with Friction
	6	Black	6628	4184169	Ball with Friction Snap	Pin with Towball
	6	Grey	2736	4211375	Ball with Cross Axle	Axle Pin with Towball
	1	Yellow	32123	4239601	Half Bush	Half Bush
	2	Red	32062	4142865	2M Cross Axle with Groove	2M Axle
	8	Grey	4519	4211815	Cross Axle 3M	3M Axle
	2	Dark Tan	6587	4566927	Cross Axle 3M with Knob	3M Axle with Stop
	2	Dark Grey	87083	4560177	Cross Axle 4M with End Stop	4M Axle with Stop
	3	Tan	99008	4666999	Cross Axle with Stop 4M	4M Axle with Middle Stop
	3	Grey	32073	4211639	Cross Axle 5M	5M Axle
	2	Dark Grey	59426	4508553	Cross Axle 5.5 with Stop 1M	5.5M Axle with Stop
	5	Black	3706	370626	Cross Axle 6M	6M Axle
	4	Dark Gray	55013	4499858	Cross Axle 8M with End Stop	8M Axle with Stop
	2	Red	32034	4234429	Angle Element 180 degrees [2]	Angle Connector #2
	4	Red	32192	4189936	Angle Element 135 degrees [4]	Angle Connector #4
	2	Grey	32039	4211553	Catch with Cross Hole	Connector with Axle Hole

Image	Qty	Color	BrickLink LDraw ID	LEGO ID	LEGO Name	Easy Name
	9	Red	32184	4128598	Double Cross Block	Double Cross Block
	14	Red	42003	4175442	Cross Block 3M	3M Cross Block
	6	Grey	48989	4225033	Beam 3M with 4 Snaps	3M Cross Block with 4 Pins
	2	Grey	87082	4560175	Double Bush 3M Ø4.9	3M Pin with Hole
	4	Grey	32138	4211888	Module Bush	2M Beam with 4 Pins
	4	Grey	32068	6013936	Steering Gear 3M	3M Cross Block, Steering
	2	Grey	92907	4630114	Cross Block/Form 2×2×2	2×2×2 Fork Cross Block
	1	Black	87408	4558692	Beam 3M Ø4.85 with Fork	Gearbox Cross Block
	1	Tan	32198	6031962	Bevel Gear z20	20z Bevel Gear
	2	Black	32269	4177430	Double Conical Wheel z20 1M	20z Double-Bevel Gear
	3	Black	32498	4255563	Double Conical Wheel z36	36z Double-Bevel Gear
	4	Grey	42610	4211758	Hub Ø11.2×7.84	Small Wheel
	4	Black	56145	4299389	Rim Wide with Cross 30/20	Large Wheel
	2	Black	50951	4246901	Tyre Low Narrow Ø14.58×6.24	Small Tire
	4	Black	44309	4184286	Tyre Normal Wide Ø43.2×22	Large Tire

Image	Qty	Color	BrickLink LDraw ID	LEGO ID	LEGO Name	Easy Name
	2	Black	53992	4502834	Caterpillar Track	Rubber Tread
	6	Red	41669	4185661	Bionicle Eye	Tooth
	1	White	61070	6015596	Right Screen Ø4.85 4×7×4	Right Mudguard
	1	White	61071	6015597	Left Screen Ø4.85 4×7×4	Left Mudguard
	3	White	64391	4547582	Right Panel 3×7	Medium Panel #4
	3	White	64683	4547581	Left Panel 3×7	Medium Panel #3
	3	White	64393	4558797	Right Panel 3×11	Long Panel #6
	3	White	64681	4558802	Left Panel 3×11	Long Panel #5
	4	White	98347	4656205	Blade with Technic Hole 1	Curved Blade
	6	Red/Grey	98568	4657296	Sword	Sword
	1	Red	85544	4544143	V-Belt Ø24	Rubber Band

Image	Qty	Color	BrickLink LDraw ID	LEGO ID	LEGO Name	Easy Name
	1	Black	53550	6024109	Magazine for Balls Ø16.5	Ball Magazine
	1	Black	54271	6024106	Shooter	Ball Shooter
	3	Red	54821	4545430	Ball Ø16.5	Ball
	1	Various	95654	6009811	MS-EV3, IR Sensor	EV3 IR Sensor
	1	Various	72156	6014051	MS-EV3, IR Beacon	EV3 Remote IR Beacon

turning the education expansion set into the retail set

If you have the Education Core set 45544 and the Education Expansion set 45560, there's a good chance that you're a schoolteacher. The following table lists the parts you would need in order to have the assortment of elements in the Retail set 31313 to build all of the robots in this book.

Image	Qty	Color	BrickLink LDraw ID	LEGO ID	LEGO Name	Easy Name
	1	Black	32316	4142135	Beam, 5M	5M Beam
	8	Black	32271	4140327	Angular Beam, 3×7	3×7 Angular Beam

Image	Qty	Color	BrickLink LDraw ID	LEGO ID	LEGO Name	Easy Name
	2	Black	32009	4111998	Double Angular Beam 3×7 45°	Double Angular Beam
	2	Black	6575	4143187	Comb Wheel	Cam
	4	Black	32293	4141300	LT Steering Gear	9M Link
	4	Grey	2736	4211375	Ball with Cross Axle	Axle Pin with Towball
	1	Tan	99008	4666999	Cross Axle with Stop 4M	4M Axle with Middle Stop
	4	Black	3706	370626	Cross Axle 6M	6M Axle
	2	Red	32192	4189936	Angle Element 135 degrees [4]	Angle Connector #4
	5	Red	32184	4128598	Double Cross Block	Double Cross Block
	2	Red	42003	4175442	Cross Block 3M	3M Cross Block
	2	Grey	87082	4560175	Double Bush 3M Ø4.9	3M Pin with Hole
	4	Grey	32068	6013936	Steering Gear 3M	3M Cross Block, Steering
	2	Grey	92907	4630114	Cross Block/Form 2×2×2	2×2×2 Fork Cross Block
	1	Black	32269	4177430	Double Conical Wheel z20 1M	20z Double-Bevel Gear
	3	Black	32498	4255563	Double Conical Wheel z36	36z Double-Bevel Gear
	4	Black	56145	4299389	Rim Wide with Cross 30/20	Large Wheel
	2	Black	50951	4246901	Tyre Low Narrow Ø14.58×6.24	Small Tire

Image	Qty	Color	BrickLink LDraw ID	LEGO ID	LEGO Name	Easy Name
	2	Black	44309	4184286	Tyre Normal Wide Ø43.2×22	Large Tire
	2	Black	53992	4502834	Caterpillar Track	Rubber Tread
	3	Red	41669	4185661	Bionicle Eye	Tooth
	1	White	61070	6015596	Right Screen Ø4.85 4×7×4	Right Mudguard
	1	White	61071	6015597	Left Screen Ø4.85 4×7×4	Left Mudguard
	1	White	64391	4547582	Right Panel 3×7	Medium Panel #4
	1	White	64683	4547581	Left Panel 3×7	Medium Panel #3
	2	White	64393	4558797	Right Panel 3×11	Long Panel #6
	2	White	64681	4558802	Left Panel 3×11	Long Panel #5
	4	White	98347	4656205	Blade with Technic Hole 1	Curved Blade
	6	Red/Grey	98568	4657296	Sword	Sword

Image	Qty	Color	BrickLink LDraw ID	LEGO ID	LEGO Name	Easy Name
	1	Black	53550	6024109	Magazine for Balls Ø16.5	Ball Magazine
	1	Black	54271	6024106	Shooter	Ball Shooter
	3	Red	54821	4545430	Ball Ø16.5	Ball
	1	Various	95654	6009811	MS-EV3, IR Sensor	EV3 IR Sensor
	1	Various	72156	6014051	MS-EV3, IR Beacon	EV3 Remote IR Beacon

HEY, IT'S MY FATHER'S CAR!
DEX, SWEETIE! DID YOU HAVE FUN?
IT WAS AWESOME!
HI, I'M...
DANNY, SURE! MY FATHER TOLD ME ABOUT YOU! I'M DEXTER'S SISTER, JENNY!
SSSHHH!
A PLEASURE!
DEXTER IS VERY TALENTED!
MY FATHER SENT ME TO SAY THAT HE WILL COVER THE RESTORATION EXPENSES OF EV3L. HE JUST WANTED A SIGN OF GOOD WILL FROM YOU!
GREAT! THANK HIM VERY MUCH FOR US!
BYE!
SEE? YOU JUST HAD TO OPEN THE DOORS TO A KID TO GET THE LAB BACK ON TRACK! WE NEED TO CELEBRATE!
LET'S GO! I'M STARVING!
QUACK!
THE END
17X1

index

E

N

O

P

R

S

The LEGO MINDSTORMS EV3 Laboratory is set in Chevin. The book was printed and bound by Edwards Brothers Malloy in Ann Arbor, Michigan. The paper is Husky 60# Opaque, which is certified by the Sustainable Forestry Initiative (SFI).

The book uses a RepKover binding, in which the pages are bound together with a cold-set, flexible glue and the first and last pages of the resulting book block are attached to the cover with tape. The cover is not actually glued to the book's spine, and when open, the book lies flat and the spine doesn't crack.

companion website

Visit *http://EV3L.com/* for the EV3 projects for the robots, errata, additional tips and tricks, and bonus models.